Nanoscience and its Applications

Nanoscience and its Applications

Alessandra L. Da Róz

Marystela Ferreira

Fábio de Lima Leite

Osvaldo N. Oliveira Jr.

ELSEVIER

William Andrew
Applied Science Publishers
elsevier.com

William Andrew is an imprint of Elsevier
The Boulevard, Langford Lane, Kidlington, Oxford, OX5 1GB, United Kingdom
50 Hampshire Street, 5th Floor, Cambridge, MA 02139, United States

Notices
Knowledge and best practice in this field are constantly changing. As new research and experience broaden our understanding, changes in research methods, professional practices, or medical treatment may become necessary.

Practitioners and researchers must always rely on their own experience and knowledge in evaluating and using any information, methods, compounds, or experiments described herein. In using such information or methods they should be mindful of their own safety and the safety of others, including parties for whom they have a professional responsibility.

To the fullest extent of the law, neither the Publisher nor the authors, contributors, or editors, assume any liability for any injury and/or damage to persons or property as a matter of products liability, negligence or otherwise, or from any use or operation of any methods, products, instructions, or ideas contained in the material herein.

Library of Congress Cataloging-in-Publication Data
A catalog record for this book is available from the Library of Congress

British Library Cataloguing-in-Publication Data
A catalogue record for this book is available from the British Library

ISBN: 978-0-323-49780-0

For information on all William Andrew publications visit our website at https://www.elsevier.com/

Working together
to grow libraries in
developing countries

www.elsevier.com • www.bookaid.org

Publisher: Matthew Deans
Acquisition Editor: Simon Holt
Editorial Project Manager: Charlotte Kent
Production Project Manager: Lisa Jones
Designer: Greg Harris

Typeset by Thomson Digital

Table of Contents

List of Contributors

Adriano Moraes Amarante
Nanoneurobiophysics Research Group, Department of Physics, Chemistry & Mathematics, Federal University of São Carlos, Sorocaba, São Paulo, Brazil

Carolina de Castro Bueno
Nanoneurobiophysics Research Group, Department of Physics, Chemistry & Mathematics, Federal University of São Carlos, Sorocaba, São Paulo, Brazil

Renata Pires Camargo
Nanoneurobiophysics Research Group, Department of Physics, Chemistry & Mathematics, Federal University of São Carlos, Sorocaba, São Paulo, Brazil

Juliana Cancino-Bernardi
Group of Nanomedicine and Nanotoxicology, Institute of Physics of São Carlos, University of São Paulo (Universidade de São Paulo—USP), São Carlos, São Paulo, Brazil

Marco Roberto Cavallari
Department of Electronic Systems Engineering, Polytechnic School of the USP, São Paulo, Brazil

Richard André Cunha
Institute of Chemistry, Federal University of Uberlandia—UFU, Uberlandia, Minas Gerais, Brazil

Daiana Kotra Deda
Nanoneurobiophysics Research Group, Department of Physics, Chemistry & Mathematics, Federal University of São Carlos, Sorocaba, São Paulo, Brazil

Fernando Josepetti Fonseca
Department of Electronic Systems Engineering, Polytechnic School of the USP, São Paulo, Brazil

Eduardo de Faria Franca
Institute of Chemistry, Federal University of Uberlandia—UFU, Uberlandia, Minas Gerais, Brazil

Luiz Carlos Gomide Freitas
Chemistry Department, UFSCAR, São Carlos, São Paulo, Brazil

Pâmela Soto Garcia
Nanoneurobiophysics Research Group, Department of Physics, Chemistry & Mathematics, Federal University of São Carlos, Sorocaba, São Paulo, Brazil

Jéssica Cristiane Magalhães Ierich
Nanoneurobiophysics Research Group, Department of Physics, Chemistry & Mathematics, Federal University of São Carlos, Sorocaba, São Paulo, Brazil

Fabio de Lima Leite
Nanoneurobiophysics Research Group, Department of Physics, Chemistry & Mathematics, Federal University of São Carlos, Sorocaba, São Paulo, Brazil

Valéria Spolon Marangoni
Group of Nanomedicine and Nanotoxicology, Institute of Physics of São Carlos, University of São Paulo (Universidade de São Paulo—USP), São Carlos, São Paulo, Brazil

Guedmiller Souza de Oliveira
Nanoneurobiophysics Research Group, Department of Physics, Chemistry & Mathematics, Federal University of São Carlos, Sorocaba, São Paulo, Brazil

Leonardo Giordano Paterno
Institute of Chemistry, Darcy Ribeiro Campus, University of Brasilia, Brasilia, Brazil

Gerson Dos Santos
Department of Electronic Systems Engineering, Polytechnic School of the USP, São Paulo, Brazil

Fabio Ruiz Simões
Institute of Marine Sciences, Federal University of São Paulo, Santos, São Paulo, Brazil

Clarice Steffens
Regional Integrated University of Upper Uruguai and Missions, Erechim, Rio Grande Do Sul, Brazil

Miguel Gustavo Xavier
Center of Biological and Nature Sciences, Federal University of Acre, Rio Branco, Acre, Brazil

Valtencir Zucolotto
Group of Nanomedicine and Nanotoxicology, Institute of Physics of São Carlos, University of São Paulo (Universidade de São Paulo—USP), São Carlos, São Paulo, Brazil

1

Basic Concepts and Principles

F.R. Simões*, H.H. Takeda**

*INSTITUTE OF MARINE SCIENCES, FEDERAL UNIVERSITY OF SÃO PAULO, SANTOS, SP, BRAZIL; **DEPARTMENT OF INTERDISCIPLINARY SCIENCES AND TECHNOLOGY, FEDERAL UNIVERSITY OF RONDÔNIA, ARIQUEMES, RONDÔNIA, BRAZIL

CHAPTER OUTLINE

1.1 Introduction

1.1.1 Understanding the Nanoscale and Nanotechnology

To understand nanoscience and nanotechnology, it is necessary to know the origin of the prefix *nano*, which is Greek and means "dwarf." One nanometer (nm) is simply 1 billionth of 1 m (1 nm = 1×10^{-9} m). For comparison, the ratio between the size of a soccer ball and the Earth is approximately the same as that between a soccer ball and a sphere of 60 carbon atoms known as a C-60 fullerene (Fig. 1.1). The Earth is approximately 100 million times larger than a soccer ball, and in turn, the ball is approximately 100 million times larger than the fullerene [1].

Several other common examples can be used to understand the "nano" scale. Fig. 1.2 compares different nanoscale materials. A human hair is approximately 100,000 nm wide, whereas a red blood cell is approximately 7,000 nm in diameter. Even smaller are typical viruses, which are between 45 and 200 nm in size. On the atomic scale, the length of a typical bond between carbon atoms and the spaces between atoms in a molecule are on the order of 0.12–0.15 nm [1–4].

Thus, structures of nanoscale materials (called nanostructures) are intermediate between the smallest structure that can be produced by man and the largest molecules of living systems. Humans' abilities to control and manipulate nanostructures, therefore,

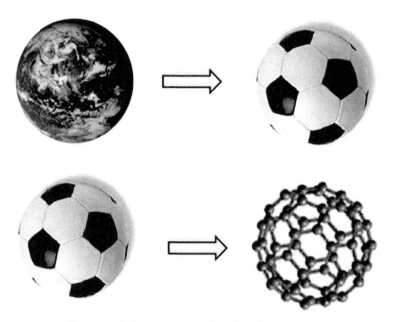

FIGURE 1.1 Illustration of the diameter ratios between the Earth and a ball, and between a ball and a C-60 fullerene.

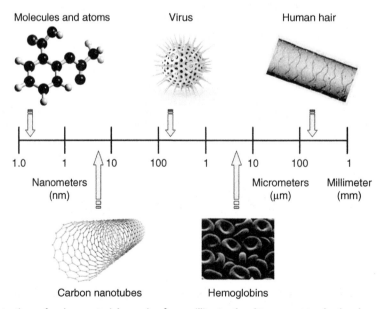

FIGURE 1.2 Illustrations of various materials ranging from millimeter *(mm)* to nanometer *(nm)* scales.

facilitate exploring novel physical, biological, and chemical properties of systems that are of intermediate size between atoms and molecules, such as, nanoscale materials.

Two standard definitions exist for the term nanotechnology: one given by the International Organization for Standardization–Technical Committee (ISO–TC) and the other given by the National Nanotechnology Initiative of the US (NNI). According to the ISO–TC, "(i) Understanding and control of matter and processes at the nanoscale, typically, but not exclusively, below 100 nanometres in one or more dimensions where the onset of size-dependent phenomena usually enables novel applications; and (ii) Utilizing the properties of nanoscale materials that differ from the properties of individual atoms, molecules, and bulk matter, to create improved materials, devices, and systems that exploit these new properties." Therefore, for a device to be considered "nanotechnological," in addition to nanometric dimensions, it must also have unique properties associated with the nanoscale [2–5]. In contrast, as defined by the NNI, nanotechnology must fall between 1 and 100 nm in size [2]. The lower limit is defined by the size of atoms, as this branch of science must construct devices from atoms and molecules. For example, one hydrogen atom has a diameter of approximately one quarter of a nanometer ($d = 0.25$ nm). The upper limit was established based on our ability to modulate properties on scales up to 100 nm and observe the resulting phenomena in larger structures that be used to generate specific devices [2]. These phenomena differentiate truly nanoscale devices from those that are simply miniaturized versions of an equivalent macroscopic device. Thus, such larger-scale devices should be considered as microtechnologies [3].

Thus, nanotechnology is used to describe molecular-scale engineering systems. More specifically, this term refers to the ability to design, construct or manipulate devices, materials and functional systems on the nanometric scale [4,5].

1.1.2 Nanoscience: History, Concepts, and Principles

Between the discovery of fire by man and the present day, tremendous advancements have occurred in science and technology, accompanied by remarkable development in the research and manufacture of new products, including new materials, pharmaceuticals, and foods. For example, very light materials with mechanical resistances exceeding that of steel have been produced as a result of advances in nanoscience, which can be defined as *the science that governs the study of nanotechnology for the development or improvement of materials based on the possibility of manipulating atoms and/or molecules with observed effects that are closely linked to the nanoscale and that have attractive physical, chemical and/or biological properties* [6].

Although a large number and variety of technologies and articles have recently been published on the topic of nanotechnology, this science has been applied and studied for a long time, albeit without knowledge of the relationship among the scale, the product and the resulting properties. In other words, man has manipulated materials at the nanometer level for a long time without understanding that the effects obtained were related to their nanoscale natures. One example is medieval glass-blowers who, using mixtures of gold nanoparticles of various sizes, produced differently colored stains for the fabrication of stained glass windows. A study by a research team at the University of Queensland [7] found that, in addition to the staining produced by gold nanoparticles, these nanoparticles also functioned as photocatalytic air purifiers: for example, when sunlight shone on the stained glass, air purification occurred. Another example is the experiment conducted by Michael Faraday (late 19th century), who synthesized gold nanoparticles [8] but did not understand their properties.

Regarding the manipulation of particles at the atomic level, scientists have investigated the nanometric world to find explanations and rationales for their theories, such as the atomic theory first proposed by Democritus in 400 BC, which was refined in 1913 by Ernest Rutherford and Niels Bohr [9]. Additionally, in 1867, James Clerk Maxwell performed an experiment known as Maxwell's demon, confirming that the second law of thermodynamics has only one statistical certainty. Briefly, in this experiment, Maxwell used a chamber containing a gas at equilibrium that was divided into two parts by a wall containing a door. When the door was opened, only the particles with higher and lower velocities could change sides, resulting in the heating of one side of the chamber and the cooling of the other [10]. Thomson (1906) and Lewis (1916) [9] developed the theory of chemical bonds (ionic and covalent bonding) to describe the formation of molecules. From 1934 to 1938, Lise Maitner, Otto Frisch, Otto Hahn, and Fritz Strassman studied the radioactive isotopes produced by bombarding uranium with neutrons (the experiment was conducted by Enrico Fermi). Based on their results, they discovered the phenomenon of nuclear fission, which releases 200 MeV of energy. This discovery led to the development of atomic bombs and nuclear power plants [11]. Another example is the nuclear fusion that occur in the Sun, n (Fig. 1.3), in which four hydrogen atoms fuse to form a helium atom, generating all the energy we observe and feel on Earth [12].

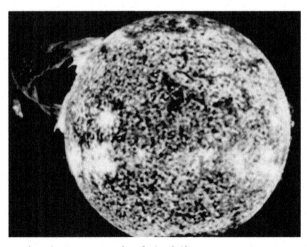

FIGURE 1.3 The Sun, a star undergoing constant nuclear fusion [12].

Many scientists have confirmed their theories or answered chemical and/or physical questions by simply observing the phenomena resulting from the atomic or molecular behaviors of materials, and thus, it can be concluded that a scientific field and its foundation develop via studies and the formulations of theories, concepts and principles based on experimental observations. Thus, defining the basic concepts and principles of nanoscience and nanotechnology would be much more difficult without the theoretical knowledge of other branches of sciences. Indeed, most of the laws, concepts and principles that govern nanoscience are essentially the same ones that govern physics and chemistry. To illustrate this, we simply review the portion of the definition of nanotechnology that states that *nanotechnology is a novel branch of science responsible for the study and development of materials at the atomic and molecular levels that have unique characteristics associated with the nanoscale.* This concept confirms that the manipulation of atoms or molecules first requires theoretical knowledge of atomic theory and chemical bonds, as indicated previously.

Nanoscience is not only related to chemical and physical knowledge. There is also a great demand for biological knowledge. Biology and biochemistry also have much to gain from nanoscientific advances because DNA, viruses, and organelles are considered nanostructures [13]. For example, the National Aeronautics and Space Administration (NASA) has studied the development of nanoparticles containing DNA repair enzymes and ligands for the recognition of damaged cells [14].

In 1959, the American physicist Richard Feynman (Fig. 1.4), in his lecture "There's plenty of room at the bottom" [15], introduced the first nanoscientific approach. Feynman explained that the entire area on the head of a pin (1/16 in.), if amplified 25,000 times, would have an area capable of housing all the pages of the Encyclopedia Britannica. He explained that the resolving power of the human eye (approximately 1,120 points per inch) corresponds to approximately the diameter of one of the tiny dots in the high-quality half-tone reproductions in the Encyclopedia. If this dot could be demagnified by over 25,000 times, it would have a diameter of 80 Å, which is sufficient space for 32 atoms of a common metal.

FIGURE 1.4 Richard Feynman [16].

In other words, Feynman explained that one such dot (1,120 of which can be seen by the human eye in 1 in. is large enough to contain approximately 1,000 atoms, and therefore, the size of each dot can be easily adjusted, as required by photoengraving. As a result, the entire contents of the Encyclopedia could fit on the head of a pin.

The term "*Nanotechnology*" was coined in 1974 by Norio Taniguchi of Tokyo University and used to describe the ability to create nanoscale materials. Until the term was formally defined, nanotechnology had not undergone major developments because it is a field that manipulates atoms and molecules, but no methods to observe and, thus, manipulate material in a controlled manner existed. However, since the invention of the first microscope, scientists have sought to amplify their ability to observe matter. Using a typical microscope with optical lenses, objects invisible to the naked eye and smaller than the wavelength of light can be observed. In contrast, with an electron microscope, it is possible to observe smaller particles with better definition, although individual atoms cannot be clearly distinguished. Then, in 1981, in the International Business Machines (IBM) laboratories in Zurich, Switzerland, Gerd Binning and Heinrich Rohrer developed a microscope known as the scanning tunneling microscope (STM), which won the Nobel Prize in Physics in 1986 and opened the door to nanotechnology and nanoscience. Fig. 1.5 shows the first commercially offered STM.

Briefly, this microscope is equipped with a very fine probe that very closely scans the sample, removing electrons and generating an image of the atomic topography on the sample surface. The Binnig and Rohrer's STM gave rise to an entire family of instruments and techniques that revolutionized our ability to visualize surfaces and materials that

FIGURE 1.5 First scanning tunneling microscope (STM) produced by "W.A. Technology of Cambridge" in 1986. *Free license Image, Creative Commons License, provided by Science Museum London, Flickr. Available from: https://www.flickr.com; https://www.flickr.com/photos/sciencemuseum/9669013645 [17].*

FIGURE 1.6 An STM image of the International Business Machines (IBM) initials [18].

previously could not be observed. Atomic force microscopy (AFM) is one example of a technique derived from STM that allowed visualizing materials that do not conduct electricity. Indeed, these novel microscopes permitted not only visualization but also manipulation of matter on the nanoscale.

One example of such manipulation is the experiment conducted by Donald M. Eigler and Erhard Schweinzer in 1989 at IBM. In their work, they manipulated 35 xenon atoms on a nickel substrate to spell out the company's initials [18] (Fig. 1.6).

Based on this experiment, many other researchers demonstrated the possibility of manipulating matter on the nanoscale. For example, researchers at the Brazilian Agricultural Research Corporation (EMBRAPA), in the Agricultural Instrumentation division, developed a method for the nanomanipulation of a compact disk's (CD's) polycarbonate surface involving mechanical modification via nanolithography using a phosphorus-doped silicon tip. They used this method to "draw" the EMBRAPA symbol and the Brazilian flag on the polycarbonate substrate in a controlled manner in an area of 10 μm × 10 μm [19] (Fig. 1.7).

More recently, in 2009, researchers at Stanford University led by Hari Manoharan wrote the initials of Stanford University (SU) in letters smaller than atoms by encoding 35 bits of information per electron [20] (Fig. 1.8).

Given these advances in atomic-scale microscopy, interest in nanoscience and nanotechnology has been steadily growing. According to Whitesides [13], there are six reasons to study nanoscience:

1. Many properties remain mysterious, such as, the operation of the flagellar motor of *E. coli* bacteria and how electrons move through organometallic nanowires.

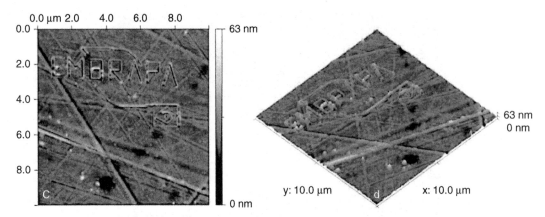

FIGURE 1.7 "EMBRAPA" and an image of the Brazilian national flag scratched onto the surface of a polycarbonate compact disk (CD) with an atomic force microscopy (AFM) tip [19].

2. Nanomaterials are relatively difficult to obtain. Unlike colloids, micelles and crystal nuclei, molecules are easily obtained and characterized. The development of chemical syntheses of colloids that are as accurate as those of molecules remains highly challenging.

3. Many nanostructures are still inaccessible, and their study may lead to the observation of new phenomena.

4. Nanostructures have many sizes in which quantum phenomena (especially quantum entanglement and other reflections of the material's wave character) are expected to occur. Observing such quantum phenomena will contribute to explaining the behaviors and properties of atoms and molecules, but they are typically masked by the classical behaviors of matter and macroscopic structures. For example, quantum dots and nanowires have been produced and found to exhibit unique electronic properties.

5. The nanometric and functional structures responsible for the primary functions of a cell represent the frontier of biology. For example, ribosomes, histones, chromatin, the Golgi apparatus, the interior structure of the mitochondria, the flagellar micromotor, the centers of photosynthetic reactions, and ATPases of cells are nanostructures that must be characterized and understood.

6. Nanoscience is the basis for the development of nanoelectronics and photonics.

The word nanotechnology is relatively new, but this field is not. It is estimated that nature has evolved on Earth for approximately 3.8 billion years, and nature includes many materials, objects and processes that function on the macroscale to the nanoscale [4]. Thus, understanding the behaviors and properties of these materials and processes may facilitate the production of nanomaterials and nanodevices. *Biomimicry*, a term derived from the Greek word *biomimesis*, was defined by Otto Schmitt in 1957 and denotes the development of biologically inspired designs that are derived or adapted from nature [4]. The term

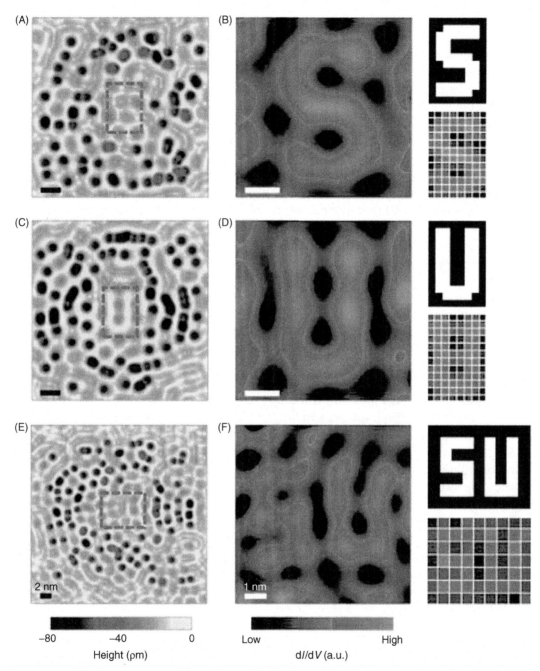

FIGURE 1.8 Stanford University's initials (SU) written using electron waves in a piece of copper and used to design a tiny hologram [20].

biomimicry is relatively new, but our ancestors have looked to nature for the inspiration and know-how to develop various devices for many centuries [21,22]. Indeed, throughout history, many objects and beings, including bacteria, plants, soil, aquatic animals, shells and spider webs, have been found to have commercially interesting properties.

Bacterial flagella rotate at more than 10,000 rpm [23] and constitute an example of a molecular biological machine. The flagella motor is driven by proton flow caused by electrochemical potential differences across the cell membrane. The diameter of the bearings is approximately 20–30 nm, and the gap is approximately ≈ 1 nm [4].

Many billions of years ago, molecules began to organize themselves into the complex structures that gave rise to life. Photosynthesis uses solar energy to support plant life. The molecular assemblies present in the leaves of plants (such as chlorophyll) take the energy from sunlight and transform it into chemical energy to power the biochemical processes of plant cells, which have processes ranging from the nanometric to the micrometric scale. This technology has been exploited and developed for solar energy applications [4].

Some natural surfaces, including plant leaves with water repellents, are known to be superhydrophobic and self-cleaning because of their roughness (arising from nanostructures) and the presence of a wax coating [24]. Using roughness to imbue surfaces with superhydrophobicity and self-cleaning properties is of interest for many applications, including windows, windshields, exterior paints, ships, kitchenware, tiles, and textiles. Superhydrophobic surfaces can also be used for energy conversion and storage [25], whereas surfaces with low wettability can reduce the friction of contacting surfaces at machine interfaces [26].

The fixation structures present on the feet of various creatures, including many insects (e.g., beetles and flies), spiders and lizards, can adhere to a variety of surfaces and be used for locomotion. These structures cling to and detach from different types of surfaces [27,28]. The dynamic adhesion capacity is called reversible adherence or smart adhesion. Common adhesives leave residues and are not reversible. Thus, replicating the characteristics of gecko feet would facilitate the development of a super adhesive polymer tape capable of clean, dry, and reversible adhesion [4]. Such a tape would have potential applications in everyday objects and high-tech applications, such as, microelectronics.

Many aquatic animals can move at high speeds through water with low drag energy. For example, most shark species move through water with high efficiency. Shark skin is fundamental for this behavior, reducing friction, and exhibiting a self-cleaning effect that removes ectoparasites from its surface [4]. These characteristics are attributable to very small structures present in shark skin, called dermal denticles, which are ridges with longitudinal grooves that result in very effective mobility through water and minimize the adherence of barnacles and algae [4].

Speedo developed a full-body, shark skin-based swimsuit, called Fastskin, for elite swimming. Furthermore, the builders of boats, ships, and aircraft have also attempted to mimic shark skin to reduce drag and minimize the fixation of organisms on the surfaces of these craft. The mucus on the skin of aquatic animals, including sharks, acts as a barrier against the osmotic salinity of sea water, protects against parasites and infections, and functions as a friction-reducing agent. Artificial fish-derived mucus products are currently used to propel crude oil through the Alaskan pipeline [4].

Shells are natural nanocomposites with laminated structures and superior mechanical properties. Spider webs are made of silk fiber with high tensile strength. The materials and structures used in these objects have led to the development of various materials and fibers with high mechanical resistance [4]. Moths have eyes with multifaceted surfaces on the nanoscale and are structured to reduce the reflection of light. Their antireflective structure inspired the development of antireflective surfaces [29].

Biological systems' self-healing abilities are highly interesting. For example, the chemical signals originating from the site of a fracture initiate a systemic response that sends agents to repair the injury. Inspired by these activities, various artificial self-structuring materials have been developed [30]. Human skin, for example, is sensitive to impact, which leads to purple discoloration of affected areas. This behavior led to the development of coatings that indicate impact-related damage [21].

Sensor arrays "mimicking" human senses, such as smell [31] and taste [32–35], consist of a set of sensors based on nanostructured materials and have been widely used in various applications, such as gas and liquid sensing, respectively [31,36–40].

Nanostructured materials are typically named according to their shapes and sizes and may take the forms of particles, tubes, wires, films, flakes and reservoirs, provided they have at least one nanoscale dimension [41,42]. One material that has been widely studied in nanotechnology is carbon nanotubes (Fig. 1.9), which are so named because their

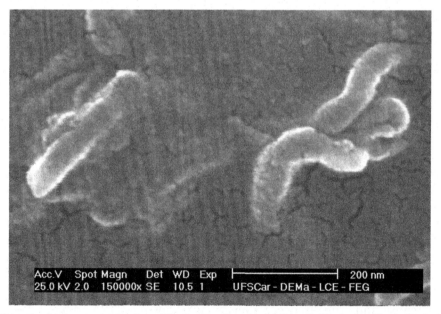

FIGURE 1.9 Image of carbon nanotubes immobilized onto a film of poly(allylamine) hydrochloride obtained with a high-efficiency electron microscope [a scanning electron microscope coupled with a field emission gun (SEM-FEG)] in the Materials Engineering Department of the Federal University of São Carlos (DEMA-UFSCar) [43].

diameters are between 1 and 100 nm, although their lengths are typically on the order of hundreds of nanometers.

There are two main modes of developing nanotechnological materials. In the bottom-up approach, materials and devices are built from molecular components that are chemically organized according to the principles of molecular recognition. In contrast, in the top-down approach, nanoscale objects are built from other, larger scale objects, without control at the atomic level [44]. Many methods using these two modes have been reported.

In the bottom-up approach, a DNA molecule may, for example, be used to build other larger and well-defined structures using DNA and other nucleic acids [4]. Another example is the self-assembly technique, which can create self-organized molecular layer films [45,46]. Additionally, as previously discussed, an AFM tip can be used as a nanoscale recording head [19].

Due to top-down approach, many "solid-state" technologies used to create silicon-based microprocessors are now able to use resources on a scale smaller than 100 nm. Other techniques can also be applied to create devices known as Nano Electro Mechanical Systems (NEMS) derived from Micro Electro Mechanical Systems (MEMS) [4,5].

Examples of NEMS include microcantilevers with integrated nanotips for STM and AFM, AFM tips for nanolithography, molecular gears used to attach benzene molecules to the outer walls of carbon nanotubes, magnetic media used in hard disk drives, magnetic tape units [4], and ion beams that can directly remove or deposit materials in the presence of precursor gases. AFM can also be used in the top-down approach for the deposition of resistive films on a substrate that is subsequently subjected to an etching process [3–5,41,42].

Based on the manipulation of materials on the nanoscale, nanoelectronics can also be used to create computer memory using individual molecules or nanotubes capable of storing bits of information, molecular switches, nanotube transistors, flat-panel nanotube displays, integrated circuits, fast-access logic gates, nanoscopic lasers, and nanotubes as electrodes in fuel cells [4].

BioMEMS or BioNEMS are micro- or nanoelectromechanical systems with biological applications and have been increasingly used commercially [5,47–51]. These devices have been applied for chemical and biochemical analyses (biosensors) for medical diagnoses (e.g., DNA, RNA, proteins, cells, blood pressure, and toxin detection and identification) [51,52] and in implantable devices for controlled drug release [53]. Biosensors have also been developed to monitor liquids and gases [54–58]. A wide variety of biosensors are based on the principles of micro-/nanofluids [55,57,58]. Indeed, micro-/nanofluid devices offer the ability to work with smaller reagent volumes and reduced reaction times and facilitate performing various analyses simultaneously [52]. Other types of biosensors include micro-/nanoarrays able to perform one type of analysis millions of times [48–52].

Micro-/nanoarrays are used in biotechnology research to analyze DNA or proteins for disease diagnosis and drug discovery. They are also known as DNA arrays and can simultaneously identify thousands of genes. They consist of microarrays of silicon nanowires with diameters of a few nanometers, which are able to selectively interact with and even detect a single biological molecule, such as a DNA or a protein, or microarrays of carbon nanotubes capable of electrically detecting glucose. Detection occurs via nanoelectronics that sense small variations in an electric signal generated by the interaction of the analyte (e.g., DNA, protein or glucose) with the micro-/nanoassay [59].

BioMEMS and BioNEMS are also being developed to minimize invasive surgical procedures, including endoscopic surgery, laser angioplasty, and microscopic surgery. Other applications include implantable drug-release devices (micro-/nanoparticles encapsulating drug molecules in functionalized shells for activity at specific locations) or silicon capsules with nanoporous membranes for controlled release [4,48,49].

Nanoscale structures assumed key roles in technological development for a variety of reasons, some of which are listed next [1].

In quantum mechanics, the wave properties of electrons inside matter are influenced by nanoscale variations. Through the nanoscale design of materials, it is possible to vary their micro- and macroscopic properties, such as their charge capacity, magnetization and melting temperature, without altering their chemical composition [1].

The systematic organization of matter on the nanoscale is a key feature of biological systems. Nanoscience and nanotechnology seek to achieve powerful combinations of biology and materials science by adding artificial nanostructured systems to living cells and creating new materials using the self-structuring properties found in nature.

On the nanoscale, the surface areas of various materials are generally much larger than their geometric volumes, which favors their use in various applications, such as, composites for drug delivery and reaction systems and the storage of both chemical energy (such as hydrogen and natural gas) and electrochemical energy (batteries and supercapacitors) [2–5].

Macroscopic systems developed from nanostructures can have much higher densities than those obtained from microstructures. Therefore, when controlling the complexity of interactions on the nanoscale, these systems may be better conductors of electricity and thereby allow the development of novel electronic devices and smaller and faster circuits with more sophisticated functions and significantly reduced power consumption [2–5].

Currently, researchers from universities and companies worldwide are developing nanomaterials to create new products and lead to technological advances in engineering, chemistry, physics, computer science, and biology. Among these numerous technological advances, some easily recognized examples include the development of faster processors, more effective drugs, lighter and more resistant materials, new charge devices (e.g., more efficient batteries) and equipment with lower energy consumption and better image definition [e.g., organic light-emitting diode (OLED) screens] [2–5].

Nanotechnology represents a new world that is finally within our reach, thanks to quantum mechanics and sufficiently advanced simulation and analysis techniques. Currently, nanotechnology has important potential applications in all fields.

Due to technological advances in visualization and the ability to manipulating matter at the atomic and molecular levels, nanotechnology has become a major research theme for the development or improvement of products. Indeed, its popularity is reflected by the abundant news items reported in the media (television, radio, newspapers, magazines, and the Internet). The term "nanotechnology" has become synonymous with quality. Thus, the nanoscience knowledge domain will be of fundamental importance for the development of new technologies.

1.1.3 Investments, Strategies, Actions, and Research in Nanotechnology

Since nanotechnology has benefitted the development or improvement of a wide variety of commercial products, the term nanotechnology has been popularized worldwide. Indeed, its use has not been restricted to the world of scientific research, and as a result, the term "nano" is sometimes improperly used to increase sales volumes or stimulate sales; therefore, "nano" products should be thoroughly reviewed [60]. As explained previously, standard definitions of "nano" exist and must be followed. Furthermore, many products have actually benefitted from nanotechnology research and have significantly influenced the global economy. According to the National Science Foundation (NSF), 1 trillion US dollars were exchanged in the nanotechnology market involved in 2015. Furthermore, Lux Research reported that in 2014, this market was valued at 2.6 trillion US dollars and that in 2012, worldwide, the total investment in this field was approximately 7.9 billion US dollars [61,62]. Table 1.1 summarizes the invested amounts up to 2015 [63].

Both private and public sectors are strongly interested in this scientific field, as demonstrated by the study conducted by Allianz, which investigated many countries' nanotechnology-development programs. Such programs are funded by and involve various organizations, ministries, government agencies, and the private sector. Countries that stand out are the United States, China, Japan, Germany, Spain, France, England, Sweden, Italy, the Netherlands, and Finland [63]. Since 2001, more than 60 countries have instituted national programs in nanotechnology [64].

To optimize these investments, long-term plans have also been studied, such as the Horizon 2020 plan (investment plan for 2014–20) launched by the European Union (EU). In this plan, after the closing of the Framework Panel 7 (FP-7) with an investment of €77 billion, an investment of €6.6 billion is destined for the Horizon 2020 program in six areas of nanotechnology: advanced manufacturing of nanotechnology, advanced materials, nanoelectronics, photonics and biotechnology. Additionally, €5 billion in investments from

Table 1.1 Countries With the Largest Investments in Nanotechnology [63]

Country	Amount Invested	Period
United States	US $15.6 billion	2001–12
Japan	€2.8 billion	2011
China	€1.8 billion	2011–15
Germany	€500 million	Per year
France	€400 million	Per year
United Kingdom	€250 million	Per year
Other European Union Countries	€100 million	Per year

public and private investors are predicted. Japan has an investment plan lasting until 2020 that provides an investment of US $1 billion per year [65].

All nanotechnological successes depend on both investments and the relationships between research and development centers (universities and public or private companies). To contextualize the importance of such relationships, we highlight the case study reported by Sá [66], who investigated the actions taken by the University of Albany (the United States) to create the first college of nanotechnology as part of a larger center for research and development with the intent of building a regional cluster for nanoelectronics research: the College of Nanoscale Science and Engineering-Albany Nanotech. For the success of this project, in addition to the invested capital, it was necessary to establish new ways of thinking and administrating. Traditional models of relationships between universities and the corporate world were dispensed with, and a new model was created. Research professors with entrepreneurial interests were highly important for this development. Thus, the University of Albany concentrated on several characteristics, including building a university campus and a technology park and establishing an agent to promote regional and industrial development [66].

This study showed the importance of not only financial investment but also new ideas, the breaking of paradigms, and strategic and specific actions, which can be critical for the success of nanotechnological development.

Another factor, which is somewhat curious and can assist in the development of nanotechnology, has nothing to do with money or commercial interests. Instead, it relates to public commitment, combining civil society, and scientific and technological research. With the emergence of new policies, at least in the United States, Australia, and Europe, public commitment has become important in deciding the path of innovation. In Europe, for example, the Responsible Research Innovation (RRI) program seeks not only to align society with science but also to make research centers accessible to society, especially regarding decision-making about relevant problems. This type of engagement between civil society and the scientific world is a form of upstream public engagement, which is the opposite of downstream public participation, in which society participates only in the final stages of technological development, for example, after important decisions have already been made and the developed product is ready to enter the market. Upstream public engagement is a public commitment allowing civil society to provide input from the beginning regarding the course that scientific and technological research should take to produce new, beneficial products [67].

As mentioned previously, the success of nanotechnology is evident and have attracted high investment. As a result, there is concern among governments and investors regarding the evaluation and justification of the impact of spending on nanotechnology. For example, in the United States and Japan, there are specialized programs for such evaluations: the Science of Science and Innovation Policy (SciSIP) and the National Institute of Science and Technology Policy (NISTEP), respectively. In addition, there is also great interest on the part of society and especially development agencies and investment sectors in the impact and quality of publications resulting from research and development [68].

Therefore, bibliometric and scientometric studies are used in the identification, quantification, qualification, classification, and frequency of publications in nanoscience and nanotechnology. Such studies have a certain level of complexity, and one way to contextualize this complexity is by discussing how such studies are conducted: searching for articles on relevant sites [for example, the Web of Science (WOS)] using keywords. In the case of nanotechnology, because of the large number of variations involving the term "nano," "nano" itself can be used as a keyword. However, the articles generated from such a search may or may be relevant to nanoscience and nanotechnology, and thus, the results must be carefully evaluated. Thus, this type of study is as complex as the actual research in nanoscience and nanotechnology, and abundant relevant articles have been published in the area. One example is the study of Porter et al. [69], who evaluated the performance of software used to search for publications on the subject of nanotechnology. In the WOS, 406,000 papers and more than 53,000 micropatents and patents were found between 1990 and mid-2006. Another assessment performed from 1990 to mid-2011 indicated that 820,000 documents could be identified in the WOS [70]. Two other studies were conducted by Shapira and Wang: one study evaluated the quantity and quality of articles published from 2008 to 2009, identifying 91,500 articles published on the subject nanotechnology, among which only 67% were informative articles [64]. The other study evaluated the impact factor and quality of published papers. That study revealed that the quality of the publishing journal and the number of citations were linked the major sponsors of the study [68]. Stopar and Bartol conducted a very restricted evaluation of the scientific output in nanoscience and nanotechnology in 2012, searching for articles in the ten journals with the most publications containing the term "nano" in the title. Journals were selected by researching the core collection (old citation index) of the WOS. The Scopus database was also evaluated, revealing the same ten journals identified in the core collection of the WOS. These 10 journals were the *Journal of Materials Chemistry, Journal of Physical Chemistry C, Applied Physics Letters, Journal of Nanoscience and Nanotechnology, Journal of Applied Physics, ACS Nano, Nano Letters, Nanotechnology, Nanoscale,* and *Materials Letters* [71].

Among all the assessments made in this study, the most striking results was that in 2012, a total of 2,700 journals published articles with the term "nano" in the title, but of these 2,700 journals, only 50 published more than 200 articles; furthermore, the ten journals selected in the study had published approximately 9,000 articles on the subject of nanoscience or nanotechnology. Another interesting result was the number of compound terms containing the term nano that were found. In ascending order, the terms most commonly found were as follows: nanoparticle, nanowire, nanotube, nanostructur*, nanocrystal, nanocompos*, nanorod, nanoscale, nanofiber, nanopor*, nanoribbon, nanosheet, nanocluster, nanosphere, nanometer, nanomaterial, nanobelt, nanocube, nanopillar, nanoplate, nanopattern, nanodot, nanoflake, nanosize, nanohybrid, nanoplatelet, nanoantenna, nanoring, nanoimprint, nanosecond, nanopowder, nanomechanic, nanocapsule, nanoshell, nanogap, nanomembrane, nanoindent* and nanoflower [71].

Finally, this study reported the number of articles in the literature published over a 1-year period and indicated how complex it is to perform this type of study. Clearly,

Table 1.2   Number of Articles Published Between 2012 and 2015 Containing the Three Most-Cited Nanoterms in 2012

Nanoterms	Number of Articles (2012–15)
Nanoparticle	18,336
Nanowire	7,348
Nanotube	36,547
Total	62,331

Source: webofknowledge.com.

abundant scientific articles are produced, as indicated by the studies cited and by a simple search of the WOS. When we searched for titles included the top three most-cited keywords containing the term "nano," according to a study by Stopar and Bartol covering 2012–15, we obtained the following results (Table 1.2)

The numbers listed previously were not subjected to scientometric or bibliometric evaluation; indeed, only a search for article titles containing the indicated nano-terms was performed. However, these results provide perspective regarding the vast number of articles in the literature. One thing is certain: all of these numbers are the result of the popularity of research in nanoscience and nanotechnology.

As shown, nanoscience and nanotechnology have provoked important changes in the political, economic, administrative, social, and scientific spheres. Given the data evaluated here, it may be concluded that more than half of the globe benefits from and participates in this new branch of science. Furthermore, financial investments increased during each studied period. Interest in this new science is attracting increasing attention from many sectors ranging from technology to health, and the dissemination of scientific research results, as reflected by the huge quantity of publications, indicates the popularity of this new scientific field.

1.1.4 Commercial Products Involving Nanotechnology

Currently, given the recent technological advances, high-resolution electron microscopes, and the ability to manipulate and manufacture materials and devices with controlled properties at the nanoscale, many companies have invested in quality improvements or producing new products.

The most easily identifiable applications are electronic devices: nanotechnology is critical to the development of various components, such as, microprocessors, digital screens, and batteries.

The production of data-storage chips begins with the wafer, which consists of a disk cut from a single silicon crystal with a diameter of at least 300 mm and is typically between 500 and 800 µm thick. Integrated circuit structures are built layer by layer on the surface of the chip using etching techniques, such as, lithography [72]. The development of the transistor (the building block of integrated circuits and microchips) was based on complementary metal–oxide semiconductors (CMOSs); transistors have existed for decades and are part of our everyday lives. The smallest of these structures was as low as 22 nm in 2015 (the "router" transistor) and is near the operating limit, approximately 5 nm, for metal–oxide semiconductor field-effect transistor technology; this limit is based on the high leakage current that should occur inside the chip and will likely be reached within the next 20 years [73]. These developments represent major technological challenges because of the manufacturing process itself and performing heat-management tests in circuits (a modern high-performance chip can dissipate heat at a density of 100 W·cm^{-2}, which exceeds the ability of a domestic kitchen hot plate) [74].

Silicon is envisaged as an important semiconductor material, but to fabricate ever smaller structures, new photoresistors must be developed. Furthermore, thin silicon oxide films become increasingly less efficient as they become thinner, and their leakage currents become very high. Thus, SiO_2 was replaced by physically thicker layers with higher dielectric constants. New semiconductors must satisfy some conditions, such as, high dielectric constants, thermal and kinetic stabilities, band shifts, good quality Si interfaces, and low defect densities. Hafnium oxides (HfO_2) have emerged as attractive oxides, typically with the incorporation of nitrogen. Ge has also been incorporated into the structure of CMOS transistors to maximize performance [75]. The origin of ferromagnetism is based on both electron charges and spin. Therefore, the miniaturization of magnetic memories has been limited not by the final size of the ferromagnetic domain but by the sensitivity of the magnetic sensors. In other words, the main limitation is not the ability to make very small storage cells but the ability to detect very small magnetic fields [74].

The influence of rotation on the electron conductivity was invoked by Nevill Mott in 1936 but remained virtually unexplored until the discovery of giant magnetoresistance (GMR) in 1988 [74,76]. The main application of spintronics (loosely defined as the field of devices in which electron spin plays a role) is the development of ultrasensitive magnetic sensors for magnetic reading memories. GMR is used for the reading and writing heads of computer hard disk drives [76].

A second type of magnetic sensor is based on the magnetic tunnel junction (MTJ), in which a very thin dielectric layer that is nonconductive under applied voltage separates the ferromagnetic layers (electrode) and the electron tunnel. The sensitivity of the magnetic field exceeds that of GMR devices. MTJ devices also have high impedance, allowing high output signal values. Unlike GMR devices, electrodes are magnetically independent and may have different critical fields to change the orientation of the magnetic moment. The first laboratory samples of MJT devices (NiFe-Al_2O_3-Co) were reported in 1995 [74].

As previously described, there are no limits on the demand for devices that are smaller, lighter, and more efficient and have lower power consumption. When merely considering the development of mobile telephone devices (cellular phones), in just 20 years, development in this segment has been remarkable. In 1993, smartphones were fiction, a distant possibility at best. At that time, at least in Brazil, cell phones were not prevalent. Furthermore, although they had existed since the 1980s, even in the 1990s, the predominant devices were expensive, large, and heavy; had antennas; and operated using analog networks.

FIGURE 1.10 Mobile devices from the early 1980s to the present day. (A) Motorola DynaTAC 8000X (1983), (B) Pager Motorola (1980–90), (C) Motorola MicroTAC 9800X (1989), (D) Motorola StarTAC (1996), (E) Apple iPhone 1 (2007), and (F) iPhone 6 Plus (2015).

Currently, cellular phones are energy-efficient machines, are fast and provide all the functions of notebook computers, and combine multiple devices into one, including a personal computer, digital camera, sound recorder, text and spreadsheet editor, gaming station, and Global Positioning System (GPS). There are also numerous manufacturers and a wide range of models and prices. The evolution of mobile phones went through a period of miniaturization, and now, with the advent of smartphones, these devices are growing again, providing increasing numbers of functions (Fig. 1.10). Recently, smart watches have also been developed that are similar to smartphones.

Another example from everyday life is the evolution of television sets and computer monitors, and the emergence of different displays in portable devices. Briefly, all of this evolution began in the mid-18th century with Charles Du Fay, who studied the electron emission from a heated metal surface, which is known as the thermoionic effect [77]. Since then, several studies on this subject were performed until it was concluded that this phenomenon was attributable to cathode rays [78]. These rays were further studied in 1897 by Joseph John Thomson [9], who found that cathode rays were simply electrons and that they behave as particles rather than waves. It should be noted that in parallel to the research of J.J. Thomson, Wilhelm Conrad Roentgen also investigated the behavior of cathode rays and discovered how the X-ray machines used in the medical field worked [79]. Finally, building on these discoveries (thermoionic effect and the nature of cathode rays), the kinescope was developed; this device was responsible for image formation in the first televisions and computer monitors [80].

The television sets and computer monitors that used cathode ray tubes for image generation have since been replaced by other devices because of the development of the

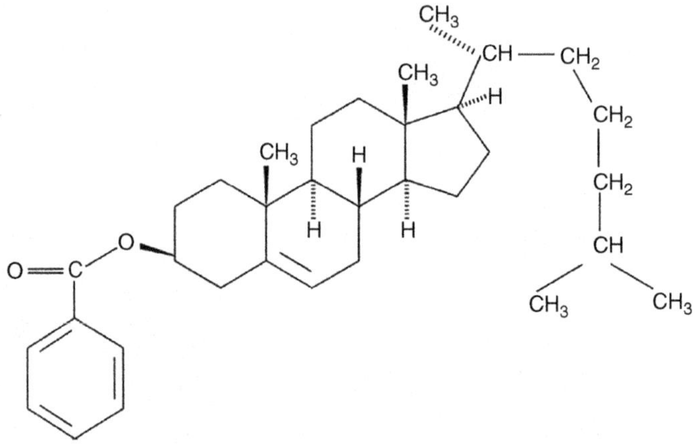

FIGURE 1.11 Structure of cholesteryl benzoate [84].

liquid crystal displays (LCDs). LCD screens comprise a combination of polymers and glass slides coated with indium tin oxide (ITO) [81]. The discovery of liquid crystals is attributed to the botanist Friederich Reinitzer in 1888 [82], when he observed the double melting point of cholesteryl benzoate (Fig. 1.11). However, the term liquid crystal was coined by Coube Lehmann, who believed that the degree of fluidity was the only difference between liquid crystals and solid crystals. However, liquid crystals are actually characterized by the degree of molecular order between the long-range orientational and positional orders of a solid crystal and the long-range disorder of isotropic liquids and gases [83].

In the Web of Knowledge database, when the title search keyword "liquid crystal displays" is used, the first works that describe this technology can be seen to date back to 1968: *Liquid Crystals—A Step Closer To Low-Voltage Displays* (anonymous author) [85] and *Reflective Liquid Crystal Television Display* [86]. However, 99,644 other works are also found (searched on September 16, 2013) that describe the development and diverse applications of LCDs. Initially, the displays that used liquid crystal materials had very limited image resolutions and were used only in wristwatches and calculators [87]. An account of their development is given by Toshiaki Fujii et al. [87], who developed a LCD with a resolution capable of reproducing not only letters and numbers but also graphics, figure patterns, and even Chinese characters. In summary, this LCD revealed the possibility of replacing cathode ray tube monitors. The results reported in that article were based on a screen tested using a computer, and the authors were believed to have developed the first LCD monitor at the time. Given the constant evolution of this technology, currently, there is competition between LCD screens and OLEDs, which were first described by two researchers from Eastman Kodak in 1987. They developed an electroluminescent device with a thickness of 135 nm that operated at low voltage (below 10 V) [88]. Indeed, currently, OLEDs are LCDs' main rivals (Fig. 1.12).

The basic structure of an OLED consists of one or more emissive (electroluminescent) polymeric layers (organic part) placed between a transparent anode and a metallic

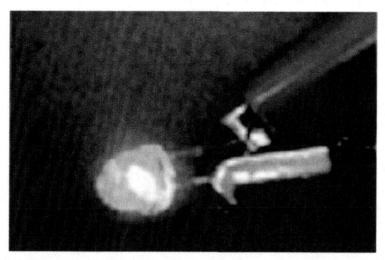

FIGURE 1.12 Organic light-emitting diode (OLED) developed to increase the power of light emission [89].

cathode. Typically, the organic layers are composed of a hole-transporting layer (HTL), an emitting layer (EL) and an electron-injection layer (EIL), and thus, a variety of architectures and desired effects can be achieved [90]. All OLEDs require at least one transparent electrode. Traditionally, ITO has been used, but the Earth's supply of indium is very limited, and at current consumption rates, it may be completely exhausted within 2–3 years. Furthermore, constant advances in integration and miniaturization are causing the effective recycling of indium from discarded components to become increasingly complicated. Thus, there is great interest in the fabrication of transparent polymer additives containing a small proportion of carbon nanotubes for conductivity.

One additional advantage of organic electronic devices is their ability to be deposited on any substrate, including flexible and robust plastic sheets. New organic light-emitting polymers are currently in the research phase and have potential for the development of thinner and flexible displays with high resolution and low power consumption [90,91] (Fig. 1.13).

In addition to the development electronic devices, nanotechnology is used in all other areas, including health, sports, clothing, automotive, and food.

A natural textile fiber, such as, cotton has an intricate nanostructure. Many of the comfortable properties of traditional textiles result from a favorable combination of chemistry and morphology. These factors are what allow the properties of natural textiles to be equal or superior to those of synthetic materials. Furthermore, nanoadditives can increase some of the properties of natural textile fibers, such as, mechanical strength, durability, heat and flame resistance, self-cleaning capability, color, and antiseptic effect.

Some textiles release useful chemical substances passively or actively. These advanced functional fabrics are often used in special applications, serving as a type of "scaffold" for cells facilitating tissue regeneration and wound sterilization to facilitate healing [74].

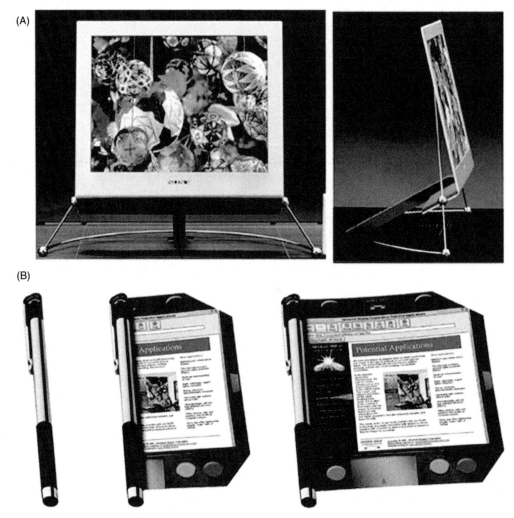

FIGURE 1.13 OLED screens: (A) 2-mm-thick OLED monitor; (B) concept for a future OLED screen, which could be rolled up like a scroll inside of a pen-sized device when not in use [91].

Dr. Robert Burrell of the University of Alberta (Canada) developed dressings with silver nanocrystalline films in 1995 [92]. While working for Westaim Corp's NUCRYST Pharmaceuticals, he invented Acticoat (Fig. 1.14), a silver-based dressing, which had antimicrobial properties and increased the healing rate. This dressing is often used in burn units and is now sold worldwide combined with common adhesive bandages. This work earned awards, including the World Union of Wound Healing Society Lifetime Achievement Award recognizing contributions to wound healing and, in 2009, the ASM Engineering Materials Achievement Award.

FIGURE 1.14 Acticoat antimicrobial adhesive [93].

Wall papers with antibacterial properties have also been developed, in which zinc oxide (ZnO) nanoparticles are incorporated into the cellulose fibers and react with the gram-positive bacteria *Staphylococcus aureus* and gram-negative bacteria *Escherichia coli*, the main causes of nosocomial infections, and the fungus *Aspergillus niger* [94].

Tennis is a good example of the integration of nanotechnology in sports. Wilson, one of the largest tennis sporting goods manufacturers in the world, manufactures balls with less-permeable rubber than traditional balls (nanoclays are mixed with the rubber to achieve this effect) and lighter, more-resistant tennis racquets produced using carbon nanotube fiber structures [95].

Eyeglasses, sunglasses, windows, and even photographic camera lenses may also include nanometric particles that function as conductive polymers and change color based on the absorption of UV light from sunlight. This mechanism is known as "photochromism," and today, it is widely commercially available.

In the automotive industry, nanotechnology has an increasingly important role in the development of new vehicles. The main overriding goal is to reduce vehicle weight without compromising other attributes, such as, safety. Therefore, researchers are especially interested in replacing the heavy metals used in components with lighter polymers that are reinforced by nanoparticulate materials or nanofiber additives. Other, more specific objectives include formulating lightweight, electrically conductive materials for use in fuel lines to prevent static electricity and abrasion-resistant paints, reducing friction, and developing lights (LED), automotive sensors, multimedia devices, and electronics [74]. Polymeric nanomaterials, such as, carbon nanotubes, have been used in bumpers with safety and resistance levels similar to those of conventional bumpers while being significantly lighter, directly influencing vehicle fuel consumption. In Brazil, Fiat, for example, developed its FCC II car with clay nanocomposites. The body is composed of fibers containing nanoclay, which is a clay with a nanoparticle-based chemical additive. This composite has characteristics similar to those of fiberglass but is lighter, cheaper, and easier to recycle [96].

Another example is the use of nanoparticles in automotive paints. Mercedes-Benz recently developed a special paint with nanometric paint bubbles that can regenerate small scratches in the paint. When a scratch occurs, these small bubbles break and release paint that covers the exposed area, improving both the protection and aesthetics of the vehicle. Additionally, a special paint has been developed that allowed a car painted yellow to change color with the application of a low current or because of the incidence of sunlight [97].

In the health field, substantial research and development has been invested in products using nanotechnology for controlled drug delivery and use in surgical catheters.

Nanomedicine is defined as the application of nanotechnology in human health. In recent decades in particular, medicine has been characterized by important technological advances, accompanied by an enormous and concomitant expansion of its ability to diagnose and cure diseases.

Molecular biology can be considered an example of conceptual nanotechnology because the structures involved are on the nanoscale. Furthermore, molecular biologists' work increasingly involves nanometrology, such as, the use of microscopic scanning probes. As mentioned earlier, biomimicry is the use of nanotechnology to artificially recreate natural nanoscale materials, devices, and systems. Since Drexler presented biology as a "proof of principle" of nanotechnology [98], a close relationship has developed between biology and nanotechnology. Thus, nanotechnology can influence the health field, and the resulting social changes have greatly impacted human health. In other words, the impact of nanotechnology on human health is not restricted to the development of drug delivery systems or devices and equipment [74].

In the pharmaceutical industry, three major developments are currently predicted: sensing, automated diagnosis, and personalized pharmaceutical products. The development of ever smaller sensors able to penetrate the body using minimally invasive procedures, such as endoscopy, is an example of the direct application of nanotechnology. Researchers are even seeking to develop such sensors as permanently implanted devices. Indeed, these devices may be able to continuously monitor physiologically relevant physicochemical parameters, such as, temperature and the concentrations of selected biomarkers. Presently, one of the greatest challenges is the automatic diagnosis of disease [74]. The third development, which is believed to be the most accessible, is the creation of custom pharmaceuticals using microfluidic mixers and reactors; however, these methods are not based on nanoscale techniques.

Prostheses and biomedical devices must be biocompatible [74]. Implants that play structural roles, such as bone substitutes, should be assimilated by the host. When assimilation fails, the implant typically becomes coated with a layer of fibrous material, which can move, causing irritation and weakening the implant's structural role. For implants in the bloodstream (arterial stents and, possibly, future sensors), the opposite behavior is required: Blood proteins should not attach to such devices. In this case, adsorption has two deleterious effects: The accumulation of protein layers may clog the blood vessel, and the adsorbed proteins may change and therefore appear foreign to the host organism, triggering inflammatory responses. Thus, to achieve the required assimilation behavior, structures with appropriate surfaces (textures) must be developed [74].

For nonimplanted medical devices, such as scalpels or needles, researchers are focusing on developments to improve their sterilization and decrease their coefficients of friction to minimize the physical force required to penetrate the skin and, thereby, minimize patients' pain [74]. These improvements can be achieved by high-precision machining and rough surface finishing at the nanoscale. Long-term implants should be designed to minimize the chance of bacterial infection. Once bacteria colonize an implant, their phenotype generally changes so that the cells become "invisible" to the immune system and antibiotics, resulting in persistent inflammation without destruction of the bacterial colony [74].

Implants with contact surfaces, such as those used in arthroplasty, typically generate particles as a result of wear, which can be relatively large and cause inflammation [74].

The oldest well-documented example of the use of nanoparticles in medicine is Paracelsus's deliberate synthesis of gold nanoparticles (called "soluble gold") as a pharmaceutical formulation [74]. The use of nanoparticles in medicine has recently become a flourishing field. Applications include magnetic nanoparticles directed by external fields to the site of a tumor and then energized by an external electromagnetic field to kill the particles' neighboring cells, nanoparticles acting as carriers for drugs and nanoparticles that can be used as markers for disease diagnosis [74]. The use of nanoparticles for drug release is being intensively researched and developed, and many products are undergoing clinical trials. A major obstacle for the successful development of new drugs is that many drugs that exhibit good therapeutic interactions with a target molecule (e.g., an enzyme) have very low solubility in water. However, these compounds can be encapsulated in nanoparticles with hydrophilic outer surfaces. Such surfaces also prevent adsorption and the triggering of an adverse immune response, such as, protein denaturation during the passage of the particle through the bloodstream [74]. Thus, this technique facilitates not only controlled drug release but also targeting these nanocapsules (Fig. 1.15) to the desired location and thus avoiding side effects in healthy organs, such as, those caused by the chemotherapeutic agents used to treat cancer [99].

Molecular biology and clinical research constitute an important market for nanoparticle. For example, nanoparticles may be useful as biomarkers: By coating them with chemical products with specific affinities for particular targets (e.g., antibodies), nanoparticles can be used to more easily map the locations of those targets using microscopy. Such particles may simply consist of heavy metals, which are easily visualized under an electron microscope, or fluorescent particles [74].

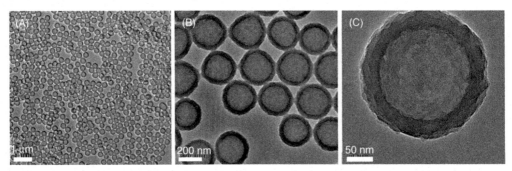

FIGURE 1.15 Images of silica nanocapsules (A–C) used in experiments investigating the treatment of tumors in rabbits obtained with a transmission electron microscope (TEM). [99].

The use of toxic materials for cosmetic purposes (e.g., applying them to the skin of the face) has a long history, as in the case of antimony salts, which were popular among the Romans. Nanotechnology contributed to creating new sunscreen formulations using smaller nanoscale particles in their emulsions. The advantages of these new formulas mainly include better spreading of the emulsions and, thus, improved skin coverage. Furthermore, upon spreading, the particles become invisible to the naked eye, and the whitish color common in older formulations is avoided. Additionally, advances in our knowledge of toxicity yielded much more benign materials, although the recent use of very fine particles (e.g., zinc oxide nanoparticles in sunscreens) has generated further concern regarding the possibility of their penetration through the outer layers of the skin or into cells and causing unknown effects [74].

Robots (microscopic or nanoscopic) are extensions of existing ingestible devices that move slowly through the gastrointestinal tract and provide information (mainly images [74]). As noted by Hogg [100], the minimum future requirements for such devices are as follows: detection (chemical), communication (receiving information and transmitting information out of the body, communicating with other nanobots), locomotion, computing (highly miniaturized electronics would be very attractive for the construction of the on-board logic circuitry), and power (it is estimated that power on the order of picowatts would be sufficient to propel a nanobot at a speed of approximately 1 mm$\cdot$s^{-1}). To be effective, these nanobots will likely need to operate in swarms, making the requirement that they be able to communicate increasingly important.

Nanoscale materials have been used to solve problems in various fields of research and development. From a simple package to a high-tech device, nanoscience is clearly of fundamental scientific importance.

1.2 Final Considerations

Nanoscience is the main tool supporting this century's scientific and technological developments. The domain of the atomic world is the main focus of this branch of science. Another key factor for the development of this field is its multidisciplinary character, as demonstrated, for example, by the numerous improvements and development of new products in technology and medicine. In this context, engineers, biologists, physicists, chemists, doctors, and many other professionals in health and technology may be involved. Additionally, given the variety of problems solved, the development of new products, and the ability to add value to existing nanotechnology products, political and economic sciences are also contributing to the development of nanotechnology worldwide. The investments in Japan, the United States, the EU and other countries are enormous, notwithstanding the substantial investments from the private sector. Thus, the nanotechnology market is expected to achieve a one-trillion-dollar valuation in the coming decades.

In scientific terms, substantial activity is occurring among scientists with regard to technical and scientific publications and mass media, and as a result, ordinary people are engaging with the scientific world. In fact, prior to nanotechnology, society was never so interested in this community. However, despite major advances and numerous problems solved by nanotechnology in health fields, some studies and research groups have expressed concern about the potentially harmful effects of nanotechnology, that is, the ability of these nanoparticles to damage health human [101–111]. Therefore, ethics and thoughtfulness are necessary in this field, particularly with respect to the disposal of nanomaterials, including laboratory or industrial waste and commercialized products, especially electronics.

The smallest unit of matter has inspired major events worldwide. Science does not stop.

List of Symbols

d	Diameter
$mm·s^{-1}$	Millimeters per second
nm	Nanometer
rpm	Rotations per minute
V	Volt
$W·cm^{-2}$	Watts per square centimeter
μm	Micrometer

References

[1] G.A. Mansori, Principles of Nanotechnology: Molecular-Based Study of Condensed Matter in Small Systems. [S.l.], World Scientific, Singapore, (2005) p 358.

[2] F. Allhoff, P. Lin, D. Moore, What is Nanotechnology and Why Does It Matter? [S.l.], John Wiley and Sons, New Jersey, NJ, (2010) pp. 3-5.

[3] S.K. Prassad, Modern Concepts in Nanotechonology, [S.l.]: [s.n.], Discovery Publishing House, New Delhi, (2008) pp. 31-32.

[4] B. Bhushan, Springer Handbook of Nanotechnology, Springer-Verlag Berlin Heidelberg, New York, (2010) 1222 p.

[5] C.P. Poole, F.J. Owens, Introduction to Nanotechnology, John Wiley & Sons, Hoboken, (2003).

[6] A. Merkoçi, Biosensing using nanomaterials, in: A. Merkoçi (Ed.), Nanoscience and Nanotechnology, Wiley Series, Hoboken, 2009, pp. XV.

[7] Laboratório de Química do Estado Sólido [Laboratory of Solid State Chemistry]. Available from: http://lqes.iqm.unicamp.br/; http://lqes.iqm.unicamp.br/canal_cientifico/lqes_news/lqes_news_cit/lqes_news_2008/lqes_news_novidades_1208.html

[8] E. Longo, L.H. Toma, Novos Materiais e Nanotecnologia [New Materials and Nanotechnology], Anais da Reunião Anual da SBPC [Annals of the 56 th Annual Meeting of the Brazilian Society for the Advancement of Science], Cuiabá. MT, 56, (2004), ISBN: 85-86957-07-0. Available from: http://www.sbpcnet.org.br/livro/56ra/banco_conf_simp/textos/ElsonLongo.htm, 2004.

[9] J.B. Russell, Química Geral [General Chemistry], São Paulo, McGraw Hill, (1982) pp. 197–251.

[10] L. Brillouin, Maxwell's demon cannot operate: information and entropy. I, J. Appl. Phys. 22 (3) (1951) 334–337.

[11] S.S. Mizrahi, Mulheres na Física: Lise Meitner [Women in Physics: Lise Meitner], Revista Brasileira de Ensino de Física 27 (2005) 491–493.

[12] R. Chang, Química Geral-Conceitos Essenciais [General Chemistry-Essential Concepts], seventh ed., McGraw-Hill, São Paulo, (2007) 929 p.

[13] G.M. Whitesides, Nanoscience, nanotechnology, and chemistry, Small 1 (2) (2005) 172–179.

[14] NASA, National Aeronautics and Space Administration. Available from: http://science.nasa.gov/science-news/science-at-nasa/2002/15jan_nano/

[15] A.D. Romig, et al. An introduction to nanotechnology policy: opportunities and constraints for emerging and established economies, Technol. Forecast. Soc. Change 74 (2007) 1634–1642.

[16] Wikipedia Free Encyclopedia. Available from: http://en.wikipedia.org/wiki/main_page; http://en.wikipedia.org/wiki/File:Feynman_at_Los_Alamos.jpg

[17] Science Museum London, Flickr. Available from: https://www.flickr.com; https://www.flickr.com/photos/sciencemuseum/9669013645

[18] Ibm, Ibm. Available from: http://www-03.ibm.com/ibm/history/exhibits/vintage/vintage_4506VV1003.html, 2013.

[19] A. Manzoli, et al., Nanomanipulação de superfície polimérica: Nanolitografia [Nanomanipulation of Polymer Surfaces: Nanolithography]. Embrapa Instrumentação Agropecuária, São Carlos—SP, 2007, pp. 4–13 (1678-0434).

[20] C.R. Moon, et al. Quantum holographic enconding in a two-dimensional electron gas, Nat. Nanotechnol. 4 (3) (2009) 167–172.

[21] Y. Bar-Cohen, Biomimetics—Biologically Inspired Technologies, [S.l.]: [s.n.], CRC Press, Florida, (2005).

[22] B. Bhushan, Biomimetics: lessons from nature—an overview, Philos. Trans. R. Soc. A-Math. Phys. Eng. Sci. 32 (2009) 1445–1486.

[23] C.J. Jones, S. Aizawa, The bacterial flagellum and flagellar motor: structure, assembly, and functions, Adv. Microb. Physiol. 32 (1991) 109–172.

[24] B. Bhushan, Y.C. Jung, Wetting, adhesion and friction of superhydrophobic and hydrophilic leaves and fabricated micro-/nanopatterned surfaces, J. Phys. D 20 (2008).

[25] M. Nosonovsky, B. Bhushan, Multiscale effects and capillary interactions in functional biomimetic surfaces for energy conversion and green engineering, Philos. Trans. R. Soc. A-Math. Phys. Eng. Sci. 367 (2009) 1511–1539.

[26] B. Bhushan, Principles and Applications of Tribology, John Wiley & Sons, New York, (1999).

[27] K. Autumn, Adhesive force of a single gecko foot-hair, Nature 405 (2000) 681–685.

[28] B. Bhushan, Adhesion of multi-level hierarchical attachment systems in gecko feet, J. Adhes. Sci. Technol. 21 (2007) 1213–1258.

[29] C.G. Bernhard, W.H. Miller, A.R. Moller, The insect corneal nipple array: a biological, broad-band impedance transformer that acts as a antireflection coating, Acta Physiol. Scand. 21 (2007) 1213–1258.

[30] J.P. Youngbloob, N.R. Sottos, Bioinspired materials for self-cleaning and self-healing, Mater. Res. Soc. Bull. 18 (2008) 732–738.

[31] J.W. Gardner, P.N. Bartlett, A brief-history of electronic noses, Sensor. Actuat. B: Chem. 18 (1994) 211–220.

[32] A. Riul, An artificial taste sensor based on conducting polymers, Biosens. Bioelectron. 18 (2003) 1365–1369.

[33] A. Riul Jr., et al. An electronic tongue using polypyrrole and polyaniline, Syn. Met. 132 (2) (2003) 109–116.

[34] Y. Vlasov, A. Legin, A. Rudnitskaya, Electronic tongues and their analytical application, Analytical and bioanalytical chemistry 373 (2002) 136–146.

[35] K. Toko, A taste sensor, Measurement science & technology 9 (1998) 1919–1936.

[36] T. Zhang, A gas nanosensor unaffected by humidity, Nanotechnology 20 (2009).

[37] R. Bogue, Nanosensors: a review of recent research, Sensor Rev. 29 (2009) 310–315.

[38] S.N. Shtykov, T.Y. Rusanova, Nanomaterials and nanotechnologies in chemical and biochemical sensors: capabilities and applications, Russ. J. Gen. Chem. 78 (2008) 2521–2531.

[39] R. Bogue, Nanosensors: a review of recent progress, Sensor Rev. 28 (2008) 12–17.

[40] A. Ramanavicius, A. Ramanaviciene, A. Malinauskas, Electrochemical sensors based on conducting polymer—polypyrrole, Electrochim. Acta. 51 (2006) 6025–6037.

[41] J. Kahn, Nanotechnology, Nat. Geo. (2006) 98–119.

[42] I. Amato, White House. Available from: http://www.whitehouse.gov/; http://www.ostp.gov/nstc/html/iwgn/iwgn.public.brochure/welcome.htm

[43] H.H. Takeda, et al. Differential pulse voltammetric determination of ciprofibrate in pharmaceutical formulations using a glassy carbon electrode modified with functionalized carbon nanotubes within a poly(allylamine hydrochloride) film, Sensor Actuat. B: Chem. 161 (2012) 755–760.

[44] Y. Gogodtsi (Ed.), Nanomaterials Handbook, Taylor & Francis-CRC Press, Philadelphia, 2006.

[45] N.I. Kovtyukhova, Layer-by-layer self-assembly strategy for template synthesis of nanoscale devices, Mater. Sci. Eng. C-Bio. S.V 19 (2002) 255–262.

[46] L.G. Paterno, L.H.C. Mattoso, O.N. De Oliveira, Ultrathin polymer films produced by the self-assembly technique: Preparation, properties and applications, Química Nova 24 (2001) 228–235.

[47] E.A. Rietman, Molecular Engineering of Nanosystems, Berlin: [s.n.], 2001.

[48] M. Okandan, P. Galambos, S.S. Mani, J.F. Jakubczak, Development of surface micromachining technologies for microfluidics and bioMEMS. Procedings of The International Society for Optcis and Photonics (SPIE), vol. 4560, Microfluidics and BioMEMS, San Francisco, CA, USA, (2001), p. 133.

[49] M.J. Heller, A. Guttman, Integrated Microfabricated Biodevices, Marcel Dekker, New York, (2001).

[50] H. Becker, L.E. Locascio, Polymer microfluidic devices, Talanta 56 (2002) 267–287.

[51] A. Van Der Berg, Lab-on-a-Chip: Chemistry in Miniaturized Synthesis and Analysis Systems, Elsevier, Amsterdam, (2003).

[52] P. Gravesen, J. Branebjerg, O.S. Jensen, Microfluidics—a review, J. Micromech. Microeng. 3 (1993) 168–182.

[53] K. Park, Controlled Drug Delivery: Challenges and Strategies, American Chemical Society, Washington DC, (1997).

[54] D. Grieshaber, Electrochemical biosensors—sensor principles and architectures, Sensors 8 (2008) 1400–1458.

[55] P. Pandey, M. Datta, B.D. Malhotra, Prospects of nanomaterials in biosensors, Anal. Lett. 41 (2008) 159–209.

[56] U. Yogeswaran, S.M. Chen, A review on the electrochemical sensors and biosensors composed of nanowires as sensing material, Sensors 8 (2008) 290–313.

[57] L.G. Carrascosa, Nanomechanical biosensors: a new sensing tool, Trac-Trend. Anal. Chem. 25 (2006) 196–206.

[58] M.N. Helmus, Nanotechnology-enabled chemical sensors and biosensors, Am. Lab. 38 (2006) 34.

[59] C. Lai Poh San, E.P.H. Yap, Frontiers in Human Genetics, World Scientific, Singapore, (2001).

[60] A.D. Roming Jr., et al. An introduction to nanotechnology policy: opportunities and constraints for emerging and established economies, Technol. Forecast. Soc. Change 74 (2007) 1634–1642.

[61] X. Liu, et al. Nanotechnology knowledge diffusion: measuring the impact of the research networking and a strategy for improvement, J. Nanopart. Res. 16 (9) (2014) 2613.

[62] F. Karaca, M.A. Öner, Scenarios of nanotechnology development and usage in Turkey, Technol. Forecast. Soc. Change 91 (2015) 327–340.

[63] I.C. Ezema, P.O. Ogbobe, A.D. Omah, Initiatives and strategies for development of nanotechnology in nations: a lesson for Africa and other least developed countries, Nanoscale Res. Lett. 9 (133) (2014) 1–8.

[64] J. Wang, P. Shapira, Funding acknowledgement analysis: an enhanced tool to investigate research sponsorship impacts: the case of nanotechnology, Scientometrics 87 (2011) 563–586.

[65] B. Bhushan, Governance, policy, and legislation of nanotechnology: a perspective, Microsyst. Technol. 21 (2015) 1137–1155.

[66] C.M. Sá, Redefining university roles in regional economies: a case study of university–industry relations and academic organization in nanotechnology, High Educ. 61 (2011) 193–208.

[67] L. Krabbenborg, H.A.J. Mulder, Upstream public engagement in nanotechnology: constraints and opportunities, Sci. Commun. 37 (4) (2015) 452–484.

[68] J. Wang, P. Shapira, Is there a relationship between research sponsorship and publication impact? An analysis of funding acknowledgments in nanotechnology papers, *PLoS One* 10 (2) (2015) 1–19.

[69] A.L. Porter, et al. Refining search terms for nanotechnology, J. Nanotechnol. Res. 10 (2008) 715–728.

[70] S.K. Arora, et al. Capturing new developments in an emerging technology: an updated search strategy for identifying nanotechnology research outputs, Scientometrics 95 (2013) 351–370.

[71] T. Bartol, K. Stopar, Nano language and distribution of article title terms according to power laws, Scientometrics 103 (2015) 435–451.

[72] A.G. Mamalis, A. Markopoulos, D.E. Manolakos, Micro and nanoprocessing techniques and applications, Nanotechnol. Percept. 1 (2005) 63–67.

[73] H. Iwai, Roadmap for 22 nm and beyond, Microelectron. Eng. 86 (7–9) (2009) 1520–1528.

[74] J.R. Ramsden, Applied Nanotechnology. [S.l.], Elsevier, Oxford, (2009).

[75] J. Robertson, R.M. Wallace, High-K materials and metal gate for CMOS applications, Mater. Sci. Eng. R 88 (2015) 1–41.

[76] M.N. Baibach, Giant magnetoresistance of (001)Fe/(001)Cr magnetic superlattices, Phys. Rev. Lett. 61 (1988) 2472–2475.

[77] E.F. De Lima, M. Foschini, M. Magini, O efeito termoiônico: Uma nova proposta experimental [The thermoionic effect: A new experimental proposal], Revista Brasileira de Ensino de Física 23 (2001) 4.

[78] J.M.F. Bassalo, A crônica da física do estado sólido: Do tudo de Geissler às válvulas a vácuo [The chronicle of solid state physics: From the Geissler tube to vacuum tubes], Revista Brasileira de Ensino de Física 15 (1993) 1–4.

[79] R. Da Silva Lima, J.C. Afonso, L.C.F. Pimentel, Raios-X: Fascinação, medo e ciência [X-rays: fascination, fear and science], Química Nova 32 (1) (2009) 263–270.

[80] M. Tolentino, R.C. Rocha-Filho, O átomo e a tecnologia [The atom and technology], Química Nova na Escola 3 (1996) 4–7.

[81] P.C. Wang, A.G. Macdiarmid, Integration of polymer-dispersed liquid crystal composites with conducting polymer thin films toward the fabrication of flexible display devices, Displays 28 (3) (2007) 101–104.

[82] A. Trokhymchuck, On Julius Planer's 1861 paper "Notiz uber das Cholestearin" in Annalen der Chemie und Pharmacie, Condens. Matter Phys. 13 (3) (2010) 1–4.

[83] I.H. Bechtold, Cristal líquido: Um sistema complexo de simples aplicação [Liquid crystals: a complex system of simple application], Revista Brasileira de Ensino de Fca 27 (2005) 333–342.

[84] B. Çaliskan, et al. EPR of gamma irradiated single crystals of cholesteryl benzoate, Radiation Eff. Defect. S. 159 (2004) 1–5.

[85] Anônimo, Liquid crystals—a step closer to low-voltage displays, Electron. Eng. 27 (1968) 17.

[86] J.A.V. Raalte, Reflective liquid crystal television display, Proc. IEEE 56 (1968) 2146–2149.

[87] T. Fujii, et al. Dot matrix LCD module for graphic display (64 × 320 dots). IEEE transactions on consumer electronics, AGOSTO CE-28 (3) (1982) 196–201.

[88] C.W. Tang, S.A. Van Slyke, Organic electroluminescent diodes, Appl. Phys. Lett. 51 (12) (1987) 913–915.

[89] C.H. Lu, et al. Output power enhancement of InGaN/GaN based green light-emitting diodes with high-density ultra-small in-rich quantum dots, J. Alloys Compd. 555 (2013) 250–254.

[90] A. Pereira, et al. Investigation of the energy transfer mechanism in OLEDs based on a new terbium β-diketonate complex, Org. Electron. 13 (1) (2012) 90–97.

[91] S.R. Forrest, The path to ubiquitous and low-cost organic electronic appliances on plastic, Nature 6986 (2004) 911–918.

[92] H.Q. Yin, R. Langford, R.E. Burrel, Comparative evaluation of the antimicrobial activity of ACTIOCOAT antimicrobial barrier dressing, J. Burn Care Rehab. 3 (1999) 195–200.

[93] B. Boonkaew, et al. Antimicrobial efficacy of a novel silver hydrogel dressing compared to two common silver burn wound dressings: Acticoat™ and PolyMem Silver(®), Burns 40 (1) (2014) 89–96.

[94] M. Jaisai, S. Baruah, J. Dutta, Paper modified with ZnO nanorods—antimicrobial studies, Beisltein J. Nanotechnol. 3 (2012) 684–691.

[95] Introduction to nanotechnology. Applications of nanotechnology. Available from: http://nanogloss.com/nanotechnology/applications-of-nanotechnology/#axzz2g1HmcoTE

[96] E. Hiroshi, Fiat FCC 2. Car and Drive-Brasil. Available from: http://caranddriverbrasil.uol.com.br/carro-moto-testes/13/artigo123553-1.asp

[97] K. Craveiro, UOL. Uol Carros. Available from: http://carros.uol.com.br/ultnot/2008/10/04/ult634u3225.jhtm

[98] K.E. Drexeler, Molecular engineering: an approach to the development of general capabilities for molecular manipulation, Proc. Natl. Acad. Sci. 78 (1981) 5275–5278.

[99] X. Wang, et al. Au-nanoparticle coated mesoporous silica nanocapsule-based multifunctional platform for ultrasound mediated imaging, cytoclasis and tumor ablation, Biomaterials 34 (2013) 2057–2068.

[100] T. Hogg, Evaluating microscopic robots for medical diagnosis and treatment, Nanotechnol. Percept. 3 (2007) 63–73.

[101] C.E. Handford, et al. Awareness and attitudes towards the emerging use of nanotechnology in the agri-food sector, Food Control 57 (2015) 24–34.

[102] L.Y.Y.L. Lu, et al. Ethics in nanotechnology: what's being done? What's missing?, J. Bus. Ethics 109 (2012) 583–598.

[103] K.H. Keller, Nanotechnology and society, J. Nanopart. Res. 9 (2007) 5–10.

[104] P.A. Schulte, et al. Occupational safety and health criteria for responsible development of nanotechnology, J. Nanopart. Res. 16 (2153) (2014) 1–17.

[105] D. Parr, Will nanotechnology make the world a better place? Trends in biotechnology, 23, n. 8. 395–398.

[106] J.J. Bang, L.E. Murr, Atmospheric nanoparticles: preliminary studies and potential respiratory health risks for emerging nanotechnologies, J. Mater. Sci. Lett. 21 (2002) 361–366.

[107] K.A.D. Guzmán, M.R. Taylor, J.F. Banfield, Environmental risks of nanotechnology: national, nanotechnology initiative funding, 2000–2004, Environ. Sci. Technol. 40 (2006) 1401–1407.

[108] R. Justo-Hanani, T. Dayan, European risk governance of nanotechnology: explaining the emerging regulatory policy, Res. Policy 44 (2015) 1527–1536.

[109] V. Türk, C. Kaiser, S. Schaller, Invisible but tangible? Societal opportunities and risks of nanotechnologies, J. Clean. Prod. 16 (2008) 1006–1009.

[110] A. Capon, et al. Perceptions of risk from nanotechnologies and trust in stakeholders: a cross sectional study of public, academic, government and business attitudes, BMC Public Health 15 (424) (2015) 1–13.

[111] Z.A. Collier, et al. Tiered guidance for risk-informed environmental health and safety testing of nanotechnologies, J. Nanopart. Res. 17 (155) (2015) 1–21.

1

Nanomaterials: Solar Energy Conversion

L.G. Paterno

INSTITUTE OF CHEMISTRY, DARCY RIBEIRO CAMPUS, UNIVERSITY OF BRASILIA,
BRASILIA, BRAZIL

CHAPTER OUTLINE

1.1 Introduction

The increase in population density and economic growth in many parts of the world since 1980 has maintained a strong pace because of the availability of 15 terawatts (TW) of energy—our current consumption—at accessible prices. However, even the most optimistic forecasts cannot ensure that this scenario will persist in the coming decades. A large part of the world's energy consumption depends on fossil fuels, especially oil. However, extraction of "cheap oil" may reach its peak in the next few years and then decline. Therefore, production of energy from alternative sources is indispensable to maintain sustainable global economic growth and the perspective of the consumption of an additional 15 TW after 2050 [1]. We must also note in this new planning effort that the alternative sources of energy production must be clean, as gas emissions from fossil fuels have negatively contributed to the quality of life on the planet because of global temperature increases and air pollution [2].

In view of the imminent collapse of the current energy production system, it is necessary to seek alternative sources of energy production. Within the current scenario of scientific and technological development and the urgent demand for 30 TW by 2050 [1], there are at least three alternative options for energy production [3]: (1) burning of fossil fuel associated with CO_2 sequestration, (2) nuclear energy, and (3) renewable energy. In option (1), emissions of greenhouse gases (GHGs), especially CO_2, can already be controlled by emission certificates known as carbon credits [4]. One ton of CO_2 is equal to one carbon credit. In practical terms, the Kyoto Protocol of 1999 establishes the maximum level of GHGs that a certain country can emit from its industrial activities. If a certain industry or country does not reach the established goals, it becomes a buyer of carbon credits equivalent to the excess emissions. By contrast, countries whose emissions are lower than the preestablished limit can sell their "excess" credits on the international market. However, this initiative is still controversial, mainly because, to many, it implies a discount on the penalty from the excessive emission of GHGs. Option (2) is very attractive because it is a clean type of energy, with high production efficiency and zero GHG emission [5]. Many European countries, as well as the United States and Japan, produce and consume electricity from nuclear plants [6]. However, the risk of accidents is constant, especially after the catastrophic events of Chernobyl (1986) [7] and Fukushima (2011) [8], and the improper use of nuclear technology for nonpeaceful purposes raises concerns. Energy production from renewable sources (3) is undoubtedly the most promising alternative [9]. The different technologies in this group, such as solar, wind, hydroelectric, and geothermal, are absolutely clean in terms of GHG emissions. Theoretically, energy production from the burning of biomass, such as wood, ethanol, and biodiesel, creates no net CO_2 emission because it is a closed cycle (the amount of CO_2 sequestered from the atmosphere during photosynthesis and transformed into biomass is the same amount released to the atmosphere when the biomass is combusted with O_2). In practical terms, the balance is slightly negative because land management, transportation, and processing use noncounterbalanced sources. Still, this balance is much less negative than the one associated with the burning of fossil fuels.

The conversion of solar energy into electricity is one of the most promising ways to produce clean and cheap energy from a renewable source [3,9,10]. Silicon-based solar cells, already manufactured in the 1950s for military and space applications, have been available for civilian use for at least three decades, mainly in buildings and for power generation at remote locations. The cost is decreasing with the increasing production scale and the corresponding reduction of the price of silicon [11]. In addition, research in the field of solar cells increased in the 1990s as a result of discoveries in the area of nanotechnology. Today, third-generation devices are being developed that use nanomaterials to convert solar light into electricity at significantly lower costs than those of conventional silicon cells [3]. Third-generation solar cells, such as organic solar cells (OSCs) and dye-sensitized solar cells (DSSCs or dye cells), are a reality, are being produced at pilot scale by small companies, and will be commercially available in the near future [12,13]. Nanotechnology research has also enabled the development of photoelectrochemical cells for artificial photosynthesis, whose efficiency is still low, but with a high potential to increase to levels that would make the cells commercially interesting [14]. The use of nanomaterials is decisive to fully achieve energy conversion in these new devices. In addition, the use of nanomaterials reduces the cost and the environmental impact of cell production to make such devices even more promising.

Initially, this chapter will present the principles of solar conversion and the operation of solar cells, with a focus on OSCs and DSSCs. Then, the role of nanomaterials in the different parts of each device and in their operation is discussed. The most recent data (until 2012) on the conversion performance of third-generation cells are provided. The last topic consists of a brief description of photoelectrochemical cells to produce solar fuels, considering that this subject is correlated to the previous ones. However, it is not the intention of this chapter to present a thorough literature review on this last subject, and the reader is encouraged to consult the references listed at the end of the chapter.

1.2 Conversion of Solar Energy Into Electricity

1.2.1 Solar Spectrum and Photovoltaic Performance Parameters

The Sun is the most abundant and sustainable source of energy available on the planet. The Earth receives close to 120,000 TW of energy from the Sun every year, an amount 10^4 larger than the current global demand [14]. The photons that reach the Earth as solar light are distributed across different wavelengths and depend on variables, such as latitude, time of day, and atmospheric conditions. This distribution, known as the solar spectrum, is shown in Fig. 1.1 [15].

The spectrum shows the solar incidence power per area per wavelength (W m^{-2} nm^{-1}), also known as irradiance, considering a bandwidth of 1 nm ($\Delta\lambda$) [16]. The terms AM0, AM1.0, and AM1.5 refer to solar spectra calculated according to different ASTM standards, appropriate for each type of application [17]. For example, spectra AM1.0 and AM1.5 are calculated according to standard ASTM G173 and are used as reference standards for terrestrial applications, whereas spectrum AM0, based on standard ASTM E 490, is used in

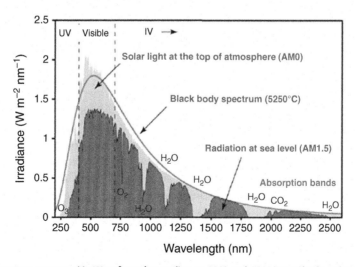

FIGURE 1.1 Solar spectrum expressed in W m^{-2} nm^{-1} according to AM0 and AM1.5 standards. *Adapted from http://org. ntnu.no/solarcells [15].*

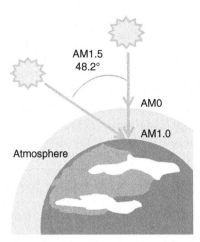

FIGURE 1.2 Schematics of the different forms of solar radiation incident on the Earth and the respective standards to calculate the solar spectrum. *Adapted from http://org.ntnu.no/solarcells [15].*

satellites. As shown schematically in Fig. 1.2, the calculation of the spectra considers specific geographic and atmospheric variables (i.e., the angle of incidence on the planet, the air density, and other parameters). The reproduction of the solar spectrum in the laboratory, according to the established standards, is fundamental to developing photovoltaic cells because it allows for comparison and certification of the performances of devices developed by different manufacturers and those still in development.

The performance of an illuminated solar cell is assessed by photovoltaic parameters, such as the power produced per illuminated cell area (P_{out}, in W·cm^{-2}), open-circuit voltage (V_{oc}, in V), short-circuit current density (J_{sc}, in mA·cm^{-2}), fill factor (*FF*), and overall conversion efficiency (η) [16]. All these parameters depend initially on the power of the light incident on the cell (P_{in}), given by Eq. 1.1 [16]:

$$P_{in} = \int_{\lambda} \frac{hc}{\lambda} \Phi_0(\lambda) d\lambda \qquad (1.1)$$

where h is the Planck constant (4.14×10^{-15} eV·s), c is the speed of light (3.0×10^8 m·s^{-1}), $\Phi(\lambda)$ is the flux of photons corrected for reflection and absorption before impacting the cell (cm^{-2} s^{-1} per $\Delta\lambda$), and λ is the wavelength of the incident light.

The open-circuit voltage corresponds to the voltage between the terminals (electrodes) of the illuminated cell when the terminals are open (infinite resistance). The short-circuit current density corresponds to the condition in which the cell's terminals are connected to a zero-resistance load. The short-circuit current density grows with the intensity of the incident light because the number of photons (and thus the number of electrons) also increases with intensity. Since the current usually increases with the active area of the solar cell, the current is conventionally expressed in terms of current density, J (current/area).

When a load is connected to a solar cell, the current decreases and a voltage is developed when the electrodes are charged. The resultant current can be interpreted as a superposition of the short-circuit current caused by the absorption of photons and a dark current caused by the voltage generated by the load that flows in the opposite direction. Considering that solar cells usually consist of a p–n (p–n junction: junction of p-type and n-type semiconductors) or D–A [D–A junction: junction of an electron donor material (D) and an electron acceptor material (A)] junction, they can be treated as diodes. For an ideal diode, the dark current density (J_{dark}) is given by [16]

$$J_{dark}(V) = J_0(e^{qV/k_BT} - 1) \tag{1.2}$$

where J_0 is the current density at 0 K, q is the electron charge (1.6×10^{-19} C), V is the voltage between the electrodes of the cell, k_B is the Boltzmann constant (8.7×10^{-5} eV·K^{-1}), and T is the absolute temperature. The resultant current can be explained as a superposition of the short-circuit current and the dark current [16]:

$$J = J_{sc} - J_0(e^{qV/k_BT} - 1) \tag{1.3}$$

The open-circuit voltage is defined at $J = 0$, which means that the currents cancel and no current flows through the cell, which is the open-circuit condition. The resultant expression is given by [16]:

$$V_{oc} = \frac{k_BT}{q} \ln\left(\frac{J_{sc}}{J_0} + 1\right) \tag{1.4}$$

The performance parameters of a solar cell are determined experimentally from a current–voltage curve ($J \times V$), schematically represented in Fig. 1.3, when the cell is subjected to standard operating conditions, in other words, illumination according to standard AM1.5, under an irradiating flux of 100 W·cm^{-2} and a temperature of 25°C.

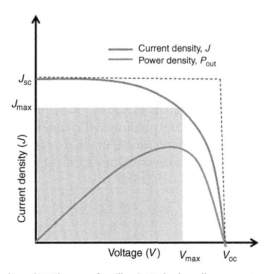

FIGURE 1.3 Current density–voltage ($J \times V$) curve of an illuminated solar cell.

The power density produced by the cell (P_{out}) (Fig. 1.3, gray area) is given by the product of the current density (J, in mA·cm^{-2}) and the corresponding operating voltage (V, in V) as per Eq. 1.5 [16]:

$$P_{out} = JV \tag{1.5}$$

The maximum power density (P_{max}) is given by

$$P_{max} = J_{max} V_{max} \tag{1.6}$$

Based on Fig. 1.3, it can be concluded that the maximum power produced by an illuminated solar cell is between $V = 0$ (short circuit) and $V = V_{oc}$ (open circuit), or V_{max}. The corresponding current density is given by J_{max}. The conversion efficiency of the cell (η) is given by the ratio between the maximum power and the incident power [16]:

$$\eta = \frac{J_{max} V_{max}}{P_{in}} \tag{1.7}$$

The behavior of an ideal solar cell may be represented by a $J \times V$ curve of rectangular shape (represented in the graph by dashed lines) where the current density produced is maximum, constant, and equal to J_{sc} up to the V_{oc} value. However, not all of the incident power is converted into energy by the cell, so in actual situations the $J \times V$ curve deviates from the ideal rectangular shape. The fill factor term (FF) is introduced to measure how close to the ideal behavior a photovoltaic cell operates. The FF is given by [16]

$$FF = \frac{J_m V_m}{J_{sc} V_{oc}} \tag{1.8}$$

By definition, $FF \leq 1$. Thus, the overall conversion efficiency can be expressed using the FF value [16]:

$$\eta = \frac{J_{sc} V_{oc} FF}{P_{in}} \tag{1.9}$$

In addition to these, another important performance parameter is the quantum efficiency, which measures how many electrons capable of performing work are generated by each incident photon of wavelength λ. The quantum efficiency is subdivided into internal and external classifications. The external quantum efficiency (EQE) measures the number of electrons collected by the electrode of the cell in the short-circuit condition divided by the number of incident photons. Also known as the incident photon to current efficiency (IPCE), its value is determined by Eq. 1.10 [16]:

$$IPCE = \frac{1240 I_{sc}}{\lambda P_{in}} \tag{1.10}$$

Since the IPCE value depends on the wavelength of the incident radiation, an IPCE versus wavelength curve corresponds to the cell's spectral response, also known as the action

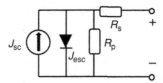

FIGURE 1.4 Equivalent circuit of a solar cell.

spectrum of the solar cell. The internal quantum efficiency (IQE) is the ratio of the number of electrons collected by the electrode of the cell in a short-circuit configuration divided by the number of photons that effectively enter the cell. This parameter does not consider all the incident photons because a portion of them is lost by reflection or absorption before they accomplish the charge separation process within the absorption layer. Both efficiencies are positive and ≤100%.

As already discussed, a solar cell can be understood as a power generator and can be represented by an equivalent circuit as in Fig. 1.4. The circuit has a solar cell, represented by the diode and its respective dark and short-circuit currents, and two resistances, one in series (R_s) and the other in parallel (R_p). The resistance in series represents the nonideal conductor behavior of the cell, whereas the resistance in parallel accounts for the leakage current inherent to any device, usually associated with insulation problems. In an ideal solar cell, $R_s = 0$, $R_p = \infty$, and the current expression initially represented by Eq. 1.3 can be expanded by including the resistances of the equivalent circuit, according to Eq. 1.11 [16]:

$$J = J_{sc} - J_0[e^{q(V+JAR_s/k_BT)} - 1] - \frac{V + JAR_s}{R_p}$$

(1.11)

where A is the active area of the cell in centimeter square.

1.2.2 Operating Principles of a Solar Cell

A typical solar cell consists of an absorbing material between two electrodes. The absorbing material can be either a semiconductor or a dye, organic, or inorganic, and it can be monocrystalline, polycrystalline, nanocrystalline, or amorphous [16]. The absorber collects (absorbs) the solar light and therefore must have a band separation energy (E_g) that matches the solar spectrum. The separation of charges into individual carriers and their transport may or may not be performed by the absorber. The electrodes are fabricated from conductive materials with different work functions, and one of them must be transparent to the incident light.

The photovoltaic conversion process can be divided into four sequential stages [16]:

1. light absorption causes an electron transition in the cell's absorbing material from the ground state to the excited state;
2. the excited state is converted into a pair of separate charge carriers, one negative and the other positive;

3. under an appropriate transport mechanism, the carriers move separately to the cell's electrical contacts; the negative carrier to the cathode and the positive carrier to the anode;

4. the electrons travel the circuit external to the cell, where they lose energy and perform useful work (i.e., to power a lamp or an engine). Then, they reach the cathode, where they recombine with the positive charge carriers and return the absorbing material to the ground state.

The mechanisms of (1) light absorption and (2) charge separation depend on the absorber's electron structure and morphology. In conventional solar cells made of inorganic semiconductors, such as silicon or gallium arsenide (GaAs) in the mono-, poly, and microcrystalline forms, absorption and separation are performed only by the absorber. The band separation energy of these semiconductors (Si: 1.1 eV; GaAs: 1.42 eV) is sufficient to collect close to 70% of the solar radiation incident on the planet. The absorbed photons promote the electrons to the conduction band and at the same time produce an equivalent number of gaps in the valence band of the absorbing material. The effect is observed both in direct-gap semiconductors, such as GaAs, and in indirect-gap semiconductors, such as Si [16]. Once formed, the charge carriers are separated in the semiconductor's junction region, which is formed at the interface of the p- and n-type doping regions (p–n junction), even when they are insulated by a third i-insulating region (p–i–n junction). The junction is obtained in the production of the absorbing film; a thin semiconductor film with a certain doping undergoes a second doping process, limited to a small depth. Thus, the final film has two different doping regions separated by an interface or junction. A built-in potential is created at the junction region because of the difference in electron affinities in each region. This potential is strong enough to separate the carriers into individual species, electron, and gap.

To better understand the operation of this type of device, Fig. 1.5 presents a schematic of the energy levels of p and n semiconductors, before and after the junction. In Fig. 1.5A, the energy levels of the insulated semiconductors are aligned and referred to vacuum, which allows visualization of the relative position of the bands and their spacing in each semiconductor. The Fermi levels of each material, E_F, are described in terms of their respective work functions, ϕ_{W1} and ϕ_{W2}. Material 1 is of the n-type, and material 2 is of the p-type. The Fermi level refers to the total chemical potential of the electrons. In Fig. 1.5B, the semiconductors are placed in contact (again, the contact is already established when the semiconductors are manufactured), and an abrupt interface is formed, where the electron affinities of each semiconductor create a step at the contact. In Fig. 1.5C, the system reaches steady state, with a single Fermi level at a certain temperature. It is important to observe that the Fermi levels of each material remain the same as they were before junction creation, considering their position relative to V (V_n and V_p) and their respective work functions. The existence of a single Fermi level for the system requires the creation of an electrostatic potential between $x = -d_1$ and $x = d_2$, and a deformation of the valence and conduction bands and the local vacuum level around the junction. The difference

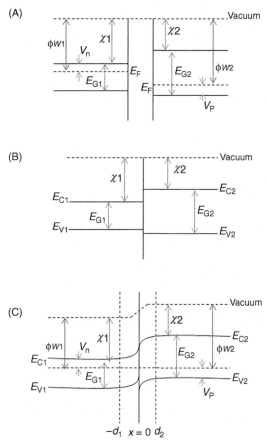

FIGURE 1.5 Electron energy levels of the n and p semiconductors. (A) Before contact, (B) during contact, and (C) after contact and in thermodynamic equilibrium. *Adapted from S.J. Fonash, Solar Cell Device Physics, second ed., Academic Press, Burlington, 2010 [16].*

of potential energy through the junction is the driving force to separate the charges generated after light absorption. The transport of the separated charges to the electrodes is performed efficiently by the absorbing film itself, which has high electron mobility [16].

In solar cells constructed from organic absorbers, such as dyes and conjugated polymers, or in cells sensitized with dyes or quantum dots, the absorption of light energy results in the formation of multiparticle excited species known as excitons [16]. An exciton is nothing more than an electron–gap pair, bonded by Coulombic attraction [18]. Therefore, they are species without charge that can move only by diffusion. The method of breaking exciton symmetry in these devices is to use donor–acceptor (D–A) architectures, in analogy to the p–n junction of traditional semiconductors [3]. The donor is the material that absorbs photons and generates excitons, and the acceptor is a material of high electron affinity. The difference in electron affinities between the donor and the acceptor results in an electric potential large enough to cause the separation of charges at the interface of both materials [3]. However, the cell design must consider whether the separation of the D and A phases, as well as the thicknesses of the absorbing layer as a whole, is compatible with the lifetime and the diffusion length of the exciton. For example, the diffusion length of the excitons in most conjugated polymers used in OSCs is between 30 and 60 nm [3,16]. Therefore, the operation of organic or dye-sensitized cells depends decisively on the control of the morphology of the interfaces at the nanoscale. The following section presents details on the structure and operation of these types of cells.

1.2.3 Organic Solar Cells

OSCs use organic semiconductors, such as small organic dye or conjugated polymer molecules, in the absorber/donor function [19]. Coordination compounds, especially phthalocyanines and metalloporphyrins, can also be used in this function, although they are not organic substances, strictly speaking [3,19]. Fullerene is the most common acceptor, although other polymers and small conjugated molecules [19], carbon nanotubes [20], and metal oxide nanoparticles [21] can also be used in this function. Figs. 1.6 and 1.7 illustrate the structural formulas of the main absorbing materials and fullerenes used to build OSCs.

The first models of this type of cell were proposed in the 1980s, with organic pigments films placed between two metal electrodes with different work functions [22]. However, the efficiency of these cells was very low (<0.1%). Years later, the use of the D–A junction in the bilayer concept, employing other dyes, such as phthalocyanines, increased the conversion efficiency to almost 1% [23]. This figure was surpassed only at the beginning of the 2000s with the introduction of bulk heterojunction solar cells based on blends of conjugated polymers and fullerenes [24,25]. Today, the best conversion efficiencies are 12% for cells based on small molecules [12b] and 9.2% for cells based on conjugated polymers [26]. Efficiency values of 15% are expected in the near future [27,28]. In addition, there is consensus on the fact that efficiency has a secondary role in the dissemination of organic cells; the most important metric is the lower watt/hour cost of these devices. Fig. 1.6 shows the structures of absorbers used in OCSs along with the photovoltaic parameters of the respective cells fabricated under optimized conditions.

The field of OSCs gained force with the introduction of conjugated polymers as absorbing material. In addition to their optoelectronic properties, conjugated polymers can be processed in solution and at room temperature by the fast printing processes used in the printing and textile industries, such as roll-to-roll and screen printing [29]. Due to this versatility, the cell manufacturing cost is even lower. In addition, the polymer films can be deposited on plastic substrates, which allow the large-scale production of flexible cells (Fig. 1.8) and, consequently, their widespread use in buildings.

Conjugated polymers are formed by a main chain of sp^2 carbon atoms, with alternate single and double bonds. Due to conjugation, the electrons of the π bonds are dislocated by the polymer chain. As a result of the different lengths of the single and double bonds (single bonds are longer than double bonds), these materials exhibit a separation between

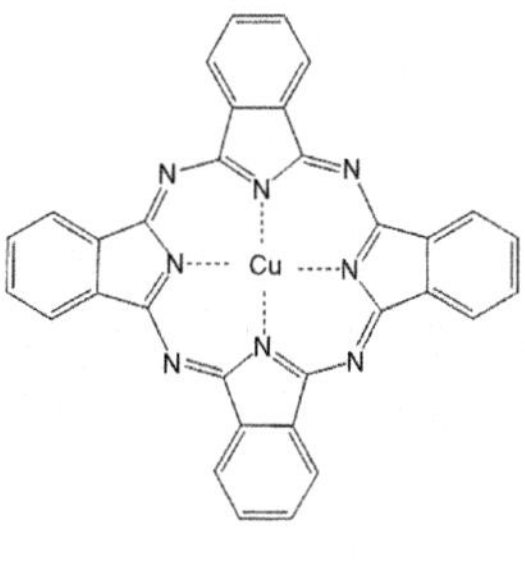

1
E_g: 3.1 eV
V_{OC}: 450 mV; J_{SC}: 2.3 mA.cm^{-2}
FF: 0.65; η: 0.95%

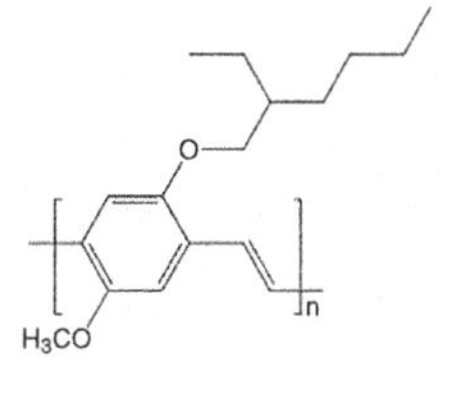

2
E_g: 2.3 eV
V_{OC}: 820 mV; J_{SC}: 5.3 mA.cm^{-2}
FF: 0.61; η: 2.5%

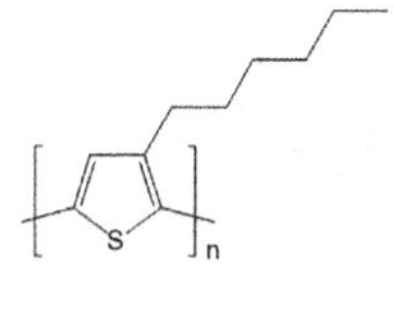

3
E_g: 2.0 eV
V_{OC}: 610 mV; J_{SC}: 11.1 mA.cm^{-2}
FF: 0.66; η: 5.0%

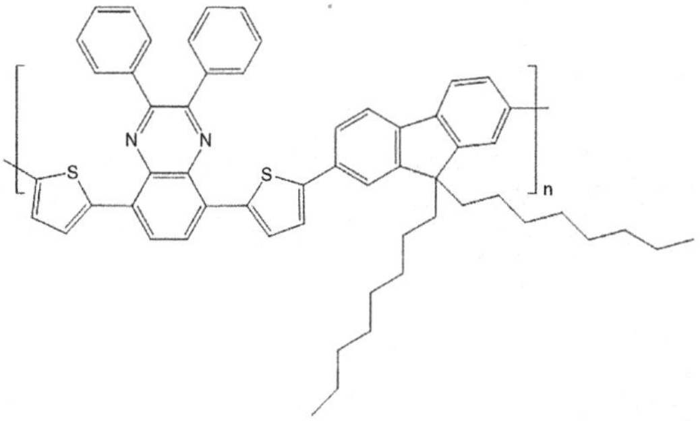

4
E_g: 1.9 eV
V_{OC}: 990 mV; J_{SC}: 9.7 mA.cm^{-2}
FF: 0.57; η: 5.5%

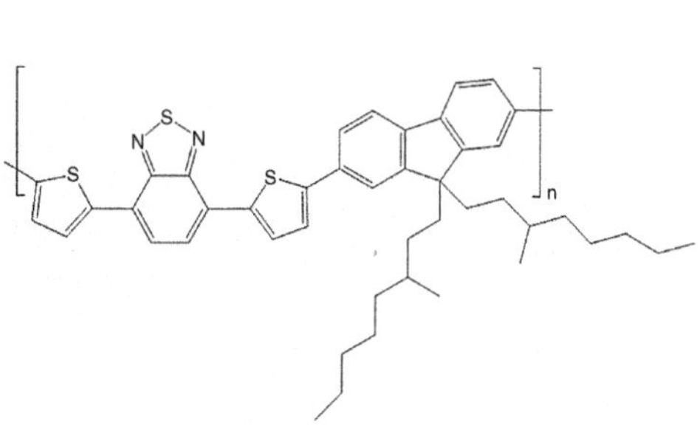

5
E_g: 1.9 eV
V_{OC}: 1000 mV; J_{SC}: 9.1 mA.cm^{-2}
FF: 0.51; η: 4.5%

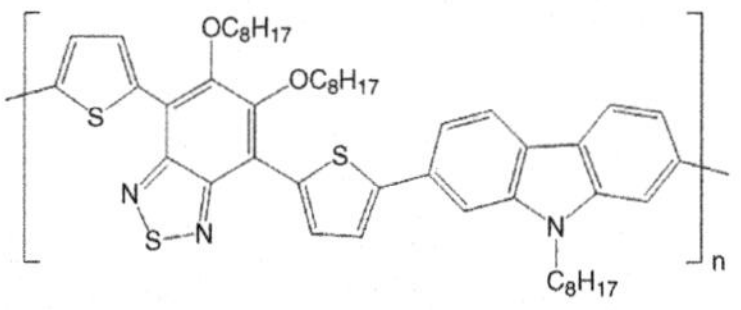

6
E_g: 1.9 eV
V_{OC}: 810 mV; J_{SC}: 9.6 mA.cm^{-2}
FF: 0.69; η: 5.4%

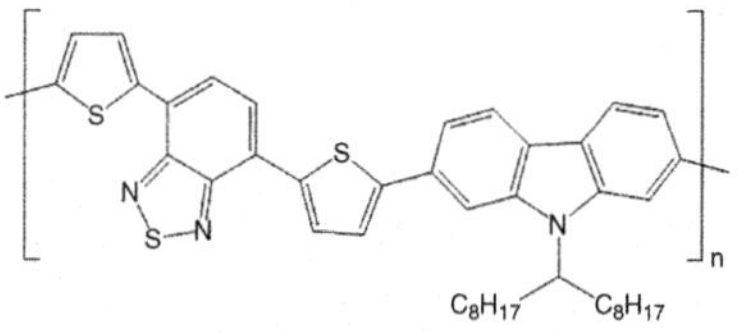

7
E_g: 1.8 eV
V_{OC}: 900 mV; J_{SC}: 12.1 mA.cm^{-2}
FF: 0.69; η: 6.8%

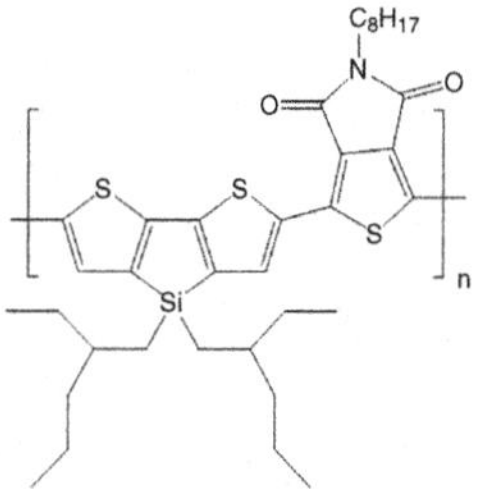

8
E_g: 1.8 eV
V_{OC}: 880 mV; J_{SC}: 12.2 mA.cm^{-2}
FF: 0.68; η: 7.3%

FIGURE 1.6 Structural formula of absorbing materials for organic solar cells (OCSs) and their respective photovoltaic parameters (for cells in optimized conditions). 1, Copper phthalocyanine; 2, MDMO-PPV; 3, P3HT; 4, NP7; 5, BisDMO-PFTDBT; 6, HXS-1; 7, PCDTBT; 8, PDTSPTBT.

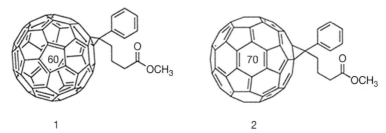

FIGURE 1.7 Structural formula of some fullerenes used in OSCs. 1, [6,6]-Phenyl-C61-butyric acid methyl ester (PC61BM); 2, [6,6]-phenyl-C71-butyric acid methyl ester (PC71BM).

FIGURE 1.8 OCS module manufactured by Konarka. *From http://www.konarka.com/*

the valence and conduction bands that characterizes them as semiconductors [30]. Conduction at levels similar to those of metals is only possible when the polymer is doped [31]. In fact, most conjugated polymers have separation energies between bands above 2.0 eV (620 nm), which considerably limits solar light absorption. Another important feature is their electron mobility (for most conjugated polymers, it is lower than 0.01 cm^2 V^{-1} s^{-1}), which is much lower than that of traditional semiconductors (e.g., the mobility of intrinsic silicon can be up to 10^4 times higher) [16]. By contrast, the optical absorption coefficient of these materials is extremely high, on the order of 10^5 cm^{-1}, so ultrathin films (thickness <100 nm) are sufficient to collect most of the incident photons and compensate for the low mobility [19].

The operation of OCSs is based essentially on the photoinduced electron transfer from the donor element to an acceptor of high electron affinity. Fig. 1.9 shows a schematic of the process, in which a conjugated polymer (donor) absorbs energy and injects the photoexcited electrons into the fullerene (acceptor) [32].

The process of electron transfer between dyes or conjugated polymers and fullerenes occurs on a much smaller time scale (on the order of femtoseconds) than that of

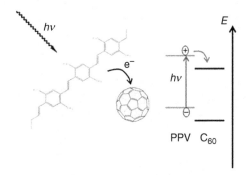

FIGURE 1.9 Schematics of the photoinduced electron transfer process from a *p*-phenylene-vinylene *(PPV)* molecule to a fullerene (C$_{60}$) and of the energy levels of each material. *Adapted from H. Hoppe, N.S. Sariciftci, Organic solar cells: an overview, J. Mater. Res. 19 (2004) 1924–1945 [19].*

competitive processes, such as photoluminescence (on the order of nanoseconds) [32,33]. The energy levels of the donor and the acceptor favor the vector transfer of the electrons. The junction region between the donor and the acceptor (D–A junction) of an OCS can be established in two ways, as illustrated in Fig. 1.10 [19]: (A) as a plane or bilayer heterojunction and (B) as a bulk heterojunction. In (A), a fullerene layer is deposited on the donor layer (small molecule or polymer) (Fig. 1.10A), and the heterojunction is established in the interface of the two layers, in the same way as in a p–n junction of an inorganic semiconductor. In Fig. 1.10B, the absorbing film is a blend of fullerene with a conjugated polymer, mixed throughout the entire volume of the film. The blend's morphology is developed on the nanoscale and provides a fundamental contribution to the photovoltaic conversion, as it significantly increases the interfacial area of the D–A junction [19].

The photovoltaic parameters of OSCs depend primarily on the energy levels of the electron donor and acceptor, as well as on the work function of the electrodes. The open-circuit voltage of an organic cell can be as high as the separation energy between the highest occupied molecular orbital (HOMO) of the donor and the lowest unoccupied molecular orbital (LUMO) of the acceptor [34]. Some studies indicate that in cells with conjugated polymers, the V_{oc} value has an inverse and linear relationship with the first reduction voltage (LUMO) of the fullerene compound (the lower the voltage is, the higher the V_{oc} of the corresponding cell) [34]. Likewise, there is a linear and direct relationship between the V_{oc} and the oxidation voltage (HOMO) of the polymer [34]. The interfaces of the absorbing layer with the electrodes also have an effect on the cell voltage. Tin oxide doped with indium (ITO) used as an anode has a work function of 4.7 eV, slightly lower than the HOMO levels of most dyes and conjugated polymers used in solar cells. To reduce the energy barrier in the electrode-active layer interface, this value can be increased by treating the ITO electrode with plasma or by depositing layers of conductive polymers of nanometric thickness, such as the poly(3,4-ethylenedioxytiophene)-poly(styrenesulfonic acid) (PEDOT:PSS) complex or the polyaniline (PANI) complex [35–37]. The cathode consists of a metal with a low work function, such as aluminum.

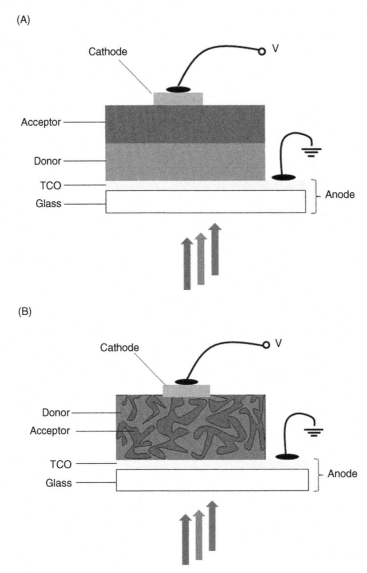

FIGURE 1.10 Schematic of an OSC for the bilayer heterojunction (A) and bulk heterojunction (B) configurations.

The current density generated by a solar cell corresponds to the number of charges collected by the electrodes, which, in the case of OSCs, is the product of three efficiencies: photon absorption (η_{abs}), exciton dissociation (η_{diss}), and the collection of carriers by the electrodes (η_{out}) [19]:

$$J_{sc} = \eta_{abs} \cdot \eta_{diss} \cdot \eta_{out} \tag{1.12}$$

The absorption efficiency depends on the absorption spectrum of the absorbent material and on optical effects, especially interference, inside the device [38]. As mentioned, the absorption coefficient of conjugated polymers, as well as that of small conjugated molecules and coordination compounds, is quite high. The OSC architecture features a relatively thick glass substrate adjacent to a highly reflective metal electrode. This combination establishes interference conditions, both constructive and destructive, that may disrupt the coincidence between the maximum amplitude of the electrooptical field and the D–A interface and, consequently, reduce the photon collection efficiency [19,38].

The dissociation efficiency of the excitons depends decisively on the interface between donor and acceptor. In bulk heterojunction OSCs (Fig. 1.10B), the morphology developed at the nanoscale must be controlled in a way that the phase segregation is minimal and the D–A interface has a dimension compatible with the exciton diffusion length [19,39,40]. Films consisting of 1:4 (m/m) MDMO-PPV:C_{60} deposited by spin coating on an ITO substrate have different morphologies when toluene or chlorobenzene is used to prepare the deposition solutions. Toluene causes the formation of globules of varied sizes up to 200 nm, possibly clustered structures that make the film surface more irregular. When chlorobenzene is used, segregation occurs at a much smaller scale, on the order of 20 nm, and the film surface appears smoother. The consequence of morphology on conversion efficiency is immediate; the efficiency of OSCs prepared using chlorobenzene is 3 times higher than that of those made with toluene [39].

The charge collection efficiency depends on the path inside the active layer that is taken by the carriers. The mobility of the materials composing the active layer has a critical role in this process. Mobility is very low in most conjugated polymers because of their disordered nature in the solid state. In these systems, the conduction mechanism occurs mainly by hopping of highly localized charge carriers [41–44]. The carrier mobility can be explained by the Poole–Frenkel model, where mobility μ is proportional to the electric field E, according to Eq. 1.13 [19]:

$$\mu(E) = \mu_0 \exp\left(\gamma\sqrt{E}\right) \tag{1.13}$$

where μ_0 is the mobility at 0 K and γ is a preexponential factor; both depend on the temperature. The electric field inside the device must be strong enough to drive the carriers to the electrodes, which requires the difference between the work functions of the electrodes to be maximized and the thickness of the active layer to be minimized (generally to less than 100 nm).

The *FF* and the overall conversion efficiency of a solar cell depend, in addition to V_{oc} and J_{sc}, on the optimization of the series and parallel resistances of the device (Section 2.1). The deposition of conductive polymers on the transparent electrode, before the active layer, reduces the contact resistance and, consequently, the R_s value [37]. The deposition of a LiF layer on the active layer, preceding the deposition of the metal cathode, increases the carrier collection by the electrode, which also contributes to reducing R_s [45]. The carrier mobility and lifetime (τ) also influence *FF* and η. The larger the $\mu \times \tau$ product is, the higher the cell efficiency.

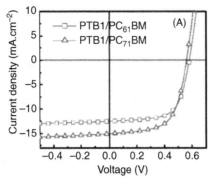
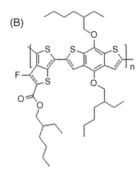

FIGURE 1.11 (A) $J \times V$ curves of OSCs subjected to AM1.5 irradiation (100 mW·cm^{-2}) prepared with two types of fullerenes as indicated, and (B) structural formula of the *PTB1* polymer used as an absorber/donor. *Reproduced with permission from Y. Liang, et al. Development of new semiconducting polymers for high performance solar cells, J. Am. Chem. Soc. 131 (2009) 56–57 [46].*

The most modern molecular engineering concept for the synthesis of conjugated polymers for OSCs is based on the preparation of monomers (and comonomers) with a modular structure. The modules are formed by a sequence of electron donating and accepting units (push–pull systems) that together with the main conjugated chain form chromophores with an absorption spectrum displaced to the red region (500–800 nm), a region with a more intense solar flux of photons [27]. Structures 4–8 illustrated in Fig. 1.6 are examples of this type of polymer. With this approach, the conversion efficiencies of the respective OSCs have reached the highest values of approximately 9.2% [27]. Fig. 1.11 shows the photovoltaic curves of OSCs that reached this efficiency record (9.2%) based on fullerenes and on the polymer absorber also shown in the figure [46].

1.2.4 Dye-Sensitized Solar Cells

The development of DSSCs began in the mid-1980s. A study revealed that colloidal TiO$_2$ particles (20 nm in diameter) could be sensitized to visible light by a ruthenium-based luminescent complex (tris(2,2′-bipyridine-4,4′-dicarboxylate) chloride of ruthenium(II)) [47]. In addition, the study found that the photoluminescence of the complex was completely extinguished when the complex was immobilized on the oxide nanoparticle, and thus, it was concluded that the complex transferred photoexcited electrons to the conduction band of the semiconductor. Years later, this principle was applied to a device, and the DSSC field has grown significantly since then [48]. Today, the overall conversion efficiency of these cells is higher than 11% [49,50].

A typical DSSC consists of a photoanode, a counter-electrode, and an electrolyte that separates the electrodes. The photoanode is constructed from a mesoporous and nanocrystalline film of a semiconductor with wide separation between bands, such as TiO$_2$, ZnO, or SnO$_2$, which is deposited on a transparent electrode, usually fluorine-doped tin oxide (FTO). The semiconductor is photosensitized by a dye, which may be a ruthenium(II) complex [51], quantum dots (CdSe, PbS) [52], synthetic organic dyes [53,54], or natural extracts [55,56]. Fig. 1.12 illustrates the structural formulas of the main dyes used in DSSCs.

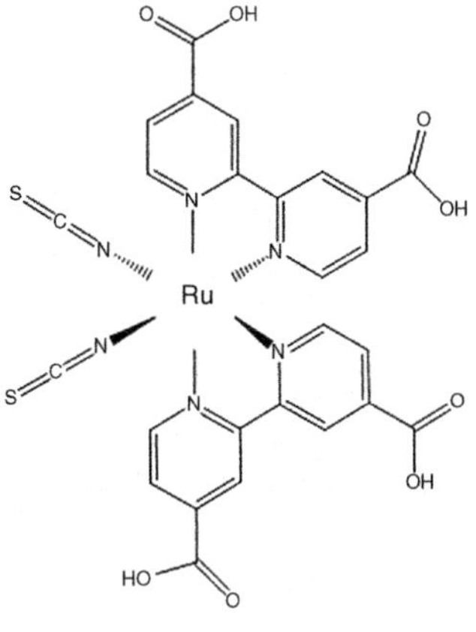

11
V_{OC}: 720 mV; J_{SC}: 18.2 mA.cm^{-2}
FF: 0.73; η: 10.0%

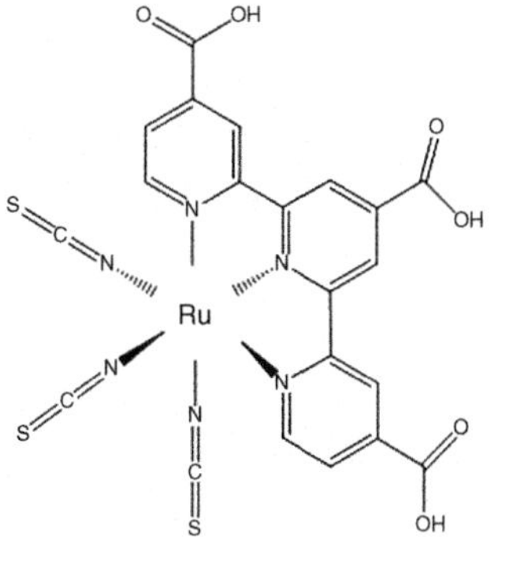

12
V_{OC}: 736 mV; J_{SC}: 20.9 mA.cm^{-2}
FF: 0.72; η: 11.1%

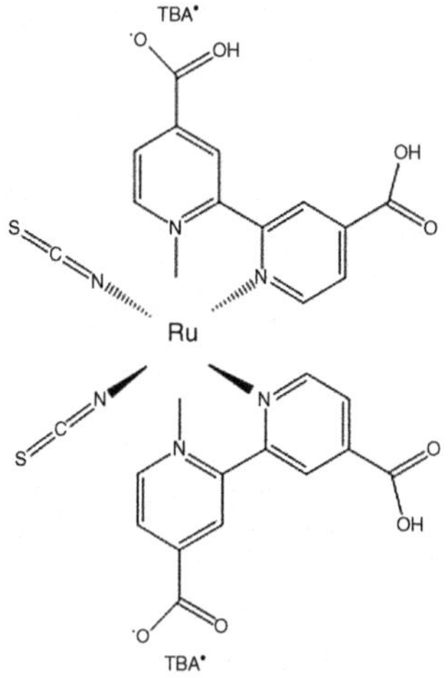

13
V_{OC}: 826 mV; J_{SC}: 17.0 mA.cm^{-2}
FF: 0.72; η: 10.1%

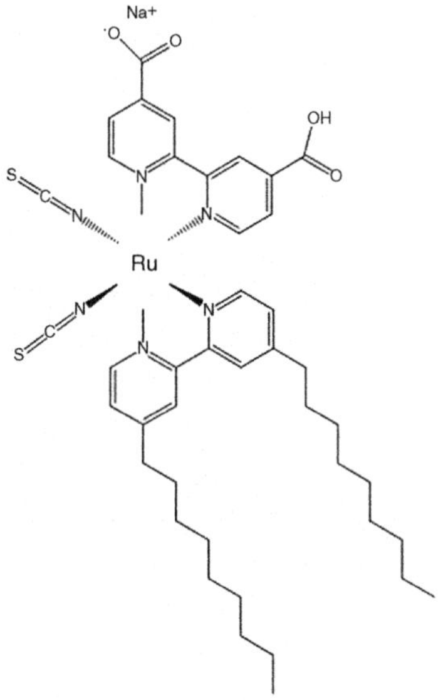

14
V_{OC}: 700 mV; J_{SC}: 15.9 mA.cm^{-2}
FF: 0.68; η: 7.2%

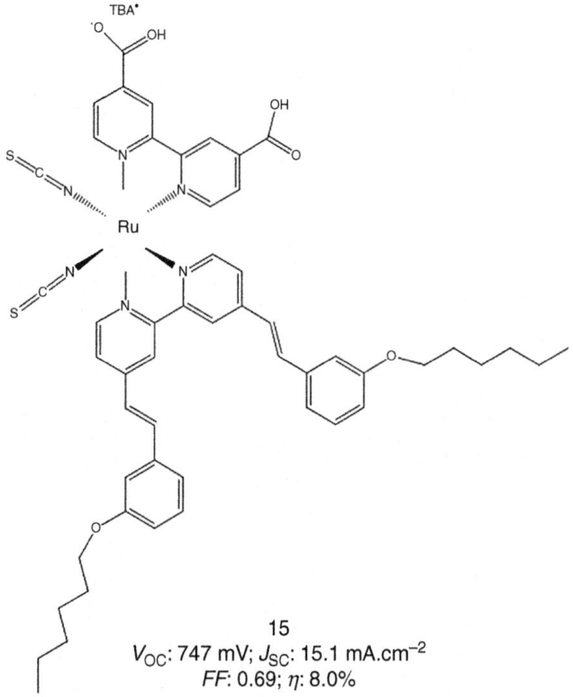

15
V_{OC}: 747 mV; J_{SC}: 15.1 mA.cm^{-2}
FF: 0.69; η: 8.0%

FIGURE 1.12 Structural formula of the main dyes used in dye-sensitized solar cells (DSSCs) and their respective photovoltaic parameters (for cells in optimized conditions). 11, N3; 12, N749; 13, N719; 14, Z907; 15, K19; 16, YD2-o-C8; 17, D205; 18, blackberry anthocyanin.

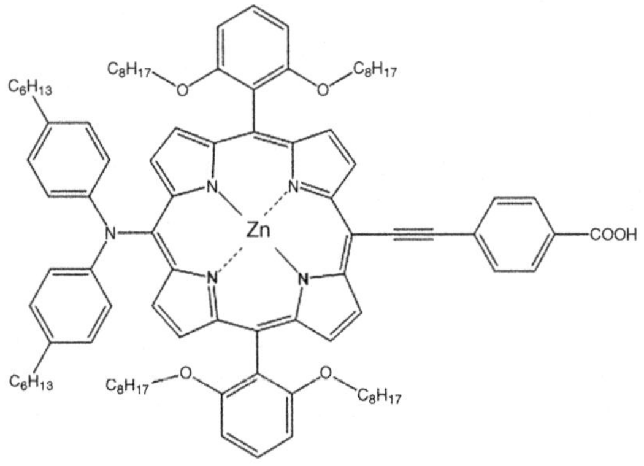

16
V_{OC}: 965 mV; J_{SC}: 17.3 mA.cm^{-2}
FF: 0.71; η: 11.9%

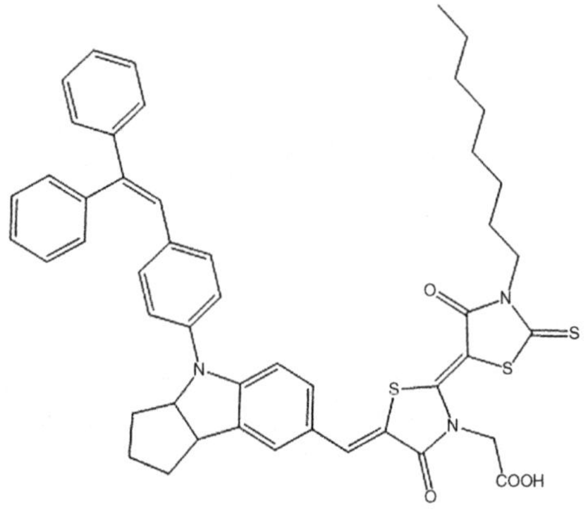

17
V_{OC}: 728 mV; J_{SC}: 13.7 mA.cm^{-2}
FF: 0.72; η: 7.2%

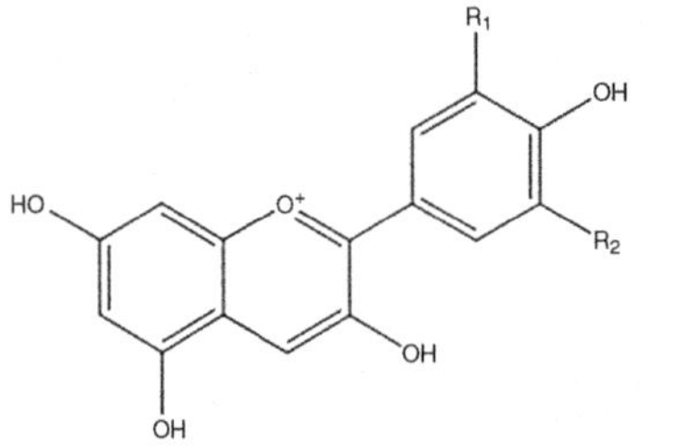

18
V_{OC}: 348 mV; J_{SC}: 15.1 mA.cm^{-2}
FF: 0.43; η: 1.15%

FIGURE 1.12 (cont.)

The electrolyte consists of a redox pair, such as I^-/I_3^-, cobalt(II/III) complexes, or TEMPO/TEMPO$^+$ (2,2,6,6-tetramethyl-1-piperidinyloxy) [51]. There is also the option of using ionic liquids [57,58], polymer electrolytes [59], and p-type semiconductors, including conjugated polymers [60,61]. The counter-electrode is formed by a transparent FTO electrode modified with a thin platinum layer. Some alternative materials can also be used, such as graphite and conductive polymers [62]. The operating principle of a DSSC and the morphology of the photoanode are schematized in Fig. 1.13A and 1.13B, respectively.

According to Fig. 1.13A, the photoexcited dye (S*) injects electrons into the semiconductor's conduction band. The photoinjected electrons are then collected by the transparent electrode to generate an anodic current in the circuit external to the cell. The photoexcited dye receives electrons from the electrolyte's redox pair and is then regenerated to its ground state. The redox pair is in turn regenerated by the electrons from the external circuit that are captured by the counter-electrode. An equivalent process occurs when the redox pair is replaced by organic semiconductors [61]. The cell operates in a cyclic and regenerative way, without any permanent chemical transformation. Differently from OSCs, in DSSCs, the energy absorption and charge transport are performed by different

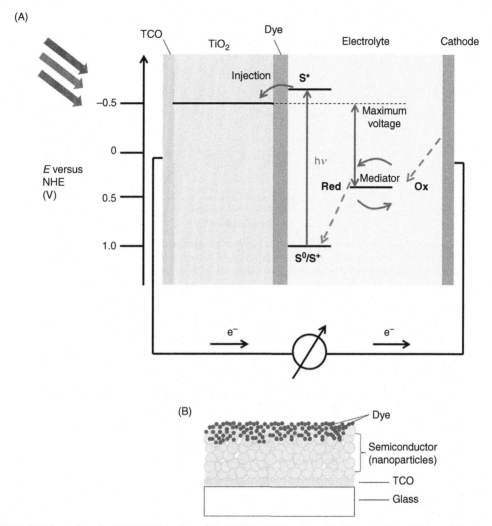

FIGURE 1.13 Schematics of a DSSC. (A) Side view of the DSSC indicating the oxidation–reduction voltages of the dye and mediator, and the position of the semiconductor's conduction band relative to the normal hydrogen electrode *(NHE).* (B) Amplified side view of the photoanode with emphasis on the semiconductor nanoparticles sensitized with dye molecules. *Adapted from M. Gratzel, Recent advances in sensitized mesoscopic solar cells, Acc. Chem. Res. 42 (2009) 1788–1798 [51].*

elements in the device, a feature that widens the options of different absorbing materials and electrodes.

The efficiency of light conversion into electricity by DSSCs depends on effective electron coupling between the excited state of the photosensitizer and the semiconductor's conduction band. This condition is met mainly by the combination of ruthenium(II)-based dyes and TiO$_2$ mesoporous films, which has the highest conversion values [63,64]. The polypyridine complexes of ruthenium(II) (structures 11–15, Fig. 1.12) strongly absorb

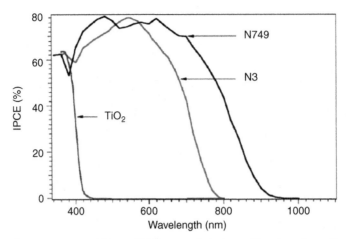

FIGURE 1.14 Incident photon to current efficiency *(IPCE)* curves of photosensitizers based on ruthenium(II) complexes **(N3, N749) and pure TiO₂ mesoporous films.** The chemical structure of the photosensitizers is shown next to the curves. *Reproduced with permission from M. Gratzel, Recent advances in sensitized mesoscopic solar cells, Acc. Chem. Res. 42 (2009) 1788–1798 [51].*

visible light because of the metal-to-ligand charge transfer (MLCT) type of transition [65,66]. In addition to the intense absorption, the energy levels of these complexes favor the electron transfer to the semiconductor, which occurs in nanoseconds to femtoseconds and, similar to OSCs, is much faster than recombination processes.

The voltage produced by an illuminated DSSC corresponds to the difference between the Fermi level of the electrons on the TiO_2 nanoparticles and the redox potential of the electrolyte (in DSSCs with semiconductors instead of a redox pair, this voltage is the chemical potential of the gaps) [51]. For a large number of photosensitizers anchored on TiO_2 nanoparticles, the driving force for electron injection is 0.15 eV [67]. However, the energy consumed to regenerate the photosensitizer is 0.6 eV. This imbalance is minimized by photosensitizers that have a redox potential closer to the redox potential of the iodide/triiodide pair (or another mediator). Another strategy to increase the efficiency of DSS-Cs is to extend the absorption range of the dye to the infrared region. The EQE of DSSCs based on two ruthenium(II) complex photosensitizers (N3 and N749) is as high as 80%, as shown in Fig. 1.14 [51]. Moreover, the spectral response of photosensitizer N749 extends to the near infrared, so an even wider range of the solar spectrum can be used by the cell. Therefore, the conversion efficiency increases 1.5-fold compared to the cell prepared with N3. As shown in Fig. 1.13, the photoelectrochemical parameters of current DSSCs are very promising; values of J_{sc} on the order of 20 mA cm², V_{oc} between 0.7 and 0.86 V, and *FF* between 0.65 and 0.8 have been found for cells under AM1.5 solar irradiation.

Ruthenium complexes are the main photosensitizers used in DSSCs. However, porphyrin-based DSSCs (structure 16, Fig. 1.12) have conversion efficiencies on the order of 12% [68]. Metal-free dyes are synthesized or extracted from natural sources to be used as photosensitizers. Synthetic organic dyes to be used in DSSCs must have electron donating and electron conducting units and anchoring groups (structure 17, Fig. 1.12) [69–72]. Natural extracts of berries, such as blackberry, raspberry, and açaí, are rich in anthocyanins, which are flavonoids that inhibit the action of free radicals and can be used as photosensitizers in DSSCs [73]. The anthocyanin molecule (structure 18, Fig. 1.12) has hydroxyl groups that can anchor to the TiO_2 surface and, when photoexcited, is able to inject electrons in the semiconductor's conduction band. The stability of a DSSC sensitized with blackberry extract was evaluated, and it was concluded that it is comparable to that of a DSSC sensitized with N3 as long as the device is adequately sealed [74].

The mesoporous and nanocrystalline layer of a semiconductor oxide (Fig. 1.13B) and the photosensitizer are key elements of the DSSC. The nanoparticles are usually produced by the sol–gel method, followed by hydrothermal growth up to a size of 20–25 nm [75]. Once obtained, the nanoparticles are dispersed in a paste prepared with a stabilizing agent, such as poly(ethylene oxide). The paste is spread on the surface of the transparent electrode, which is then placed in an oven for sintering. The sintering process creates a connection between the particles that ensures electrical contact throughout the entire mesoscopic structure. The pores' mesoscopic structure significantly increases the surface area of the semiconductor film, which increases the amount of photosensitizer absorbed. Consequently, the light absorption capacity of the mesoporous system is much higher than that of a compact film. TiO_2 in anatase form is the most widely used semiconductor in DSS-Cs. The TiO_2 particles in this form have a bipyramidal shape, with facets exposed to the (101) direction, which is the direction with the lowest surface energy [51]. Fig. 1.15 shows the morphology of a TiO_2 mesoporous film used in DSSCs. In general, the thickness of a

FIGURE 1.15 Scanning electron microscopy micrograph of a TiO₂ mesoporous film. *Reproduced with permission from A.O.T. Patrocinio, et al., Layer-by-layer TiO₂ films as efficient blocking layers in dye-sensitized solar cells, J. Photochem. Photobiol. A 205 (2009) 23–27 [78].*

semiconductor's mesoporous layer is 5–10 µm. In the dark, this semiconductor is an insulator because the separation energy between the bands is very high (3.2 eV). However, when electrons are injected into a TiO$_2$ nanoparticle during DSSC operation, the electron concentration can increase to up to 10^{17} cm^3, corresponding to a conductivity of 10^{-4} S cm^{-1} [67]. The charge transport mechanism inside the material is still not entirely understood. Nevertheless, there is consensus that the photoinjected electrons are caught in traps in the semiconductor, and once these traps are filled, the electrons begin to move freely by diffusion. The electron diffusion coefficient is not fixed but rather increases with light intensity. In addition, electron transport is accompanied by the movement of cations in the electrolyte, whose role is to avoid the formation of a noncompensated spatial charge that would certainly hinder the movement of electrons through the semiconductor [67]. The mesoporous film also has a terminal layer (~5 µm) of larger particles (400–500 nm) used to increase the optical path and the light collection in the red region [67].

In addition to the photosensitizer dye and the semiconductor film, a DSSC is composed of other parts: the electrolyte and the counter-electrode. The electrolyte has a support role in the operation of the DSSC and is responsible for the regeneration of the photosensitizer to its ground state. Depending on the aggregation state of the electrolyte, it can be classified as a liquid, solid, or gel [61]. A liquid electrolyte is usually prepared with a redox pair, such as iodide/triiodide, dissolved in acetonitrile. In many cases, a solvent mixture is prepared to facilitate dispersion through the mesoporous layer and to passivate empty sites on the semiconductor surface. From a technological point of view, two important issues related to the use of liquid electrolytes must be considered. First, the photosensitizer's regeneration by the redox pair occurs mainly by diffusion, which is more effective when the electrolyte is less viscous. Second, the use of a liquid solvent requires effective sealing of the device to avoid solvent leakage and evaporation. Some alternatives have been proposed and tested, such as solid and gel polymer electrolytes, such as poly(ethylene oxide) impregnated with iodide/triiodide, organic conductors, such as PEDOT:PSS and spiro-OMeTAD, or even mixtures of a liquid solvent with an ionic liquid. Ionic liquids are being investigated as candidates for the partial or total substitution of organic solvent-based electrolytes in DSSCs [58]. The low vapor pressure of ionic liquids is ideal for outdoor applications, such as solar panels in buildings. However, the diffusivity of the redox pair in ionic liquids is low, which requires studies to optimize the electrolyte composition. In addition, there is one photosensitizer better adapted to each type of mediator system or electrolyte, which makes the choice of the mediator even less straightforward.

The DSSC performance is significantly reduced by recombination reactions [76]. In the case of liquid electrolyte DSSCs, the reaction between the photoinjected electrons and I$_3$ occurs preferably at the FTO/semiconductor interface because of the permanent contact of the electrolyte with the conductive electrode. However, this reaction can be inhibited by a compact layer placed between the electrode and the semiconductor layer that physically blocks the contact of the I$_3$ ions of the electrolyte with the photoinjected electrons in the electrode [77]. This layer is usually made of the same semiconductor used in the mesoporous layer, although it is thinner, on the order of 100–200 nm. The self-assembly technique seems to be a promising method for deposition of the blocking layer because of its high degree of control of the film's thickness and morphology at the nanoscale and its low execution cost [78,79]. Basically, the technique consists of the successive and alternate immersion of a solid substrate in suspensions of anionic and cationic materials. This procedure results in the spontaneous adsorption of ultrathin layers, whose thicknesses are regulated by the conditions of the suspensions (i.e., concentration, pH, and ionic strength) [80]. Blocking layers composed of TiO_2 cationic particles (5 nm in diameter) and the polyanion poly(styrene sulfonic acid) PSS on FTO electrodes formed prior to the deposition of the TiO_2 mesoporous layer increased the overall conversion efficiency of an N3-based DSSC by 30% [78]. In addition, it was confirmed that the self-assembled layer not only blocks recombination but also improves the electrical contact between the electrode and the mesoporous layer, and both effects are responsible for the improvement in cell performance [81].

1.3 Photoelectrochemical Cells for the Production of Solar Fuels

Solar cells are devices that convert solar energy directly into electricity; therefore, they represent a feasible alternative to produce clean and cheap energy. However, one might ask how to produce energy at night? The Sun is an intermittent energy source, and it is estimated that only 6 h of the total solar irradiation are available for conversion to other useful forms of energy. Therefore, for solar energy to be a primary energy source, it must be associated with effective means of storage. To address this issue, we look at nature and how approaches have emerged naturally to produce and store energy from the Sun. The answer is photosynthesis. Briefly, photosynthesis can be described as the process through which plants absorb and convert solar energy, together with CO_2 and H_2O, into chemical energy in the form of substances with high energy content, such as carbohydrates [82]. Far from being science fiction, actual ways to artificially reproduce photosynthesis are being researched today. This strategy would allow the conversion of solar energy into high-energy products that could be used as fuel in power plants or to generate electricity in fuel cells during the night or at any time of the day.

Direct conversion of solar light into fuels (solar fuels) consists of transforming solar energy into energy stored in chemical bonds with the aid of a photocatalyst [14]. Solar fuels are an attractive solution toward energy sustainability because they can store large amounts of energy for a virtually unlimited time. The two most researched artificial photosynthesis techniques are based on the water-splitting reaction to produce molecular hydrogen (reaction 1.I) and the CO_2 reduction reaction to produce methanol or hydrocarbons (reaction 1.II):

$$2H_2O + 4hv \rightarrow 2H_2 + O_2 \ (\Delta G° = 4.92 \text{ eV}, n = 4) \tag{1.I}$$

$$2H_2O + CO_2 + 8h\nu \rightarrow CH_4 + 2O_2 \ (\Delta G^\circ = 10.3 \text{ eV}, n = 8) \tag{1.II}$$

where n is the number of transferred electrons. Both reactions are thermodynamically unfavorable and involve a large number of transferred electrons. Thus, it is necessary to develop a photocatalytic system able to provide the electrochemical potential required to drive the reactions and allow efficient conversion of solar energy into chemical energy. Although water electrolysis is a well-established technology, it depends on the use of platinum as an electrocatalyst to accelerate the hydrogen evolution and water oxidation reactions. Platinum is a noble and expensive metal that does not represent a sustainable source. Therefore, it is necessary to develop lower cost photocatalytic systems that can be produced from sustainable sources.

A photoelectrochemical cell for artificial photosynthesis consists essentially of a pair of electrodes, with one or both made of a photoactive material, and an electrolyte responsible for completing the path between the electrodes and the external circuit [14]. The electrode made of the photoactive material acts as a photocatalyst; that is, it absorbs and converts solar energy at the oxidation–reduction potentials necessary to perform the water splitting or CO_2 reduction reactions. Different semiconductor materials can be used as photocatalysts in the water splitting reaction, as illustrated in Fig. 1.16. Hematite (α-Fe_2O_3) in particular is very promising. Its theoretical light-into-hydrogen conversion efficiency is 16.8% [83]. Pure TiO_2 is an excellent photocatalyst when excited with UV light [84], but it must be dye-photosensitized for the visible and infrared ranges. Photocatalytic systems based on metal complexes [85–87] and biocatalysts [88,89] (enzymes, especially hydrogenases and dehydrogenases) have also been tested for both water splitting and CO_2 reduction. Noble metals and their oxides (e.g., Pt, Pd, RuO_x, and IrO_x) are generally used as cocatalysts to accelerate the O_2 and H_2 evolution reactions. However, high cost and low abundance limit the large-scale production of photoelectrochemical cells with this type of electrocatalyst [14].

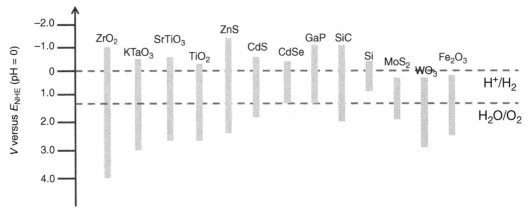

FIGURE 1.16 **Energy levels of inorganic semiconductors in aqueous medium of pH 0.** The *horizontal dashed lines* in *red* represent the oxidation potential of water (1.23 V) and the reduction potential of the proton (0 V). *Adapted from P.D. Tran, et al., Recent advances in hybrid photocatalysts for solar fuel production, Energy Environ. Sci. 5 (2012) 5902–5918 [14].*

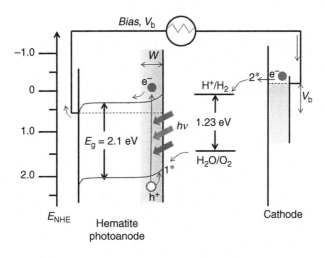

1*: $2OH^- + 2h^+$ $H_2O + ½ O_2$

2*: $2H_2O + 2e^-$ $H_2 + 2OH^-$

FIGURE 1.17 Schematic of a photoelectrochemical cell for water oxidation based on a hematite photoanode. The energy levels of the semiconductor and the water oxidation and H^+ reduction reactions are indicated in relation to the potential of the NHE. *Adapted from K. Sivula, et al., Solar water splitting: progress using alfa-Fe₂O₃ photoelectrodes, ChemSusChem 4 (2011) 432–449 [83].*

One of the most researched photoelectrochemical cell prototypes for the production of hydrogen from water photodecomposition uses photoanodes manufactured with hematite nanostructured films [83]. Hematite is the most stable crystalline form of iron oxide. It exhibits semiconductor behavior with a band separation energy of 2.1 eV, is stable at operating conditions, and is available cheaply. A schematic of a hematite-based photoelectrochemical cell and its operating principle during water photooxidation is shown in Fig. 1.17. The energy levels of hematite's valence and conduction bands, the water oxidation potential, and the proton reduction potential are also represented. When light is absorbed, the electrons are promoted to hematite's conduction band and the remaining gaps in the valence band are used in the O_2 evolution reaction (reaction 1.I). In the second stage, the electrons in the photoanode's conduction band are transferred to the external circuit until reaching the cathode, where they reduce protons to generate H_2 (reaction 1.II). The electrolyte medium may be an aqueous solution of a base, such as a sodium hydroxide solution of pH ~ 13.

As shown in Fig. 1.18, the level of the hematite conduction band is not sufficient to reduce the proton to hydrogen. This requires the application of an overvoltage in the cell for photooxidation to begin. The overvoltage can be supplied by a photovoltaic cell connected in series to the photooxidation cell [83]. Hematite has other limitations that must be considered: a low absorption coefficient that requires thick films (400–500 nm) for full light absorption, low conductivity of the major carriers, and a very short diffusion length (L_d = 2–4 nm) of the minor carriers. Nanotechnology has been applied to overcome these limitations to produce hematite photoanodes with ideal performance. The overvoltage (up to 1.0 V) required to begin the water splitting process with hematite photoanodes is due to surface effects of the material. First, the presence of oxygen vacancies and the crystalline disorder create electron states in the middle of the forbidden band that act as traps for the gaps. In addition, the oxygen evolution reaction promoted by the hematite is very slow. The application of a passivating layer, such as Al_2O_3, and a layer of cobalt(II) or amorphous cobalt(II) phosphate eliminates the surface states and catalyzes the oxygen evolution reaction, respectively, to reduce overvoltage to close to 0.1–0.3 V [83]. The control of the morphology and the use of dopants such as Si significantly improve light absorption and the electrode's conductivity. Hematite films doped with Si and of columnar morphology, with "cauliflower-shaped" nanostructured columns, currently have the highest conversion efficiencies of solar light into hydrogen (2.1%) [90]. Fig. 1.19 shows a micrograph and the photoelectrochemical response of a hematite photoanode with this morphology.

Fig. 1.18B shows that at 1.23 V (water oxidation potential), the photocurrent density is 2.3 mA cm^{-2}, the highest value achieved to date [90]. The functionalization of the electrode's surface with cobalt ($Co(OH)_2$) increases the photocurrent to 2.7 mA cm^{-2} and alters the initiation of photooxidation by approximately 80 mV. Other studies have also demonstrated that nanostructured films of pure hematite are potentially promising as long as the growth of the columnar nanostructures has a preferential crystallographic orientation, under which charge transport is more effective [91]. Thus, it is believed that the development of nanotechnology will result in new methods to increase the conversion efficiency of hematite-based cells.

Coordination compounds based on ruthenium, cobalt, and nickel can be used to sensitize semiconductor electrodes and as cocatalysts of the O_2 evolution reaction. One of these complexes [(4,4′-((HO)$_2$P(O)CH$_2$)$_2$bpy)$_2$Ru^{II}(bpm) Ru^{II}(Mebimpy)(OH$_2$)]$^{4+}$, where bpy = 2,2′-bipyridine, bpm = 2,2′-bipyrazine, and Mebimpy = 2,6-bis(1-methylbenzimidazol-2-yl)pyridine), whose structure immobilized on an electrode is shown in Fig. 1.19, has different "modules" for each task, conceived to mimic the biological system of photosynthesis [87,92,93]. The group that is bonded directly to the electrode is the chromophore responsible for sensitization. Under the action of light, it injects electrons into the semiconductor, which are transferred to the external circuit up to the cathode. The more external group is the catalytic center that oxidizes the water molecules to O^{2+} and H^+. In addition, the electrons removed in this reaction are transferred to the chromophore part through a "bridge" (2,2′-bipyrazine unit). The electrons that reach the cathode are utilized in the reduction reaction, $2H^+ + 2e^- \rightarrow H_2$. Electrochemical studies show that the catalytic system is stable for up to 28,000 cycles without signs of deterioration [93]. However, under longer operation times and because of the environmental conditions, which are usually very oxidizing, the organic ligands of the complex degrade. These systems are

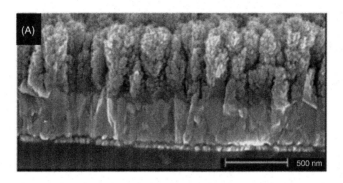

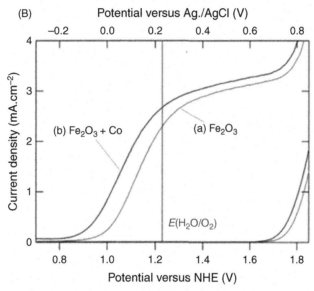

FIGURE 1.18 (A) Cross-section of a hematite film grown by chemical vapor deposition on an fluorine-doped tin oxide (FTO) substrate. (B) $J \times V$ curves of the hematite film in the dark and under simulated illumination (AM1.5) without modification and modified with cobalt, as indicated. *Reproduced with permission from A. Kay, et al., New benchmark for water photooxidation by nanostructured α-Fe2O3 films, J. Am. Chem. Soc. 128 (2006) 15714–15721 [90].*

not competitive for large-scale applications today, but they enable some of the photosynthesis reactions to be applied in a controlled manner.

Cobaloxime, a cobalt(III)-based complex, is another efficient catalyst of the H_2 evolution reaction, especially in strongly acidic media and organic solutions. The stability of this family of complexes is still low when immobilized on substrates to prepare photoelectrodes [14,85,86]. To date, only a few photoanodes based on cobalt complexes have been reported to enable the water splitting reaction. One potential candidate is based on an ITO electrode sensitized with perylene (an n-type organic semiconductor) and then covered with a second layer of cobalt phthalocyanine. When operated together with a platinum

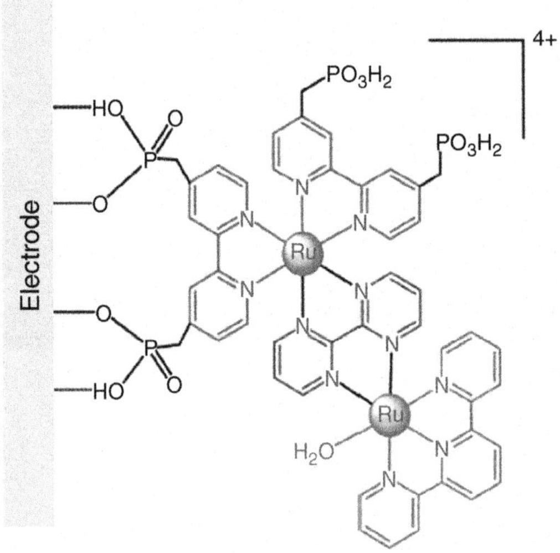

FIGURE 1.19 Structure of the complex [(4,4'-((HO)$_2$P(O)CH$_2$)$_2$bpy)$_2$Ru^{II}(bpm)Ru^{II}(Mebimpy)(OH$_2$)]$^{4+}$, where bpy = 2,2'-bipyridine, bpm = 2,2'-bipyrazine, and Mebimpy = 2,6-bis(1-methylbenzimidazol-2-yl)pyridine), immobilized on an electrode. *Reproduced with permission from J.J. Concepcion, et al., Making oxygen with ruthenium complexes, Acc. Chem. Res. 42 (2009) 1954–1965 [92].*

contraelectrode, this photoanode is able to split water (pH 11) in the presence of visible light under a voltage of 0.4 V (Ag/AgCl) and a yield of 3500 cycles/h [86].

1.4 Conclusions and Perspectives

Solar cells and photoelectrochemical cells for artificial photosynthesis are concrete solutions to current and future global energy demands and an alternative to the scarcity and environmental impact problems associated with fossil fuels. However, coordinated efforts among different segments of society—the scientific community, industry, and government—are still necessary to implement scientific discoveries and technological advances for the service of the population.

In terms of science and technology, conversion of solar light into useful energy is based on an established sequence of events—light absorption and charge transfer, separation and transport—that occur at very short times and at an atomic and molecular scale. Therefore, the role of nanotechnology is fundamental because nanomaterials can be used as construction blocks to assemble organized systems with virtually unlimited capabilities for solar light capture and charge transport. Thanks to a large number of semiconductors (both organic and inorganic) and new properties that manifest at the nanoscale, as well as the possibility for combination with molecular systems to increase light collection efficiency, third-generation solar cells (OSCs and DSSCs) are promising prospects to convert solar light into electricity. In addition, the possibility of having flexible (or even disposable) devices, fabricated with cheaper materials in much simpler manufacturing systems, is an exclusive feature of these new cells. However, materials and devices with higher efficiency and stability still need to be developed for these cells to be effectively introduced into the market. The development of solar cells based on crystalline silicon was considered impractical in the past because of their high cost. However, the fact that silicon has ideal properties for photovoltaic conversion cannot be ignored, and silicon-based solar cells are now readily available. The same path must be followed in the case of third-generation solar cells.

Regarding artificial photosynthesis and the production of solar fuels, this field is still in an early development stage, but it looks very promising because of the discoveries with third-generation cells together with the consolidation of the photochemistry of coordination compounds and semiconductors. Although conversion efficiencies are still modest, the breakthroughs in cells based on hematite photoelectrodes are encouraging, especially considering that hematite is an abundant and low-cost material. The discoveries in the field of the photochemistry of coordination compounds go beyond energy production. Molecular photocatalytic systems have provided a single and elegant platform to study biological processes whose full understanding will certainly result in the development of more efficient energy conversion devices.

List of Symbols

DSSCs	Dye-sensitized solar cells
EQE	External quantum efficiency
FF	Fill factor
FTO	Fluorine-doped tin oxide
GHGs	Greenhouse gases
HOMO	Highest occupied molecular orbital
IPCE	Incident photon to current efficiency
ITO	Indium-doped tin oxide
J_{sc}	Short-circuit current density
LUMO	Lowest occupied molecular orbital
MDMO-PPV	Poly[2-methoxy-5-(3′,7′-dimethyloctyloxy)-1,4-phenylenevinylene]
MLCT	Metal-to-ligand charge transfer
OSCs	Organic solar cells
PANI	Poly(aniline)
PC61BM	[6,6]-Phenyl-C61-butyric acid methyl ester
PC71BM	2, [6,6]-Phenyl-C71-butyric acid methyl ester
PEDOT:PSS	Poly(3,4-ethylenedioxytiophene)-poly(styrenesulfonic acid)
Pout	Power produced per illuminated area
PPV	Poly(p-phenylene vinylene)
PSS	Poly(styrenesulfonic acid)
spiro-OMeTAD	N2,N2,N2′,N2′,N7,N7,N7′,N7′-octakis(4-methoxyphenyl)-9,9′-spirobi[9H-fluorene]-2,2′,7,7′-tetramine
TCO	Transparent conductive oxide
TEMPO	2,2,6,6-Tetramethyl-1-piperidinyloxy
V_{oc}	Open circuit voltage
η	Overall conversion efficiency

References

[1] N.S. Lewis, D.G. Nocera, Powering the planet: chemical challenges in solar energy utilization, Proc. Natl. Acad. Sci. USA 103 (2006) 15729–15735.

[2] M.E. Mann, et al., Global-scale temperature patterns and climate forcing over the past six centuries, Nature 392 (1998) 779–787.

[3] P.V. Kamat, Meeting the clean energy demand: nanostructure architectures for solar energy conversion, J. Phys. Chem. C 111 (2007) 2834–2860.

[4] http://www.institutocarbonobrasil.org.br

[5] Key World Energy Statistics, International World Agency, 2012.

[6] (a) http://en.wikipedia.org/wiki/Nuclear_power_in_the_United_States; (b) http://pt.wikipedia.org/wiki/Energia_nuclear_no_Japao

[7] http://pt.wikipedia.org/wiki/Acidente_nuclear_de_Chernobil

[8] http://pt.wikipedia.org/wiki/Acidente_nuclear_de_Fukushima_I

[9] M.S. Dresselhaus, I.L. Thomas, Alternative energy technologies, Nature 414 (2001) 332–337.

[10] N.S. Lewis, et al., Basic Energy Sciences Report on Basic Research Needs for Solar Energy Utilization, Office of Science, U.S. Department of Energy, Washington, DC, 2005, http://www.sc.doe.gov/bes/reports/files/SEU_rpt.pdf

[11] http://www.downtoearth.org.in/content/falling-silicon-prices-shakes-solar-manufacturing-industry

[12] (a) http://www.polyera.com; (b) http://www.heliatek.com; (c) http://www.plextronics.com; (d) http://www.solarmer.com; (e) http://www.konarka.com

[13] (a) http://www.g24i.com; (b) http://www.dyesol.com; (c) http://www.m-kagaku.co.jp; (d) http://www.solaronix.com

[14] P.D. Tran, et al., Recent advances in hybrid photocatalysts for solar fuel production, Energy Environ. Sci. 5 (2012) 5902–5918.

[15] http://org.ntnu.no/solarcells

[16] S.J. Fonash, Solar Cell Device Physics, second ed., Academic Press, Burlington, 2010.

[17] http://www.astm.org/standards

[18] C. Kittel, Introduction to Solid State Physics, eighth ed., John Wiley & Sons, New Jersey, 2005.

[19] H. Hoppe, N.S. Sariciftci, Organic solar cells: an overview, J. Mater. Res. 19 (2004) 1924–1945.

[20] A.F. Nogueira, et al., Polymer solar cells using single-wall carbon nanotubes modified with thiophene pendant groups, J. Phys. Chem. C 111 (2007) 18431–18438.

[21] S. Luzzati, et al., Photo-induced electron transfer from a dithienothiophene-based polymer to TiO_2, Thin Solid Films 403 (2002) 52–56.

[22] D. Wohrle, D. Meissner, Organic solar cells, Adv. Mater. 3 (1991) 129–138.

[23] C.W. Tang, Two-layer organic photovoltaic cell, Appl. Phys. Lett. 48 (1986) 183–185.

[24] G. Yu, et al., Polymer photovoltaic cells: enhanced efficiencies via a network of internal donor-acceptor heterojunctions, Science 270 (1995) 1789–1791.

[25] C.Y. Yang, A.J. Heeger, Morphology of composites of semiconducting polymers mixed with C60, Synth. Met. 83 (1996) 85–88.

[26] Z. He, et al., Enhanced power conversion efficiency in polymer solar cells using an inverted device structure, Nat. Photon 6 (2012) 591–595.

[27] L. Bian, Recent progress in the design of narrow bandgap conjugated polymers for high efficiency organic solar cells, Prog. Polym. Sci. 37 (2012) 1292–1331.

[28] R.F. Service, Outlook brightens for plastic solar cells, Science 332 (2011) 293–303.

[29] F.C. Krebs, et al., Product integration of compact roll-to-roll processed polymer solar cell modules: methods and manufacture using flexographic printing, slot-die coating and rotary screen printing, J. Mater. Chem. 20 (2010) 8994–9001.

[30] A.J. Heeger, Nobel Lecture: semiconducting and metallic polymers: the fourth generation of polymeric materials, Rev. Mod. Phys. 73 (2001) 681–700.

[31] A.G. Macdiarmid, Synthetic metals: a novel role for organic polymers, Synth. Met. 125 (2001) 11–22.

[32] N.S. Sariciftci, et al., Photoinduced electron transfer from a conducting polymer to buckminsterfullerene, Science 258 (1992) 1474–1476.

[33] C.J. Brabec, et al., Tracing photoinduced electron transfer process in conjugated polymer/fullerene bulk heterojunctions in real time, Chem. Phys. Lett. 340 (2001) 232–236.

[34] M.C. Scharber, Design rules for donors in bulk-heterojunction solar cells—towards 10% energy-conversion efficiency, Adv. Mater. 18 (2006) 789–794.

[35] C.C. Wu, et al., Surface modification of indium tin oxide by plasma treatment: an effective method to improve the efficiency, brightness, and reliability of organic light-emitting devices, Appl. Phys. Lett. 70 (1997) 1348–1350.

[36] C. Ganzorig, M. Fujihira, Chemical modification of indium tin-oxide electrodes by surface molecular design, in: S.C. Moss (Ed.), Organic Optoelectronic Materials (Mater. Res. Soc. Symp. Proc. 708), Warrendale, PA, 2002, p. 83

[37] J.C.B. Santos, et al., Influence of polyaniline and phthalocyanine hole-transport layers on the electrical performance of light-emitting diodes using MEH-PPV as emissive material, Thin Solid Films 516 (2008) 3184–3188.

[38] L.A.A. Pettersson, et al., Modeling photocurrent action spectra of photovoltaic devices based on organic thin films, J. Appl. Phys. 86 (1999) 487–489.

[39] S.E. Shaheen, et al., 2.5% efficient organic plastic solar cells, Appl. Phys. Lett. 78 (2001) 841–843.

[40] H. Hoppe, et al., Nanoscale morphology of conjugated polymer/fullerene based bulk-heterojunction solar cells, Adv. Funct. Mater. 14 (2004) 1005–1011.

[41] P.W.M. Blom, et al., Electric-field and temperature dependence of the hole mobility in poly(*p*-phenylenevinylene), Phys. Rev. B 55 (1997) 656–659.

[42] L. Bozano, et al., Temperature- and field-dependent electron and hole mobilities in polymer light emitting diodes, Appl. Phys. Lett. 74 (1999) 1132–1134.

[43] V.D. Mihailetchi, et al., Electron transport in a methanofullerene, Adv. Funct. Mater. 13 (2003) 43–46.

[44] C. Melzer, et al., Hole transport in poly(phenylenevinylene)/methanofullerene bulk-heterojunction solar cells, Adv. Funct. Mater. 14 (2004) 865–870.

[45] C.J. Brabec, et al., Effect of LiF/metal electrodes on the performance of plastic solar cells, Appl. Phys. Lett. 80 (2002) 1288–1290.

[46] Y. Liang, et al., Development of new semiconducting polymers for high performance solar cells, J. Am. Chem. Soc. 131 (2009) 56–57.

[47] J. Desilvestro, et al., Highly efficient sensitization of titanium dioxide, J. Am. Chem. Soc. 107 (1985) 2988–2990.

[48] B. O'regan, M. Gratzel, A low-cost, high efficiency solar cell based on dye sensitized colloidal TiO_2 films, Nature 335 (1991) 737–740.

[49] Y. Chiba, et al., Dye sensitized solar cells with conversion efficiency of 11,1%, Jpn. J. Appl. Phys. Part 2 45 (2006) L638–640.

[50] J.M. Kroon, et al., Nanocrystalline dye sensitized solar cells having maximum performance, Prog. Photovolt. 15 (2007) 1–18.

[51] M. Gratzel, Recent advances in sensitized mesoscopic solar cells, Acc. Chem. Res. 42 (2009) 1788–1798.

[52] I. Robel, et al., Quantum dot solar cells: harvesting light energy with CdSe nanocrystals molecularly linked to mesoscopic TiO_2 films, J. Am. Chem. Soc. 128 (2006) 2385–2393.

[53] S. Ito, et al., High-conversion-efficiency organic dye-sensitized solar cells with a novel indoline dye, Chem. Commun. 10 (2008) 5194–5196.

[54] G. Zhang, et al., High efficiency and stable dye-sensitized solar cells with an organic chromophore featuring a binary π-conjugated spacer, Chem. Commun. 16 (2009) 2198–2200.

[55] A.S. Polo, N.Y.M. Iha, Blue sensitizers for solar cells: natural dyes from Calafate and Jaboticaba, Sol. Energy Mater. Sol. Cells 90 (2006) 1936–1944.

[56] A.O.T. Patrocinio, et al., Efficient and low cost devices for solar energy conversion: efficiency and stability of some natural-dye-sensitized solar cells, Synth. Met. 159 (2009) 2342–2344.

[57] M. Gratzel, Conversion of sunlight to electric power by nanocrystalline dye-sensitized solar cells, J. Photochem. Photobiol. A Chem. 164 (2004) 3–14.

[58] Y. Bai, et al., High-performance dye-sensitized solar cells based on solvent-free electrolytes produced from eutectic melts, Nat. Mater. 7 (2008) 626–630.

[59] A.F. Nogueira, et al., Dye-sensitized nanocrystalline solar cells employing a polymer electrolyte, Adv. Mater. 13 (2001) 826–830.

[60] Y. Saito, et al., Photo-sensitizing ruthenium complexes for solid state dye solar cells in combination with conducting polymers as hole conductors, Coord. Chem. Rev. 248 (2004) 1469–1478.

[61] B. Li, et al., Review of recent progress in solid-state dye-sensitized solar cells, Sol. Energy Mater. Sol. Cells 90 (2006) 549–573.

[62] T. Kitamura, et al., Improved solid-state dye solar cells with polypyrrole using a carbon-based counter electrode, Chem. Lett. 10 (2001) 1054–1055.

[63] A.S. Polo, et al., Metal complex sensitizers in dye-sensitized solar cells, Coord. Chem. Rev. 248 (2004) 1343–1361.

[64] R. Argazzi, et al., Design of molecular dyes for application in photoelectrochemical and eletrochromic devices based on nanocrystalline metal oxide semiconductors, Coord. Chem. Rev. 248 (2004) 1299–1316.

[65] J.P. Paris, W.W. Brandt, Charge transfer luminescence of a ruthenium(II) chelate, J. Am. Chem. Soc. 81 (1959) 5001–5002.

[66] K.W. Hipps, G.A. Crosby, Charge transfer excited states of ruthenium(II) complexes. III. Electron-ion coupling model for dπ* configurations, J. Am. Chem. Soc. 97 (1975) 7942–7948.

[67] M. Gratzel, Solar energy conversion by dye-sensitized photovoltaic cells, Inorg. Chem. 44 (2005) 6841–6851.

[68] A. Yella, et al., Porphyrin-sensitized solar cells with cobalt (II/III)–based redox electrolyte exceed 12 percent efficiency, Science 334 (2011) 629–633.

[69] K. Hara, et al., Novel conjugated organic dyes for efficient dye-sensitized solar cells, Adv. Funct. Mater. 15 (2005) 246–252.

[70] Z.S. Wang, Molecular design of coumarin dyes for stable and efficient organic dye-sensitized solar cells, J. Phys. Chem. C 112 (2008) 17011–17017.

[71] X.R. Song, D-π-A type organic photosensitizers with aromatic amine as electron-donating group— application in dye-sensitized solar cells, Prog. Chem. 20 (2008) 1524–1533.

[72] D. Kuang, Organic dye-sensitized ionic liquid based solar cells: remarkable enhancement in performance through molecular design of indoline sensitizers, Angew. Chem. Int. Ed. 47 (2008) 1923–1927.

[73] Q. Dai, J. Rabani, Photosensitization of nanocrystalline TiO_2 films by anthocyanin dyes, J. Photochem. Photobiol. A Chem. 148 (2002) 17–24.

[74] A.O.T. Patrocinio, N.Y.M. Iha, Em busca da sustentabilidade: células solares sensibilizadas por extratos naturais [Toward sustainability: solar cells sensitized by natural extracts], Quim. Nov. 33 (2010) 574–578.

[75] M.K. Nazeeruddin, et al., Conversion of light to electricity by Cis-X_2-bis(2,2'-bipyridyl-4,4'-dicarboxylate)ruthenium(II) charge transfer sensitizers ($X = Cl^-$, Br^-, I^-, CN^-, SCN^-) on nanocrystalline TiO_2 electrodes, J. Am. Chem. Soc. 115 (1993) 6382–6390.

[76] A.J. Frank, et al., Electrons in nanostructured TiO_2 solar cells: transport, recombination and photovoltaic properties, Coord. Chem. Rev. 248 (2004) 1165–1179.

[77] E. Palomares, et al., Control of charge recombination dynamics in dye sensitized solar cells by the use of conformally deposited metal oxide blocking layers, J. Am. Chem. Soc. 125 (2003) 475–482.

[78] A.O.T. Patrocinio, et al., Layer-by-layer TiO_2 films as efficient blocking layers in dye-sensitized solar cells, J. Photochem. Photobiol. A 205 (2009) 23–27.

[79] A.O.T. Patrocinio, et al., Filmes de óxidos semicondutores automontado [Self-assembled semiconductor oxide films], PI0802589-4 A2, deposit date 08/14/2008.

[80] L.G. Paterno, et al., Filmes poliméricos ultra-finos produzidos pela técnica de automontagem: preparação, propriedades e aplicações [Ultra-thin polymer films produced by the self-assembly technique: preparation, properties and applications], Quim. Nov. 24 (2001) 228–235.

[81] A.O.T. Patrocinio, et al., Role of polyelectrolyte for layer-by-layer compact TiO_2 films in efficiency enhanced dye-sensitized solar cells, J. Phys. Chem. C 114 (2010) 17954–17959.

[82] J. Whitmarsh, The photosynthetic process, in: G.S. Singhal, et al., (Ed.), Concepts in Photobiology: Photosynthesis and Photomorphogenesis, Kluwer Academic Publishers, Boston, 1999, pp. 11–51.

[83] K. Sivula, et al., Solar water splitting: progress using alfa-Fe_2O_3 photoelectrodes, ChemSusChem 4 (2011) 432–449.

[84] A. Fujishima, K. Honda, Electrochemical photolysis of water at a semiconductor electrode, Nature 238 (1972) 37–38.

[85] A.J. Esswein, D.G. Nocera, Hydrogen production by molecular photocatalysis, Chem. Rev. 107 (2007) 4022–4047.

[86] V. Artero, et al., Splitting water with cobalt, Angew. Chem. Int. Ed. 50 (2011) 7238–7266.

[87] J.J. Concepcion, et al., Chemical approaches to artificial photosynthesis, PNAS 109 (2012) 15560–15564.

[88] E. Reisner, et al., Visible light-driven H_2 production by hydrogenases attached to dye-sensitized TiO_2 nanoparticles, J. Am. Chem. Soc. 131 (2009) 18457–18466.

[89] T.W. Woolerton, et al., CO_2 photoreduction at enzyme-modified metal oxide nanoparticles, Energy Environ. Sci. 4 (2011) 2393–2399.

[90] A. Kay, et al., New benchmark for water photooxidation by nanostructured α-Fe_2O_3 films, J. Am. Chem. Soc. 128 (2006) 15714–15721.

[91] R.H. Gonçalves, et al., Magnetite colloidal nanocrystals: a facile pathway to prepare mesoporous hematite thin films for photoelectrochemical water splitting, J. Am. Chem. Soc. 133 (2011) 6012–6019.

[92] J.J. Concepcion, et al., Making oxygen with ruthenium complexes, Acc. Chem. Res. 42 (2009) 1954–1965.

[93] W. Song, et al., Making solar fuels by artificial photosynthesis, Pure Appl. Chem. 83 (2011) 749–768.

2

Supramolecular Systems

M.L. de Moraes*, L. Caseli**

*INSTITUTE OF SCIENCE AND TECHNOLOGY, FEDERAL UNIVERSITY OF SÃO PAULO, SÃO PAULO, SÃO PAULO, BRAZIL; **INSTITUTE OF ENVIRONMENTAL, CHEMICAL AND PHARMACEUTICAL SCIENCES, FEDERAL UNIVERSITY OF SÃO PAULO, SÃO PAULO, SÃO PAULO, BRAZIL

CHAPTER OUTLINE

2.1 General Concepts of Supramolecular Systems

A supramolecular system is an assembly of molecular subunits that are organized via intermolecular interactions that can be ionic or covalent. These interactions include hydrogen bonds, metallic coordination, hydrophobic interactions, Van der Waals forces and ionic interactions [1].

Supramolecular systems can be classified into two types:

1. macrocycle compounds that are formed via cyclic chemical structures and are linked by covalent bonds (e.g., compounds derived from porphyrin and phthalocyanine); and
2. systems with molecular self-aggregation, such as, micelles and lipid vesicles.

There are some interesting technological and scientific applications of supramolecular systems, such as, molecular recognition, catalysis, biomimetic systems, and transport of substances [2–4].

Many macrocycles, as well as self-aggregated disperse systems, have the capability to recognize and "capture" substances and, in many cases, to carry them to strategic locations. In general, molecular recognition is not an exclusive property of supramolecular systems. Molecular recognition was first described for biomacromolecules, such as enzymes, and later for nucleic acids. Hermann Emil Fischer [5], was one of the pioneers to identify the systems capable of recognizing molecules in 1902. He described the interactions between an enzyme and its substrate using the "key and lock" model. However, his comprehension of structure was very superficial. At the beginning of the 20th century, Van der Waals forces (noncovalent molecular interactions) improved the understanding of protein structure and of several biological processes.

In this context, an important advance for elucidating molecular recognition processes was the assumption that noncovalent bonds were fundamental for maintaining many of the aforementioned structures. The possibility of synthesizing structures capable of recognizing molecules or, at least, of "capturing" them using noncovalent interactions came when Charles J. Pedersen described the synthesis of crown ethers in the 1980s [6], which improved the understanding of self-organized structures in solutions, such as, micelles and liposomes.

The 1987 Nobel Prize in Chemistry was awarded to Donald J. Cram, Jean-Marie Lehn, and Charles J. Pedersen for their work in developing high selectivity molecules of the "host-guest" type. This can be considered as an acknowledgement of the importance of supramolecular chemistry. Later, other researchers started to develop molecular machinery with self-organized structures and complex sensors using construction blocks, such as, fullerenes and dendrimers.

The self-assembly concept must be highlighted for self-aggregation systems. Self-assembly involves the construction of systems in which the molecules are driven to assemble structures via noncovalent interactions. Thus, self-assembly can be subdivided into intermolecular and intramolecular. Intramolecular assembly involves macromolecules, such as, DNA and proteins. Intermolecular assembly involves the interaction of a certain number of smaller molecules for building larger structures, such as, micelles, vesicles, membranes, and liquid crystals.

2.2 Molecular Recognition

Molecular recognition is a specific interaction between a host molecule and a complementary guest molecule, which results in a host-guest complex. The designation of which substance is a "host" and which one is a "guest" is typically arbitrary. Molecules are capable of recognizing each other via noncovalent interactions. The main applications of this field are the construction of molecular sensors and catalysis.

In the case of template-driven synthesis, self-assembly can be used with reactive species to organize a system for a chemical reaction. Intermolecular interactions may facilitate the desired chemical interaction by minimizing the effect of side reactions. The molecule used as a template can be removed after the reaction.

The systems that are formed via molecular printing follow this principle. In these systems, the host is built from small molecules using a species of interest as a template. Therefore, the molecule (typically a polymer) is synthesized around the template molecule. The model is removed after the construction. This leaves only the host with a region in its chain (i.e., a cavity) with a molecular geometry that is equal or similar to the geometry of the molecule that was removed. The construction model of the cavity can subtly differ from the guest-molecule, which is a molecule of interest that is to be molecularly recognized later.

Some supramolecular recognition systems are considered "bioinspired" or "biomimetic" because they try to copy some functions of biological systems. This includes the production of biosensors or high-efficiency catalytic systems.

2.3 Self-Organized Systems

2.3.1 Surfactants and Detergents

Surfactants are amphiphilic substances. These molecules have one hydrophobic end and one hydrophilic end (Fig. 2.1) and are capable of changing the surface and interface tensions of a liquid. This effect occurs because of the strong adsorption of surfactant molecules at the air-water interface that reduces its surface energy, which is a consequence of the Gibbs theory [7].

Amphiphilic molecules have different properties that depend on the composition of their hydrophilic parts, which can range from hydroxyl to more complex groups, and their hydrophobic parts that may have saturated, unsaturated, or ramified chains. Surfactants are classified according to their hydrophilic part. Specifically, surfactants are classified as cationic, anionic, nonionic, and amphoteric or zwitterionic. The latter type has both positive and negative charges at different locations. Table 2.1 shows some surfactants with different charge types.

Additionally, surfactants are classified according to their function or application (i.e., detergents, foaming agents, conditioning agents, bactericides, moisturizers, emollients, dispersants, and solubilizing agents).

Detergents are surfactants that remove oily or greasy materials from surfaces, such as fabric fibers or human skin, and disperse these dirty materials in an aqueous medium. In

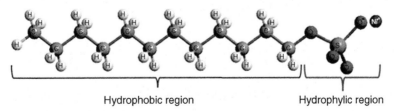

FIGURE 2.1 Amphiphilic molecule—sodium dodecyl sulfate (SDS). Hydrophobic and hydrophilic regions are composed of a hydrocarbon chain and a sulfate group, respectively.

Table 2.1   Some Example of Surfactants

Surfactant	Application
Anionic	
SDS	Solubilize proteins Protein electrophoresis cosmetics cleanliness
Cationic	
CTAB	Protein electrophoresis Synthesis of nanoparticles DNA extraction
Nonionic	
PEG	Pharmaceutical products Controlled drug delivery Eye lubricant
Zwitterionic	
DPPC	Controlled drug delivery Mimic cell membranes

CTAB, Cetyltrimethylammonium bromide; DPPC, dipalmitoylphosphatidylcholine; PEG, polyethylene glycol; SDS, sodium dodecyl sulfate.

biochemistry, detergents also solubilize a large range of species, such as, membrane proteins for biophysical analysis.

Oily materials are removed via detergent adsorption on the oily or greasy surface, which increases its contact angle. Thus, the grease and oil droplets are easily removed using mechanical action, and the surfactant, which is adsorbed around the droplet surface, stabilizes the oily material in an aqueous solution (Fig. 2.2) [8].

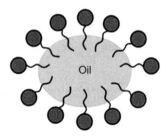

FIGURE 2.2 Adsorbed detergent around the surface of an oil droplet.

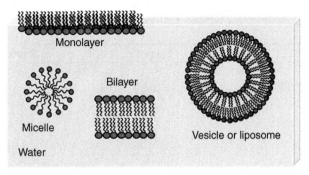

FIGURE 2.3 Lipid aggregates in aqueous medium.

2.3.1.1 Lipid Aggregates

Amphipathic lipids upon contact with water form lipid aggregates. Thus, the hydrophilic region of the molecules remains in contact with water, and the hydrophobic part aggregates itself via hydrophobic interactions. This molecular ordering results in a higher thermodynamic stability. [9] Different types of aggregates can be formed (e.g., monolayers at the air-water interface, bilayers, micelles, vesicles, or liposomes) depending on the composition and shape of the lipid in aqueous medium (Fig. 2.3).

The lipid aggregate structure can be predicted by the critical packing parameter (*PP*), which considers the volume, length of the hydrocarbon tail (hydrophobic region), and the polar head area (hydrophilic region). Thus, the critical *PP* can be calculated using Eq. (2.1):

$$PP = \frac{V}{al_c} \tag{2.1}$$

where, V is the hydrophobic volume, a is the surface area occupied by the polar region of the amphiphilic molecule at the water–air interface, and l_c is the hydrophobic region length.

The *PP* can be used as a guide of the surfactant architecture. The typical values and their corresponding structures are shown in Table 2.2.

The *PP* is useful for predicting the structure of a lipid aggregate. However, the structure can be changed through the addition of an electrolyte, such as ionic cosurfactant, as well as by temperature change or by the insertion of ramified or unsaturated chains.

The control of architecture of these structures has a broad potential in many biochemistry areas, such as, controlled drug delivery, catalysis, and solubilization of membrane proteins of low-solubility species. Some of these applications will be discussed in Section 2.3.6.

2.3.2 Micelles: Formation and Critical Micelle Concentration

Micelle aggregate formation occurs due to the shape and size of a surfactant and due to its concentration. The micelle structure forms only if the surfactant or monomer concentration is equal or higher than the critical micelle concentration (CMC) (Fig. 2.4). If the surfactant concentration is below the CMC, micelle formation does not occur, and the solution will only contain the surfactant monomers. [10]

Table 2.2   Use of the Critical Packing Parameter for Predicting the Lipid Aggregate Structure.

PP = V/al_c	Shape	Structure
PP ~ 1 Example: phosphatidylethanolamine		Bilayers
PP < 1 Example: single phospholipids		Micelles
PP > 1 Example: phosphatidylcholine		Vesicles or liposomes

PP, Packing parameter.
Source: Adapted from R.M. Pashley, M.E. Karaman, Applied Colloid and Surface Chemistry, J. Wiley, Chichester, West Sussex, England; Hoboken, N.J, 2004 [7].

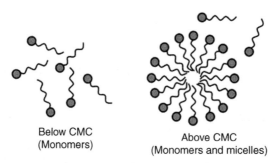

Below CMC
(Monomers)

Above CMC
(Monomers and micelles)

FIGURE 2.4 Dependence of micelle formation on the surfactant concentration.

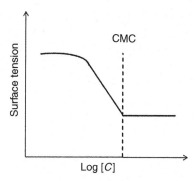

FIGURE 2.5 Typical graph of surface tension versus surfactant concentration for obtaining the critical micelle concentration *(CMC)*.

The main method for determining surfactant CMC is through surface tension measurement. However, other methods, such as electrical conductivity, spectroscopy, electrochemistry, light scattering and viscosity [11–13], are also available.

It is possible to determine the CMC from the curve of surface tension versus surfactant concentration. Additionally, this curve defines a surface excess (Γ) parameter, which is associated with the surfactant excess at the surface at any cross-section inside the solution. [8,14,15] This parameter is determined through the Gibbs adsorption equation (Eq. 2.2):

$$\Gamma = -\frac{1}{2RT}\left(\frac{\partial \gamma}{\partial \ln C}\right)_T \tag{2.2}$$

where γ is the surface tension, C is the surfactant concentration, R is the gas constant and T is the absolute temperature.

Fig. 2.5 shows the behavior of the curve of surface tension versus the logarithm of the surfactant concentration (log C). When C is well below the CMC, the increase of the surfactant volume does not result in a significant increase of surface tension or adsorption density. When the surfactant concentration is close to the CMC, the surface tension decreases abruptly, and there are changes in conductivity and turbidity. When the surfactant concentration reaches the CMC, $d\gamma/d\ln C = 0$, micelle aggregates are formed, the surface becomes completely charged, and no additional alteration of surface tension is observed.

Additionally, micelle aggregates may occur in nonpolar solvents. In this case, inverted or reverse micelles are formed. Reverse micelles have the hydrophilic region of the surfactant inside the aggregate and the hydrophobic region toward the nonpolar medium (Fig. 2.6)

2.3.3 Lipid Bilayers and Vesicular Systems

A lipid bilayer is another type of lipid aggregate that can be formed in an aqueous medium, in which two single layers form a bidimensional sheet. The hydrophobic tails of each single layer interact with each other and exclude water, while the hydrophilic heads interact with

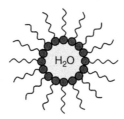

FIGURE 2.6 Typical reverse micelle structure.

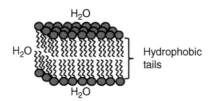

FIGURE 2.7 Lipid bilayer structure.

water in each bilayer surface (Fig. 2.7). Since the hydrophobic region at the ends of a bilayer is in contact with water, the bilayer sheet is relatively unstable, and it spontaneously turns into other lipid aggregates that are spherical structures called vesicles or liposomes.

Liposomes or vesicles were first described by Alec D. Bangham in 1964 [15], and since then, they have raised the interest of many researchers. Because their structure mimics a cell membrane, liposomes have been used as membrane models for physical and chemical studies, as drug carriers to target cells, to solubilize and incorporate biomolecules and as barriers to viral and bacterial invasions.

In general, vesicles are formed because of the tension applied at the hydrophobic–hydrophilic interface. This causes molecules to group to minimize contact of the hydrocarbon elements with the aqueous phase. The equilibrium is reached via steric repulsion among the hydrophilic groups and the entropic repulsion among acyl chains of single layers. The formed spheres capture water and create an aqueous compartment inside lipid bilayers, which is separated from the medium via a hydrophobic barrier (Fig. 2.8). The

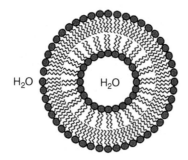

FIGURE 2.8 Structure of a vesicle or liposome.

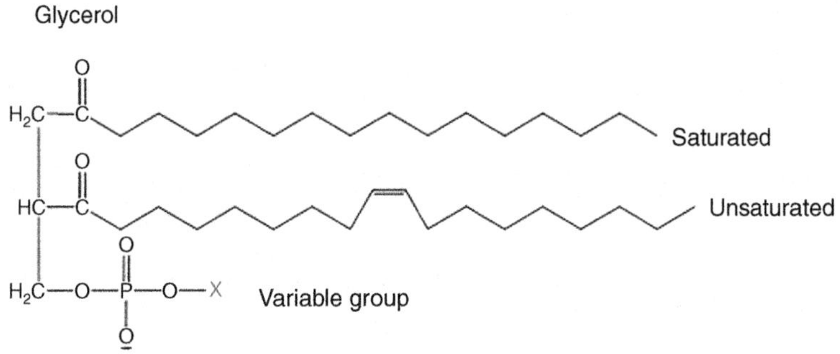

FIGURE 2.9 General phospholipid structure.

lipid bilayer mimics biological or cell membranes that consist of an approximately 3-nm-thick lipid bilayer that contains proteins. Proteins that are inserted in the cell membrane bilayer help with molecular diffusion between the medium and the inside of the bilayer. In the case of vesicles or liposomes produced in laboratories, a bilayer made of pure lipids is essentially nonpermeable to polar solutes. A lipid bilayer has a flexible structure with the ability to change its shape without losing its integrity. This property is a consequence of noncovalent interactions among the lipids in a bilayer.

The lipids that constitute a bilayer (i.e., phospholipids) may vary both in the size of the hydrocarbon part and of the hydrophilic group, as shown in Table 2.3. The hydrophobic group consists of a fatty acid chain [i.e., a carboxylic group (—COOH)] linked to an alkyl chain of 4–28 carbon atoms that can generally be saturated or unsaturated] and a polar head group with a positive, negative or zwitterionic charge (Fig. 2.9).

Another property of phospholipids is the phase transition temperature or critical temperature (T_c) at which, depending on the phospholipid composition, the membrane passes from a crystalline to a fluid phase [16]. At conditions close to physiologic temperatures (20–40°C), when the lipids have 12–14 carbon atoms in the hydrophilic tail and are not unsaturated, the membrane will most likely be in a crystalline phase (i.e., the gel phase). If the hydrophobic tail is unsaturated, it is probable that the bilayer structure will be in a fluid phase, which favors the rotation of molecules in the bilayer (Table 2.3). Thus, the membranes that are more fluid tend to allow the solute to flow through the bilayer of the aqueous compartment to the external medium and vice versa.

2.3.4 Preparation Method of Liposomes and Multilayer Systems

When lipid bilayers are hydrated, they separate during agitation and self-organize into spherical structures to form multilamellar vesicles (MLVs). Once MLVs are formed, it is possible to reduce the size and quantity of lamellas by introducing either sonic (sonication) or mechanical (extrusion) energy to the system. In general, liposomes are prepared using the following stages: [4,17,18]

Table 2.3 Properties of Phospholipids

Name (abbreviation)	Fatty Acid Chain	Transition Temperature Tc (°C)	Charge in pH 7.4
DLPC	12:0	−1	0
DMPC	14:0	23	0
DPPC	16:0	41	0
DSPC	18:0	55	0
DOPC	18:1	−20	0
DMPE	14:0	50	0
DPPE	16:0	63	0
DOPE	18:1	−16	0
DMPA	14:0	50	−1.3
DPPA	16:0	67	−1.3
DOPA	18:1	−8	−1.3
DMPG	14:0	23	−1
DPPG	16:0	41	−1
DOPG	18:1	−18	−1
DMPSD	14:0	35	−1
DPPSD	16:0	54	−1
DOPSD	18:1	−11	−1

DL, Dilauryl; DM, dimyristoyl; DO, dioleyl; DP, dipalmitoyl; PA, phosphatidic acid; PC, phosphatidylcholine; PE, phosphatidylethanolamine; PG, phosphatidylglycerol; PSD, phosphatidylserine.

Stage 1: Dissolution of lipids

Phospholipids are dissolved in an organic solution, typically chloroform or a chloroform–methanol mixture, to obtain a homogeneous solution. The choice of solvent depends on the phospholipid composition.

Stage 2: Lipid film formation

Organic solvent is removed to form a film on the flask wall. The solvent can be evaporated using the compressed air or nitrogen flow.

Stage 3: Formation of MLVs

The dry lipid film is hydrated via the addition of an aqueous medium followed by agitation. The hydration temperature must be above the phase transition temperature during the entire hydration process. Hydration time depends on the composition and phospholipid structure. In general, 1 h of agitation is enough for most phospholipids. The hydration product is an MLV. Its structure resembles an onion, with bilayers or lamellas separated by a water layer.

Stage 4: Formation of unilamellar vesicles (SUVs)

SUVs can be obtained via sonication or extrusion of the MLV suspension.

2.3.4.1 Sonication

The breakage of MLVs with sonic energy (i.e., sonication) produces SUVs with diameters between 15 and 150 nm. The diameter depends on the composition, lipid concentration, temperature, and sonication time. Sonication conditions influence the breakage

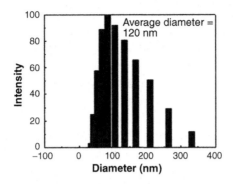

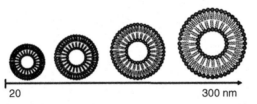

FIGURE 2.10 Size distribution measured via light scattering of DPPG liposomes prepared using a titanium tip sonicator. *Adapted from M.L. Moares, M.S. Baptista, R. Itri, V. Zucolotto, O.N. Oliveira, Immobilization of liposomes in nanostructured layer-by-layer films containing dendrimers. Mater. Sci. Eng. C, 28 (4) (2008) 467–471 [19].*

reproducibility. Therefore, the average diameter of SUVs that are obtained via sonication may vary between the preparations. In addition, SUVs with a small diameter, between 15 and 50 nm, have a high degree of curvature that makes them unstable. Due to high degree of curvature, fusion among SUVs occurs spontaneously to form larger vesicles. This occurs when the solution is stored below the phase transition temperature.

The following equipment is commonly used for sonication: ultrasound baths and sonicators with titanium tips. Fig. 2.10 shows the size distribution measured via light scattering of liposomes that were prepared using a titanium tip sonicator.

2.3.4.2 Extrusion
In the extrusion technique, MLVs are forced through a filter (i.e., a polycarbonate membrane with a defined pore size) to obtain particles with diameters close to pore diameters. In addition, extrusion must be performed above the phase transition temperature of the lipid. This technique results in a more defined size distribution. For example, a membrane with a 100-nm pore size has an average size distribution of 120–140 nm.

2.3.5 Biomedical Applications of Liposomes

Phospholipid vesicles or liposomes have been widely explored since the 1970s [20]. Liposomes have been used as model membrane systems, especially to mimic the processes of adsorption and incorporation of biomolecules as well as drug carriers to target cells [21–23]. The use of liposomes to treat diseases is favored because of the biocompatibility and capability to stabilize proteins and soluble and nonsoluble drugs, in addition to the versatility of lipid composition, which may help to guide the drug to a specific organ.

As a result, different formulations have been investigated. Thus, this technique became potentially useful to treat certain diseases. Many formulations are already commercially available. Liposome systems for treating cancer and antibacterial and antifungal diseases have attracted significant attention regarding controlled drug delivery because of the potential for minimizing side effects. Doxil, a doxorubicin, is an example of a liposome system for systemic administration of chemotherapy against cancer.

In addition, immobilized liposomes in solid substrates have been explored for the controlled release of drugs in patches [24,25] and to build biosensors [26]. Drug release from patches is a promising application because the transdermal delivery system occurs using diffusion across skin layers, which allows continuous drug administration. In this case, the use of liposome systems is more viable because of its higher skin diffusivity compared to the majority of pure drugs. [27]

Geraldo and coworkers described the production of layer-by-layer (LbL) films made with liposomes that incorporate ibuprofen, which is an antiinflammatory drug. The dipalmitoylphosphatidylcholine (DPPC), dipalmitoylphosphatidylglycerol (DPPG), and palmitoyl-2-oleoyl-sn-glycero-3-phosphoglycerol (POPG) liposomes containing ibuprofen were assembled using the alternate adsorption of generation 4 polyamidoamine dendrimer (PAMAM). The release of ibuprofen incorporated in liposomes was slower than that of the free drug.

Additionally, LbL films with liposomes were explored by Xavier et al. [25] to release aloin, a component of *Aloe vera* with scar healing properties. The authors investigated aloin encapsulated in liposomes and immobilized in LbL films with a polyelectrolyte. The aloin release was followed from the solutions and LbL films with liposomes of different phospholipid compositions using fluorescence spectroscopy. After comparing different phospholipid compositions of the systems, the authors concluded that the main factors that control the release are the electrostatic interactions that involve charged phospholipids and the drug. Thus, these interactions can be estimated in self-assembled films that break the ground for new systems for the controlled delivery of drugs. Fig. 2.11 represents the release of aloin that is incorporated in liposomes and is immobilized in self-assembled films with the insertion of a polyelectrolyte.

Liposomes that were wrapped in LbL films in suspension were explored for the development of vaccines [28] and as drug release vehicles [29,30].

Liposomes that were wrapped in LbL films (*layersomes*) were studied by Harde et al. [28] to develop a bivalent vaccine. Layersomes were prepared using phospholipid hydration in the presence of toxoids, followed by the LbL assembly of polyacrylic acid (PAA) and polyallylamine hydrochloride (PAH) polyelectrolytes. The formulation was stable in biomimetic medium and kept its structural integrity and preserved the biological activity and conformational stability of toxoids.

Chen et al. [29] encapsulated paclitaxel, which is used in cancer treatment, in liposomes wrapped in multilayers of PAA and chitosan to improve the stability of the drug

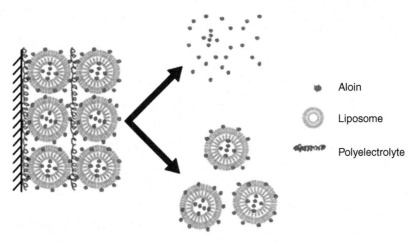

FIGURE 2.11 Schematic of aloin release after its incorporation in liposomes and immobilization in self-assembled films.

release system. The formulation had good stability in gastrointestinal fluids and was stable after 6 months of storage both at 4°C and at ambient temperature. The release of paclitaxel in vitro in the chitosan-PAA-liposome-paclitaxel formulation had a retention time of approximately 3 h, which is shorter than that of liposome-paclitaxel. Furthermore, compared to the formulation without chitosan, the formulation that contained chitosan showed an increase of cytotoxicity induced by paclitaxel in a culture medium of human cervical cancer cells.

Additionally, liposomes have been used in the construction of immunosensors [26]. In this case, the bioreceptor [i.e., the specific biomolecule (an enzyme, an antibody or a peptide] is incorporated in the phospholipid vesicle and is immobilized in solid substrates. The stability of the bioreceptor is maximized when it is encapsulated in liposomes before its immobilization, compared to a bioreceptor that is immobilized directly on the substrate. This occurs because the encapsulated bioreceptor has its secondary structure induced, in addition to being immobilized by noncovalent bonds and kept free in a microspace, which reduces denaturation and/or structural changes during immobilization. The schematic of biorecognition and structural organization of the bioreceptor encapsulated in liposomes and immobilized in thin films is shown in Fig. 2.12. The response of the electrochemical immunosensor shows that there is a change in the current when the antibody interacts with the bioreceptor.

2.3.6 Cell Membranes: Chemical Composition and Physicochemical Properties

Many self-organized systems are used as models of biological membranes. Therefore, it is important to know basic aspects of the composition, structure, and properties of a cell membrane.

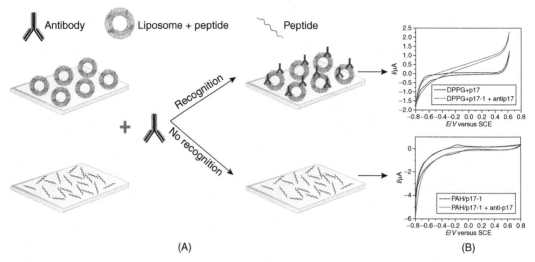

 Antibody   Liposome + peptide   Peptide

(A)   (B)

FIGURE 2.12 (A) Schematic of an immunosensor with a bioreceptor immobilized directly on the substrate, and when a bioreceptor is incorporated in liposomes. (B) Antigen/antibody biorecognition response of the electrochemical immunosensor. *Adapted from Petri et al., Toward preserving the structure of the antigenic peptide p17-1 from the HIV-1 p17 protein in nanostructured films, J. Nanosci. Nanotechnol. 11 (8) (2011) 6705–670 [26].*

A cell membrane is a bilayer that separates the inside of all cells from the external environment [31,32]. It is selectively permeable to substances that enter or leave through it, controls their flow to the inside and the outside of the cell and protects the cell against external forces. Chemically speaking, it is formed by lipids that constitute the functional skeleton of a membrane (Fig. 2.13) and may or may not have proteins incorporated to it. Cell membranes are involved in a series of cellular process (e.g., cellular adhesion, ionic conductivity, and cellular signaling) and serve as a surface for attaching different extracellular structures such as the cell wall and intracellular cytoskeleton. [12] Cell membranes can be artificially assembled using different techniques, including lipid vesicles.

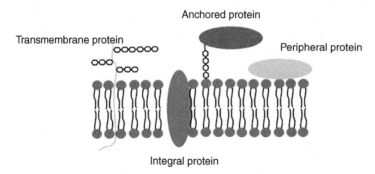

FIGURE 2.13 Schematic of a cell membrane lipid bilayer showing integral, transmembrane, peripheral, and anchored proteins.

The most fundamental model for explaining the structure of a cell membrane is the "Fluid mosaic model," which was developed by Singer and Nicolson (1972) [33] and replaced a previous model by Davson and Danielli (1935) [34]. In this model, the cell membrane can be considered as a bidimensional liquid in which lipid and protein molecules are spread laterally. Although lipid bilayers are capable of forming bidimensional liquids without the presence of other substances, the plasmatic membrane also has a large number of proteins that influence the physical and structural properties of the membrane. Some specific agglomerates of proteins and lipids make the cell membrane laterally heterogeneous. An example of this is the existence of lipid rafts, which are relatively condensed surface dominions in the membrane that are surrounded by a relatively fluid matrix. The composition of these rafts is specific and is formed specifically by cholesterol, sphingolipids and specific proteins, such as, immunoglobulins and alkaline phosphatases. In addition, the inner and outer layers of the membrane are chemically different [35].

Lipid bilayers can be formed via physical self-assembly (without covalent interactions). Lipids can be classified as amphipathic or amphiphilic molecules because they have a polar region with an affinity for water (a hydrophilic head) and a nonpolar region with a low affinity for water (a hydrophobic tail). These terms are used because the nonpolar region—a long alkyl chain (approximately 10–20 carbon atoms)—resembles a "tail," while the relatively smaller hydrophilic region resembles a "head."

A cell membrane primarily consists of a thin layer of lipids that spontaneously combine in a way that the hydrophobic tails are isolated from the surrounding polar fluid. Similarly, the hydrophilic heads tend to associate with the intracellular (cytosolic) and extracellular faces of the resulting bilayer. This forms a continuum that originates a relatively spherical lipid bilayer. Van der Waals and electrostatic forces, hydrogen bonds, and noncovalent interactions contribute to lipid bilayer formation.

In general, hydrophobic interactions are the driving force for forming lipid bilayers. Additionally, entropy (S) increases because of the higher mobility of hydrophobic tails that are protected from aqueous environments and thermodynamically contributes to the bilayer formation process, as per the Gibbs equation at constant pressure and temperature:

$$\Delta G = \Delta H - T \Delta S \tag{2.3}$$

where G is the Gibbs free energy, H is the enthalpy, T is the absolute temperature, and S is the entropy. From this equation, it can be seen that an increase in the entropy of the system favors the spontaneity of lipid bilayer formation and reduces G. Lipid bilayers are generally impermeable to ions and polar molecules. The arrangement of hydrophilic heads and hydrophobic tails of the lipid bilayer may impede the diffusion of polar solutes (e.g., amino acids, nucleic acids, carbohydrates, proteins, and ions) through the membrane but usually allows passive diffusion of hydrophobic molecules. This provides the cell with a capability to control the movement of these substances using transmembrane protein complexes, such as, pores and channels. A certain material can be incorporated into the membrane through the fusion of intracellular vesicles with the membrane (exocytose). However, if a membrane is continuous with a tubular structure that is made of the membrane material, the tube material can be continuously dragged into the membrane. In addition, there are molecular exchanges between the lipid and aqueous phases.

Additionally, there are variety of lipids inside the cell, such as, main membrane phospholipids and glycolipids (i.e., phosphatidylcholine, phosphatidylethanolamine, phosphatidylinositol and phosphatidylserine). A cell membrane consists of three classes of amphipathic lipids: phospholipids, glycolipids, and sterols. The amount of each depends on the cell type. However, in most cases, phospholipids are the most abundant class. The chains that originate from fatty acids in phospholipids and glycolipids typically have an even number of carbon atoms that can be saturated or unsaturated. The length and degree of unsaturation of the fatty acid chains influence membrane fluidity. Thus, unsaturated lipids, due to the angular torsion of hydrophobic chains, impede the regular packing of the chains, reduce the melting temperature and increase membrane fluidity. Another consequence of fluidity diminution is the increase in lateral membrane compressibility. In other words, the membrane withstands a high compression or decompression rate without a significant increase in its lateral pressure. This indicates a low lateral elasticity. In addition, surface viscosity can be increased by the presence of chains with torsion or ramifications.

Therefore, membrane composition determines its fluidity, surface elasticity, and viscosity properties. It is important to highlight that a membrane is heterogeneous in two aspects: (1) if we compare the internal and external layers, and (2) if we compare different regions of the same layer (e.g., aggregates of the "lipid raft" type that contain the characteristic lipid and protein composition). Lipid rafts and caveolae are examples of cholesterol-rich microdomains in the cell membrane. This molecule adapts well to irregular spaces between the hydrophobic tails of the membrane lipids, which provides a condensation and reinforcement effect to the cell membrane structure.

Thus, cell membranes can be studied as models. In addition to liposomes, Langmuir films (1917) [36] are also useful models. They are made of thin films oriented as monolayers at the liquid–air interface, as shown in Fig. 2.14. These films are formed when organic solutions of amphiphilic materials are spread at the air–water interface. If the material is insoluble in water and the solvent is volatile, a film of monomolecular thickness can be formed at the interface. Usually, the substance must spread uniformly at the interface to form a stable film. Thus, the film can be characterized in different ways, using tensiometric,

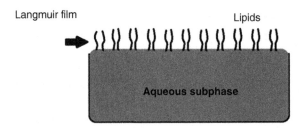

FIGURE 2.14 Schematic of a Langmuir film that is mounted on an aqueous subphase.

volumetric, microscopic, and spectroscopic measurements. The most classical character-ization approach is via surface pressure measurement (π), defined as:

$$\pi = \gamma_0 - \gamma \qquad (2.3)$$

where γ_0 is the surface tension of pure water and γ is the surface tension in the presence of the Langmuir film.

Langmuir films have been used in the interaction of membrane enzymes [37,38] as well as in molecular recognition processes [39] or in the interaction of drugs with cell mem-branes [40,41].

2.4 Multicyclic Supramolecular Systems

Ions or molecules that are close to the central atom are called binders. Binders are gener-ally linked to the central atom via a coordinated covalent bond. Specifically, a pair of non-bonded electrons in an empty metallic orbital is donated, which is said to be coordinated with the atom.

The coordination number and the number of ligands bonded to the metal define the ligand structure, which is determined by the interactions between the p and s orbitals of the ligands and the d orbitals of the metallic ions. Specifically, lanthanides and actinides, in addition to transition metals, tend to have high coordination numbers because their d and f orbitals are filled or semifilled with electrons.

Therefore, different resulting structures are the consequences of the coordination number. The geometries vary and depend on the coordination number and on the ligand.

In the particular case of macrocycle systems, such as, porphyrins and phthalocyanines, which are shown in Fig. 2.15, many of their properties are associated with the capability of recognizing molecules in their internal cavities. Furthermore, in those cases, transition

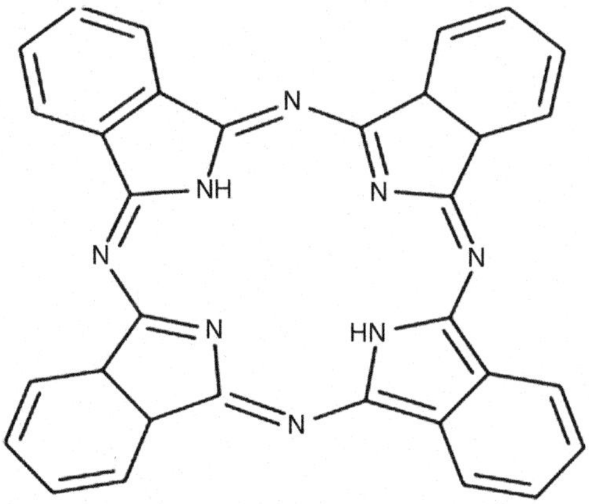

FIGURE 2.15 Chemical structure of phthalocyanine (right).

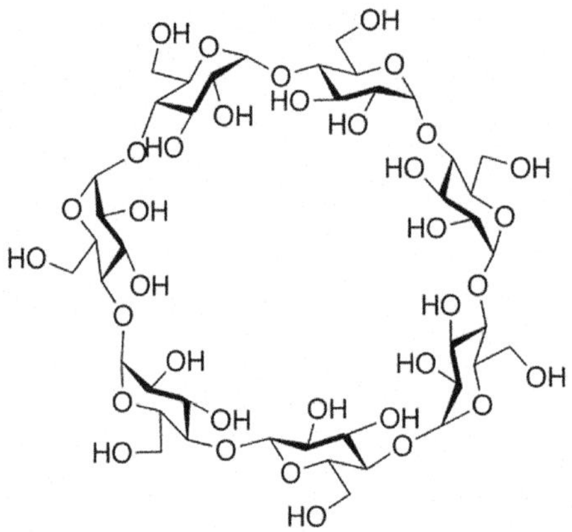

FIGURE 2.16 Chemical structure of a cyclodextrin.

metals can covalently bond to the free electrons of nitrogen atoms of those molecules to form coordinated inorganic complexes. The classical example is porphyrin. It is a prosthetic (nonpolypeptide) element inside the hemoglobin protein, which is present in blood. Porphyrin has the capability of bonding to iron ions in the bloodstream and participating in the process of oxygen transport to biological tissues.

Many other more complex macrocycle compounds, such as, cyclodextrins (Fig. 2.16), act in more specific cases of molecular recognition. These compounds are cyclic oligosaccharides (sometimes called cycloamyloses) that are used in the pharmaceutical industry to increase the solubility of drugs in water. The active molecules are inserted in the molecule cavity and are transported to their performance location. Thus, the molecules are protected from oxidation and electromagnetic radiation, enabling their controlled release. Many studies in the literature have shown these applications for cyclodextrins [41–44].

In all examples that are mentioned in this study, there is a target substance on which the supramolecular system must act, which involves covalent and ionic interactions as well as intermolecular forces. Therefore, chemical and noncovalent interactions are key for these processes. Particularly, the study of noncovalent interactions is crucial for understanding many biological processes that involve cell structure and other biological interfaces. Biological systems are often the inspiration for supramolecular research. [2,45]

Thus, supramolecular chemistry and molecular self-assembly have been applied to develop new materials. Large-dimension structures can be readily obtained using small molecules that require fewer synthetic steps.

Catalysis is one of the applications of supramolecular chemistry. An example is template-driven synthesis, which is a special case of supramolecular catalysis. Other examples are encapsulating systems such as micelles. More examples can be found in recent bibliographic reviews.

References

[1] P.A. Gale, J.W. Steed, Supramolecular Chemistry: From Molecules to Nanomaterials, EUA, Hoboken, (2012).

[2] Z.G. Zhang, Fabrication of novel biomaterials through molecular self-assembly, Nat. Biotechnol. 21 (2003) 1171–1178.

[3] Ariga Katsuhiko, J.P. Hill, M.V. Lee, A. Vinu, R. Charvet, S. Acharya, Challenges and breakthroughs in recent research on self-assembly, Sci. Technol. Adv. Mater. 9 (2008) 014109.

[4] M. Lehn, Perspectives in supramolecular chemistry—from molecular recognition towards molecular information-processing and self-organization, Angew. Chem. Int. Ed. (English) 29 (1990) p1304–1319.

[5] E. Fischer, A. Speier, Darstellung der Ester, Chemische Berichte 28 (1895) p.3252–3258.

[6] C.J. Pedersen, The discovery of crown ethers, Chem. Scr. 28 (1988) 229–235.

[7] R.M. Pashley, M.E. Karaman, Applied Colloid and Surface Chemistry, J. Wiley, Chichester, West Sussex, England; Hoboken, N.J, (2004).

[8] A.W. Adamson, A.P. Gast, Physical Physical Chemistry of Surfaces, Wiley, New York, (1997).

[9] J.N. Israelachvili, Intermolecular and Surface Forces, Academic Press, Burlington, MA, (2011).

[10] B. Lindman, H.M. Wennerström, Topics in current chemistry, Springer-Verlag, Berlin, 87 (1980) 1–87.

[11] W. Al-Soufi, L.P. Mercedes Novo, A model for monomer and micellar concentrations in surfactant solutions: application to conductivity, NMR, diffusion, and surface tension data, J. Colloid Interface Sci. 370 (2012) 102–110.

[12] A.B. Mandal, B.U. Nair, D. Ramaswamy, Determination of the critical micelle concentration of surfactants and the partition coefficient of an electrochemical probe by using cyclic voltammetry, Langmuir 4 (3) (1988) 736–739.

[13] K. Khougas, Z. Gao, A. Eisenberg, Determination of the critical micelle concentration of block copolymer micelles by static light scattering, Macromolecules 27 (1994) 6341–6346.

[14] C.-H. Chang, E.I. Franses, Adsorption dynamics of surfactants at the air/water interface: a critical review of mathematical models, data, and mechanisms, Colloids Surf. A: Physicochem. Eng. Asp. 100 (1995) 1–45.

[15] A.D. Bangham, R.W. Horne, Negative staining of phospholipids and their structural modification by surface-active agents as observed in the electron microscope, J. Mol. Biol. 8 (5) (1964) 660–710.

[16] J.R. Silvius, Thermotropic phase transitions of pure lipids in membranes and their modification by membrane proteins, in: P.C. Jost, O.H. Griffith (Eds.), Lipid-Protein Interactions, vol. 2, Wiley, New York, NY, 1982, pp. 235–281.

[17] S.C. Basu, M. Basu, Liposome Methods and Protocols, Humana Press, Totowa, N.J, (2002).

[18] F. Szoka, Comparative properties and methods of preparation of lipid vesicles (liposomes), Ann. Rev. Biophys. Bioeng. 9 (1980) 467–508.

[19] M.L. Moares, M.S. Baptista, R. Itri, V. Zucolotto, O.N. Oliveira, Immobilization of liposomes in nanostructured layer-by-layer films containing dendrimers, Mater. Sci. Eng. C 28 (4) (2008) 467–471.

[20] J. De Gier, C.W.M. Haest, J.G. Mandersloot, L.L.M. Van Deenen, Valinomycin-induced permeation of 86Rb+ of liposomes with varying composition through the bilayers, Biochim. Biophys. Acta. 211 (1970) 373–375.

[21] V.P. Torchilin, Recent advances with liposomes as pharmaceutical carriers, Nat. Rev. Drug Discov. 4 (2005) 145–160.

[22] W.T. Al-Jamal, K. Kostarelos, Liposomes: from a clinically established drug delivery system to a nanoparticle platform for theranostic nanomedicine, Acc. Chem. Res. 44 (2011) 1094–1104.

[23] A.I. Rahis-Uddin, K. Salim, M.A. Rather, W.A. Wani, A. Haque, Advances in nano drugs for cancer chemotherapy, Curr. Cancer Drug Targets 11 (2011) 135–146.

[24] V.P.N. Geraldo, et al. Immobilization of ibuprofen-containing nanospheres in layer-by-layer films, J. Nanosci. Nanotechnol. 11 (2) (2011) 1167–1174.

[25] A.C.F. Xavier, M.L. de Moraes, M. Ferreira, Immobilization of aloin encapsulated into liposomes in layer-by-layer films for transdermal drug delivery, Mater. Sci. Eng. C 33 (3) (2013) 1193–1196.

[26] Petri, et al. Toward preserving the structure of the antigenic peptide p17-1 from the HIV-1 p17 protein in nanostructured films, J. Nanosci. Nanotechnol. 11 (8) (2011) 6705–6709.

[27] T. Tanner, R. Marks, Delivering drugs by the transdermal route: review and comment, Skin Res. Technol. 14 (2008) 249–260.

[28] H. Harde, K. Siddhapura, A.K. Agrawal, S. Jain, Development of dual toxoid-loaded layersomes for complete immunostimulatory response following peroral administration, Nanomedicine 10 (2015) 1077–1091.

[29] M.X. Chen, B.K. Li, D.K. Yin, J. Liang, S.S. Li, D.Y. Peng, Layer-by-layer assembly of chitosan stabilized multilayered liposomes for paclitaxel delivery, Carbohydr. Polym. 13 (2014) 298–304.

[30] S. Jeon, C.Y. Yoo, S.N. Park, Improved stability and skin permeability of sodium hyaluronate-chitosan multilayered liposomes by layer-by-layer electrostatic deposition for quercetin delivery, Colloids Surf. B: Biointerfaces 129 (2015) 7–14.

[31] H. Lodish, A. Berk, L.S. Zipursky, et al., Molecular Cell Biology, fourth ed., W.H. Freeman New York, USA, 2007.

[32] F.M. Goñi, The basic structure and dynamics of cell membranes: an update of the Singer–Nicolson model, Biochim. Biophys. Acta. 1838 (2014) 1467–1476.

[33] S.J. Singer, G.L. Nicolson, The fluid mosaic model of the structure of cell membranes, Science 175 (1972) 720–731.

[34] J.F. Danielli, H. Davson, A contribution to the theory of permeability of thin films, J. Cell. Comp. Physiol. 5 (1935) 495–508.

[35] F.G. Zalduondo, D.a. Rintoul, J.C. Carlson, W. Hansel, Luteolysis-induced changes in phase composition and fluidity of bovine luteal cell membranes, Proc. Natl. Acad. Sci. USA 79 (1982) 4332–4336.

[36] I. Langmuir, The constitution and fundamental properties of solids and liquids. II. Liquids, J. Am. Chem. Soc. 39 (1917) 1848–1906.

[37] L. Caseli, et al. Enzymatic activity of alkaline phosphatase adsorbed on dimyristoylphosphatidic acid Langmuir–Blodgett films, Colloids Surf. B Biointerfaces 25 (2002) 119–128.

[38] L. Caseli, et al. Effect of molecular surface packing on the enzymatic activity modulation of an anchored protein on phospholipid Langmuir monolayers, Langmuir 21 (9) (2005) 4090–4095.

[39] L. Caseli, et al. Probing the interaction between heparan sulfate proteoglycan with biologically relevant molecules in mimetic models for cell membranes: a Langmuir film study, Biochim. Biophys. Acta. Biomembranes 1818 (2012) 1211–1217.

[40] N. Hussein, et al. , Surface chemistry and spectroscopy studies on 1,4-naphthoquinone in cell membrane models using Langmuir monolayers, J Colloid Interface Sci. 402 (2013) 300–306.

[41] Pascholati, et al. , The interaction of an antiparasitic peptide active against African sleeping sickness with cell membrane models, Colloids Surf. B Biointerfaces 74 (2009) 504–540.

[42] D. Ma, H.B. Zhang, Y.Y. Chen, J.T. Lin, L.M. Zhang, New cyclodextrin derivative containing poly(L-lysine) dendrons for gene and drug co-delivery, J. Colloid Interface Sci. 405 (2013) 305–311.

[43] M.C. Hamoudi, J. Saunier, C. Gueutin, E. Fattal, A. Bochot, Beads made of alpha-cyclodextrin and soybean oil: the drying method influences bead properties and drug release, Drug Dev. Ind. Pharm. 39 (2013) 1305–1314.

[44] T.S. Anirudhan, D. Dilu, S. Sandeep, Synthesis and characterisation of chitosan crosslinked-β-cyclodextrin grafted silylated magnetic nanoparticles for controlled release of Indomethacin, J. Magn. Magn. Mater. 343 (2013) 149–156.

[45] B. Alberts, A. Johnson, J. Lewis, et al. Molecular Biology of the Cell, fourth ed., Garland Science, New York, USA, 2002.

2
Nanoelectronics

M.R. Cavallari, G. Santos, F.J. Fonseca
DEPARTMENT OF ELECTRONIC SYSTEMS ENGINEERING, POLYTECHNIC SCHOOL OF THE USP, SÃO PAULO, BRAZIL

2.1 Organic Materials for Nanoelectronics: Insulators and Conductors

Organic materials that have potential application in electronic products are small molecules (e.g., oligomers) or polymers whose structure is composed predominantly of carbon. The element carbon has six electrons distributed in levels $1s^2\ 2s^2\ 2p^2$; that is, it has four electrons in the highest level and median electronegativity (a tendency to share electrons rather than transfer them, which would correspond to covalent bonding). The sp^2 hybridization is characteristic of organic semiconductors. This state is shown in Fig. 2.1A for an ethylene molecule (or ethene). In this molecule, each carbon atom is hybridized as follows: σ bonds are formed on the plane defined by the carbon and hydrogen atoms with electrons strongly located between the nuclei of the atoms; however, π bonds also exist off the plane that enables a highly delocalized character.

The π-conjugated molecules are characterized by alternating single and double bonds, with some examples shown in Fig. 2.1B–E. The bonds formed by overlapping of the p_z atomic orbitals (i.e., off the plane) create molecular orbitals ligands (π) and antiligands (π^*) [1]. Electrons of the orbital ligand have less energy, and therefore, the orbital is

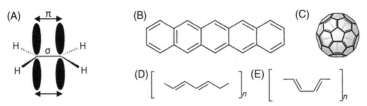

FIGURE 2.1 (A) Carbon sp² hybridization in the ethylene molecule. Organic materials can be divided into small molecules [(B) pentacene, (C) fullerene] and polymers [monomer of (D) *trans*-, (E) *cis*-polyacetylene].

known as the highest occupied molecular orbital (HOMO). Given that π bonds are typically weaker, these electrons are easily excited to the upper level and are free to move along the molecule, benefiting from the overlap of the p_z orbitals with neighboring atoms (delocalization of π orbitals). This higher energy level is known as the lowest unoccupied molecular orbital (LUMO). When the energy difference between the HOMO and LUMO levels is much greater than the thermal energy ($k_B T$, where k_B is Boltzmann's constant and T is temperature) a forbidden band (i.e., bandgap) arises. In such a case, one can compare the HOMO to the valence band (VB) and the LUMO to the conduction band (CB) in an inorganic semiconductor or insulator.

Conjugated polymers exhibit electrical conductivity that can be typical of both insulators and conductors, according to chemical or natural modifications and the degree of doping. The idea of using polymers for their conductive properties appeared as recently as 1977 with the discoveries of Shirakawa et al. [2], in which *trans*-polyacetylene (Fig. 2.1D) doped with iodine exhibited conductivity of 10^3 S cm^{-1}. The explanation of this discovery enabled Heeger, McDiarmid, and Shirakawa to win the Nobel Prize in Chemistry in 2000. Since then, interest in synthesizing other organic materials that exhibit this property has increased, and other polymers with the π-electron structure, such as those shown in Fig. 2.2 [polyaniline (PAni), polypyrrole (PPy), polythiophene (PT), polyfuran (PFu), poly(*p*-phenylene) (PPP), and polycarbazole (PCz)], have been synthesized and tested in electronic devices [3,4].

FIGURE 2.2 **Chemical structure of monomers from the principal semiconducting polymers.** (A) Polyaniline (PAni), (B) polypyrrole (PPy), (C) polythiophene (PT), (D) polyfuran (PFu), (E) poly(*p*-phenylene) (PPP), and (F) polycarbazole (PCz).

The transport of charge carriers may be limited by injection because of the height of the potential barrier at the interfaces between different materials. When the electric field is sufficiently elevated or the height of the barriers is not significant, transport is dominated by the special charge accumulated on the organic semiconductor due to transport limitations [5]. In this case, the limiting factor of the electric current is the effective mobility of charge carriers (μ), defined by the ratio between the velocity of the carriers and the applied electric field. Greater mobilities than those found in hydrogenated amorphous silicon have been obtained for pentacene [6] and fullerene [7] ($\mu \sim 6 \text{ cm}^2 \text{ V}^{-1} \text{ s}^{-1}$), and although these values are still lower than those of monocrystalline silicon ($\mu \sim 10^2 \text{ cm}^2 \text{ V}^{-1} \text{ s}^{-1}$) [8], they are sufficient for the manufacturing of electronic products.

One of the advantages of organic semiconductors in relation to inorganic semiconductors is the possibility of synthesizing materials by modulating the bandgap and energy levels. This modulation generally provides materials with different optical properties and the ability to absorb and emit photons with wavelengths both inside and outside the visible range. In their simplest form, these devices, based on thin organic films, are composed of stacked layers positioned between two electrodes with high potential for application in light-emitting diodes (LEDs) [9], and even photovoltaics—organic solar cells (OSCs) [10]. Organic materials with these properties are promising for applications in information displays, such as mobile devices and large-scale screens, including their use in ambient illumination and flexible and transparent screens. Passive devices, such as resistors [11], capacitors [12], and diodes [13] can also be constructed. In addition to these devices, organic thin film transistors (OTFTs) [14], key elements for the implementation of electronic circuits, can be highlighted. If employed as field effect transistors (FETs) they can be used for various purposes, such as logic gates, processors, and memory [15,16]. Application of OTFTs in radio frequency identification devices (RFIDs) promises to revolutionize the automatic identification systems both in commercial applications and in industry [17]. Polymeric electronic memories are being investigated as an interesting alternative because of their compatibility with electronic circuits containing polymeric devices and miniaturization toward the nanoscale [18].

Among the conductive materials, PAni [19] (Fig. 2.2A), poly(3,4-ethylenedioxythiophene) doped with poly(styrenesulfonic acid) (PEDOT:PSS) [20], graphene sheets [21], and single-walled carbon nanotubes can be highlighted (Fig. 2.3). Graphene enables current modulation by an electric field, a typical behavior of semiconductors, but has greater potential for use in chemical sensors and electrodes. In turn, as shown in Fig. 2.3C, nanotubes can be separated into three categories with respect to the angle at which the graphene sheet is rolled: armchair, zigzag, or chiral. The rolling angle defined by the vector $c_h = na_1 + ma_2$ (n,m), where n and m are integers, produces electrical behavior typical of conductors or semiconductors. All armchair structures exhibit metallic properties, whereas the other two structures may have semiconducting or metallic properties depending on the diameter of the nanotube [22,23]. Some specific electrical properties of these materials can be exploited in the construction of devices different from inorganic semiconductors, such as molecular memories [15], oscillator monolayers [24], smart windows [25], artificial muscles [26], and biosensors [27].

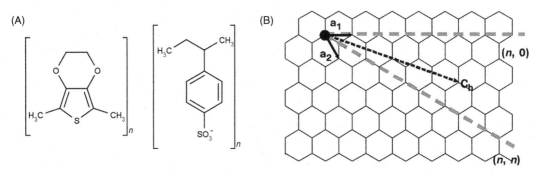

FIGURE 2.3 Chemical structure of the principal organic conductors. (A) PEDOT:PSS and (B) graphene. Carbon nanotubes are obtained by rolling the graphene sheets in the direction $c_h = (n,m)$. The chiral structure is limited by the zigzag and armchair cases defined in turn by the vectors $(n,0)$ and (n,n), respectively. According to this representation, the angle between the zigzag structure and c_h is negative. *Adapted from T.W., Odom, et al. Atomic structure and electronic properties of single-walled carbon nanotubes. Nature 391 (1998) 62–64 [22].*

Organic electronic systems offer the advantage of being lightweight and flexible, and also cover large areas at a reduced cost [28]. Manufacturing techniques are diverse, but there is a tendency to produce integrated circuits by direct printing of all system components, dispensing with expensive and complex techniques, such as thermal oxidation and photolithography [29]. Nanoelectronics permit the development of products whose mechanically flexible component provides solutions of lower weight and greater biocompatibility because they are near to the body, as is the case for portable electronics [28]. In this context, one of the possible applications is to reduce the cost of integrated circuits in the disposable sensor industry [30].

Furthermore, it appears that currently, there is a demand for components with applications in environmental [31] and food [32] monitoring, the application of drugs [33] the detection of chemical and biological weapons [34], and the fabrication of electronic tongues and noses [35]. In medical diagnostics in situ [36], diseases at an early stage can be detected through the recognition of specific biomolecules chains (deoxyribonucleic acid, DNA). Based on the compatibility of semiconductors with biomolecules and living cells, one can expect the integration of biomedicine and automation via advanced cybernetics [27].

2.1.1 Techniques for Making Organic Films

Electrochemical polymerization is the polymerization reaction performed directly on the electrodes. This was the first technique used to form a semiconductor layer employed in the fabrication of organic transistors [37,38]. This technique is usually associated with the formation of disordered films of low mobility; the first organic thin film transistor (TFT) was polythiophene and exhibited $\mu \approx 10^{-5}$ cm^2 V^{-1} s^{-1} [37].

Spin coating, one of the deposition techniques of polymeric films, is widely used in this field because of its low cost, simplicity, and capability to produce homogeneous films with adequate thickness control [39]. Application of the solution is performed on the substrate, which undergoes rotation; the resulting thickness is a function of the frequency and rotation time. Alternatively, organic materials can be cast by the liquid phase deposition technique, which consists of applying the solution to the fixed substrate. Due to longer time required for drying the deposited layer, this technique provides the material more time to organize and thus achieve higher crystallinity. In contrast, it is more difficult to control the thickness of the formed layer, which is usually in the micron range [40].

Vacuum thermal evaporation is used to obtain films of small organic molecules that can be challenging to prepare in solution form. It provides high thickness control on the nanometer scale and high film purity, as well as a high degree of ordering depending on the growth rate [41]. Films obtained by this technique tend to be highly ordered such that microcrystalline devices incorporating pentacene [6] and fullerene [7] may provide greater performance than those of hydrogenated amorphous silicon. Alternatively, a monocrystal can be obtained from lamination or transfer of a mold to the substrate [42–44]. Crystals of high purity and molecular ordering can be obtained under vacuum by physical vapor transport in a horizontal tube.

An alternative technique for the formation of nanometric films of small molecules or conjugated polymers is self-assembly or electrostatic self-assembly, which uses two ionic species to form successive fine film bilayers [45]. This method is highlighted because it is low cost and simple, can be performed at ambient temperature and permits precise control of the structure and film thickness.

Currently, printing techniques have progressed to the industry stage for the manufacture of commercial devices. Notable techniques include inkjet, screen [46], roll-to-roll [47], and microcontact printing (μCP) [48,49]. These techniques are capable of providing resolution down to 2 μm. Microcontact printing consists of the transfer (stamping) of conductive material (e.g., nanoparticles and PEDOT:PSS) in the process of additive binding of the mold to the substrate. As shown in Fig. 2.4, the process can be divided into two stages:

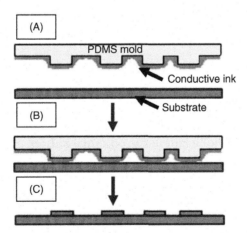

FIGURE 2.4 Schematic diagram of the microcontact printing process (μCP). (A) inking of the mold with conductive ink; (B) stamping of the conductive ink of the mold onto the substrate; and (C) substrate with the pattern of electrodes transferred from the mold.

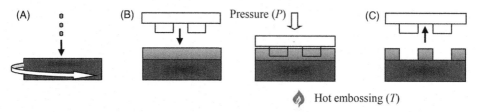

Hot embossing (*T*)

FIGURE 2.5 Hot embossing lithography. (A) spin coating of the photoresist on the substrate; (B) embossing or imprinting the pattern of the mold with heat; and (C) postbaking to cure the transferred structures.

(1) inking, which involves the formation of a conductive ink layer on the surface of a mold, followed by (2) stamping or transfer of this conductive layer onto a substrate.

Microcontacts can be alternatively obtained by a subtractive process known as imprint lithography [50]. Developed by Stephen Chou et al. [51], hot embossing, shown in Fig. 2.5, uses a polymer that can be thermally modified to conform to the relief of the mold. Under compression, this polymer is heated above its glass transition temperature (T_g) to flow through the channels of the stamp and then cooled down to solidify the patterned structures.

2.2 Process of Charge Transport in Organic Devices

Organic materials have a structure characterized on the atomic scale by disorganization, in which the charge carriers are delocalized in their molecules (or along the conjugation length of the polymer chains), where the rate-limiting step in transport is intermolecular (or between-chain) hopping of these carriers [52]. This structure, in the absence of crystallinity, inherently leads to a high density of defects and therefore a mobility (μ) of 10^{-7}–10^{-3} cm^2 V^{-1} s^{-1} depending on the electric field and temperature. In the literature, the two main models of charge carrier injection in organic semiconductors are Fowler–Nordheim [53] or Schottky thermionic emission [54]. In the latter, the electric current (I) can be described as a function of the applied electric potential (V) generically as:

$$I(V) = AA^{**}T^2 \left(e^{\frac{-\phi_B}{k_B T}} \right) \left(e^{\frac{q(V_a - IR_s)}{\eta k_B T}} \right) \left(1 - e^{\frac{-q(V_a - IR_s)}{k_B T}} \right) \tag{2.1}$$

where A is the geometric area of the electrical contact of the device, A^{**} is Richardson's constant, T is the temperature, q is the electron charge, R_s is the series resistance, k_B is Boltzmann's constant, ϕ_B is the height of the potential barrier, and η is the factor of ideality. Under ideal conditions, such that R_s is negligible and the ideality factor is equivalent to one, Eq. 2.1 is reduced to describe the behavior of the diffusion current due to thermionic emission at the interface between the metal and the semiconductor. Another factor to be taken into consideration is the Richardson–Schottky thermionic emission, which defines the reduction of the potential barrier at the interface between the metal and semiconductor with the applied electric field. Thus, the current density ($J = I/A$) associated with this model is [8]

$$J = A^{**}T^2 e^{\frac{-\left(\phi_B - \beta\sqrt{F}\right)}{k_B T}}$$

$$A^{**} = \frac{4\pi q m^* k^2}{h^3} \qquad (2.2)$$

$$\beta = \sqrt{\frac{q^3}{4\pi\varepsilon_r\varepsilon_0}}$$

with the constants, where h is the Plank constant, ε_r is the relative dielectric constant, ε_0 is the electric vacuum permittivity, and m^* is the effective electron mass. The Fowler–Nordheim mechanism, in turn, consists of a tunneling model through the triangular potential barrier, governed by [8,53]

$$J = \frac{q^3 F^2}{8\pi h \phi_B} e^{\frac{-8\pi\sqrt{2m^*}\phi^{3/2}}{3qhF}} \qquad (2.3)$$

Both models are suitable for inorganic semiconductors but limited for organic semiconductors, which have disordered structures and a high defect density. Electrical conduction in the volume of these semiconductors is represented by space charge limited current (SCLC) models, as proposed by Rose and Lampert [54]. This theory considers the effect of traps on the displacement of the charge carriers; that is, it is assumed that there is a current limitation due to charge carriers trapped [trapped charge limited (TCL)] by defects or impurities. Initially, taking into account a low electrical potential applied and disregarding any injection barriers, the expectation is that mobility is low and governed by an ohmic behavior of thermally generated free carriers (n_0). In this case, electrical potential and current density are linearly related according to:

$$J = \frac{q\mu n_0 V}{d} \qquad (2.4)$$

where μ is the mobility and d is the thickness of the thin film used in the structure with the electron transport function. If n_0 is negligible in relation to the injected charge carriers, there is a predominance of the SCLC mechanism, as

$$J = \frac{9}{8}\frac{\mu\varepsilon_r\varepsilon_0 V^2}{d^3} \qquad (2.5)$$

When the trap density is high, the quadratic dependency is only maintained if trapping of charge carriers in discrete energy levels or shallow traps occurs. In this condition, μ in Eq. 2.5 is modified by $\mu_{eff} = \theta\mu$, where $\theta = n/(n + n_t)$ represents the ratio between the density of free carriers (n) and the total density including those trapped (n_t). If they have an energy distribution, the traps will gradually be filled with an increase of the electric field; that is, θ is dependent on the electric field, and the electric current will increase with dependence higher than quadratic until all traps are filled. This behavior characterizes the TCL current regime; the accompanying equation shown in Eq. 2.6 corresponds to an exponential distribution $h(E) = (H/k_B T_c)\exp(E/k_B T_c)$, especially for deep traps, where N_v (or N_c) is the effective density of states in the valence (or conduction) band [5,55]

$$J = \mu q N_c \left(\frac{\varepsilon_r \varepsilon_0 m}{qH(m+1)}\right)^m \left(\frac{2m+1}{m+1}\right)^{m+1} \frac{V^{m+1}}{d^{2m+1}}$$ (2.6)

Assuming that the mobility is dependent on the electric field, an analytical solution exists that takes into account an arbitrary dependence on this factor and disregards the presence of traps. An analytical approximation was obtained by Murgatroyd [56] and incorporates the dependence of Poole–Frenkel mobility according to

$$\mu(F) = \mu_0 \exp\left(\beta\left(\sqrt{F} - \sqrt{F_0}\right)\right)$$ (2.7)

where μ_0 is the mobility when $F = F_0$, which depends on F, and β is related to the system disorder. This parameter is often observed in disordered molecular materials, doped polymers and in most π-conjugated polymers [57]. The current density in this case is part of the free SCLC model of charge carrier trapping multiplied by the Poole–Frenkel mobility according to

$$J \approx \frac{9}{8} \frac{\mu_0 \varepsilon_r \varepsilon_0 V^2}{d^3} \exp\left(0.89\beta\sqrt{\frac{V}{d}}\right)$$ (2.8)

In summary the organic devices and their electrical response, it is necessary to define if the electrical current is limited by the injection of charge carriers or by a spatial charge. In the latter case, it should be noted that the distribution of traps is in discrete or exponential levels through the dependence of μ on the electric field (F) and temperature (T).

2.3 Organic Thin Film Transistors

2.3.1 Structure of the TFT

Electronics technology has evolved rapidly over the last decades, and the trend is that the devices used in modern life are increasingly faster and compact. Currently, almost all electronic devices are based on extrinsic monocrystalline semiconducting silicon, whose conventional incorporation into FETs uses silicon substrates covered by silicon dioxide (SiO_2) obtained by thermal growth. Designed and patented by Lilienfeld in 1930 [58] but first manufactured in 1960 in the Bell laboratories by D. Kahng and M. Atalla [59], the FET transistor is miniaturized and integrated with millions of other transistors to implement highly complex and sophisticated circuits. However, for greater integration and consequently smaller size as predicted by Moore's law [60], there is the possibility of overheating and interference (cross-talk) between components, which affects the overall performance. A potential solution for overcoming such limitations is to develop new materials and manufacturing technologies to perform the same functions commonly performed by conventional semiconductors.

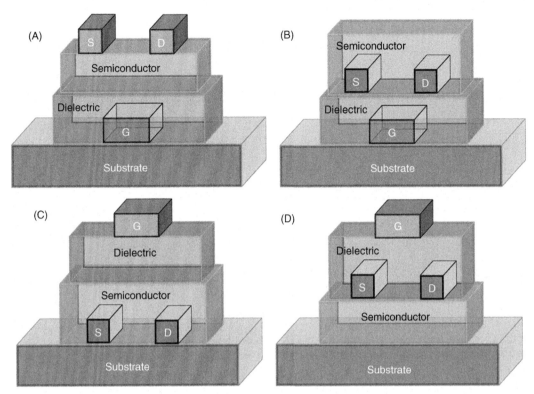

FIGURE 2.6 Possible structure of organic thin film transistors (OTFTs) according to the relative position of the gate *(G)*, source *(S)* and drain electrodes *(D)*: bottom gate (A) top contact (staggered) or (B) bottom contact (coplanar); top gate (C) bottom contact or (D) top contact. The width *(W)* and length *(L)* of the channel between the source and drain are shown in (A).

OTFTs are characterized as a type of FET constructed by the deposition of organic layers on low cost substrates, such as glass and plastics. This type of device involves a wide range of applications, such as flexible displays [28], radio frequency identification tags [61,62], memories [63], textile electronics [64], and chemical, biological, and pressure sensors [27,65,66]. The bottom-gate geometry (Fig. 2.6A–B) currently adopted in research laboratories uses a highly doped silicon substrate covered with high-quality thermal SiO_2, easily found on the market [67]. In this structure, the gate electrode is on the dielectric, and electrical contacts of the source and drain are formed between the insulating and semiconducting layers. The current trend in hybrid transistors with such a structure is to deposit a dielectric layer with a high dielectric constant (high-k) because this allows for defining the gate electrode and contact channels for the implementation of circuits, in addition to functioning at a reduced operating voltage compared to SiO_2 ($\varepsilon_r = 3.9$). Cavallari et al. demonstrated the potential of silicon oxynitride ($\varepsilon_r = 4.5$) formed by plasma enhanced chemical vapor deposition (PECVD) for TFTs of the semiconductor poly (2-methoxy-5-(3,7-dimethyloctyloxy)-1,4-phenylenevinylene) (MDMO-PPV) operating at ± 20 V on

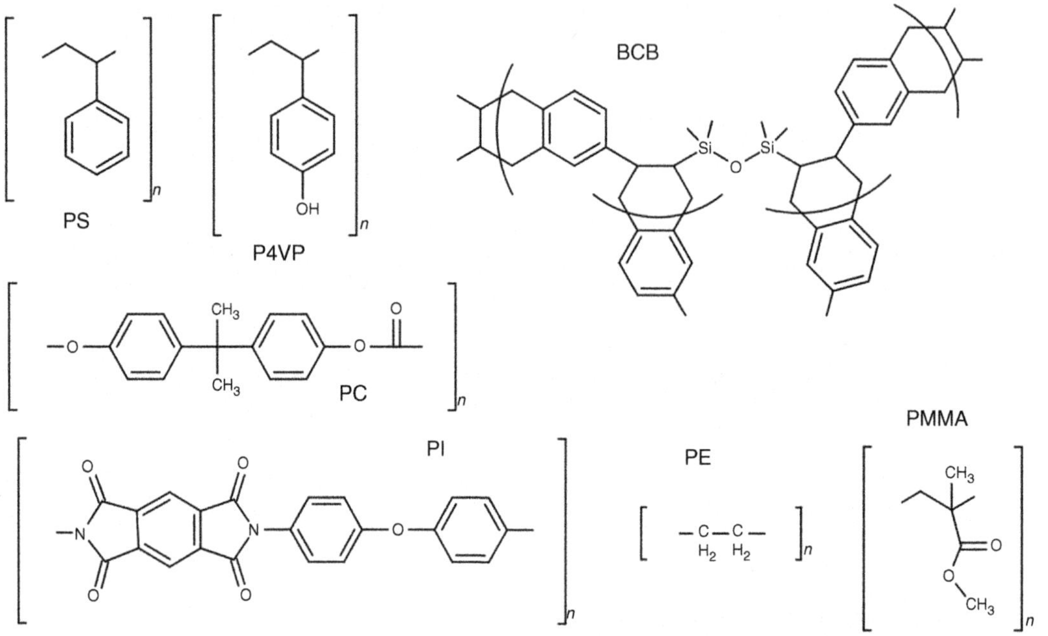

FIGURE 2.7 Organic insulators for modifying the dielectric semiconductor interface in OTFTs. polymethylmethacrylate *(PMMA)*, polyethylene *(PE)*, divinyltetramethyldisiloxane-bis (benzocyclobutene) *(BCB)*, polycarbonate *(PC)*, polystyrene *(PS)*, polyimide *(PI)*, and poly(4-vinylphenol) *(P4VP)*.

Si [68]. In turn, Zanchin et al. employed titanium oxynitride ($\varepsilon_r = 30$) obtained by RF magnetron sputtering to produce TFTs of poly(3-hexylthiophene) (P3HT) that operated at ±3 V on glass [69].

It was recently demonstrated that organic dielectrics are promising materials for the construction of OTFTs, as they (1) can be processed via solutions, (2) produce thin films on transparent glass and plastic substrates (with the possibility of being flexible), (3) are suitable for the construction of photosensitive OTFTs in the field of optoelectronics because of the optical transparency of these films, and (4) may possess a high dielectric constant up to 18 [70,71]. An immediate opportunity to use organic dielectric layers appears in top-gate OTFTs (Fig. 2.6C–D) if the process of the deposition of organic dielectrics does not damage or remove the lower layers. The main organic dielectrics addressed in the literature are shown in Fig. 2.7.

2.3.2 Modeling of the Characteristic Curves

Most OTFTs found in the literature have the structure of an FET, in which conduction occurs in the channel between the source and drain electrodes, and is modulated by the gate voltage (V_{GS}) [72,73]. The selection of the metal–insulator–semiconductor (MIS) structure is due to extensive use of metal–oxide–semiconductor FET (MOSFET) technology in the digital circuits market and because it is the structure most used in amorphous silicon TFTs [74].

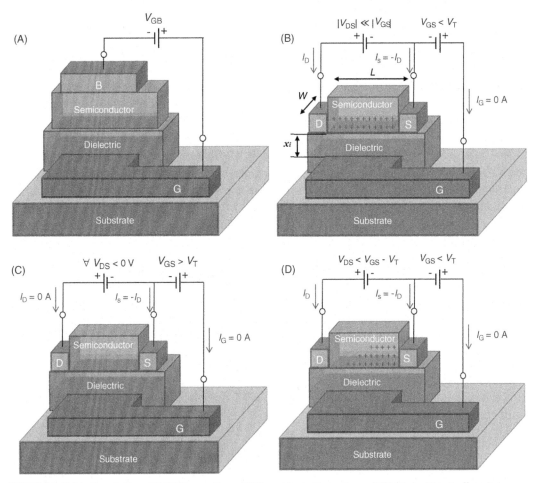

FIGURE 2.8 Field effect devices. (A) MOS capacitor; p-FET transistor in the regimes of (B) triode; (C) cut-off; and (D) saturation.

The core structure of the device is the MIS capacitor of Fig. 2.8A, whose gate electrode controls the charge density (n_Q) in the transistor channel and therefore its conductance. The density of accumulated charges in the triode regime [75] can be described by:

$$n_Q = \frac{C_i (V_{GS} - V_T)}{q}$$

$$C_i = \frac{\varepsilon_r \varepsilon_0}{x_i}$$

(2.9)

is the capacity per area, ε_r where q is the elementary charge, V_T is the threshold voltage, $C_i = x_i$ is the relative dielectric constant, ε_0 is the electrical permittivity of the vacuum, and x_i is the thickness of the dielectric film. Unlike the enrichment and depletion FETs whose conduction channel is formed by inversion, OTFTs function by the accumulation of charges in the channel. Furthermore, the semiconductor at the interface with the dielectric does not need to be doped for the formation of the channel because a channel is formed that permits the passage of current between the source and drain for a nonzero V_{DS}, even if $V_{GS} = 0$ (Fig. 2.8B). Thus, a positive threshold voltage (V_T) is defined that decreases the concentration of holes in the dielectric/semiconductor interface and eliminates current flow in the transistor (Fig. 2.8C).

Given that the concentration of charges available for transport in the channel depends on the voltage difference between the terminals of the MIS capacitor, what would happen if the source voltage was maintained at zero as the drain voltage became increasingly negative (note that in a p-TFT, the operating voltages are negative)? In this case, a non-homogeneous charge distribution in the channel is observed, and when $V_{GS} - V_{DS} > V_T$, the concentration of holes is minimal near the drain electrode, leading to competition between two phenomena that occur simultaneously:

1. an increasingly negative drain voltage would increase the current circulating between the source and drain electrodes;
2. choking of the channel around the drain due to the low concentration of holes would tend to limit the current in the channel (a phenomenon known as pinch-off). This situation leads to saturation of the current that would ideally remain constant (Fig. 2.8D).

Characteristic curves for I_D versus V_{DS} for a p-TFT of P3HT obtained experimentally are shown in Fig. 2.9A. When $V_{DS} > V_{GS} - V_T$, the OTFT operates in the triode region and the drain current is described by [74]

$$I_D = \mu C_i \frac{W}{L} V_{DS}\left[(V_{GS} - V_T) - \frac{V_{DS}}{2}\right] \tag{2.10}$$

where μ is the carrier mobility, W is the channel width, and L is the channel length. It is observed that at low drain voltages, that is, $|V_{DS}| << 2|V_{GS} - V_T|$, the current exhibits a linear behavior (or ohmic) as a function of V_{DS}:

$$I_D \approx \mu C_i \frac{W}{L} V_{DS}(V_{GS} - V_T) \tag{2.11}$$

The parameters μ and V_T can be obtained from the transconductance of the channel in this regime [74]:

$$g_m = \left.\frac{\partial I_D}{\partial V_{GS}}\right|_{V_{DS} = \text{constant}} = \mu C_i \frac{W}{L} V_{DS} \tag{2.12}$$

where the intersection of the approximate straight line with the x axis (V_{GS0}) is equal to

$$V_{GS0} = V_T + \frac{V_{DS}}{2} \tag{2.13}$$

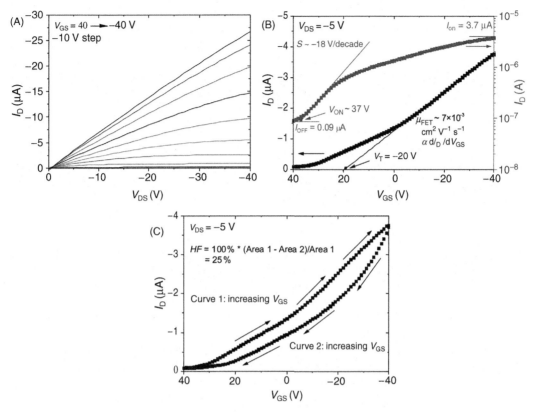

FIGURE 2.9 Characteristic curves of p-transistors with 50 nm of P3HT deposited from 4.3 mg mL^{-1} chloroform by spin coating on a substrate of p$^+$ – Si (190 nm) with W = 1.1 mm, and L = 10 μm. (A) I_D versus V_{DS}, (B) I_D versus V_{GS}, and (C) hysteresis of I_D versus V_{GS}.

Considering that $V_{DS} < V_{GS} - V_T$, the transistor operates in the saturation region as described by Eq. 2.14:

$$I_D = \frac{1}{2}\mu C_i \frac{W}{L}(V_{GS} - V_T)^2 \tag{2.14}$$

Mobility can be extracted through the first derivative of $\sqrt{I_D}$ in relation to V_{GS}:

$$\frac{\partial \sqrt{I_D}}{\partial V_{GS}} = \sqrt{\frac{1}{2}\mu C_i \frac{W}{L}} \tag{2.15}$$

where the intersection of the approximate line with the x axis (V_{GS1}) is equivalent to V_T (Fig. 2.9).

The same model can be applied to n-OTFTs, taking into account that the operating voltages are positive. As in p-TFTs, the threshold voltage (V_T) is opposite to the working voltage (and hence is negative), and the increase in channel conductivity is given by the accumulation of negative charges (or electrons) at the semiconductor/dielectric interface.

Analyzing the curve of I_D versus V_{GS} in Fig. 2.9B, it is possible to observe two operating regimes: (1) the cut-off region in which $V_{GS} > V_{on}$ (the characteristic voltage for a given p-transistor) and the current $I_D = I_{off} \approx 0$, and (2) subthreshold, in which I_D depends exponentially on V_{GS}. In this regime, it is possible to define the subthreshold slope (S), which corresponds to the increase of $|V_{GS}|$ needed to increase $|I_D|$ by a factor of ten. It is expected that both S and V_T tend to zero to reduce the operating voltage of the electronic circuits, typically approximately 5 V. From the operating current (I_{on}) defined for an arbitrary pair of V_{DS} and V_{GS}, the current modulation can be calculated from the relationship between I_{on} and I_{off}. As in the case of amorphous silicon TFTs, nonidealities, such as hysteresis and current leakage by the dielectric ($I_{leakage}$) adversely affect the device performance [76,77]. Hysteresis is bistable operation of the transistor current and appears as a difference in I_D during ascending and descending scans of V_{DS} or V_{GS}. An illustration of these techniques is shown in Fig. 2.9.

In an organic transistor, the mobility of charge carriers is more properly assumed to depend on the overdrive voltage (i.e., the difference between V_{GS} and V_T) [78] according to

$$\mu = k(V_{GS} - V_T)^{\gamma} \tag{2.16}$$

where k contains information on the morphology of the film, mainly related to the ease of hopping between sites, and γ is related to the width of an exponential distribution of states (DOS) by $\gamma = 2(T_c/T-1)$. A possible explanation is given by the variable-range hopping model proposed by Vissenberg and Matters [79], in which carriers participate in the current flow only when they are excited to an energy level termed the transport energy. If the carrier concentration is high, the initial average energy is close to the transport energy, which reduces the required activation energy and thus increases mobility considerably. Alternatively, according to the multiple trapping and release and transport model [80], only one fraction of the charge induced by the gate contributes to current flow in the channel. The remaining part continues to be trapped in an exponential tail of localized states. Since the ratio between free carriers and traps is greater in high-injection conditions, the mobility increases with increasing gate voltage. In both cases, the drain current I_D in the triode regime becomes:

$$I_D = C_i \frac{W}{L} k(V_{GS} - V_T)^{\gamma+1} V_{DS} \tag{2.17}$$

In the saturation regime, the current becomes:

$$I_D = \frac{1}{\gamma+2} C_i \frac{W}{L} k(V_{GS} - V_T)^{\gamma+2} \tag{2.18}$$

The presence of disorder in the thin film semiconductor generally involves the occurrence of Poole–Frenkel phenomena, in which the dependence of the field effect mobility and the longitudinal electric field can be described by Eq. 2.7 [56]: the relationship between β and γ can be derived from Eqs. 2.3 and 2.11 [81].

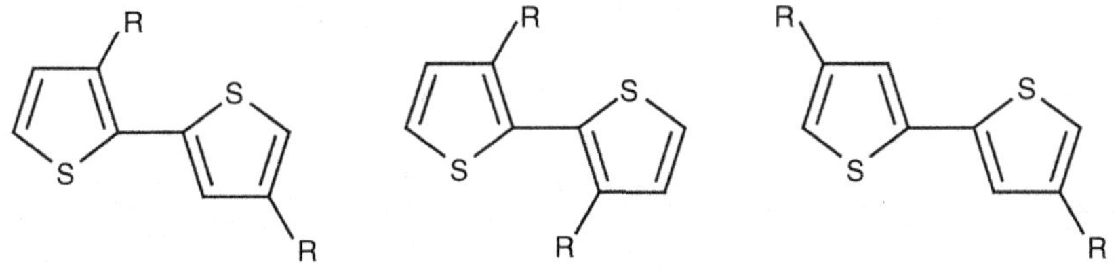

Head-to-tail (HT)   Head-to-head (HH)   Tail-to-tail (*TT*)

FIGURE 2.10 Regioregularities in poly(3-hexylthiophene).

Elevated contact resistances (R_s) decrease the effective voltage applied to the transistor and consequently the performance of devices. In high-mobility organic semiconductor TFTs, that is $\mu > 10^{-1}$ cm^{2}V^{-1}s^{-1}, the value of R_s can be estimated from the resistance of the device in the triode regime (R_{on}) and the resistance of the channel in the triode regime ($R_{channel}$):

$$R_{on} = \frac{V_{DS}}{I_D} = R_{channel} + R_S = \frac{L}{\mu_i C_i W (V_{GS} - V_{T,i})} + R_S \tag{2.19}$$

where μ_i and $V_{T,i}$ are intrinsic parameters of the material [82].

2.3.2.1 Field Effect Mobility

Organic semiconductors present mobility dependent on morphology and thus on the electric field and temperature. For this reason, the evolution in performance of these devices is intrinsically related to the study of the influence of each layer and fabrication conditions for their operation. In fact, the first polythiophene FET obtained by electrochemical polymerization exhibited field effect mobilities of only 10^{-5} cm^2 V^{-1} s^{-1}. The regularity in the arrangement of radicals in the molecule, as in poly(3-alkylthiophenes) (P3AT), can greatly affect the morphology and thus the charge transport in a film semiconductor [83]. These radicals can be incorporated into the polymer chain at three different regioregularities: head-to-head (HH), tail-to-tail (TT), and head-to-tail (HT), as shown in Fig. 2.10. A regiorandom polymer has both configurations in the same chain, and these are randomly distributed, whereas a regioregular polymer has only one of these regioregularities in their chains. The latter tend to have a lower bandgap energy, and higher order and crystallinity in the solid state, in addition to increased electroconductivity [84]. In 1996, Bao et al. reported remarkable results in regioregular P3HT FETs deposited by casting: $\mu \approx 0.015$–0.045 cm^2 V^{-1} s^{-1} [40].

The final morphology of the film is related to the applied solvent. For example, Bao et al. observed that P3HT precipitated during evaporation of the solvent tetrahydrofuran (THF), resulting in a nonuniform and discontinuous film. Electron microscopy revealed the presence of lamellar micrometric and granular nanometric crystals. Using solvents, such as 1,1,2,2-tetrachloroethane and chloroform, the FET mobility increased from 6.2×10^{-4} to c. 5×10^{-2} cm^2 V^{-1} s^{-1} [40]. Comparatively, films from derivatives of

Table 2.1   A Comparative Table Between the First Flexible Microprocessor From An Organic Semiconductor and the Intel 4004 on Rigid Silicon

Microprocessor	Plastic	Intel 4004
Year of production	2011	1971
Number of transistors	3381	2,300
Area	1.96×1.72 cm^2	3×4 mm^2
Wafer diameter	6 in.	2 in.
Substrate	Flexible	Rigid
Number of pins	30	16
Voltage source	10 V	15 V
Power consumption	92 µW	1 W
Operational speed	40 operations s^{-1}	92,000 operations s^{-1}
P-type semiconductor	Pentacene	Silicon
Mobility	~0.15 cm^2 V^{-1} s^{-1}	~450 cm^2 V^{-1} s^{-1}
Operation	Accumulation	Inversion
Technology	5 µm	10 µm
Bus width	8 bit	4 bit

poly(*para*-phenylenevinylene) (PPV), important for application in OLEDs and OSCs, have mobilities on the order of 10^{-5}–10^{-4} cm^2 V^{-1} s^{-1} [85,86]. The fact that these materials have an amorphous structure and that transport of charge carriers in films of these materials occurs by hopping limits the mobility [87].

Monocrystalline or highly ordered polycrystalline films deposited by vacuum sublimation exhibit field effect mobilities greater than 0.1 cm^2 V^{-1} s^{-1} [88]. This performance allows the manufacture of many flexible commercial devices, such as the first microprocessor on plastic presented in Table 2.1.

The silicon Intel 4004 microprocessor, whose performance is similar to that of pentacene, was produced in 1971, approximately 2 years after the arrival of humans on the moon on July 20, 1969 [15]. These small molecules, of great interest to the scientific community because of the high mobility of charge carriers, have been modified to allow for processing in the liquid phase. Such is the case of pentacene (0.1–0.5 cm^2 V^{-1} s^{-1}) [89] and polythienylenevinylene (0.22 cm^2 V^{-1} s^{-1}) [90], for which precursor materials were synthesized for their conversion into polymers during film treatment postdeposition. However, it should be noted that grain boundaries in polycrystalline films can play a key role in transistor performance, as they act as traps to charge carriers [91].

Finally, it should be noted that the interface between the gate dielectric and the semiconductor also plays an important role in charge transport, possibly representing up to three orders of magnitude depending on the dielectric used [92]. The SiOH groups present on the surface of SiO$_2$ act as electron traps. The passivation of these groups may occur by the deposition of self-assembled monolayers prior to the deposition of the semiconductor (Fig. 2.11A–B). In 2005, Chua et al. [93] verified a significant increase in the current in an n-type TFT of F8BT (Fig. 2.11C) when treating SiO$_2$ with monolayers of octadecyltrichlorosilane

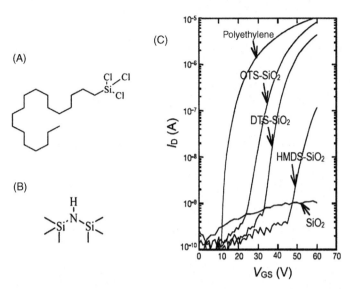

FIGURE 2.11 Self-assembled monolayers for treating the surface of SiO_2: (A) OTS; (B) HMDS; (C) drain current (I_D) as a function of the gate voltage (V_{GS}) in an F8BT n-TFT for different treatments with self-assembled siloxanes or a polyethylene passivating film (buffer). *Adapted from H. Ohnuki, et al. Effects of interfacial modification on the performance of an organic transistor based on TCNQ LB films. Thin Solid Films 516 (2008) 2747–2752 [93].*

(OTS), hexamethyldisilazane (HMDS), or decyltrichlorosilane (DTS). According to Ohnuki et al. [94], these self-assembled films chemically react with SiOH groups, passivating surface defects and reducing the concentration of traps. A similar effect can be produced when using an organic insulator. Chua et al. in work dating from 2005, used polyethylene and divinyltetramethyldisiloxane-bis(benzocyclobutene) (BCB) together with calcium electrodes and demonstrated that it is possible to manufacture n-OTFTs and that the mobility of the electrons can be greater than that of the holes (until then, the opposite was believed). In 2008, Benson et al. [95] demonstrated complementary metal oxide semiconductor (CMOS) technology with organic transistors using pentacene TFTs obtained by replacing the dielectric with polymethylmethacrylate (PMMA), polycarbonate (PC), (polystyrene) PS, (polyimide) PI, or poly(4-vinylphenol) (P4VP).

In summary, when seeking to overcome the threshold $\mu \leq 10^{-1}$ cm^2 V^{-1} s^{-1} in polymers, in the first decade of the 21st century, high mobility polymer, such as those shown in Fig. 2.12, were synthesized. The obtained p-semiconductors are mostly derivatives of polythiophene, such as poly[5,5-bis(3-dodecyl-2-thienyl)-2,2-bithiophene] (PQT-12) [96,97] and poly(2,5-bis(3-hexadecylthiophen-2-yl)thieno[3,2-*b*]thiophene (pBTTT) [98–100]. Among the n-types are poly(benzobisi midazobenzophenanthroline) (BBL) [101] and [*N,N*-9-bis(2-octyldodecyl)naphthalene-1,4,5,8-bis(dicarboximide)-2,6-diyl]-*alt*-5,59-(2,29-bithiophene) (P(NDI2OD-T2)) [102–104]. Finally, also pursued was the synthesis of ambipolar semiconductors, such as diketopyrrolopyrrole–benzothiadiazole copolymer (PDPP-TBT), studied by Ha et al. [105].

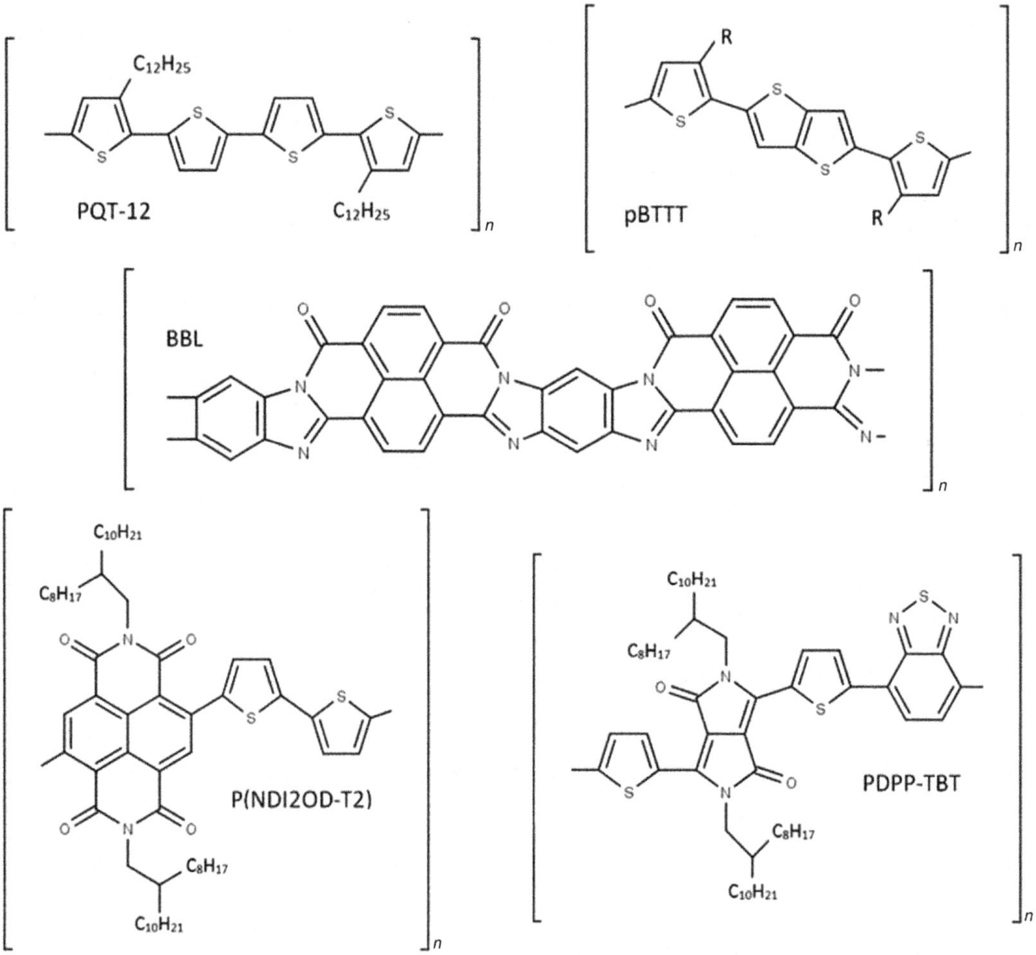

FIGURE 2.12 Chemical structure of high mobility polymers. PQT-12, pBTTT, BBL, P(NDI2OD-T2), and PDPP-TBT.

In parallel, the effort for the development of small high mobility molecules, shown in Fig. 2.13, including those that can be processed in liquid medium, should be highlighted. In this case, the main p-semiconductor are acenes, such as tetracene [106] *n*-heteropentacenes (HP) [107] and rubrene [44,106,108,109] as well as the already mentioned thiophenes, such as *trans*-1,2-di[thieno[3,2-*b*][1]benzothiophenic-2-]ethylene (DTBTE) [110] dinaphtho-[2,3-*b*:2′,3′-*f*]thieno[3,2-*b*]-thiophene (DNTT) [43,111,112], and 2,7-dioctyl[1] benzothieno[3,2-*b*][1]benzothiophene (C_8-BTBT) [113]. Small acene molecules can be processed in solution by adding branches or radicals, which make them soluble and cause them to react with the adjacent molecule of the same semiconductor, resulting in the formation of a monocrystal. This is the case for 6,13-bis[triisopropylsilylethynyl] (TIPS) pentacene, crystallized from solution by heat treatment at 120°C under a 100 sccm flow of forming gas for 4 h [106]. The main n-type molecules are tetracyanoquinodimethane (TCNQ) [44,114]

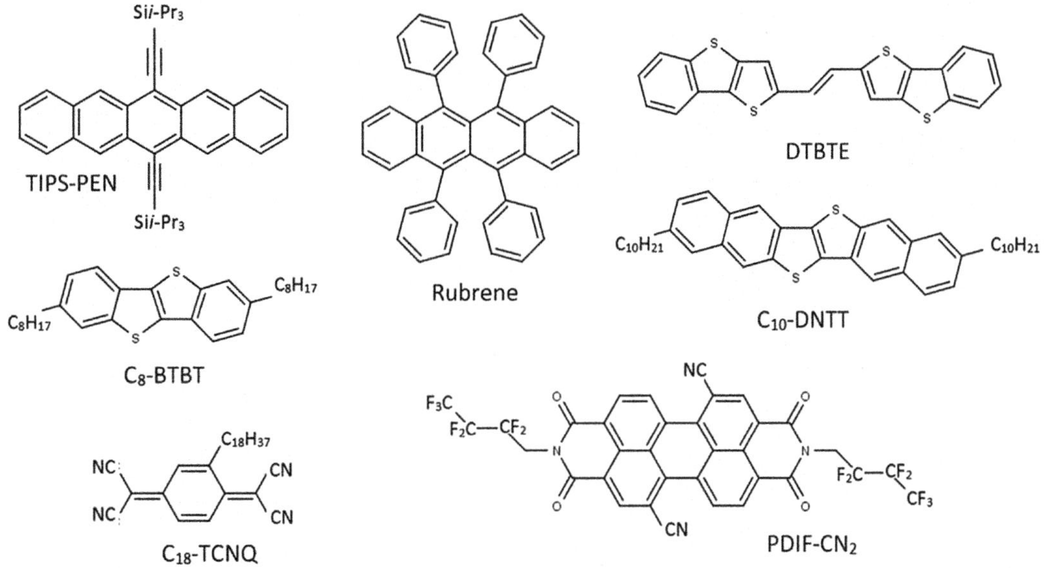

FIGURE 2.13 Chemical structure of small high mobility organic molecules. TIPS-pentacene, rubrene, DTBTE, C_{10}-DNTT, C_8-BTBT, TCNQ, and PDIF-CN$_2$ [118].

and N,N''-bis(n-alkyl)-(1,7 and 1,6)-dicyanoperylene-3,4:9,10-bis(dicarboximide) (PDIR-CN$_2$) [115,116]. In the class of ambipolar semiconductors, reduced graphene oxide (RGO) is highlighted [31,117]. Developments in the performance of organic semiconductors applied in FETs are presented in Table 2.2 via a comparative table.

2.3.2.2 Organic TFT-Based Sensors

The susceptibility of organic semiconductors to operating conditions and impurities is a limiting factor in the efficiency of photovoltaic and electroluminescent devices [organic light-emitting diodes (OLEDs)] but has found direct application in chemical sensors. Materials, such as acenes [120,121], oligothiophenes [122], polythiophenes [123], and poly(phenylenevinylene) [124] are semiconductors successfully used in this field. In the specific case of OTFT-based sensors, inorganic oxides (e.g., SiO$_2$, a high-dielectric constant hafnium oxide) [125] and cross-linked polymers [e.g., poly(4-vinylphenol)] [126,127] are primarily employed as insulators in academic laboratories. Conductive polymers, such as PAni [19] and PEDOT:PSS [20], have recently been used as electrodes, although gold remains the most widely used source and drain metal. Rigid silicon or flexible polymer substrates covered with a thin film conductor [127,128] are generally used as the gate electrode and substrate.

OTFT-based sensors rely on the interaction between chemical or biological analytes and the semiconductor active layer. These analytes may be specific to chemical modification and the incorporation of recognition sites into the transistor structure. However, alteration of the deposition conditions of each layer, a consequence of this modification,

Table 2.2   Comparative Table of Organic Semiconductors Applied in TFTs

Semiconductor	Deposition Technique[a]	Substrate	Dielectric[b]	μ (cm² V⁻¹ s⁻¹) μ_h	μ_e	V_T (V)	$I_{on/off}$	Year	References
PQT-12	S	Si	SiO₂/OTS-8	0.18	—	−5	10^7	2005	[97]
pBTTT	S	Si	SiO₂/OTS	0.12	—	−4	—	2007	[98]
P3HT:TCNQ	S	Si	SiO₂/OTS	0.01	—	2,6	10^4	2008	[114]
P3HT:RGO	S	Si	SiO₂	8×10^{-3}	—	−5	10^4	2011	[21]
BBL	S	Si	SiO₂	—	10^{-3}	15	10^5	2011	[101]
P(NDI2OD-T2)	S	Glass or PEN	PMMA	—	0.2–0.3	8	10^{4-6}	2011	[102]
PDPP-TBT	S	Si/SiO₂/OTS	D139	0.53	0.58	—	—	2012	[105]
Tetracene	E	—	Parylene-N	1	—	0	—	2010	[106]
TIPS-PEN	S	—	Parylene-N	0.05	—	0	—	2010	[106]
Rubrene	L	—	PE	4.7	—	0	10^6	2011	[44]
RGO	S	Si	SiO₂	1	0.2	0	10^2	2008	[117]
DTBTE	E	Si	SiO₂/OTS	0.5	—	−25	10^6	2011	[110]
DNTT	L	Si	SiO₂/CYTOP	8.3	—	0	10^8	2009	[43]
C₁₀-DNTT	S	Si	SiO₂	11	—	10	10^7	2011	[111]
C₈-BTBT	S	Si/SiO₂	Parylene-C	31.3	—	−10	10^7	2011	[113]
TCNQ	S	Si	SiO₂/HMDS	—	4.4	−3	10^2	2008	[94]
PDIF-CN₂	L	Si	SiO₂/PMMA	—	6	−5–5	10^4	2009	[115]

TFTs, Thin film transistors.

[a]S, Processed in solution (e.g., spincoatinge casting); EV, thermal evaporation; L, lamination of monocrystals.

[b]CYTOP, Commercial amorphous fluoropolymer [119].

affects the device response signal because it is intrinsically related to the quality of interfaces in the device (i.e., insulator/semiconductor and electrode/semiconductor) [129] and the nanomorphology of the layers [130,131]. Direct interactions between the analyte and the active layer may occur via charge transfer or doping, resulting in a change in conductivity. Furthermore, analytes can be adsorbed and then diffuse through the grain boundaries, introducing new energy states located in the film (i.e., charge transport traps), thus increasing the resistance to charge transport. This effect is observed in the alteration of the mobility of charge carriers (μ) in a semiconductor film. The accumulation of analytes at the interface with the insulator, in turn, can change the local electric field distribution at this interface and thus the conductivity of the channel. In this case, the main parameter monitored is the threshold voltage (V_T). In both cases, the impact is in the variation of the current circulating in the channel (ΔI_{DS}) [132].

One of the principal means of analyte interaction with organic materials is in the vapor phase. Pentacene is sensitive to environmental conditions, such as relative humidity [133], oxygen [134], and ozone (O_3) [135]. An 80% decrease of the saturation current for TFTs of pentacene can be observed as a result of an increase from 0% to 30% humidity [133]. In turn, Crone et al. mapped the detection of volatile organic species by a set of organic TFT-based sensors [136]. A total of 16 analytes in the vapor phase were studied by monitoring the ΔI_{DS} in 11 organic semiconductors. The sensitivity of the devices was extrapolated to

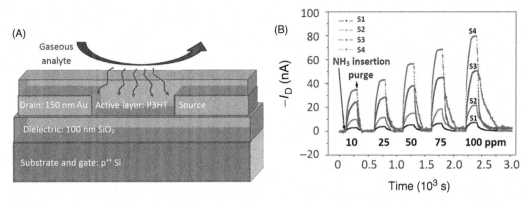

FIGURE 2.14 (A) Schematic diagram of the OTFT-based gas sensor: gold on a thermal oxide forms the source and drain electrodes; the active layer is obtained by spin coating. (B) Response array of sensors I_D to the insertion of 10–100 ppm of gaseous ammonia into the chamber. *Adapted from K.-H. Kim, S.A. Jahan, E. Kabir, A review of breath analysis for diagnosis of human health. TrAC 33 (2012) 1–8 [139].*

approximately 1 ppm via electrical circuit-based analysis techniques [137]. These studies demonstrated that the analyte interacts with the semiconductor via van der Waals, permanent dipole–dipole or hydrogen bond electrostatic interactions. Chang et al. demonstrated the applicability of polythiophene derivatives deposited by printing in the detection of organic solvents, such as alcohols, acids, aldehydes, and amines [138]. The material of the active layer is principally altered by the insertion of functional groups along the polymer chain or at its ends. The use of polythiophene P3HT in electronic noses enables the detection of ammonia [139], a biomarker of diseases, such as infection in the gastrointestinal tract by the bacteria *Helicobacter pylori*, uremia, or renal insufficiency and hepatic cirrhosis [140]. Jeong et al. observed the dependence of μ, V_T, $I_{on/off}$, and S of devices, as shown in Fig. 2.14A, on the ammonia concentration in parts per million (ppm). The response of I_D over time as a function of the concentration in ppm is shown in Fig. 2.14B for four different values of W/L (identified by $S1$ to $S4$ in the legend).

Analyte detection in liquid medium can be accomplished using the ion-sensitive field effect transistor (ISFET), invented in 1970 by Bergveld [141]. OTFTs are generally considered incompatible with water because of uncertainty regarding their stability at high operating voltages and the possibility of physical delamination. In 2000, however, the ability of OTFTs to respond to ions in solution was demonstrated by Bartic et al. [142]. A pH meter was fabricated on a silicon wafer using an active layer of regioregular P3HT and a silicon nitride (Si_3N_4) dielectric. Since then, efforts have been made to commercialize these sensors. In 2003, Gao et al. integrated a reference electrode [143]. Two years later, Loi et al. fabricated an ISFET on the flexible film Mylar [30]. In 2008, Roberts et al. manufactured ISFETs of 5,5-bis-(7-dodecyl-9H-fluoren-2-yl)-2,2-bithiophene (DDFTTF) operating at 1 V [126]. These TFTs without encapsulation exceeded 10^4 cycles of operation (V_{GS} of 0.3 at -1 V and $V_{DS} = -0.6$ V) without significant changes in their electrical parameters. Applicability was demonstrated in sensors of glucose, trinitrobenzene, cysteine, and methylphosphonic

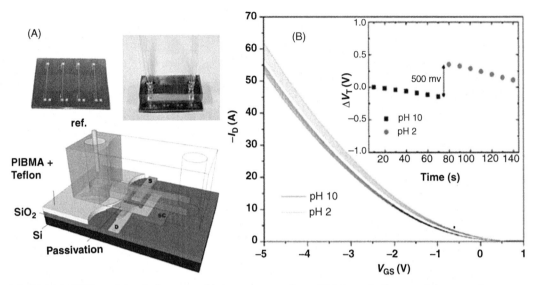

FIGURE 2.15 OTFT-based chemical sensor stable in aqueous medium. (A) Schematic diagram of the sensor in the aqueous flow cell, and (B) the current I_{DS} depends on the pH of the solution. *Adapted from N.T. Kalyani, S.J. Dhoble, Organic light emitting diodes: energy saving lighting technology—a review. Renow. Sust. En. Rev. 16 (2012) 2696–2723 [144].*

acid for concentrations on the order of parts per billion (ppb). An example of microfluidic integration with organic electronics for pH detection is shown in Fig. 2.15. In this case, stability of the polytriarylamine transistor (PTAA) processed by Spijkman et al. in 2010 is guaranteed by the presence of a bottom gate of SiO_2 and a top gate of Teflon and polyiso-butylmethacrylate (PIBMA) [144].

2.4 Organic Light–Emitting Diodes

2.4.1 Structure of Thin Films in OLEDs and Typical Materials Used

OLEDs [145] are fabricated in the form of a simple thin film structure by two electrodes and a conjugated polymer, with suitable energy levels for the emission of wavelengths in the visible region [146–148]. Among the conjugated polymers best known and currently studied are those derived from PPV [149], polyfluorene (PF), and polyvinylcarbazole (PVK) [150]. These materials emit wavelengths associated with the primary colors reddish-orange, greenish-yellow and blue, respectively. The emission spectrum of these materials is directly related to the bandgap that separates the HOMO from the LUMO, a characteristic equivalent to that of the inorganic semiconductor whose nomenclature is defined as the VB and CB, respectively [151]. Due to this separation, charge carriers can receive external power via high-energy photons or even by applying a potential difference to form a species denominated exciton. This exciton represents an electron–hole pair in an excited energy state whose radioactive decay is in a wavelength band characteristic of each material.

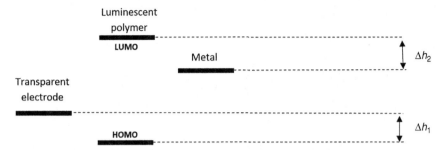

FIGURE 2.16 Diagram of the energy levels considering a transparent electrode, a luminescent polymer, and a stacked metal in which Δh_1 refers to the existing barrier height between the working function of the transparent electrode and the highest occupied molecular orbital *(HOMO)* level of the polymer and Δh_2 to the barrier height between the lowest unoccupied molecular orbital *(LUMO)* level and the working function of the metal.

Joining of the materials used with functions of emission and conduction, when in thermal equilibrium, can produce energy barriers to the transport of charge carriers; as a result, recombination occurs outside the appropriate location for light emission, and loss may occur by nonradiative decay or lattice vibration [152]. Thus, auxiliary materials are usually employed to facilitate the reduction of such energy barriers. In a thin film structure, two classes of materials studied and applied for this purpose are termed hole transport layers (HTL) [153] and electron transport layers (ETL) [154]. Materials, such as PEDOT:PSS, MTDATA, and TPD are exploited for the hole transporting function; in contrast, materials, such as hydroxyquinoline aluminum (Alq3) and 2-(4-biphenylyl)-5-(4-*tert*-butylphenyl)1,3,4-oxadiazol (butyl-PBD) are used for the electron transport function. This classification arises from study of the energy levels of these materials, which provide appropriate and preferential transport of a specific type of charge carrier. Materials with HTL and ETL functions minimize the barrier heights both between the work function of the transparent electrode and the HOMO level of the polymer (Δh_1), and between the LUMO level and the working function of the metal (Δh_2), shown in Fig. 2.16.

2.4.2 Electrooptic Characterization of OLEDs

The main parameters that characterize the performance of an OLED or LEC are power efficiency (η_{WW}), power luminous efficiency (η_P), luminous efficiency (h_L), and external quantum efficiency (h_{EQE}). Calculation of the power efficiency (η_{WW}) [155], also known in the literature as the wall plug efficiency [156], requires the optical power [P_{opt} (W)], which is determined as follows:

$$P_{opt} = \int_0^{\infty} P_{OLED}(\lambda)d\lambda$$

(2.20)

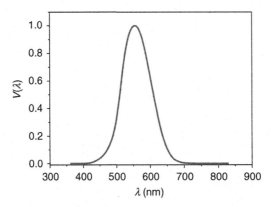

FIGURE 2.17 Photopic response function $V(\lambda)$ according to the CIE 1924 with a standard accuracy of 1 nm. *Adapted from Luminous efficiency graphic, http://www.cvrl.org/cie.htm [158].*

The electric power $P_{opt} = VI$. Thus, we obtain the dimensionless power or wall plug efficiency:

$$\eta_{WW} = \frac{P_{opt}}{P_{ele}} = \frac{\int_0^\infty P_{OLED}(\lambda)d\lambda}{VI} \tag{2.21}$$

Applications of these devices as information displays or ambient lighting must take into account the perception of light based on the sensitivity of the human eye. Thus, for the total luminous flux [Φ_{tot} (lm)], we have:

$$\Phi_{tot} = \int_0^\infty K(\lambda)P_{OLED}(\lambda)d\lambda \tag{2.22}$$

where $K(\lambda) = 683V(\lambda)$ (lm W^{-1}) and $V(\lambda)$ is the photopic dimensionless response function. Fig. 2.17 shows the standard of the Commission Internationale de l'Eclairage (CIE) [157] of 1924 for $V(\lambda)$, whose maximum point [$V(\lambda) = 1$] occurs for the yellowish-green wavelength at 555.17 nm.

Therefore, the power luminous efficiency [η_P (lm W^{-1})] is defined by the relationship between the luminous flux and electrical power according to [152]:

$$\eta_P = \frac{\Phi_{tot}}{P_{ele}} = \frac{\int_0^\infty K(\lambda)P_{OLED}(\lambda)d\lambda}{VI} \tag{2.23}$$

In turn, calculation of the current luminous efficiency (η_L) requires the concepts of luminous intensity (i) and luminance (L_{OLED}). From a circular flux of light with a radius (r) of an infinitesimal area dA generated by a light source, as shown in Fig. 2.18, it is possible to assume for a distance (x), such that $x \gg r$, the point light source hypothesis [152].

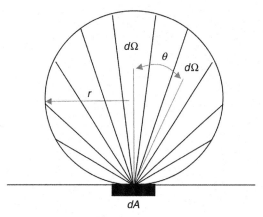

FIGURE 2.18 Circular scattering representative of light scattering.

The detection of this signal (stereo-radian photon flux) is defined as the luminous intensity (i), where its scale is a function of the polar (θ) and azimuth angles (φ) such that $d\Omega$ is the infinitesimal variation of the emission angle. Thus, the total luminous flux emitted is defined according to:

$$\Phi_{\text{tot}} = \int_0^{2\pi} \int_0^{\theta/2} i(\theta,\varphi)\operatorname{sen}\theta\, d\varphi\, d\theta \tag{2.24}$$

For simplification purposes, the photon flux is usually considered only as a function of θ, whose reference is 90 degree in relation to the surface. In this case, the approximation of a Lambertian surface [159] is used, corresponding to an intensity of $i(\theta) = i_0\cos(\theta)$. Thus, the total luminous flux is defined as

$$\Phi_{\text{tot}} = 2\pi \int_0^{\pi/2} i_0 \cos\theta \operatorname{sen}\theta\, d\theta \tag{2.25}$$

Eq. 2.25 for a Lambertian surface has a solution of πi_0; that is., $i_0 = 1$ candela (cd) in the normal direction, which is equivalent to $\Phi_{\text{tot}} = \pi$ lumen (lm). In practice, optical characterization begins by measuring the luminance [L_{OLED} (cd m^{-2})] defined by

$$L_{\text{OLED}} = \frac{i(\theta = 0°)}{A} \tag{2.26}$$

Thus, the luminous efficiency [η_L (cd A^{-1})] [157] can be defined as the ratio between L and the current density ($J = I/A$) according to

$$\eta_L = \frac{L}{J} \tag{2.27}$$

The power luminous efficiency [η_P (lm W^{-1})] can be calculated from η_L, assuming the emission of an electroluminescent device, such as a Lambertian emitter, according to

$$\eta_P = \frac{\pi}{V}\eta_L \qquad (2.28)$$

Another figure of merit that can be calculated is the external quantum efficiency (EQE), which is the ratio between the number of photons obtained and the amount of charge carriers injected into the system. Considering the integrated angle spectrum of electroluminescence (optical power) [$P_{OLED}(\lambda)$], whose emission is a function of a given wavelength (λ), the number of photons generated in unit time (N_F) is given by [152]

$$N_F = \int_0^\infty \frac{P_{OLED}(\lambda)}{hc}\lambda d\lambda \qquad (2.29)$$

where c is the speed of light and h is Planck's constant. In turn, the number of charge carriers passing through the device subject to current I in unit time is given by $N_{PC} = I/q$, where q is the electron charge. Therefore, EQE, which is dimensionless, can be obtained by [152]

$$\frac{N_F}{N_{PC}} = \frac{e}{hcI}\int_0^\infty \frac{P_{OLED}(\lambda)}{hc}\lambda d\lambda \qquad (2.30)$$

In the calculation of EQE, the photopic response function $V(\lambda)$ and η_L can be taken into account according to [152]

$$\eta_{EQE} = \frac{\pi e \lambda_D}{K(\lambda_D)hc}\eta_L \qquad (2.31)$$

which can be approximated using the numerical constant values given by

$$\eta_{EQE} = \frac{3.7\times10^3 \lambda_D}{V(\lambda_D)}\eta_L \qquad (2.32)$$

in which the dependence of h_{EQE} is observed relative to the dominant wavelength (λ_D), the photopic response function $V(\lambda)$ and η_L. The value of λ_D is obtained directly from the chromaticity diagram of CIE, using, for example, the pattern 1931, as defined by the colorimetry theory, which is not covered in this chapter [152].

Fig. 2.19 shows the photographic register of a high-performance OLED, whose thin film is composed of a polymeric matrix and a rare earth complex, produced in the laboratory and polarized with voltages of 10, 15, 20, and 30 V. The same figure presents the CIE 1931 chromaticity diagram with coordinates related to the polarization imposed on the OLED.

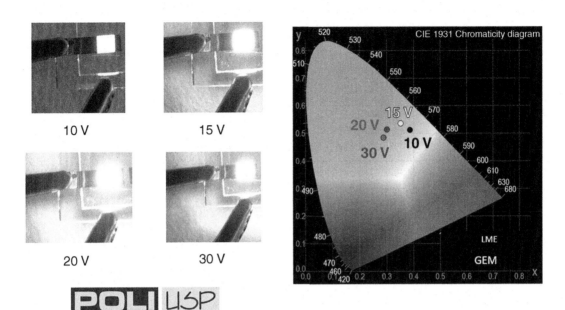

FIGURE 2.19 Photographic record of an organic light-emitting diode (OLED) produced in the laboratory and chromaticity diagram with coordinates related to the polarization of the OLED.

List of Symbols

ε_r	Relative dielectric constant
ε_0	Vacuum permittivity (F m^{-1})
η	Ideality factor in the Schottky thermionic emission model
η_{EQE}	External quantum efficiency
η_L	Luminous efficiency (cd A^{-1})
η_P	Power efficiency of current (lm W^{-1})
η_{WW}	Power or wall plug efficiency
θ	Polar angle (rad)
λ_D	Dominant wavelength (m)
μ	Effective mobility of charge carriers (cm^2 V^{-1} s^{-1})
ϕ	Azimuthal angle (rad)
φ_B	Height of the potential barrier (eV)
Φ_{tot}	Total luminous flux (lm)
Ω	Emission angle (rad)
A	Geometric area of the electrical contact device
A^{**}	Richardson's constant (A m^{-2} K^{-2})
C_i	Capacitance per area (F m^{-2})
c	Speed of light (m s^{-1})
d	Thickness of the thin film (m)
F	Electric field
h	Planck's constant (Js)
I	Electric current (A)

I_D	Current between the source and drain in the transistor (A)
$I_{leakage}$	Leakage current of the transistor dielectric (A)
I_{off}	Cut-off current of the transistor (A)
I_{on}	Operating current of the transistor (A)
$I_{on/off}$	Current modulation of the transistor
i	Luminous intensity or stereo-radian photon flow
k_B	Boltzmann's constant (J K^{-1})
L	Channel length of the transistor (m)
L_{OLED}	Luminance (cd m^{-2})
m^*	Effective electron mass (kg)
N_c	Effective density of states in the conduction band (1 cm^{-3})
N_F	Number of photons generated per unit time (1 s^{-1})
N_{PC}	Number of charge carriers crossing the device
N_v	Effective density of states in the valence band (1 cm^{-3})
n	Density of free carriers (1 cm^{-3})
n_Q	Charge density in the transistor channel (1 cm^{-2})
n_t	Density of trapped carriers (1 cm^{-3})
n_0	Free thermally generated carriers (1 cm^{-3})
P_{ele}	Electric power (W)
P_{OLED}	Optical power in a wavelength (W)
P_{opt}	Optical power (W)
q	Charge of the electron or element (C)
r	Radius (m)
$R_{channel}$	Channel resistance in the triode transistor regime (Ω)
R_{on}	Resistance in the triode transistor regime (Ω)
R_s	Series resistance (Ω)
T	Temperature (K)
T_g	Glass transition temperature of polymers (K)
V	Electric potential (V)
V_{DS}	Voltage between the source and drain of the transistor (V)
V_{GS}	Voltage between the gate and drain of the transistor (V)
V_T	Threshold voltage of the transistor (V)
x_i	Thickness of the dielectric film in the transistor (m)
W	Width of the transistor channel (m)
μCP	Microcontact printing

References

[1] M. Sampietro, Curso de Elettronica a semiconduttori organici: Principi, dispositivi ed appli-cazioni. Aula Aspetti base dei semiconduttori organici, Politecnico di Milano, Milão, Itália. Maio 2009. http://home.dei.polimi.it/sampietr/ESO/index.html

[2] H. Shirakawa, et al., Synthesis of electrically conducting organic polymers: halogen derivatives of polyacetylene, (CH)x, J. Chem. Soc. Chem. Commun. (1977) 578, doi: 10.1037/C39770000578.

[3] T.A. Skotheim, Handbook of Conducting Polymers, vols. 1 and 2, Marcel Dekker, New York, 1986.

[4] K.S.V. Srinivasan, Macromolecules: New Frontiers, vols. 1 and 2, Allied Publishers, New Delhi, 1998.

[5] K.C. Kao, W. Hwang, Electrical Transport in Solids, Pergamon Press, Oxford, 1981.

[6] T.W. Kelley, et al., High performance organic thin film transistors, Mater. Res. Soc. Symp. Proc. 771 (2003) L6.5.1.

[7] T.H.B. Singh, et al., Enhanced mobility of organic field-effect transistors with epitaxially grown C60 film by in-situ heat treatment of the organic dielectric, Mater. Res. Soc. Symp. Proc. 871E (2005) I4.9.1–I4.9.8.

[8] S.M. Sze, Physics of Semiconductor Devices, second ed., John Wiley & Sons, New York, 1981, 880 p.

[9] S. Reineke, et al., White organic light-emitting diodes with fluorescent tube efficiency, Nature 459 (2009) 234–238.

[10] S.H. Park, et al., Bulk heterojunction solar cells with internal quantum efficiency approaching 100%, Nat. Photon. 3 (2009) 297–302.

[11] P. Stallinga, General concept, Electrical Characterization of Organic Electronic Materials and Devices, John Wiley & Sons Ltd, Chichester, West Sussex, United Kingdom, 2009, pp. 1–38 (Chapter 1).

[12] S.H. Jeong, C.K. Song, M. Yi, Capacitance enhancement in the accumulation region of C-V characteristics in metal-insulator-semiconductor capacitors consisting of pentacene and poly(4-vinylphenol), Appl. Phys. Lett. 94 (2009) 183302.

[13] K. Myny, et al., An integrated double half-wave organic Schottky diode rectifier on foil operating at 13.56 MHz, Appl. Phys. Lett. 93 (2008) 093305.

[14] H. Yan, et al., A high-mobility electron-transporting polymer for printed transistors, Nature 457 (2009) 679–686.

[15] K. Myny, et al., An 8-bit, 40-instructions-per-second organic microprocessor on plastic foil, IEEE J. Sol. State Circ. 47 (1) (2012) 284–291.

[16] L. Ma, et al., Nanometer-scale recording on an organic-complex thin film with a scanning tunneling microscope, Appl. Phys. Lett. 69 (1996) 3762.

[17] K. Myny, et al., Plastic circuits and tags for 13.56 MHz radio-frequency communication, Solid State Electron. 53 (2009) 1220–1226.

[18] Q.-D. Ling, et al., Polymer electronic memories: materials, devices and mechanisms, Prog. Polym. Sci. 33 (2008) 917–978.

[19] K.S. Lee, et al., High-resolution characterization of pentacene/polyaniline interfaces in thin-film transistors, Adv. Funct. Mater. 16 (2006) 2409–2414.

[20] H.S. Kang, et al., Electrical characteristics of pentacene-based thin film transistor with conducting poly(3,4-ethylenedioxythiophene) electrodes, J. Appl. Phys. 100 (2006) 064508.

[21] A. Liscio, et al., Charge transport in graphene–polythiophene blends as studied by Kelvin Probe Force Microscopy and transistor characterization, J. Mater. Chem. 21 (2011) 29242931.

[22] T.W. Odom, et al., Atomic structure and electronic properties of single-walled carbon nanotubes, Nature 391 (1998) 62–64.

[23] R.H. Baughman, A.A. Zakhidov, W.A. De Heer, Carbon Nanotubes—the route toward applications, Science 297 (5582) (2002) 787–792.

[24] R. Toniolo, C. Lepienski, I. Hümmelgen, Organic electronic pulse generator, Electron. Lett. 40 (2004) 566–567.

[25] C. Zhang, et al., Alkali metal crystalline polymer electrolytes, Nat. Mater. 8 (2009) 580–584.

[26] T.F. Otero, M. Broschart, Polypyrrole artificial muscles: a new rhombic element. Construction and electrochemomechanical characterization, J. Appl. Electrochem. 36 (2) (2006) 205–214.

[27] A.N. Sokolov, M.E. Roberts, Z. Bao, Fabrication of low-cost electronic biosensors, Mater. Today 12 (9) (2009) 12–20.

[28] Y. Inoue, et al., Low-voltage organic thin-film transistors on flexible plastic substrates, Mat. Res. Soc. Symp. Proc. 736 (2003) D4.2.1.

[29] O. Yokoyama, Active-matrix full color organic electroluminescent displays fabricated by ink-jet printing, Optronics 254 (2003) 119.

[30] A. Loi, I. Manunza, A. Bonfiglio, Flexible, organic, ion-sensitive field-effect transistor, Appl. Phys. Lett. 86 (2005) 103512.

[31] K.S. Johnson, et al., Chemical sensor networks for the aquatic environment, Chem. Rev. 107 (2007) 623–640.

[32] F. Liao, C. Chen, V. Subramanian, Organic TFTs as gas sensors for electronic nose applications, Sens. Actuat. B 107 (2005) 849–855.

[33] L. Zhu, et al., Flavour analysis in a pharmaceutical oral solution formulation using an electronic-nose, J. Pharm. Biomed. Anal. 34 (2004) 453–461.

[34] I. Voiculescu, et al., Micropreconcentrator for enhanced trace detection of explosives and chemical agents, IEEE Sens. J. 6 (2006) 1094–1104.

[35] L. Wang, et al., Nanoscale organic and polymeric field-effect transistors as chemical sensors, Anal. Bioanal. Chem. 384 (2006) 310–321.

[36] D.J. Macaya, et al., Simple glucose sensors with micromolar sensitivity based on organic electrochemical transistors, Sens. Actuat. B 123 (2007) 374–378.

[37] A. Tsumura, H. Koezuka, T. Ando, Macromolecular electronic device: field-effect transistor with a polythiophene thin film, Appl. Phys. Lett. 49 (1986) 1210.

[38] H. Koezuka, A. Tsumura, T. Ando, Field-effect transistor with polythiophene thin film, Synth. Met. 18 (1987) 699–704.

[39] C.J. Lawrence, The mechanics of spin coating of polymer films, Phys. Fluids 31 (1988) 2786–2795.

[40] Z. Bao, A. Dodabalapur, A.J. Lovinger, Soluble and processable regioregular poly(3-hexyl-thiophene) for thin film field-effect transistor applications with high mobility, Appl. Phys. Lett. 69 (1996) 4108–4110.

[41] M. Kitamura, T. Imada, Y. Arakawa, Organic light-emitting diodes driven by pentacene-based thin-film transistors, Appl. Phys. Lett. 83 (2003) 3410.

[42] J. Takeya, et al., Field-induced charge transport at the surface of pentacene single crystals: a method to study charge dynamics of two-dimensional electron systems in organic crystals, J. Appl. Phys. 94 (2003) 5800.

[43] S. Haas, et al., High-performance dinaphtho-thieno-thiophene single crystal field-effect transistors, Appl. Phys. Lett. 95 (2009) 022111.

[44] H.T. Yi, et al., Vacuum lamination approach to fabrication of high-performance single-crystal organic field-effect transistors, Adv. Mater. 23 (2011) 5807–5811.

[45] G. Decher, J.-D. Hong, Buildup of ultrathin multilayer films by a self-assembly process: I. consecutive adsorption of anionic and cationic bipolar amphiphiles, Makromol. Chem. Macromol. Symp. 46 (1991) 321–327.

[46] Z. Bao, et al., High-performance plastic transistors fabricated by printing techniques, Chem. Mater. 9 (1997) 1299–1301.

[47] J.A. Rogers, et al., Printing process suitable for reel-to-reel production of high-performance organic transistors and circuits, Adv. Mater. 11 (1999) 741–745.

[48] Y. Xia, G.M. Whitesides, Soft lithography, Angew. Chem. Int. Ed. 37 (1998) 550.

[49] D. Li, L.J. Guo, Micron-scale organic thin film transistors with conducting polymer electrodes patterned by polymer inking and stamping, Appl. Phys. Lett. 88 (2006) 063513.

[50] L. Fortuna, et al., On the way to plastic computation, IEEE Circuits Syst. Mag. 8 (3) (2008) 6–18.

[51] S.Y. Chou, P.R. Krauss, P.J. Renstrom, Imprint of sub-25 nm vias and tren-ches in polymers, Appl. Phys. Lett. 67 (21) (1995) 3114.

[52] W. Brütting, et al., Device physics of organic light-emitting diodes based on molecular materials, Org. Electron. 2 (2001) 1–36.

[53] R.H. Fowler, L. Nordheim, Electron emission in intense electric fields, Proc. R. Soc. Lond. 119 (781) (1928) 173–181.

[54] P.E. Burrows, et al., Relationship between electroluminescence and current transport in organic heterojunction light-emitting devices, J. Appl. Phys. Lett. 79 (1996) 7991–8006.

[55] M.A. Lampert, P. Mark, Current Injection in Solids, Academic Press, New York, 1970.

[56] P.N. Murgatroyd, Theory of space-charge-limited current enhanced by Frenkel effect, J. Phys. D Appl. Phys. 3 (1970) 151.

[57] P.M. Borsemberg, D.S. Weiss, Organic Photoreceptors for Imaging Systems, Marcel Dekker, New York, 1993.

[58] J.E. Lilienfeld, Methods and apparatus for controlling electric current. US Patent 1,745,175. Applied on October 8, 1926, granted on January 18, 1930.

[59] D. Kahng, M.M. Atalla, IRE Solid-State Devices Research Conf, Carn-egie Institute of Technology, Pittsburgh, PA, 1960.

[60] G. Moore, Cramming more components onto integrated circuits, Electron. Mag. 38 (1965) 114–117.

[61] D. Voss, Cheap and cheerful circuits, Nature 407 (2000) 442–444.

[62] R. Rotzoll, et al., Radio frequency rectifiers based on organic thin-film transistors, Appl. Phys. Lett. 88 (2006) 123502.

[63] M.F. Mabrook, et al., A pentacene-based organic thin film memory transistor, Appl. Phys. Lett. 94 (2009) 173302.

[64] M. Maccioni, et al., Towards the textile transistor: assembly and characterization of an organic field effect transistor with a cylindrical geometry, Appl. Phys. Lett. 89 (2006) 143515.

[65] J. Mabeck, G. Malliaras, Chemical and biological sensors based on organic thin-film transistors, Anal. Bioanal. Chem. 384 (2005) 343–353.

[66] C. Bartic, G. Borghs, Organic thin-film transistors as transducers for (bio) analytical applications, Anal. Bioanal. Chem. 384 (2006) 354–365.

[67] T.B. Singh, N.S. Sariciftci, Progress in plastic electronics devices, Annu. Rev. Mater. Res. 36 (2006) 199–230.

[68] M.R. Cavallari, et al., PECVD silicon oxynitride as insulator for MDMO-PPV, J Integr. Circ. Syst. 5 (2010) 116–124.

[69] V.R. Zanchin et al., Low voltage organic devices with high-k TiOxNy and PMMA dielectrics for future application on flexible electronics, 26th Symposium on Microelectronics Technology and Devices (SBMicro 2011), 2011, João Pessoa (Paraíba), Microelectronics Technology and Devices—SBMicro 2011, vol. 39, Electrochemical Society Transactions, New Jersey, USA, 2011, pp. 455–460.

[70] H. Klauk, et al., Pentacene organic transistors and ring oscillators on glass and on flexible polymeric substrates, Appl. Phys. Lett. 82 (2003) 4175.

[71] J. Lee, et al., Flexible semitransparent pentacene thin-film transistors with polymer dielectric layers and NiOx electrodes, Appl. Phys. Lett. 87 (2005) 023504.

[72] C.D. Dimitrakopoulos, et al., Low-voltage organic transistors on plastic comprising high-dielectric constant gate insulators, Science 283 (1999) 822–824.

[73] Y.-Y. Lin, et al., Stacked pentacene layer organic thin-film transistors with improved characteristics, IEEE Electron. Dev. Lett. 18 (12) (1997) 606–608.

[74] J.A. Martino, M.A. Pavanello, P.B. Verdonck, Transistor MOS, Electrical Characterization of MOS Technology and Devices (Caracterização elé-trica de tecnologia e dispositivos MOS), Pioneira Thomson Learning, São Paulo, 2003, pp. 75–106.

[75] M.-H. Yoon, et al., Gate dielectric chemical structure-organic field-effect transistor performance correlations for electron, hole, and ambipolar organic semiconductors, J. Am. Chem. Soc 128 (2006) 12851–12869.

[76] A. Bonfiglio, F. Mameli, O. Sanna, A completely flexible organic transistor obtained by a one-mask photolithographic process, Appl. Phys. Lett. 82 (20) (2003) 3550–3552.

[77] L. Fumagalli, et al., Al_2O_3 as gate dielectric for organic transistors: charge transport phenomena in poly-(3-hexylthiophene) based devices, Org. Electron. 9 (2008) 198–208.

[78] D. Natali, L. Fumagalli, M. Sampietro, Modeling of organic thin film transistors: effect of contact resistances, J. Appl. Phys. 101 (2007) 014501.

[79] M.C.J.M. Vissenberg, M. Matters, Theory of the field-effect mobility in amorphous organic transistors, Phys. Rev. B 57 (1999) 12964–12967.

[80] G. Horowitz, M.E. Hajlaoui, R. Hajlaoui, Temperature and gate voltage dependence of hole mobility in polycrystalline oligothiophene thin film transistors, J. Appl. Phys. 87 (2000) 4456.

[81] L. Fumagalli, et al., Dependence of the mobility on charge carrier density and electric field in poly(3-hexylthiophene) based thin film transistors: effect of the molecular weight, J. Appl. Phys. 104 (2008) 084513.

[82] J. Zaumseil, K.W. Baldwin, J.A. Rogers, Contact resistance in organic transistors that use source and drain electrodes formed by soft contact lamination, J. Appl. Phys. 93 (2003) 61176124.

[83] H. Sirringhaus, et al., Two-dimensional charge transport in self-organized, high-mobility conjugated polymers, Nature 401 (1999) 685–688.

[84] I. Osaka, R.D. Mccullough, Advances in molecular design and synthesis of regioregular polythiophenes, Acc. Chem. Res. 41 (2008) 1202–1214.

[85] M.R. Cavallari, et al., Methodology of semiconductor selection for polymer thin-transistors based on charge carrier mobility, J. Mater. Sci. Eng. Adv. Technol. 4 (2011) 33–56.

[86] C.A. Amorim, et al., Determination of carrier mobility in MEH-PPV thin-films by stationary and transient current techniques, J. Non-Cryst. Sol. 358 (2012) 484–491.

[87] M.R. Cavallari, et al., Time-of-flight technique limits of applicability for thin-films of conjugated polymers. 2011 MRS Fall Meeting, 2012, Boston (USA). MRS Proceedings—Symposium U Charge Generation/Transport in Organic Semiconductor Materials, Cambridge University Press, Cambridge (UK), pp. 1402–1403.

[88] J.G. Laquindanum, H.E. Katz, A.J. Lovinger, Synthesis, morphology, and field-effect mobility of anthradithiophenes, Am. Chem. Soc. 120 (1998) 664–672.

[89] S.R. Saudari, P.R. Frail, C.R. Kagan, Ambipolar transport in solution-deposited pentacene transistors enhanced by molecular engineering of device contacts, Appl. Phys. Lett. 95 (2009) 023301.

[90] H. Fuchigami, A. Tsumura, H. Koezuka, Polythienylenevinylene thin-film transistor with high carrier mobility, Appl. Phys. Lett. 63 (1993) 1372.

[91] J.H. Schon, B. Batlogg, Modeling of the temperature dependence of the field-effect mobility in thin film devices of conjugated oligomers, Appl. Phys. Lett. 74 (2) (1999) 260–262.

[92] W.R. Salaneck, S. Stafström, J.L. Brédas, Conjugated Polymer Surfaces and Interfaces, Cambridge University Press, Cambridge, 1996, pp. 50–71 (Chapters 4 and 5).

[93] L.-L. Chua, et al., General observation of n-type field-effect behaviour in organic semiconductors, Nature 434 (2005) 194–199.

[94] H. Ohnuki, et al., Effects of interfacial modification on the performance of an organic transistor based on TCNQ LB films, Thin Solid Films 516 (2008) 2747–2752.

[95] N. Benson, et al., Electronic states at the dielectric/semiconductor interface in organic field effect transistors, Phys. Stat. Sol. (a) 205 (3) (2008) 475–487.

[96] A.C. Arias, et al., All jet-printed polymer thin-film transistor active-matrix backplanes, Appl. Phys. Lett. 85 (2004) 3304.

[97] Y. Wu, et al., Controlled orientation of liquid-crystalline polythiophene semiconductors for high-performance organic thin-film transistors, Appl. Phys. Lett. 86 (2005) 142102.

[98] M.L. Chabinyc, et al., Effects of the surface roughness of plastic-compatible inorganic dielectrics on polymeric thin film transistors, Appl. Phys. Lett. 90 (2007) 233508.

[99] N.C. Cates, et al., Tuning the properties of polymer bulk heterojunction solar cells by adjusting fullerene size to control intercalation, Nano Lett. 9 (2009) 4153–4157.

[100] J. Rivnay, et al., Structural origin of gap states in semicrystalline polymers and the implications for charge transport, Phys. Rev. B 83 (2011) 121306.

[101] A.L. Briseno, et al., n-Channel polymer thin film transistors with long-term air-stability and durability and their use in complementary inverters, J. Mater. Chem. 21 (2011) 16461–16466.

[102] K.-J. Baeg, et al., High speeds complementary integrated circuits fabricated with all-printed polymeric semiconductors, J. Polym. Sci. B Polym. Phys. 49 (2011) 62–67.

[103] J. Rivnay, et al., Drastic control of texture in a high performance n-type polymeric semiconductor and implications for charge transport, Macromolecules 44 (2011) 5246–5255.

[104] Z. Chen, et al., Naphthalenedicarboximide- vs perylenedicarboximide-based copolymers. Synthesis and semiconducting properties in bottom-gate N-channel organic transistors, J. Am. Chem. Soc. 131 (1) (2009) 8–9.

[105] T.-J. Ha, et al., Charge transport and density of trap states in balanced high mobility ambipolar organic thin-film transistors, Org. Electron. 13 (2012) 136–141.

[106] B. Lee, et al., Origin of the bias stress instability in single-crystal organic field-effect transistors, Phys. Rev. B 82 (2010) 085302.

[107] Z. Liang, et al., Soluble and stable *N*-heteropentacenes with high field-effect mobility, Adv. Mater. 23 (2011) 1535–1539.

[108] J. Takeya, et al., Very high-mobility organic single-crystal transistors with in-crystal conduction channel, Appl. Phys. Lett. 90 (2007) 102120.

[109] W. Xie, C.D. Frisbie, Organic electrical double layer transistors based on rubrene single crystals: examining transport at high surface charge densities above 10^{13} cm^{-2}, J. Phys. Chem. C 115 (2011) 14360–14368.

[110] H. Chen, et al., Synthesis and characterization of novel semiconductors based on thieno[3,2-*b*][1]benzothiophene cores and their applications in the organic thin-film transistors, J. Phys. Chem. C 115 (2011) 23984–23991.

[111] K. Nakayama, et al., Patternable solution-crystallized organic transistors with high charge carrier mobility, Adv. Mater. 23 (2011) 1626–1629.

[112] M.A. Mccarthy, et al., Low-voltage, low-power, organic light-emitting transistors for active matrix displays, Science 332 (2011) 570–573.

[113] H. Minemawari, et al., Inkjet printing of single-crystal films, Nature 475 (2011) 364–367.

[114] L. Ma, et al., High performance polythiophene thin-film transistors doped with very small amounts of an electron acceptor, Appl. Phys. Lett. 92 (2008) 063310.

[115] A.S. Molinari, et al., High electron mobility in vacuum and ambient for PDIF-CN2 single-crystal transistors, J. Am. Chem. Soc. 131 (2009) 2462–2463.

[116] S. Ono, et al., High-performance n-type organic field-effect transistors with ionic liquid gates, Appl. Phys. Lett. 97 (2010) 143307.

[117] G. Eda, G. Fanchini, M. Chhowalla, Large-area ultrathin films of reduced graphene oxide as a transparent and flexible electronic material, Nat. Nanotech. 3 (2008) 270–274.

[118] H. Ohnuki, et al., Effects of interfacial modification on the performance of an organic transistor based on TCNQ LB films, Thin Solid Films 516 (2008) 2747–2752.

[119] Cytop® amorphous fluoropolymer—technical information, Agc Chemicals Europe, Reino Unido, http://www.agcce.com/CYTOP/TechInfo.asp

[120] J.E. Anthony, Functionalized acenes and heteroacenes for organic electronics, Chem. Rev. 106 (2006) 5028–5048.

[121] M.L. Tang, et al., Ambipolar, high performance, acene-based organic thin film transistors, J. Am. Chem. Soc. 130 (2008) 6064–6065.

[122] A.R. Murphy, J.M.J. Frechet, Organic semiconducting oligomers for use in thin film transistors, Chem. Rev. 107 (2007) 1066–1096.

[123] S.C.B. Mannsfeld, et al., Highly efficient patterning of organic single-crystal transistors from the solution phase, Adv. Mater. 20 (2008) 4044–4048.

[124] C. Tanase, et al., Optimization of the charge transport in poly(phenylenevinylene) derivatives by processing and chemical modification, J. Appl. Phys. 97 (2005) 123703.

[125] O. Acton, et al., π-σ-Phosphonic acid organic monolayer/sol-gel hafnium oxide hybrid dielectrics for low-voltage organic transistors, Adv. Mater. 20 (2008) 3697–3701.

[126] M.E. Roberts, et al., Water-stable organic transistors and their application in chemical and biological sensors, Proc. Natl. Acad. Sci. USA 105 (2008) 12134–12139.

[127] C. Yang, et al., Low-voltage organic transistors on a polymer substrate with an aluminum foil gate fabricated by a laminating and electropolishing process, Appl. Phys. Lett. 89 (2006) 153508.

[128] T. Someya, et al., A large-area, flexible pressure sensor matrix with organic field-effect transistors for artificial skin applications, Proc. Natl. Acad. Sci. USA 101 (2004) 9966–9970.

[129] C.-a. Di, et al., Interface engineering: an effective approach toward high-performance organic field-effect transistors, Acc. Chem. Res. 42 (10) (2009) 1573–1583.

[130] J. Locklin, Z. Bao, Effect of morphology on organic thin film transistor sensors, Anal. Bioanal. Chem. 384 (2006) 336–342.

[131] M.M. Ling, Z.N. Bao, Thin film deposition, patterning, and printing in organic thin film transistors, Chem. Mater. 16 (2004) 4824–4840.

[132] L. Torsi, et al., Multi-parameter gas sensors based on organic thin-film-transistors, Sens. Actuat. B 67 (2000) 312–316.

[133] Z.-T. Zhu, et al., Humidity sensors based on pentacene thin-film transistors, Appl. Phys. Lett. 81 (2002) 4643–4645.

[134] D.M. Taylor, et al., Effect of oxygen on the electrical characteristics of field effect transistors formed from electrochemically deposited films of poly(3-methylthiophene), J. Phys. D Appl. Phys. 24 (1991) 2032–2038.

[135] R.B. Chaabane, et al., Influence of ambient atmosphere on the electrical properties of organic thin film transistors, Mater. Sci. Eng. C 26 (2006) 514–518.

[136] B. Crone, et al., Electronic sensing of vapors with organic transistors, Appl. Phys. Lett. 78 (2001) 2229–2231.

[137] B.K. Crone, et al., Organic oscillator and adaptive amplifier circuits for chemical vapour sensing, J. Appl. Phys. 91 (2002) 10140–10146.

[138] J.B. Chang, et al., Printable polythiophene gas sensor array for low-cost electronic noses, J. Appl. Phys. 100 (2006) 014506.

[139] J.W. Jeong, et al., The response characteristics of a gas sensor based on poly-3-hexylithiophene thin-film transistors, Sens. Actuat. B Chem. 146 (2010) 40–45.

[140] K.-H. Kim, S.A. Jahan, E. Kabir, A review of breath analysis for diagnosis of human health, TrAC 33 (2012) 1–8.

[141] P. Bergveld, Development, operation, and application of the ion-sensitive field-effect transistor as a tool for electrophysiology, IEEE Trans. Biomed. Eng. 19 (1972) 342–351.

[142] C. Bartic, et al., Organic-based transducer for low-cost charge detection in aqueous media, IEDM Tech. Digest (2000) 411–414, doi: 10.1109/IEDM.2000.904343.

[143] C. Gao, et al., A disposable polymer field effect transistor (FET) for pH measurement, Proc. of IEEE Transducers 2 (2003) 1172–1175.

[144] M.J. Spijkman, et al., Dual-gate organic field-effect transistors as potentiometric sensors in aqueous solution, Adv. Funct. Mater. 20 (2010) 898–905.

[145] N.T. Kalyani, S.J. Dhoble, Organic light emitting diodes: energy saving lighting technology—a review, Renew. Sustainable Energy Rev. 16 (2012) 2696–2723.

[146] Burroughes, et al., Light-emitting diodes based on conjugated polymers, Nature 347 (1990) 539.

[147] W. Brütting, S. Berleb, A.G. Mückl, Devices physics of organic light-emitting diodes based on molecular materials, Org. Electron. 2 (2001) 1–36.

[148] L.S. Hung, C.H. Chen, Recent progress of molecular organic electroluminescent for materials and devices, Mater. Sci. Eng.: R: Reports 39 (2002) 143–222.

[149] D.D.C. Bradley, et al., Polymer light emission: control of properties through chemical structure and morphology, Opt. Mater. 9 (1998) 1–11.

[150] S.I. Ahn, et al., OLED with a controlled molecular weight of the PVK (poly(9-ninylcarbazole) formed by a reative ink-jet process, Org. Electron. 13 (2012) 980–984.

[151] B. Geffroy, P. Roy, C. Prat, Organic light-emitting diode (OLED) technology: materials, devices and displays technologies, Polym. Int. 55 (2006) 572–582.

[152] G. Santos, Study of organic light emitting devices employing complexes of rare earth and transition metals (Estudo de dispositivos orgânicos emissores de luz empregando complexos de terras raras e de metais de transição), PhD Thesis, Politechnical University of Sao Paulo (Escola Politécnica da Universidade de São Paulo), 2008. http://www.teses.usp.br/teses/disponiveis/3/3140/tde-16042009-144608/

[153] A. Nesterov, et al., Simulation study of the influence of polymer modified anodes on organic LED performance, Synth. Met. 130 (2002) 165–175.

[154] V.I. Adamovich, et al., New charge-carrier blocking materials for high efficiency OLEDs, Org. Electron. 4 (2003) 77–87.

[155] G. Santos, et al., Organic light emitting diodes with europium (III) emissive layers based on b-diketonate complexes: the influence of the central ligand, J. Non-Cryst. Sol. 354 (2008) 2897–2900.

[156] S.R. Forrest, D.D.C. Bradley, M.E. Thompson, Measuring the efficiency of organic light emitting devices, Adv. Mat. 15 (2003) 1043–1048.

[157] G. Santos, et al., Opto-electical properties of single layer flexible electroluminescence device with ruthenium complex, J. Non-Cryst. Sol. 354 (2008) 2571–2574.

[158] Luminous efficiency graphic, http://www.cvrl.org/cie.htm

[159] Diffuse reflectors—essentials of reflections and table of contents, Gigahertz-optik, p. 123. http://www.gigahertz-optik.de/pdf/catalogue/Diffuse_Reflectors.pdf

3

Electrochemical Synthesis of Nanostructured Materials

F. Trivinho-Strixino*, J.S. Santos*, M. Souza Sikora**

*FEDERAL UNIVERSITY OF SÃO CARLOS, CENTER FOR SCIENCES AND TECHNOLOGY FOR SUSTAINABILITY, SOROCABA, SÃO PAULO, BRAZIL; **FEDERAL TECHNOLOGICAL UNIVERSITY OF PARANÁ, PATO BRANCO, PARANÁ, BRAZIL

CHAPTER OUTLINE

3.1 Introduction

Nanotechnology associated with the design of surfaces for fabricating nanostructures and new materials has caught the attention of many researchers and has become a topic of intense scientific interest. The great progress observed in this area is a direct result of the modern advances in device miniaturization and the development of specialized instrumentation, which make it possible to visualize objects and surfaces with nanometric resolution. As a convention, nanostructured materials are those possessing structures smaller than 100 nm in any of their dimensions, including particle size, grain size, and layer width and thickness [1,2]. From a technological viewpoint, so–called *nanomaterials* have greater application potential and versatility than conventional materials, with micrometric and larger sizes. This is a consequence of a series of modifications in their physicochemical, mechanical, optical, and electronic properties, resulting from the miniaturization process [1–4], and has raised great interest in the study and synthesis of these types of materials in recent years.

A large number of sophisticated nanomaterials have been fabricated on various types of surfaces using electrochemical techniques. The electrochemical method presents a series of advantages with respect to traditional chemical methods: ease of fabrication, low cost, homogeneity in the samples, control over the material properties, and good reproducibility. This method allows the fabrication of metallic [4] and bimetallic [5–8] nanomaterials and nanostructures; alloys [9,10], metallic oxides [11–13], etc. The main applications for these materials are in photocatalysis [14], electrocatalysis for fuel cells [15,16], photoelectric [17] and photovoltaic [18] devices, heterogeneous catalysis [16,19], thermoelectric energy converters [20], electrodes for batteries [21], and sensors [22–24].

Several experimental techniques may be employed for electrochemical synthesis, but theoretically, this process is based on the same classical foundations associated with reactions in conductor electrodes immersed in solutions containing ions [25]. It can also be performed in all types of materials (metals, oxides, and polymers), given that the "fuel" normally employed for the synthesis is the electron, which can reduce or oxidize atoms, ions, or molecules present in the solution or placed on a conductor substrate (electrode). This substrate may be part of the synthesis, or support the synthesized material. In the latter case, the synthesized material can be separated from substrate in a second stage, and it can be used for other applications of interest.

The great challenge for someone who wishes to perform electrochemical synthesis is mastering all the fundamental aspects of electrochemistry. Thus, it is imperative that whoever wants to benefit from the advances of this technique have a solid knowledge of the foundations of electrochemistry and materials science. The goal of this chapter is to survey the field of electrochemistry with a focus on the production of nanoscale metals and oxides. We do not aim to cover all of the fundamental aspects of electrochemistry and the disciplines that derive from it. Therefore, the reader eager to achieve a deeper understanding on the topic should consult further references. Detailed discussions can be found in the publications cited in this chapter, as well as in textbooks on electrochemistry and materials science.

The reader will find the content of this chapter divided into the fundamental aspects of electrochemistry, the electrodeposition of metals, and anodic oxidation. In the specific content of electrochemical synthesis of anodic oxides the focus is the synthesis of aluminum and titanium anodic oxides with high morphological control at the nanoscale, and the manipulation of the properties of the materials formed.

3.2 Fundamental Aspects of Electrochemistry

3.2.1 Faradaic and Non-Faradaic Processes

Two types of electrochemical processes can occur in a conductor substrate submerged in a solution containing ions. One of them comprises charge/electronic transfer reactions through the surface between the conductor substrate and the electrolyte. In this case, the transfer of electrons through the interface promotes oxidation or reduction processes, known as *faradaic* processes. By contrast, under certain boundary conditions, this same interface may contain a region of potentials where no redox reactions take place. Consequently, in this region, it may occur physical adsorption processes that modify the chemical molecular structure of the electrode/electrolyte interface, and which depend on the applied potential and the composition of the electrolytic solution. These processes are described as *non-faradaic.*

In faradaic processes, Faraday's law states that the amount of product formed or reagent consumed by an electric current is stoichiometrically equivalent to the number of electrons provided. For example, in the case of a positive electric current (electrons removed from the electrode) going through an aluminum metallic plate submerged in an acid aqueous solution, electrons flow through that interface in the opposite direction to the applied current. Simultaneously, aluminum ions are transferred to the acid electrolyte, according to the reactions:

$$2\,Al_{(s)} + 3\,H_2O_{(l)} \rightarrow Al_2O_{3(s)} + 6\,H^+_{(aq.)} + 6e^- \tag{3.1}$$

$$Al_2O_{3(s)} + 3\,H^+_{(aq.)} \rightarrow 2\,Al^{3+}_{(aq.)} + 3\,OH^-_{(aq.)} \tag{3.2}$$

If the solution is composed of an acid that does not dissolve the oxide formed on metallic surface, only the reaction described by Eq. 3.1 will be favored. According to Faraday, the magnitude of the effect of an electrochemical reaction is the same in both the metallic surface and the electrolytic solution, and it is a function only of the amount of electricity crossing the system. This means that the amount (in grams) of species oxidized or reduced on the electrode surface is proportional to the number (in moles) of electrons involved. The charge corresponding to 1 mol of electrons is denominated 1 Faraday (F), and mathematically it corresponds to:

$$F = Ne \tag{3.3}$$

where F is Faraday's constant, N is Avogadro's number, and e corresponds to the elementary charge. Substituting these values into Eq. 3.3, we obtain a good approximation for the value of $F = 96{,}485$ C per mol of electrons. Therefore, in an electrodic process kept at constant current I, the variation in the theoretical mass of the oxidized or reduced material, dm, can be calculated as a function of the charge variation in the process, dQ, according to the following equation:

$$\left(\frac{dm}{dQ} \right)_I = \frac{MM}{z \cdot F} \tag{3.4}$$

where MM is the molecular mass of the species being oxidized or reduced, z is the amount (in moles) of electrons involved in the process, and F is the Faraday constant. Thus, it is

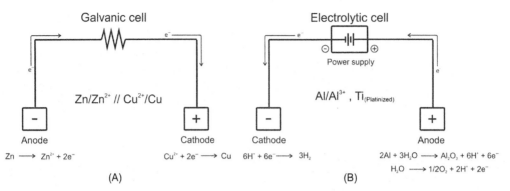

FIGURE 3.1 Schematic representation of (A) galvanic and (B) electrolytic cells. *Adapted from F. Trivinho-Strixino, J.S. Santos, M.S. Sikora, Síntese Eletroquímica de Materiais Nanoestruturados [electrochemical synthesis of nanostructured materials], In: A.L. Da Róz, F.L. Leite, et al. (Eds.), Nanoestruturas: Princípios e Aplicações, vol. 1, Elsevier, Rio de Janeiro, 2015, pp. 63–110 (Chapter 3) [26].*

ideally possible to predict the theoretical deposited mass of a metallic ion being reduced in a conductor substrate under faradaic processes.

3.2.2 Electrochemical Cells—Types and Definitions

Electrochemical cells presenting a faradaic current flowing through its electrodes are classified as galvanic or electrolytic cells. In a galvanic cell, spontaneous reactions occur in the electrodes when they are externally connected through a conductor (Fig. 3.1A). These reactions are commonly employed in the conversion of chemical energy to electric energy. By contrast, the chemical reactions in electrolytic cells are affected by the application of an external voltage larger than the open–circuit potential of the cell (Fig. 3.1B).

In general, the term *electrolysis* is given to the process associated with the chemical changes accompanied by faradaic reactions over the electrode of interest, in contact with an electrolytic solution. In electrochemistry, by definition, the reactions that take place at the cathode are reduction reactions, and those occurring at the anode are oxidation reactions. In the former case, the transport of electrons from the electrode interface to the species in solution generates a cathodic current. In the latter case, the transport of electrons from the species in solution to the electrode interface generates an anodic current. In an electrolytic cell, the cathode is negative with respect to the anode, and, in a galvanic cell, the cathode is positive with respect to the anode.

In an electrochemistry experiment, the electrolytic cell is inserted into an electrochemical circuit consisting of a voltage or current source, metallic connectors, an ohmic resistor, an ammeter and a voltmeter, and a switch. Fig. 3.2 shows the schematic of a simple electrochemical system, where S corresponds to the voltage or current source, R is the ohmic resistor, V is the voltmeter, A the ammeter, C is the electrochemical cell, and K is the circuit on/off switch.

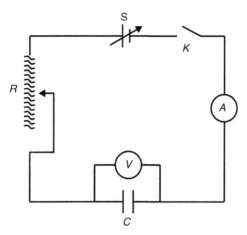

FIGURE 3.2 Electrochemical circuit commonly used in electrochemistry experiments. *Adapted from F. Trivinho-Strixino, J.S. Santos, M.S. Sikora, Síntese Eletroquímica de Materiais Nanoestruturados [electrochemical synthesis of nanostructured materials], In: A.L. Da Róz, F.L. Leite, et al. (Eds.), Nanoestruturas: Princípios e Aplicações, vol. 1, Elsevier, Rio de Janeiro, 2015, pp. 63–110 (Chapter 3) [26].*

A simplified way of visualizing an electrochemistry experiment consists of imagining a system that responds to a given perturbation. When a given excitation function (e.g., a potential step) is applied to the electrochemical cell, a response function to that perturbation is recorded (e.g., the produced current variation as a function of time), keeping all other system variables constant. Thus, the main goal of an electrochemistry experiment is to obtain information (thermodynamic, kinetic, analytic, etc.) from the observation of perturbation and response processes, which may contribute to the construction of appropriate models, and the comprehension of the fundamentals aspects of the investigated system. This allows a better control of the variables influencing the electrochemical synthesis.

In electrochemical experiments, a typical electrochemical cell is formed by a vessel commonly made of glass where the electrolyte is inserted. The electrodes are then immersed in this solution. The cell can be covered by an external and isolated glass wall jacket connected to a thermostatic bath to maintain the electrolyte temperature constant. Other ways of developing the cell and controlling the electrolyte temperature during the process can also be employed. Generally, three types of electrodes are used in this type of experiment: working, reference, and auxiliary electrodes (the last one is also called counter electrode). The working electrode is the conductor substrate where the chemical reaction of interest takes place. The working electrode potential is measured in relation to the reference electrode, which must present a known constant potential with respect to the standard hydrogen electrode within the potential window analyzed. An auxiliary or counter electrode is necessary for the flow of electric and ionic current; it closes the electrical circuit of the system and allows for the passing of a current through the external system. A magnetic stirrer can also be used to keep constant the transport process of species at

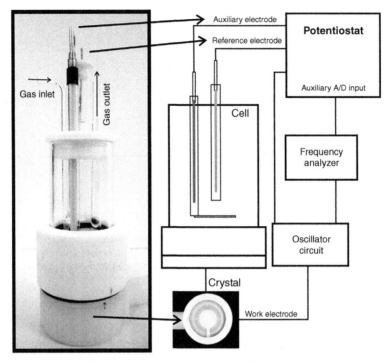

FIGURE 3.3 Experimental configuration of a conventional three-electrode electrochemical cell used in measurements coupled to the electrochemical quartz–crystal microbalance (EQCM). *Adapted from F. Trivinho-Strixino, J.S. Santos, M.S. Sikora, Síntese Eletroquímica de Materiais Nanoestruturados [electrochemical synthesis of nanostructured materials], In: A.L. Da Róz, F.L. Leite, et al. (Eds.), Nanoestruturas: Princípios e Aplicações, vol. 1, Elsevier, Rio de Janeiro, 2015, pp. 63–110 (Chapter 3) [26].*

the electrode/electrolyte interface. A bubbler for N_2 input and a saturator can also be employed for removal of the oxygen dissolved in the electrolyte, and the maintenance of an inert gaseous phase inside the cell. This type of cell configuration is commonly used in electrodeposition reactions, which are reactions where the ions present in the electrolyte are deposited over the substrate surface by means of a reduction reaction. A typical cell used in these types of reactions is shown in Fig. 3.3.

By contrast, in anodization (anodic oxidation) reactions, which are characterized by the growth of an oxide film on the working electrode surface due to oxidation of the metallic substrate, the reference electrode is often not used. In such cases, one considers the potential difference between the working electrode and the counter electrode. This approximation can be taken once the potential difference between the electrodes is very high (exceeding 40 V) and there is no reference electrode that behaves as an ideal unpolarizable electrode, under the experimental conditions in which anodization is usually carried out. However, there are books that describe experimental configurations for adequately measuring the working electrode potential with respect to the reference electrode, in anodization reactions where the potential range allows the use of these electrodes [27].

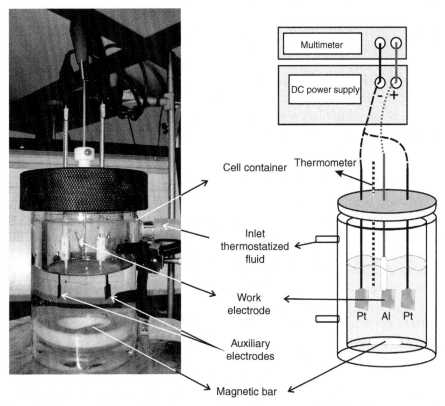

FIGURE 3.4 Schematic representation of a electrochemical cell used for anodization. *Adapted from F. Trivinho-Strixino, J.S. Santos, M.S. Sikora, Síntese Eletroquímica de Materiais Nanoestruturados [electrochemical synthesis of nanostructured materials], In: A.L. Da Róz, F.L. Leite, et al. (Eds.), Nanoestruturas: Princípios e Aplicações, vol. 1, Elsevier, Rio de Janeiro, 2015, pp. 63–110 (Chapter 3) [26].*

Furthermore, the use of two counter electrodes placed parallel and symmetrically by the working electrode to maintain a homogeneous electric field at the faces of the electrode being anodized is common in this type of experiment. Fig. 3.4 shows a picture of this type of cell used for anodization.

The fundamental aspects of these two widely used electrochemical processes for the synthesis of nanostructured materials, namely, electrodeposition and anodic oxidation, are described in the following sections.

3.3 Synthesis of Nanostructured Films by Electrodeposition

Electrochemical reactions can easily be controlled by the potential applied to the electrolytic cell. These reactions take place at the interface between the electrode and the electrolyte and are stimulated by the gradients formed by the chemical and electric potentials of the species involved [28]. One of early models used for describing the electrode/electrolytic interface is the Helmholtz model [28,29]. According to this model, the positive and nega-tive species can be represented as a simple parallel–plate capacitor. Due to the simplistic nature of this model, other models were proposed for description of the charge distribu-tion of the electrical double layer, as Gouy–Chapman and Stern models [28]. The Gouy–Chapman model considers the influence of the electrolyte and the applied potential on the capacitance of the double layer; consequently, the double layer is diffuse with a variable thickness. The Stern model is a combination of the previous models: the double layer is formed by a compact layer close to the electrode, and a diffuse layer that extends into the solution. Fig. 3.5 exhibits a refined model of the metal/electrolyte interface, considering the adsorption of fully and partially solvated ions on metal surface [29].

Fig. 3.5A illustrates the adsorption process of ionic species at the electrode surface. The potential variation as a function of the distance from the electrode surface, located at $x = 0$, is shown in Fig. 3.5B. In this figure, the plane parallel to the electrode located at a distance that passes through the center of the adsorbed ion is called the inner Helmholtz plane, whereas the parallel plane situated at a distance x_2, passing through the center of the fully dissolved ion, is called the outer Helmholtz plane. The distance between the electrode/solution interface at $x = 0$ and the outer Helmholtz plane is called the electrical double layer and is represented schematically in the electrochemical cell of Fig. 3.5C.

During the electrodeposition, the working electrode is negatively polarized with respect to the reference electrode, so the positively charged metallic ions present in the electrolytic solution are attracted toward the metallic substrate and reduced to their metallic form at the electrode surface. The result is the formation of a compact film, dense and strongly adhered to the substrate, whose thickness can be monitored by the amount of charged consumed.

To control the process of electrodeposit fabrication at the nanoscale, it is necessary to know the fundamentals of the processes of nucleation, film growth, and kinetics of elec-trochemical reactions. These processes are influenced by a series of factors, such as the electrochemical technique used, the type of substrate, temperature, composition, and concentration of the electrolyte.

The first stage of metallic film formation is the nucleation of the metal, which tends to occur preferentially at regions of electrode surface containing defects and imperfections, such as grain contour regions, holes, inclusions, oxide layers, and adsorbed molecules [3]. Consequently, the presence of adsorbed impurities and the crystallographic structure and orientation of the substrate tend to influence the nucleation process, altering the char-acteristics of the formed film. Therefore, an adequate choice of substrate where the film will be deposited is of great importance. Among the materials most commonly used as working electrodes in electrodeposition are platinum [30–32], gold [33,34], copper [35,36], stainless steel [37–39], vitreous carbon [39–41], ITO [42,43], and FTO [44]. These materials exhibit chemical stability, good resistance to corrosion, ease to clean up surface, and pos-sible functionalization of the surface for various applications.

The electrochemical reaction can be controlled through four types of mechanisms: charge-transfer reaction, diffusion, chemical reaction, and crystallization [45]. The

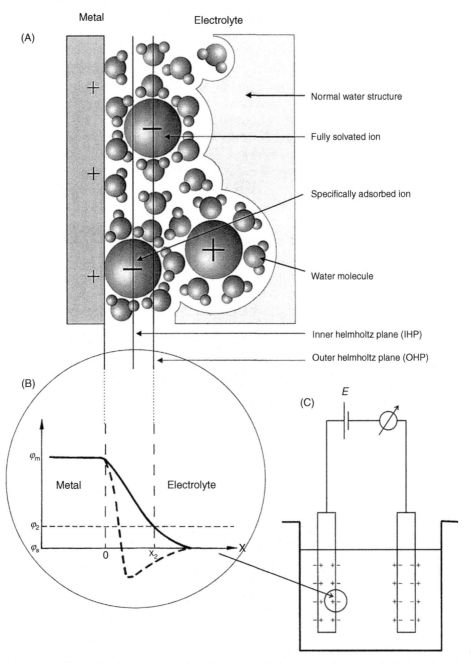

FIGURE 3.5 The metal/electrolyte interface: (A) Adsorption process of the ionic species at the electrode surface. (B) Potential variation, φ, as a function of the distance, X. (C) Schematic representation of the electrical double layer in the electrochemical cell. [29]

charge-transfer reaction involves the transfer of charge carriers, such as ions and electrons, through the electrical double layer. In the case of diffusion, the species consumed or formed during the reaction are transported from the solution to the substrate interface and vice versa. Chemical reactions correspond to homogeneous reactions in solution or heterogeneous reactions at the electrode surface and do not involve charge transfer. Thus, they are not affected by the applied potential. In crystallization, atoms are incorporated into or removed from the crystal lattice of the metal being formed.

To select an electrolyte, one must consider the composition of the electrolytic bath, the pH, and the presence of additives since these factors can strongly affect the structure of the electrolyte/substrate interface, the charge, and mass transfer kinetics, as well as the kinetics of secondary chemical reactions that may occur simultaneously with the process. During the electrodeposition of certain metals, such as, cobalt, nickel, and iron, acid solutions containing sulfur and chlorine ions are commonly utilized to avoid passivation of the metal [46–48]. One example of additive use is during the electrodeposition of Co and Ni films and their alloys, which are normally carried out in an acid saline solution containing the metallic ion. Among the additives most used in the electrolyte in this process are boric acid and saccharin. In the former case, boric acid is used as a buffer during the electrodeposition of cobalt and nickel films. This prevents pH variation near the electrode surface and inhibits the formation of hydroxylated species that can precipitate together with the metallic film, forming a porous structure [49,50] and making the material unusable for technological applications. On the other hand, saccharin is used as an additive during electrodeposition of the Ni–Co alloy to decrease the internal tension and to reduce the grain size of the electrodeposits [48]. In the study performed by Mascaro et al. [44], glycerol was used as an additive to improve the crystallinity and to reduce the number of defects on Cu, In, Ga, and Se films deposited on the FTO substrate in acid medium.

The nucleation and crystallization processes are represented in Fig. 3.6. Nucleation (Fig. 3.6A) is a process characterized by the following stages: (i) diffusion of the ions onto the electrolyte/substrate interface, (ii) partial desolvation, (iii) adsorption in the substrate surface, and (iv) complete desolvation and nucleation [51]. Fig. 3.6B displays a possible ion attachment sites during the crystallization stage on a metallic surface with crystalline plane dislocation. In this case, when the ion at position (a) approaches the substrate, it can be incorporated into the following sites: on the crystal lattice, as an ad-ion (b), on a step (c), or in a corner (d) [51].

Another important variable that affects the electrodeposition process and the electrodeposited film properties is the electrolyte temperature, which interferes with the reaction speed, the diffusion processes, and the viscosity of the medium. However, it should be emphasized that in contrast with other methods used to fabricate nanostructured materials, electrochemical synthesis does not require high temperatures. Nanostructured films obtained by electrodeposition are generally produced at room temperature, although records can be found in the literature where such materials have been synthesized at temperatures ranging from 20–90°C [9,52,53].

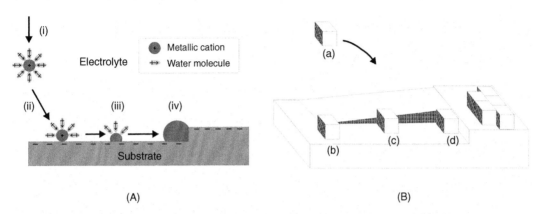

FIGURE 3.6 Schematic representation of the (A) nucleation and (B) crystallization processes on a surface with crystal plane dislocation. *Adapted from F. Trivinho-Strixino, J.S. Santos, M.S. Sikora, Síntese Eletroquímica de Materiais Nanoestruturados [electrochemical synthesis of nanostructured materials], In: A.L. Da Róz, F.L. Leite, et al. (Eds.), Nanoestruturas: Princípios e Aplicações, vol. 1, Elsevier, Rio de Janeiro, 2015, pp. 63–110 (Chapter 3) [26].*

Among the electrochemical techniques commonly used for electrodeposition are cyclic voltammetry, chronoamperometry, chronopotentiometry, and pulsed methods [45,54], which are chosen according to the type and properties of the desired material. Cyclic voltammetry is a potentiodynamic method, which involves the simultaneous variation of voltage and current. It consists of a triangular potential scanning within a specific interval and at a constant speed, while the current is monitored. Chronoamperometry is a potentiostatic method in which the voltage is kept constant, then the reactions tend to be limited by the applied voltage, and the observed response is the current variation as a function of time. Chronopotentiometry is a galvanostatic method in which the applied current density is fixed and the potential variation is observed during the process. Pulsed methods involve the application of current or voltage pulses over a specific time interval. The frequency of pulses can be controlled.

The film nucleation processes and growth kinetics will change depending on the technique used, leading to the formation of nanostructured electrodeposits with distinct properties and characteristics. Notably, these techniques do not provide morphological details of the film surface area and thickness, which can be monitored by complementary analysis techniques, such as, scanning tunneling microscopy (STM) and atomic force microscopy (AFM) [45,54]. As an example, Fig. 3.7A–B present STM and AFM images, respectively, of PbS films obtained by potentiostatic techniques under different experimental conditions. The corresponding rugosity variations in certain sample regions can be observed.

The electrochemical quartz-crystal microbalance (EQCM) is another technique that can be coupled with electrochemical techniques. This is an in situ method that provides information such as mass variation during the experiment, obtained from changes in oscillation frequency of the Pt- or Au-coated quartz crystal, generally used as a working

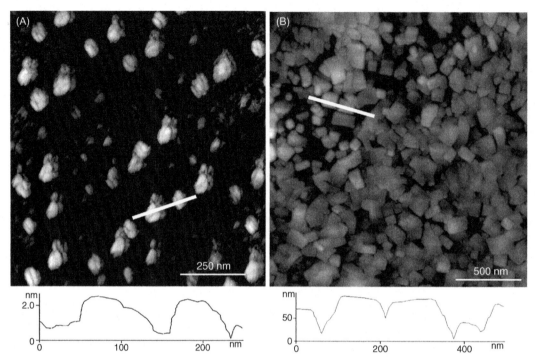

FIGURE 3.7 Scanning tunneling microscopy (STM) (A) and atomic force microscopy (AFM) (B) images of the PbS surface on Au, obtained through potentiostatic deposition under the following conditions: (A) 6 min at 30°C and (B) 30 min at 90°C. The rugosity profiles corresponding to the film sections marked in white are shown at the bottom of each image [9].

electrode in this system [46,49]. Fig. 3.8 shows the results of cyclic voltammetry, illustrating the Co deposition process, followed by its oxidation and the corresponding mass variation on the electrode, as a function of the potential during experiments carried out at different pH values.

In addition to the formation of single-constituent metallic films, the electrodeposition technique also facilitates the formation of nanostructured films containing more than one type of metal, such as alloys [9,10,20,48,55] and multilayered films [15,16,56–58]. An alloy consists of a mixture of different compounds miscible in various proportions, which can be classified as a solid solution. On the other hand, the multilayered films correspond to a series of thin layers with thicknesses of a few nanometers, constituted by different metals. Due to small sizes of the alternating metallic layers, the resultant material acquires interesting properties that would not be observed if the same material were synthesized in larger dimensions [15,57]. Fig. 3.9 presents a cross-sectional image of a multilayered film obtained by transmission electron microscopy (TEM). The multiple layers are obtained through repetition of 10-nm-thick Cu and Ni nanolayers.

In the case of alloys, the electrolytic bath is composed of different salts containing the metallic ions that will be simultaneously deposited. As an example, in the codeposition

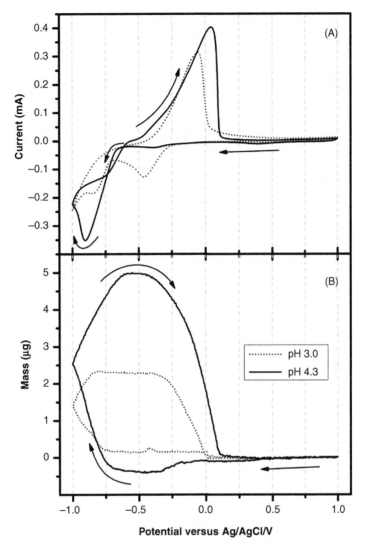

FIGURE 3.8 (A) Cyclic voltammogram obtained in a solution containing 0.01 mol L^{-1} CoSO$_4$ + 0.2 mol L^{-1} Na$_2$SO$_4$ and (B) mass variation as a function of potential. Scanning rate = 50 mV s^{-1}. *Adapted from F. Trivinho-Strixino, J.S. Santos, M.S. Sikora, Síntese Eletroquímica de Materiais Nanoestruturados [electrochemical synthesis of nanostructured materials], In: A.L. Da Róz, F.L. Leite, et al. (Eds.), Nanoestruturas: Princípios e Aplicações, vol. 1, Elsevier, Rio de Janeiro, 2015, pp. 63–110 (Chapter 3) [26].*

process of nickel and cobalt ions for the formation of a Ni–Co alloy, Qiao et al. [48] used a solution containing salts of cobalt sulfate and nickel chloride, and boric acid as an additive. The final composition of the alloy is controlled by the ratio of the metallic ions in solution and the applied current. Comparing the alloys to their constituent materials in pure form, the alloys present certain properties that their individual constituents do not. Furthermore, an improvement in the mechanical, optical, and magnetic properties can be

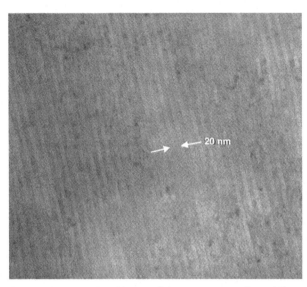

FIGURE 3.9 Transmission electron microscopy (TEM) image of the cross section of a multilayered Cu and Ni film; each layer is 10 nm thick [58] .

observed in the mixed material. Notably, it is not necessary that all constituents of an alloy be metals. Significant amounts of nonmetallic elements are also used in certain types of nanostructured alloys, such as Bi_2Se_3 [20], CoNiP [55], PbS [9], and TiN [59] which are used in thermoelectric materials, magnetic devices, and optoelectronic devices.

In contrast with the process of alloy production, during the formation of multilayered films by electrodeposition, different cations that will form each layer are deposited separately, such that either a single electrolytic bath or two different and subsequent electrolytes can be used. In the former case, each layer is formed in the same electrolyte by controlling the applied voltage. This method can be used when the difference between the reduction potentials of the ionic species is large, and the electrolytes are similar. For example, nickel and copper can be deposited in acidic solutions containing both sulfide and sulfamate ions, and even then, the difference between their electrode potentials is 0.5 V. In the latter case, the layers are formed in different electrolytes, with the working electrode being removed from one solution after the formation of the first layer, and subsequent transferred to another electrolyte, where the deposition of a new layer will take place on top of the first one. This method is used when the reduction potentials of the different ions are very similar, and simultaneous deposition may occur. However, this approach may become impractical if the number of layers deposited is large.

Independently of the electrodeposition method, most multilayered films are formed by metallic compounds; they display a crystalline structure and preferential crystallographic orientation [56]. The first works about electrochemical synthesis of multilayers in the literature describe the magnetic properties of the systems Cu-Ni, Co-P, Co-W, and Co-Pt [56]. Recently, there has been great interest in the synthesis of these types of materials for applications in fuel cells, using the systems Pt/Ir/Pt [15], Pt/Rh/Pt [16], and Pt/Bi/Pt [57] as substrate for the oxidation of small organic molecules, such as methanol. In comparison with traditional Pt electrodes, which are normally used for this type of application, the electrooxidation of methanol is much faster in the Pt/Ir/Pt multilayered films than for Pt electrodes, although both systems present the same electroactive area. This indicates a significant increase in the catalytic process efficiency [15]. Another interesting characteristic of this method is its great economy in the amount of reagent used to achieve a desired property. In the case of Pt, a significantly expensive noble metal, the use of this multilayer film offers great economic advantages.

In addition to metallic films, it is also possible to obtain metallic oxides by electrodeposition of the metal onto the substrate, followed by its oxidation. A negative potential is initially applied for the reduction of the ions in the solution; then, the polarization is inverted, applying a positive potential for the oxidation of the material previously deposited on the surface. An example of this method is the formation of zinc oxide on gold electrodes, obtained from the prior reduction of the Zn^{4+} ions present in a solution of $ZnCl_2$ and KCl, and followed by the oxidation of metallic Zn into ZnO [60]. However, the most common form of electrochemical synthesis of oxides is the oxidation of the working electrode itself. This method, known as anodization, will be described in the following section.

3.4 Oxide Formation by Anodization of Valve Metals

The term "valve metal" has been used for a long time, but eventually, some confusion related to the nomenclature emerged in the literature because of the types of diodes used as "thermionic valves." In fact, there was technological interest in developing alternatives based on semiconductor oxides for the already-discovered silicon-oxide-based devices. However, valve metal oxides exhibited high densities of structural defects that discouraged their application for electronic properties similar to those of silicon oxide. Therefore, it is more accurate to state that the nomenclature "valve metals" comes from their characteristic property as a group: the tendency to allow the passing of current in only one direction of the polarization. This is analogous to a hydraulic valve, whose flow can go only in one direction.

Some metals, including Ta, Nb, Al, Zr, Ti, Hf, Bi, Sb, and W, can be classified as valve metals: when they are cathodically polarized in the appropriate electrolytic solution, the current flows easily, whereas when it is polarized anodically, the current falls until it reaches a stationary state value [61]. Its surface is always covered by a thin oxide layer with a thickness of only a few nanometers. This layer is denominated native oxide, given that it forms spontaneously when the metal is exposed to a medium containing O_2 or H_2O [62]. Other metals such as Be, Mg, Si, Ge, Ti, and U also form an insulating oxide layer under specific boundary conditions when anodically polarized, displaying very similar characteristics to those observed in valve metals. These metals present a tendency to form a highly resistive oxide film that protects the substrate when anodically polarized. The current decay during anodic polarization is due to the formation and growth of these oxides; this process is called anodic oxidization or anodization.

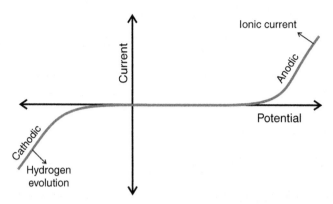

FIGURE 3.10 Current versus voltage profile for a valve metal on which an oxide layer has already formed over the electrode. *Adapted from F. Trivinho-Strixino, J.S. Santos, M.S. Sikora, Síntese Eletroquímica de Materiais Nanoestruturados [electrochemical synthesis of nanostructured materials], In: A.L. Da Róz, F.L. Leite, et al. (Eds.), Nanoestruturas: Princípios e Aplicações, vol. 1, Elsevier, Rio de Janeiro, 2015, pp. 63–110 (Chapter 3) [26].*

Fig. 3.10 shows a plot of the current as a function of potential for a valve metal with a previously formed oxide layer. The current increases rapidly in the direction of negative potentials, where the processes that allow the passing of current through the electrode/electrolyte interface are mostly related to and limited by the hydrogen evolution as water electrolysis. The behavior in this cathodic region depends on the duration of polarization, the method of preparation, and the nature of the oxide film. By contrast, the current going through the film during anodic polarization is small and cannot be reproduced until the voltage region where the ionic current begins to dominate over the electronic current. After this potential, the ionic current increases rapidly with the potential. Normally, the electronic current causes oxygen generation once the potential reaches values for which this process becomes thermodynamically favorable. Given that in most cases of valve metal anodization, the electric field is sufficiently large (10^6–10^8 V m^{-1}) [63,64] to allow for preferential current passage of an ionic current, this behavior becomes time-dependent, since the thickness of the oxide film is continuously changing.

The characteristics of anodic oxide films are determined by the experimental conditions employed, such as the anodization regime, electrolyte composition and concentration, temperature, presence of dopant salts, microstructure and quality of the metal, previous surface treatment, and stirring of the solution. Regarding the anodization regime, there are two methods that can be employed for oxide production: potentiostatic and galvanostatic.

In the potentiostatic regime, the potential is kept constant during the process, and the observed response is the current variation as a function of time. In this case, the system energy is kept constant such that the processes tend to be limited by the applied potential. The current through the system tends to decrease up to a stationary value. This current decrease is related to the formation of an compact oxide layer over the electrode that increases the local electrical field intensity. Thus, the kinetic energy of the process is not held constant. Eventually, the observed growth rate of the oxide is so small that the film thickness can be considered to have reached a limit value. The total current associated with this stage may or may not be small, given that it corresponds to the sum of the ionic and electronic currents. Therefore, the limit thickness hypothesis is acceptable in most cases. The idea of a limit thickness that is "proportional" to the applied voltage was transformed into a parameter commonly found on synthesis of anodic oxides, expressed in angstroms per volt ($\mathring{A}\,V^{-1}$), and is related to a fully-formed film. This parameter may have practical applications during the experimental parameters selection for the synthesis, however its limitations must be known. Parallel electrochemical reactions that simultaneously occur during the formation of oxides are not considered, such as, oxygen generation and release or the combustion of organic compounds from the electrolyte. These reactions are known to consume charge during the anodization and to alter the kinetics constants of the overall process.

In the galvanostatic method, the oxide is grown by application of a constant current density, and the observed response is the potential as a function of time. In this method, it is possible to monitor oxide growth by observing the different processes occurring during synthesis. The growth rate is kept constant and the system is free to reach a stationary state characterized by a plateau in the potential versus time curve. As mentioned previously, the metallic substrate always presents an inhomogeneous thin film of native oxide, formed as soon as the metal is exposed to an oxygen-rich atmosphere. In this context, the heterogeneities of the oxide film play an important role because the electric field is inversely proportional to the film thickness. In other words, the oxidation reaction of the metal occurs locally, and the number of active areas is proportional to the applied current. Hence, in this constant current method, oxidation processes are more difficult to control and prevent local disruption of the formed oxide layer over the electrode.

The films produced by these two methods present significantly different characteristics. In the potentiostatic regime, the beginning of the process is marked by an increase in the electric current due to the metal oxidation. Nonetheless, as the oxide thickness increases, the current decreases until reaching a stationary value. If the electrolyte used is able to dissolve the oxide, there may be a competition between the oxide formation and dissolution reactions, leading to the formation of a highly ordered structure if a narrow window of specific experimental parameters are employed. This type of anodization is often applied in the synthesis of oxides with highly organized structures, such as porous anodic alumina (PAA), commonly used in templates for nanostructures, and TiO_2 nanotubes, widely utilized in photocatalysis.

By contrast, in the galvanostatic method, the potential grows almost linearly at the beginning of the process, due to the formation of a barrier oxide film. The high electric field (approximately 10^6–$10^8\,V\,m^{-1}$) achieved during the anodic oxidation of these metals exceeds the critical values for electric breakdown in dielectric materials [63]. Therefore, the probability of a dielectric breakdown during the growth of these oxides is high. The dielectric breakdown observed during anodization under high electric field is also known as electrolytic breakdown. Thus, after reaching a critical thickness value, the electrolytic breakdown of the film takes place, characterized by potential oscillations resulting from localized oxide destruction and reconstruction processes [63]. Under those conditions, the formation of electric discharges on the electrode surface due to the intense localized electric field is also observed. If the anodization time is relatively long, the potential fluctuations tend to stabilize, and the potential begins oscillating about a mean potential value, reaching a steady-state in which the rates of oxide film dissolution and reconstruction processes become constant. With this method, it is possible to obtain a porous crystalline material, although with a low pore ordering degree. Consequently, there is little interest in using this material for technological applications requiring a large available surface area, and this type of material is more commonly used as coatings.

A more detailed description of these anodization methods, synthesis conditions, influence of experimental parameters on the oxide film properties, and applications will be presented in the following sections.

3.4.1 Anodization Under High Electric Field

The process for obtaining anodic oxides by anodization under high electric field has received different names throughout time. Currently, the most common terms used are "Hard Anodization," [65] "Spark Anodization," [66,67] "Micro-Arc Oxidation," [68,69] and "Plasma Electrolytic Oxidation" [70,71]. Normally, the term "Hard Anodization" is applied to the anodization of metals under high electric fields without reach the dielectric breakdown potential. This method is widely utilized for obtaining thick anodic oxides with self-organized structures, such as PAA and TiO_2 nanotubes films. By contrast, the terms "Spark Anodization," "Micro-Arc Oxidation," and "Plasma Electrolytic Oxidation" refer to the anodization technique in which the oxide growth reaches the dielectric breakdown potential. In this case, the energy released is dissipated as heat and light that can be observed as electric discharges on the electrode surface. Depending on the emission characteristics and the experimental conditions, the electric discharges are classified as glow, spark, microdischarge, microarc, or arc [70–73]. This is the origin of the different names usually found for this technique.

Fig. 3.11 shows current versus voltage curves corresponding to the different types of electric discharge observed in the anodization process [64]. According to Yerokhin et al. [64], the current versus voltage profile for type-A metals is observed in metals that do not form passive films, whereas the type-B profile is characteristic of metals that undergo passivation, such as, valve metals. Both systems obey Faraday's law at the beginning of the process; however, when the potential increases, a deviation from this behavior can be observed in both systems. The current increase with the potential is accompanied by the formation of electric discharges in the material. Nonetheless, for type-A metals, this current increase is limited by the gas evolution reaction along the surface of the metal, and the electrode is enveloped in plasma. A current drop due to the stabilization of the cloud surrounding the electrode is observed, and the electrical arcs are formed after the system reaches E_3. For type-B metals, the behavior is more complex. Initially, there is formation of the passive film, also known as a barrier film. However, after reaching the potential E_4, if the electrolyte is able to dissolve the oxide being formed, the dissolution process begins to compete with the oxide growth reaction, resulting in the formation of a porous oxide film.

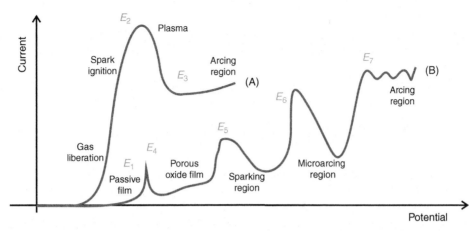

FIGURE 3.11 Types of electric discharge observed in PEO processes (A) in the near-electrode area and (B) in the dielectric film on the electrode surface. *Adapted from A.L. Yerokhin et al., Plasma electrolysis for surface engineering, Surf. Coat. Technol. 122 (2–3) (1999) 73–93 [64].*

After the potential E_5, the electric field reaches a critical value, and there is a rupture of the oxide due to the ionization processes by impact or tunneling [64]. At this point, electric discharges (*sparks*) begin to appear along the entire surface of the electrode until reaching the potential E_6, at which point the longest-lasting electric microarcs begin to emerge on top of the electrode. After the potential E_7, once the system reaches the electric arc regime, the electric microarcs become larger, more intense and last longer and may have destructive effects on the oxide film (*thermal cracking*) [64].

Due to their characteristic electric, thermal, and wear resistance [74–76], materials fabricated using this technique have a variety of applications, mainly in the aerospace, textile, electronic, automotive, and biomedical industries [64,77–80]. These materials present various morphologies, which strongly depend on the experimental growth conditions of these films. A porous structure may form if the electrolyte used can dissolve the oxide being formed in an active oxide dissolution steady state. The diameter of these pores may vary from the nanometric to the micrometric scale.

Another remarkable characteristic of materials fabricated using this technique is the crystallization of the oxides films during their growth, forming crystallites with different sizes, depending on the anodization conditions employed [67,81]. The reason of film crystallization in PEO process has been extensively discussed in the literature [63,70,82–85]. There are two main theories that aim to explain the crystallization process of these materials. The first one ascribes crystallization to the compressive stress generated by the oxide growth [82,83]. The second one states that the electric discharges are responsible for the localized heating of the films, which would be the factor responsible for the oxide crystallization in regions neighboring the discharges sites [70,84,85].

With the advancement of experimental technologies, such as high-time-resolution signal acquisition, high-speed cameras, and optical emission spectroscopy equipment, extensive development has been observed in the research, classification, and understanding of these phenomena [66,86]. In spectroscopic studies, Hussein et al. [85] observed a wide variation in the values of the plasma temperature of discharge events, ranging from $4,500 \pm 450$K to $10,000 \pm 1,000$K. Dunleavy et al. [87] attributed the wide temperature range to the existence of two distinct regions of plasma: a central nucleus at high temperature ($\sim 16,000 \pm 3,500$K) with a high electronic density ($N_e \sim 5 \times 10^{17}$ cm^{-3}), and a surrounding region extending in the direction towards the electrolyte, with a lower temperature ($\sim 3,000$–$4,000$K) and electronic density ($N_e \sim 10^{15}$ cm^{-3}). Computer modeling techniques have also been applied to investigate these phenomena. For example, Sikora et al. [88] used finite-element methods to simulate the effects of heat propagation from an electric discharge produced by a rupture event at a TiO_2 pore and found that the heat propagation area is proportional to the *spark* temperature and the oxide thickness.

The anodization process under conditions of electrolytic breakdown (PEO) can be monitored through the galvanostatic anodization curves described by the potential variation as a function of time or applied charge. As an example, Fig. 3.12 presents the galvanostatic anodization curve of titanium in phosphoric acid, where the variation of the potential as a function of applied charge can be observed. At the beginning of the process (Region 1), a linear increase in the potential with the applied charge ($q = I \times t$) can be observed, up until the beginning of the electrolytic breakdown of the film. The potential at which these oscillations begin is known as the breakdown potential [63]. Electrolytic breakdown is characterized by the appearance of the potential oscillations, resulting from localized rupture processes, growth, and dissolution of the oxide. In Region 2, the amplitude of the oscillations increases, and there is a decrease in the angular coefficient

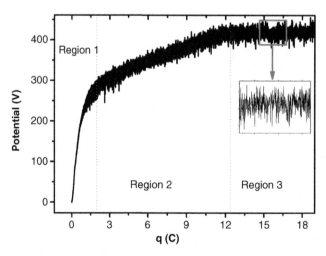

FIGURE 3.12 Galvanostatic anodization curve for titanium in 0.5 mol L^{-1} H$_3$PO$_4$ obtained at 20 mA cm^{-2} and $T = 20°$C.
Adapted from F. Trivinho-Strixino, J.S. Santos, M.S. Sikora, Síntese Eletroquímica de Materiais Nanoestruturados [electrochemical synthesis of nanostructured materials], In: A.L. Da Róz, F.L. Leite, et al. (Eds.), Nanoestruturas: Princípios e Aplicações, vol. 1, Elsevier, Rio de Janeiro, 2015, pp. 63–110 (Chapter 3) [26].

(dV/dq) of the anodization curve. In Region 3, the potential oscillations stabilize around a mean value.

The initial voltage increase is attributed to the enlargement of oxide thickness, which increases the resistive characteristics of the oxide layer. A high resistance of oxide film increases the potential and the local electric field. The potential at the oxide/electrolyte surface is given by [70]:

$$E = E_0 - IR \tag{3.5}$$

where E_0 corresponds to the potential difference applied externally; R represents all the resistances in series within the system (contacts, electrolyte, film resistance, etc.); and I is the current applied to the system. Given that the current is kept constant during the entire process, the potential rises due to the increase in film resistance. This growth refers to the thickness increase or the decrease in the metallic area exposed to the material, as follows [70]:

$$R = \frac{\rho l}{A} \tag{3.6}$$

where ρ is the specific resistivity of the film, l is the thickness, and A is the surface area of the material. After this region of linear potential increase, where a high potential value is reached, potential oscillations can be observed. In this process, the appearance of spark discharges is observed at the electrode surface. This phenomenon can be observed on the surface of the titanium electrode during anodization in phosphoric acid, as can be seen in Fig. 3.13.

The potential oscillations are described in the literature as being a result of the film destruction and reconstruction processes [63]. As previously discussed, the potential increases due to the film thickening, therefore the angular coefficient dV/dq can also be described as the oxide film growth rate (anodization rate). In steady-state, the dissolution

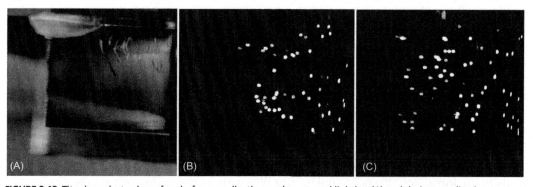

FIGURE 3.13 Titanium electrode surface before anodization under external lighting (A) and during anodization in 0.5 mol L^{-1} H$_3$PO$_4$, 20 mA cm^{-2} and $T = 20°C$, in the absence of lighting. The images shown in (B) and (C) were acquired in Region 2 of the anodization curve (Fig. 3.12) *Adapted from F. Trivinho-Strixino, J.S. Santos, M.S. Sikora, Síntese Eletroquímica de Materiais Nanoestruturados [electrochemical synthesis of nanostructured materials], In: A.L. Da Róz, F.L. Leite, et al. (Eds.), Nanoestruturas: Princípios e Aplicações, vol. 1, Elsevier, Rio de Janeiro, 2015, pp. 63–110 (Chapter 3) [26].*

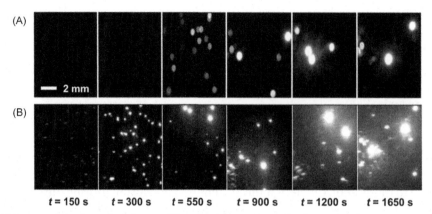

FIGURE 3.14 Images of the electrode surface at different galvanostatic anodization stages of Zr at 20 mA cm^{-2} in solutions of (A) 0.5 mol L^{-1} H$_3$PO$_4$ [1] and (B) 0.5 mol L^{-1} H$_2$C$_2$O$_4$. *Adapted from J.S. Santos et al., Characterization of electrical discharges during spark anodization of zirconium in different electrolytes, Electrochim. Acta 130 (2014) 477–487 [66].*

and growth processes of oxide film reach a constant diffusion rate due to the equivalence of the formation and dissolution rates, remaining in this condition until the end of the experiment. Depending on the metal and the experimental conditions, changes in the anodization rate, the amplitude of the potential oscillations, the oxide electrolytic rupture potential, and the spatial pattern of the electric discharges can be observed. Fig. 3.14 shows images of the Zr electrode surface during galvanostatic anodization at 20 mV cm^{-2} in phosphoric acid and oxalic acid, where changes in the spatial distribution, sizes, and intensity of the electric discharges can be observed during the anodization.

The materials obtained through the PEO technique display different morphologies. A porous structure may be formed if the oxide is soluble or partially soluble in the electrolyte because the channels for electric charge propagation constitute the preferential sites for pore formation. If the oxide is not soluble in the electrolyte, the formation of a compact oxide film with fissures may occur due to the film breakdown [63]. In addition, the incorporation of electrolyte anions into the film affects the electrolytic breakdown process [89] and can also modify the morphology of the material. In acidic electrolytes, for example, the anionic incorporation tends to be favored, independently of the value of the applied current. This intensifies the breakdown process and leads to the formation of a porous structure if the oxide is soluble in the electrolyte.

In neutral and alkaline electrolytes, the oxide dissolution mechanism changes depending on the value of the applied current, and a homogeneous dissolution may occur. This leads to the formation of a pore-free barrier film with fractures throughout the surface. In porous films, depending on the experimental conditions, the pore size can change from a few nanometers up to a few micrometers. Fig. 3.15 shows micrographs obtained by scanning electron microscopy (SEM) of TiO$_2$ films prepared via PEO in phosphoric acid. The SEM images were obtained at different stages of the experiment.

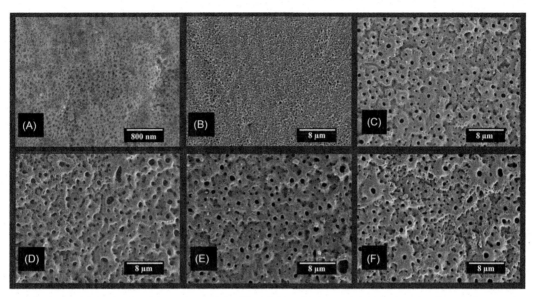

FIGURE 3.15 SEM micrographs of the surfaces of TiO_2 films obtained by PEO in 0.3 mol L^{-1} H$_3$PO$_4$, after application of the following charges: (A) 1.8 C, (B) 4.6 C, (C) 25 C, (D) 108 C, (E) 180 (C), and (F) 216 C [63].

One of the advantages that the PEO technique has over low-voltage anodization is the formation of a thick film with high mechanical resistance and crystallinity. These films are widely used as coatings in technological applications and as protection against corrosion. By contrast, materials obtained using this technique display a low degree of ordering and a variety of pore sizes at the same sample; for this reason, these oxide films prepared under high electric field are rarely explored in applications that demand well-ordered structures with high superficial area as photocatalysis or template formation. Nonetheless, one study can be found in the literature describing the photocatalytic properties of TiO_2 in films produced by PEO from titanium in a phosphoric acid solution [67]. In that study, the authors observed the influence of morphology and the microstructure on the photocatalytic activity of the oxide, by analysis of the degradation of methyl orange dye. Their results demonstrated that when the number of pores was high the photocatalytic activity improved as a consequence of the increase in surface area. Also, the most relevant result revealed that the crystallite size strongly influenced the photocatalytic activity of TiO_2 films since the grain boundary conditions acted as electron-hole recombination centers. This demonstrated that the photocatalytic activity is proportional to the sample crystallite size. The authors presents in this study the viability of using TiO_2 films produced via PEO as an alternative to the traditional TiO_2 nanotubes, which are normally used for this purpose but require a longer synthesis time and a subsequent thermal treatment for the film crystallization. A more detailed description of the electrochemical synthesis of TiO_2 nanotubes will be presented in Section 3.4.3

One way of controlling the electric microdischarges to avoid superheating and localized destruction of the oxide film structure in the galvanostatic regime, is the use of an alternating [90] or a pulsed current [85,91]. On the potentiostatic regime, by contrast, this control can be achieved by the application of a pulsed voltage. The choice of potential values depends on the characteristics of the desired oxide [92–94]. These methods, known also as *pulsed anodization* [93], constitute an alternative method because the potentiostatic method—mostly used to produce nanotubes and nanopores—requires a long anodization time, whereas the galvanostatic method, although faster, produces disordered porous structures and, in some cases, the local disrupture of the oxide. Choosing an adequate pulse sequence, it is possible to control the composition, microstructure, and morphology of the oxide film.

3.4.2 Self-Organization in Anodic Oxides: PAA

Aluminum anodization is a widely explored process of great commercial impact. It represents one of the most important methods used to synthesize ordered nanostructures that exhibit an array of hexagonal cells containing nanopores in their centers [95] (Fig. 3.16). In general, aluminum anodization promotes the formation of two types of Al_2O_3 (alumina) films: barrier-type oxide films and porous oxide films. The main factor that determines this type of growth on an Al surface is strongly related to the nature of the electrolyte being used [27,96,97]. Basically, the formation of barrier-type aluminum oxide is enabled by boric acid, ammonium borate, ammonium tartrate, phosphate aqueous solution, tetraborate in ethylene glycol medium, perchloric acid and ethanol, and organic electrolytes such as citric, succinic, and glycolic acid electrolytic solutions [97–100]. By contrast, obtaining porous oxide film requires the use of strongly acidic electrolytes; the oxide must be an effective solubility in the acid solution. Thus, the production of porous anodic alumina (PAA) is associated with an active and localized dissolution effect where the pores will originate.

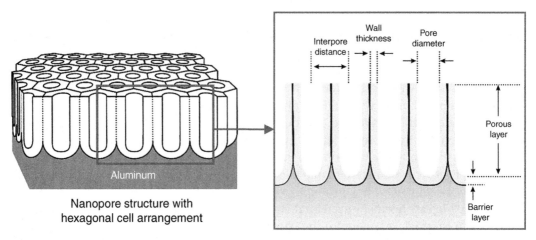

Nanopore structure with
hexagonal cell arrangement

FIGURE 3.16 Schematic representation of the hexagonal configuration of the self-organized porous anodic alumina (PAA) cells, grown on aluminum surface. *Adapted from F. Trivinho-Strixino, J.S. Santos, M.S. Sikora, Síntese Eletroquímica de Materiais Nanoestruturados [electrochemical synthesis of nanostructured materials], In: A.L. Da Róz, F.L. Leite, et al. (Eds.), Nanoestruturas: Princípios e Aplicações, vol. 1, Elsevier, Rio de Janeiro, 2015, pp. 63–110 (Chapter 3) [26].*

Normally, phosphoric acid or oxalic acid aqueous solutions are employed, although other electrolytes with the same described characteristics can also be used [95]. Another important factor for obtaining highly ordered layers of porous alumina is the anodization time. Anodization carried out during long periods allows the process to reach a steady-state where the oxide formation/dissolution processes stabilize. This promotes the growth of deep and well-organized pores [95].

The self-organized PAA obtained by Al anodization can be represented schematically by a compact array of hexagonal cells, with pores inside the cells, as illustrated in Fig. 3.16. The high ordering degree of the nanostructures and their morphology are described by geometric parameters, such as pore diameter, wall thickness, barrier layer thickness, and distance between pores (cell diameter). A uniform pore diameter—a factor that is strongly dependent on the experimental conditions of anodization—can vary from a few to several hundred nanometers. The depth of the pores' parallel channels can exceed 100 µm, depending on the synthesis conditions [101]. Compared with other nanostructured fabrication techniques, this is one of the features that make PAA one of the most desired nanostructures with a high pore depth/diameter ratio, as well as a high nanopore density that can be obtained.

The growth of the oxide layer occurs at the metal/oxide interface at the pore base and involves the conversion of a preexisting barrier-type surface through a porous oxide layer. The pore formation mechanism can be briefly described as follows [101]: during growth of the porous oxide, a thin and compact barrier layer located at the pore base and in contact with the oxide/electrolyte interface is continuously dissolved by the action of the increasing local electric field. This process results from the increased resistance to the passage of excess current through the oxide formed. Simultaneously, there is a continuous formation of a new barrier oxide layer at the metal/oxide interface. The film growth in the steady-state requires equilibrium between the oxide growth rate and its dissolution rate at the bottom of the pores. In practice, what occurs is the consumption of aluminum oxide at the nanopore base simultaneously with the formation of additional oxide at the metal/oxide interface. The result is the erosion of the aluminum metal being consumed to form the oxide, and the penetration of the channels inside the oxide, originated by an active dissolution at the oxide bottom. In materials science, this phenomenon can be classified as a *top-down* design technology. Nonetheless, the origin of the self-organization is related to the precise control of the oxide formation and dissolution processes, so that the terms *bottom-up* and *top-down* can both be used in those cases, respectively. Various published works present proposals for mechanisms for the formation of self-organized anodic alumina hexagonal molecules [63,98,101–106].

Self-organized aluminum oxide cells can be easily fabricated by aluminum anodization under potentiostatic or galvanostatic regimes [101]. However, other anodization regimes, such as the hybrid pulsed methods, are also described in the literature and present advantages [93,107–109]. A typical growth transient curve for this oxide, analyzing the current or potential as a function of anodizing time, can be observed in Fig. 3.17. When a constant current is applied during the growth of porous alumina (constant current plot in Fig. 3.17), the measured potential increases linearly with time up to a maximum value.

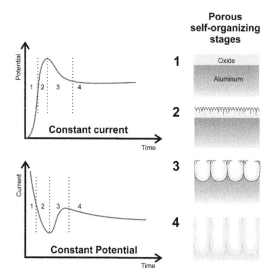

FIGURE 3.17 Anodization curves illustrating the growth kinetics of anodic alumina under galvanostatic (constant current) and potentiostatic (constant potential) regimes, and related stages of porous oxide growth: **(1)** barrier film growth, **(2)** formation of pore precursors, **(3)** pore formation, and **(4)** pore growth. *Adapted from F. Trivinho-Strixino, J.S. Santos, M.S. Sikora, Síntese Eletroquímica de Materiais Nanoestruturados [electrochemical synthesis of nanostructured materials], In: A.L. Da Róz, F.L. Leite, et al. (Eds.), Nanoestruturas: Princípios e Aplicações, vol. 1, Elsevier, Rio de Janeiro, 2015, pp. 63–110 (Chapter 3) [26].*

Then, it gradually decreases until reaching a stationary potential. During the initial period (stage 1 in Fig. 3.17), the potential increase is associated with the linear growth of a highly resistive compact film (barrier film). Subsequently, the appearance of individual penetration channels occurs in the barrier layer (pore precursors, stage 2 in Fig. 3.17). The barrier film breaks at the maximum potential (stage 3 in Fig. 3.17), and the porous structure begins to form. Finally, the steady-state growth of the porous alumina continues (stage 4 in Fig. 3.17), and the potential becomes constant [101].

Currently, the potentiostatic method is preferred over the galvanostatic method for the fabrication of highly organized PAAs with precise control of the pore diameter. The current density decreases rapidly at the beginning of the PAA potentiostatic formation process (constant potential plot in Fig. 3.17), and immediately afterward, a minimum is reached. A linear increase of current follows up to a local maximum. After this maximum, the current smoothly decreases until the steady-state region, where the formation of the porous oxide proceeds (Fig. 3.17). When the system reaches this stage at which the value of the differential dj/dt decreases, the pore formation enters the steady-state regime. The self-ordering pore formation process directly depends on the anodization conditions. In general, the onset of pore formation, i.e., the current minimum, decreases with the increase in electric field. Therefore, the current minimum is usually low for anodization performed at high potential. The decrease in the observed current minimum is also verified in concentrated acid solutions [101].

Most anodic films are directly dependent on the applied potential. In addition, during oxide growth in the steady-state regime, both the current and potential measured during potentiostatic and galvanostatic anodization remain constant. Therefore, the anodic film morphology can be controlled by the anodization parameters. According to Sulka [101], the concentration and pH of the electrolyte, the potential, the anodization time, and the pore-widening time at the end of anodization, are parameters that directly influence the pore diameters in alumina films grown in the potentiostatic regime.

There are, in general, two PAA fabrication methods that can be used as templates: nanoindentation and two-step anodization (Fig. 3.18) [95,101]. It is also possible to apply lithographic methods [110]. In the first method, the anodization is performed on top of a prepatterned plate previously marked by nanoindentation, in order to guide and form a perfectly ordered network of nanopores during anodization. The second method consists of growing the oxide film on the metal during a sufficiently long time interval to mark the aluminium surface under the pore base. After finishing the first anodization, the formed oxide layer is chemically removed from substrate. Then, the film is anodized again under either the same or different conditions as the first anodization step. An aluminum metallic structure with marks at the nanoscale emerges after the oxide removal This process aids and orients the formation of self-organized nanopores in the second anodization stage (Fig. 3.18).

The fabrication of alumina membranes with self-organized nanopores obtained from aluminum anodization is a multistage process. It involves the pretreatment of the substrate, the anodization process, and subsequent posttreatments related to the nanomanipulation of the templates, aiming toward the desired application. Thus, various aspects related to the adequate manipulation of nanoporous films must be verified to separate the oxide membrane from the metallic substrate without damaging it. One can also subsequently open the pore base to turn the structure into a nanosieve that can be used to fabricate a mold or other nanostructured materials.

The quality and purity of the aluminum substrates and the previous treatment of the surface significantly influence the self-organized pore formation process. The structure of the preexisting oxide film on the metal surface, which can be formed by exposure to air during chemical or electrochemical treatment, depends on the type of pretreatment employed. During aluminum anodization, the nucleation process of the self-organized pores depends on the defects present on the metal surface, such as imperfections and grain boundaries. The surface pretreatment procedure must have as a goal reducing the presence of these structural and surface defects. Thus, the ideal substrate for initiating the formation of self-organized nanopores must be high purity aluminium and must also have been previously processed by a thermal treatment to remove surface stress and increase the grain size.

The pretreatment procedure consists of a thermal treatment known as *annealing* of the aluminum metal, in the absence of oxygen and at high temperatures to relax the metal crystalline structure and remove structural stress [101]. In some cases, the *rapid thermal annealing* technique can be employed, which consists of applying a thermal

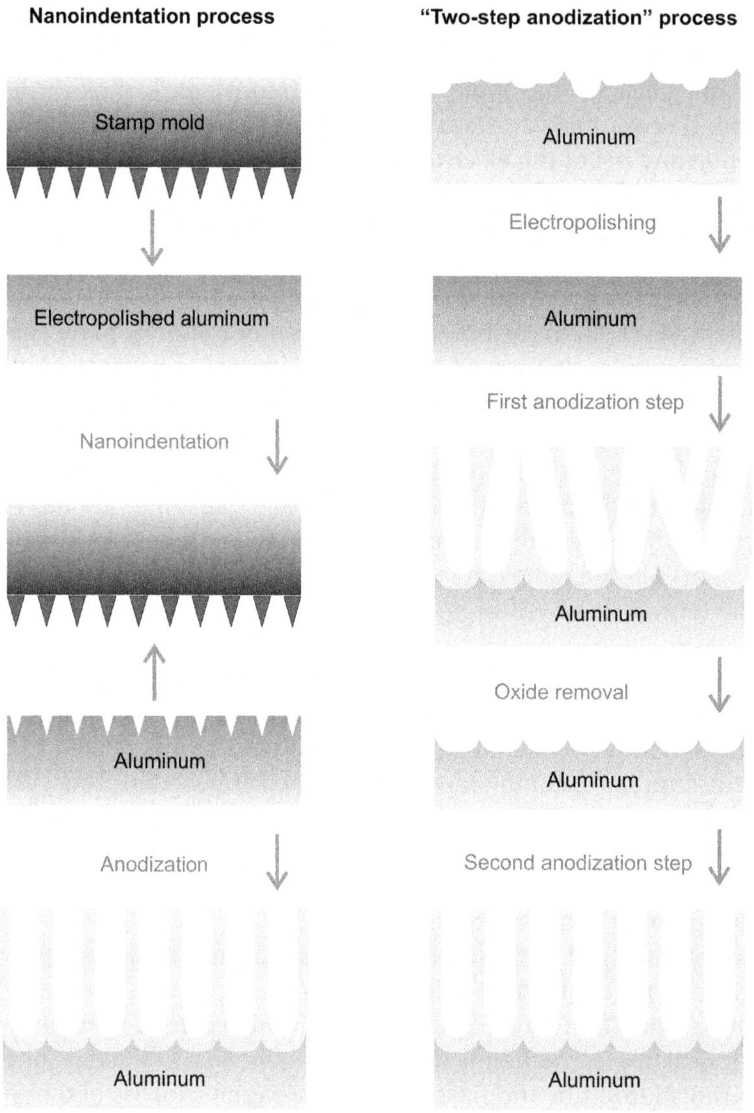

Nanoindentation process

"Two-step anodization" process

FIGURE 3.18 Illustration of the two most common processes for the fabrication of nanoporous alumina. *Adapted from F. Trivinho-Strixino, J.S. Santos, M.S. Sikora, Síntese Eletroquímica de Materiais Nanoestruturados [electrochemical synthesis of nanostructured materials], In: A.L. Da Róz, F.L. Leite, et al. (Eds.), Nanoestruturas: Princípios e Aplicações, vol. 1, Elsevier, Rio de Janeiro, 2015, pp. 63–110 (Chapter 3) [26].*

shock to the sample using high-powered incandescent lamps for a brief period. Next, the samples are appropriately purified to remove any organic substance. Among the various solvents utilized for cleaning the substrate, acetone and ethyl alcohol are the most commonly employed. After this stage, the substrates undergo a mechanical, chemical, or electrochemical (electropolishing) polishing procedure to produce flat and mirror-finished metallic substrates, depending on the desired application. Several tables can be found in Sulka et al. [101] describing examples of experimental conditions for the pretreatment of the aluminium surfaces.

As previously described, the two-step anodization consists of the usual anodization process, with the previous choice of experimental conditions (potential or current, temperature, and concentration/composition of the electrolyte), and the subsequent chemical dissolution of the oxide formed in the first anodization step. In the chemical attack for the oxide removal, a concave triangular profile is formed at the substrate base, where the formed pores used to be (Fig. 3.18). This structural profile will serve as a natural orientation for the formation of the oxide layer during the second anodization. Normally, this second step is carried out under the same conditions as the first anodization. Finally, using adequate chemical or physical methods [111], the porous structure can be separated from the substrate base to enable the desired manipulation of the membrane. Also, the pores diameter can be increased or their closed bases can be opened, thereby opening the membrane on both sides. The general scheme of the procedure for the fabrication of PAAs and their possible applications are illustrated in Fig. 3.19.

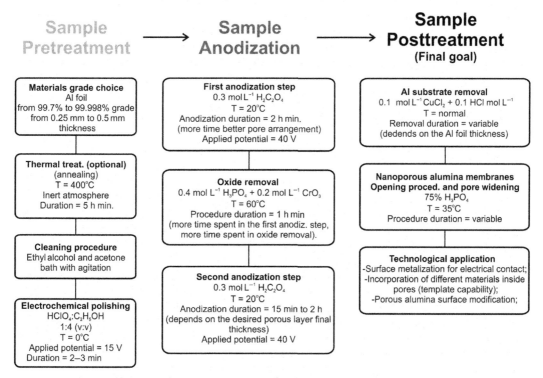

FIGURE 3.19 Schematic illustrating the experimental stages for obtaining anodic self-organized alumina for different applications.

The pore-opening process and the chemical erosion of the barrier oxide at its aluminium base depend on the chemical dissolution rate of anodic alumina. This procedure depends on the concentration and temperature of the electrolyte. Aqueous solutions of H_2SO_4, $H_2C_2O_4$, and H_3PO_4 are normally employed, and it is also possible to add small amounts of CrO_3, which increases the oxide dissolution rate. The procedure may be conducted in the same the electrolytic cell. Depending on the conditions used, the dissolution rate ranges from 0.02 to 4.45 nm min^{-1} [101].

The separation of the nanoporous oxide from its aluminum base can be performed using an electrochemical method, in which inverted potential pulses are applied [93,107,109,112]. However, the most commonly used process is a chemical removal of aluminum [101]. This method consists of exposing the metal layer to chemical substances that oxidize the metal but not the oxide. The most used chemical substances are $HgCl_2$, copper (II) sulfates or chlorides, and mixtures of Br_2 and ethanol. The solution temperature can also be varied between 5 and 40°C, as well as the exposure time of the metallic layer.

Surface images obtained by SEM of PAA samples prepared under galvanostatic and potentiostatic regimes are illustrated in Fig. 3.20 and Fig. 3.21, respectively. All samples were prepared using two-step anodization method, although under different conditions of electrolyte concentration, temperature, and growth regime.

The two-step anodization method is quite interesting from the practical viewpoint and useful when one does not have the experimental apparatus for nanoindentation. Nevertheless, to obtain self-organized structures with high morphologic homogeneity, a long anodization time is necessary for the first step, where the aluminum base is marked. For example, to produce a sample containing self-organized porous layers, with a thickness from 1 to 3 µm using the galvanostatic method, using oxalic acid as electrolyte and temperature of 5°C, approximately 4 h are necessary for the experiment. This period is

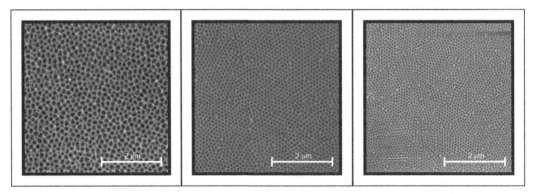

FIGURE 3.20 Micrographs showing a top view of PAA films, obtained by two-step anodization method in the galvanostatic regime. *Adapted from F. Trivinho-Strixino, J.S. Santos, M.S. Sikora, Síntese Eletroquímica de Materiais Nanoestruturados [electrochemical synthesis of nanostructured materials], In: A.L. Da Róz, F.L. Leite, et al. (Eds.), Nanoestruturas: Princípios e Aplicações, vol. 1, Elsevier, Rio de Janeiro, 2015, pp. 63–110 (Chapter 3) [26].*

 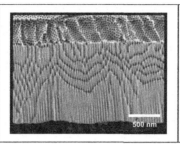 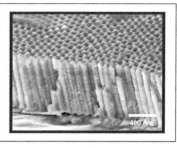

FIGURE 3.21 Micrographs showing a lateral view of PAA films, obtained by two-step anodization method in the potentiostatic regime. *Adapted from F. Trivinho-Strixino, J.S. Santos, M.S. Sikora, Síntese Eletroquímica de Materiais Nanoestruturados [electrochemical synthesis of nanostructured materials], In: A.L. Da Róz, F.L. Leite, et al. (Eds.), Nanoestruturas: Princípios e Aplicações, vol. 1, Elsevier, Rio de Janeiro, 2015, pp. 63–110 (Chapter 3) [26].*

distributed as follows: 2 h of anodization in the first step, 30 min for the removal oxide, and 1.5 h for the second anodization. The process is limited by the diffusional transport of the oxygenated species from the electrolytic solution to the metal/oxide interface, concentrated in the pores channels. This limits the oxide growth rate (2–6 μm h^{-1}) and introduces a maximum value of 20 for the depth/diameter ratio [65].

A solution to this problem arose with the development of the so-called "Hard Anodization" method. In this method, aluminum substrates were anodized in oxalic acid in the high-field regime and in the potentiostatic mode, with applied potential values ranging from 100 to 150 V for oxalic acid electrolytes. Three-dimensional arrays of regular nanopores can be obtained through this method and utilized in nanotechnology applications such as photonic crystals, microfluidics, and template for nanowire and nanotube fabrication with diameter control [113]. However, to suppress the electrolytic breakdown produced by the high associated electric field, the existence on the aluminum substrate of a previously formed barrier-oxide layer larger than 400 nm is necessary before applying large anodization potentials (>100 V). This oxide layer can be growth using mild anodization conditions, for example, performing a potentiostatic anodization, applying 40 V in 0.3 mol L^{-1} oxalic acid for 5–10 min. Then, in sequence, the applied potential must be slowly increased up to values between 100 and 150 V at a rate of 0.5–0.9 V s^{-1} and be maintained at these values until the end of anodization. The hard anodization is accompanied by intense release of heat, which must be efficiently removed using an adequate experimental apparatus (Fig. 3.22).

Thus, "Hard Anodization" method offers great advantages compared with "Mild Anodization" method. This method accelerates the nanostructure fabrication of PAA films due to the high growth rate of the self-organized oxide arrays and it allows the fabrication of these arrays in larger areas and using only one side of the metal. This aspect increases its technological relevance. Some authors [65,114] have demonstrated the advantages of utilizing a hard anodization cell, as opposed to a conventional cell using the mild anodization mode. It is possible to obtain PAA films with aspect ratio 20 times larger than the conventional method using the same anodization time. This means pores with depths up to 110 μm,

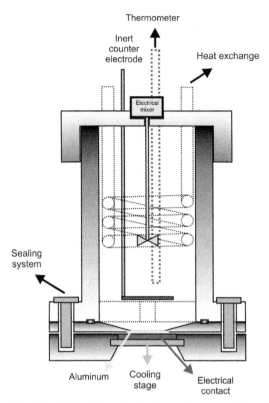

FIGURE 3.22 Illustration of a cell used in the "Hard Anodization" method. *Adapted from F. Trivinho-Strixino, J.S. Santos, M.S. Sikora, Síntese Eletroquímica de Materiais Nanoestruturados [electrochemical synthesis of nanostructured materials], In: A.L. Da Róz, F.L. Leite, et al. (Eds.), Nanoestruturas: Princípios e Aplicações, vol. 1, Elsevier, Rio de Janeiro, 2015, pp. 63–110 (Chapter 3) [26].*

when compared with depths of 4 μm in average obtained by conventional anodization. Thus, the fabrication efficiency gains for self-organized structures are well established. Considering the same anodization time, the hard anodization technique allows obtaining self-organized alumina structures with much larger thicknesses and pore diameters.

The PAAs can also be modified by incorporation of specific ions. In addition to emitting in blue [115–121], these materials are particularly promising for use as a host material [118,122,123]. They have also been employed in the preparation of waveguides and in the fabrication process of nanoscopic luminescent devices that facilitate controlling the optical properties of the material at the molecular level. Applications of PAAs are countless. We can find studies that explore the advantages related to the porous alumina itself, as well as optical and mechanical properties [11,13,120,124–129]. Other studies report the use as templates to fabricate other nanostructured materials [101,128,130], and applications in biological systems [131–136], polymeric systems [137], and semiconductor devices [128].

3.4.3 Electrochemical Synthesis of TiO$_2$ Nanotubes

Since Gong and coworkers [138] electrochemically synthesized TiO$_2$ nanotubes for the first time in 2001, this material has been broadly investigated for applications in photocatalysis [14,139], photoelectric [17], and photovoltaic devices [18,140], sensors [22,141], and immunosensors [24]. This great interest is due to their large surface/volume ratio; consequently, the TiO$_2$ nanotube formation mechanism has been extensively investigated, as well as the experimental parameters involved in its preparation. The description of the electrochemical synthesis of other types of oxide nanotubes can be found in the literature, with materials such as ZrO$_2$ and WO$_2$ [142]. However, due to the intrinsic properties of TiO$_2$, as well as its catalytic power [143] and biocompatibility with live tissues [144,145], this oxide has been widely studied, and it presents great advantages for technologic applications. TiO$_2$ nanotubes presenting pore sizes between 22 and 110 nm, width between 200 and 6,000 nm, and wall thickness between 7 and 34 nm are easily obtained via electrochemical methods by controlling the synthesis parameters [18,146]. In this section, we will present a brief review of the experimental conditions for the preparation of TiO$_2$ nanotubes and the influence of these parameters on the architecture of the produced nanomaterial.

As previously described, the formation of porous films using electrochemical methods is directly related to the solubility of the oxide in the electrolyte used. Anodic TiO$_2$ films can display either a porous structure with low spatial organization and a pore diameter measured in micrometers and nanometers or a highly organized structure in the form of nanotubes with well-defined walls, depending on the experimental conditions employed in the synthesis. In electrolytes in which the films present low solubility, such as phosphoric or sulfuric acid, or even alkaline media containing phosphates or aluminates [75], one generally obtains compact or porous structures with low spatial organization [67,147]. By contrast, highly organized structures in the form of nanotubes with well-defined walls can be obtained through anodization in media containing fluoride [14,75,138,143,144,148,149], or even perchlorate or chloride [142].

The electrochemical synthesis of TiO$_2$ nanotubes is generally carried out employing the galvanostatic or the potentiostatic method in electrolytes in which the oxide presents high solubility. Under these conditions, the potential reached by the galvanostatic system is smaller than the oxide breakdown potential (Section 4.1). In the potentiostatic method, a potential lower than the breakdown potential is applied, for the purpose of forming a homogeneous and self-organized structure, free of heterogeneities promoted by the phenomenon of electrolytic breakdown.

There is also the titanium oxidation technique using a pulsed charge, where the efficiency of the current related to the oxide formation is approximately 100%, eliminating parallel reactions such as O$_2$ evolution, commonly observed in potentiostatic and galvanostatic methods. In addition, according to Poznyak et al. [148], the formed material exhibits better electrocatalytic activity than the titanium oxide obtained through the galvanostatic method.

Considering potentiostatic anodization, it is observed that at the beginning of anodization, there is an abrupt potential drop due to the formation of a high-resistance compact oxide film that reduces the current intensity. In the sequence, a current increase due to the oxide dissolution process is observed, and then, it occurs the random growth of the nanoporous structure. The competition between the increase in oxide thickness and the formation of self-organized nanotubes (local film dissolution) promotes the slow decrease of the current intensity. Regular fluctuations can be observed, which are related to the growth and dissolution of the oxide film formed.

The first synthesis of TiO_2 nanotubes films were carried out in a hydrofluoric acid medium. According to Regonini et al. [150], these films consisted the first generation of TiO_2 nanotubes. The presence of fluoride in the reaction medium produces the partial dissolution of the formed film and promotes the formation of titanium dioxide nanotubes after the system achieves dynamic equilibrium between the formation and dissolution processes of the oxide layer. The chemical dissolution of TiO_2 occurs because of the formation of the soluble complex TiF_6^{2-}, and the dynamic equilibrium between the formation and dissolution rates is reached after a long anodization period [151]. The formation mechanism of TiO_2 nanotubes has been thoroughly discussed in the literature and is based on the existing proposals for PAA growth, considering the appropriate differences between the systems [152–158].

In potentiostatic anodization conducted at low potentials, an analogous morphology to porous alumina structures [159] is frequently observed. With the increased potential, a particulate surface is obtained. For potentials above 10 V, the particulate appearance is lost, with the initial formation of nanotubes [18]. If anodization is performed for a long period, there will be a uniform growth of the titanium nanotube array with an insulating barrier film in the lower portion of the films, separating the nanometric array from the titanium electrode, similarly to the barrier film observed in nanoporous alumina. At high potentials, the highly organized structure is lost and replaced by a randomly porous structure, due to the breakdown of the previously formed nanotubular structure [160].

The effect of the applied potential on TiO_2 nanotube formation has been investigated by several researchers. According to Yahia et al. [161], the applied potential can strongly influence the nanotube growth and their properties. These researchers [161] observed a linear dependence of the barrier film thickness with the applied potential within the interval from 20 to 50 V. Similarly, by analysis of the applied voltage and the anodization time, Matykina et al. [160] confirmed a barrier layer increase proportional to this voltage. Furthermore, the authors [160] also observed the incorporation of fluoride ions into the barrier film, proportional to the potential. A percentile of 4–6% was detected in the films grown at 20 V, whereas up to 13% of fluoride was identified in the films grown at 60 V. In another study, Alivov et al. [162] demonstrated the growth of TiO_2 nanotubes in a wide potential range (10–240 V) by electrolyte modification. According to the authors [162], the resistivity of the formed films was found to be proportional to the applied potential, with an observed increase of eight orders of magnitude for the films obtained at 240 V, with respect to those obtained at 10 V. In addition, at potentials lower than 60 V, well-ordered

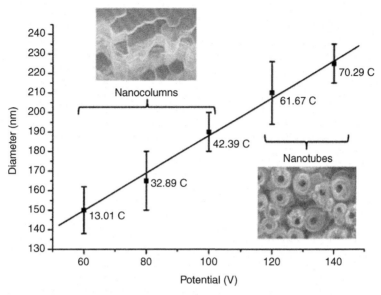

FIGURE 3.23 Nanostructure diameter versus anodization potential at 10°C. The insets display SEM micrographs of the nanocolumn and nanotube structures of TiO_2 films [163].

structures were obtained for a solution of NaF + glycerol + H_2O, with a NaF concentration of 0.7%, and for media containing NH_4F + ethylene glycol + H_2O, in different proportions of NH_4F (0.5–1%) and water. For potentials higher than 60 V, the same result was obtained by diluting the solution to 0.1%. According to the authors, the increase in the water-electrolyte ratio promotes an increase in the nanotube growth rate, which modifies the morphology and increases the resistivity of the films.

Complex structures, such as hexagonal nanostructures formed at high potentials and low temperatures, have been reported by Ruff et al. [163] Fig. 3.23 reveals that the transition from a nanocolumn structure to nanotubes is caused by modification of the applied potential, among other factors. The authors also suggest that the formation of the nanocolumns results from the enlargement of the inner nanotube walls due to intensification of the oxide formation reaction, when compared with the dissolution reaction.

The pH of the electrolyte is also an important factor that affects the morphology of TiO_2 films. At pH = 1.0, well-defined nanotubes are produced with a length and diameter proportional to the anodization potential. The formation of a conical structure may occur after the formation of the nanometric array due to the increase of pore base [164]. According to Mor [18], the optimal pH range for obtaining relatively long nanotubes is between 3.0 and 5.0. However, a reduction in the acidity and the use of salts containing fluoride promote a decrease in the oxide dissolution rate, and it is possible to obtain nanotubes with lengths between 2 and 4 μm. [165] The oxide synthesis in aqueous medium with fluoride salt solutions forms the second generation of TiO_2 nanotubes [150].

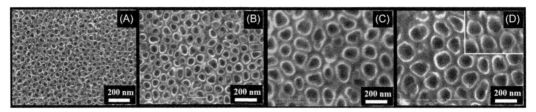

FIGURE 3.24 SEM images of TiO$_2$ nanotubes obtained in a mixture of methanol 50% vol. + HF 1% mass, under the following current densities: (A) 10 mA cm^{-2}, (B) 15 mA cm^{-2}, (C) 20 mA cm^{-2}, (D) 30 mA cm^{-2}. *Adapted from S. Kaneko et al., Fabrication of uniform size titanium oxide nanotubes: impact of current density and solution conditions, Scr. Mater. 56 (5) (2007) 373–376 [166].*

Kaneko et al. [166] performed galvanostatic anodizations in a solution containing methanol and HF in order to study the effect of methanol and the influence of the current density on TiO$_2$ nanotube formation. Their results demonstrated that an increased current density promotes an enlargement of nanotube diameter and the formation of channels between them, as can be observed in Fig. 3.24. The changes in the nanotube diameter is promoted by the increase of dissolution rate and of magnitude of the electric field. In addition, the authors concluded that adding methanol to the electrolyte eliminated the current oscillations related to the gaseous evolution and produced a spacing between the nanotubes. This led to obtaining standardized nanomaterials with high surface area.

The synthesis of nanotubes in organic media containing fluoride salts and small amounts of water comprises the third and fourth generations of TiO$_2$ nanotubes [150]. Nanotubes grown under these conditions can display thicknesses of 100–1000 μm [146,165,167–169], and are more homogeneous than nanotubes of the first and second generations. Studying viscous electrolytes, such as glycerol or ethylene glycol containing NaF, Macak and Schmuki [167] observed the formation of self-organized TiO$_2$ nanotubes with smooth walls and lengths in micrometers. According to the authors, diffusion is limited in viscous electrolytes, causing local acidification in the pore, which promotes a high current efficiency compared with aqueous electrolytes. Due to the simplicity of the synthesis and reproducibility of the TiO$_2$ nanotubes in organic media, great attention is currently being paid to these synthesis conditions.

Fourth-generation nanotubes are synthesized at the same electrolytic medium. However, the anodization parameters, such as anodizing time, applied potential, and concentration of fluoride ions, are carefully controlled to obtain a highly homogeneous and well-organized structure [150].

TiO$_2$ nanotubes formed during the electrochemical synthesis from first to fourth generation are amorphous. A thermal treatment known as annealing is required for their conversion to crystalline oxide. After annealing, the titanium dioxide can exist in three different crystalline phases: anatase (tetragonal), rutile (tetragonal), and brookite (orthorhombic) [150]. Generally, the anatase phase is predominant after the thermal treatment of this material. This characteristic, the large surface area, makes TiO$_2$ nanotubes excellent photocatalysts [14]. In absence of fluoride in the electrolyte, there is usually no formation

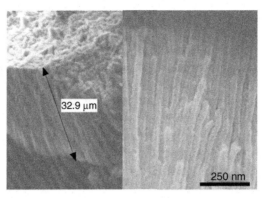

FIGURE 3.25 SEM micrographs of the lateral view of TiO$_2$ nanotubes prepared in 0.1 mol L^{-1} HClO$_4$ at 30 V for 60 s. *Adapted from R. Hahn, J.M. Macak, P. Schmuki, Rapid anodic growth of TiO$_2$ and WO$_3$ nanotubes in fluoride free electrolytes. Electrochem. Commun. 9 (5) (2007) 947–952 [142].*

of a tubular structure, but the formation of a porous structure caused by the electrical discharges at the surface during the oxide growth [67]. Studies carried out in media where TiO$_2$ is poorly soluble, such as sulfuric acid [84,167] and phosphoric acid [67], show a transition from amorphous oxide to the crystalline anatase phase during the film growth.

In potentiostatic studies performed in electrolytes containing perchlorate and chloride ions, Hahn et al. [142] observed that, contrary to common sense, there is formation of a tubular array in the absence of fluoride, as shown in Fig. 3.25. Furthermore, this structure is obtained in a relatively short time. However, since the formation of this structure is induced by local breakdown—that is, by the formation of pitting corrosion [142]—the array does not display rigorous self-organization, such as that obtained in fluoride medium.

Temperature is another experimental parameter of great importance in the growth of nanotubular TiO$_2$ arrays. A reduction of the electrolyte temperature tends to increase the wall thickness of the pore. As result, the formed material presents a sponge-like structure [18]. According to Prida et al. [151], low temperatures inhibit the formation of self-organized TiO$_2$ nanotubes, favoring the formation of a randomly disordered structure due to the internal growth of the nanotubes. This effect is promoted by the direct influence of temperature on the oxide solubility in the electrolyte. The decrease in temperature reduces the solubility and alters the viscosity of the solution, promoting disequilibrium in the steady-state reached between the oxide formation and dissolution reactions. This way, a new steady-state is achieved. Consequently, there is an increase in the film resistivity and the formation of a structure with a higher or lower degree of ordering, depending on the electrolyte and the temperature used.

The effect of temperature becomes more pronounced in the growth of homogeneous nanotubes in viscous electrolyte, since the viscosity is inversely proportional to the temperature. According to Macak and Schmuki [167], the high temperature promotes the enlargement of the diameter and length of the tubes; and the limiting factor of nanotube growth is the diffusion of the reagents to the inside of the pore (barrier oxide/electrolyte interface) or of the reaction products to the outside of the tubes.

Despite the experimental aspects of the electrochemical synthesis of TiO_2 nanotubes being widely disseminated and discussed in the literature, great effort has been recently observed to improve the optical properties of TiO_2 for applications as photocatalysts under sunlight irradiation. The absorption band of TiO_2 can be shifted to the visible light region by the incorporation of other species into the oxide. In addition, the insertion of doping ions can promote different effects to improve the photocatalytic properties of TiO_2, such as a bandgap reduction, increased crystallization of the material, and increased surface area [170–172]. Considering the method of fabrication of these nanostructures by electrochemical techniques, the introduction of dopants into the oxide crystal lattice—a process known as anodic doping—enables the formation of doped anodic oxides with various interesting properties. The main characteristics of this process will be discussed in the next section.

3.4.4 Modification of the Oxide Properties by Anodic Doping

As previously discussed, the growth of anodic films in valve metals is determined by a high-field mechanism [63]. This means that the ion transport inside the oxide, which depends exponentially on magnitude of the electric field, is the determinant stage of the oxide-forming reaction. Consequently, a linear increase of the barrier-oxide thickness with the electrode potential in the galvanostatic regime is observed. During this process, impurities in the solution can migrate into the film and be incorporated into the oxide crystal lattice during its growth. According to some authors [89,173], the incorporation of anions into the electrolyte and their ability to release electrons for the conduction band are the main factors that cause anodic oxide film breakdown. In general, the adsorption of ions from the electrolyte will be preferred in positively polarized electrodes and will compete with the OH^- anions generated during electrolysis of the water.

In this sense, the modification of the oxide properties can be performed by introduction of dopant salts into the electrolyte, using the technique known as anodic doping [12,174–177]. In this technique, the doping ions present in the electrolyte can migrate into the film and be incorporated into the oxide crystal lattice during the anodization reactions. Various works in the literature describe the doping of oxides by chemical methods [178–185]; however, the doping process by electrochemical techniques, as the anodic doping method, is seldom explored.

A dopant can be defined as an atom that is originally not part of the crystal structure of the material. It can be intentionally introduced as an impurity to modify the properties of the material in question. In the case of doping by substitution at the reticule, the dopant atom can have a different number of valence atoms than the atom it substitutes, which enables forming extra energy levels and modifying the material properties. The type of dopant used and its location in the host structure (substitutional and interstitial defects, vacancies, etc.) can produce different effects during the capture of electrons and/or holes at the surface, or during charge transfer. This way, the dopant nature exerts a great influence in the properties and characteristics of the resulting oxide.

The dopant atoms can be of two types: acceptors and donors. Acceptor dopants are ions that present a lower valence than the ions of the host material. By contrast, donor dopants are ions with a higher valence than those of the host material. An example of doping with acceptors is the case of ZrO_2 doped with Eu^{3+} atoms [174], where the Eu^{3+} ions substitute the Zr^{4+} ion in the oxide crystal lattice, altering the defect density in its microstructure. The acceptor species presents an effective negative charge, represented by the notation Eu'_{Zr}. The introduction of acceptor species tends to add balancing holes to the structure. The inclusion of lower-valence impurities will influence the stoichiometry and the charge balance of the host material, so they must be counterbalanced by a series of defects. Acceptor doping can be done by addition of interstitial cations, anionic vacancies, and holes (h^+) [186].

An example of donor dopant is the doping of ZrO_2 with Nb^{5+} ions [12]. In this case, the Nb^{5+} ions substitute the Zr^{4+} ion in the oxide matrix. The donor species have a positive effective charge ($Nb^{\bullet}_{Zr}$), and their introduction to the oxide matrix produces counterbalanced electrons in the structure. Such electrons also affect the material stoichiometry, which is balanced by defects. Donor doping can be performed through cationic vacancies, interstitial anions, and electrons (e^-).

The ZrO_2 in its most abundant crystalline phase (monoclinic [81]) have a large bandgap of approximately 5.2 eV and can present localized electronic states inside the bandgap, normally associated with irregularities of the crystal lattice or the presence of structural defects. This characteristic makes it an optimal candidate for use in photophysical and photochemical processes [19,187,188]. Furthermore, this oxide presents low phonon energy, i.e., electronic recombination processes that may take place during decay are mostly radioactive processes. Furthermore, nonradioactive recombination or vibrational processes are less probable due to the "harder" nucleus of Zr, which enables its use as a host in materials with luminescent properties. Thus, rare-earth elements can be incorporated into this oxide and have their luminescence signals intensified, creating probes of technological interest [178,188]. One example of this application was verified by Trivinho-Strixino et al. [174], who produced zirconium oxides doped with Eu^{3+} ions, using the anodic doping technique. Fig. 3.26 presents the photoluminescence spectra of pure ZrO_2 and of ZrO_2 doped with Eu^{3+}. A wide emission band is observed between 350 nm (3.5 eV) and 740 nm (1.7 eV), presenting two shoulders centered at 3.1 eV and 2.7 eV, as well as a maximum at 2.3 eV. These bands are related to emission in the oxide matrix, most likely originating from the structural defects introduced during the oxide film growth [19,63,189–192]. The photoluminescence spectrum of the ZrO_2 sample doped with Eu^{3+} exhibits emission lines corresponding to the transition from the Eu(III) $^5D_0$ level to various sublevels of the $^7F_{0,\,1,\,2,\,3,\,4}$ ground state (e_{max}= 2.09 eV, 2.07 eV, 2.02 eV, 1.98 eV, and 1.74 eV, respectively) [193], where the emission line with e_{max} = 2.02 eV ($^5D_0 \rightarrow {}^7F_2$ transition) is the most intense. In addition, Fig. 3.26 also illustrates that the emission lines of Eu^{3+} are asymmetric, which may be related to the fact that these ions are incorporated into the host matrix (ZrO_2) at

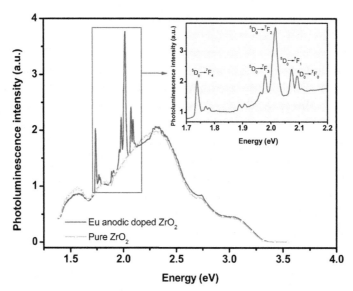

FIGURE 3.26 Emission spectrum of ZrO_2 doped with Eu^{3+} ions prepared in an electrolytic solution containing 0.04 mol L^{-1} H_3PO_4 + 0.05 mol L^{-1} citric acid + 5.3 × 10^{-4} mol L^{-1} $EuCl_3$, I = 16 mA cm^{-2}, T = 20°C. Laser λ_{exc} = 325 nm (3.85 eV) and T = 4.7K. Inset: zoom-in of the Eu^{3+} ion emission lines.

substitutional defects of the crystal, causing an increase in the lattice microdeformation (inset in Fig. 3.26). Since the absorption region for the Eu^{3+} ion is smaller than the excitation of laser light at 325 nm (3.85 eV), it is assumed that the intense emissions of the europium are a consequence of the electronic energy transfer from the ZrO_2 host matrix, which are superimposed onto the electronic states of the Eu^{3+} ions in the energy range below 3.85 eV [174].

Other dopants such as Ca^{2+} and Nb^{5+} ions modify the microstructural properties of ZrO_2. The zirconia films produced by galvanostatic anodization under PEO conditions crystallize predominantly in the monoclinic phase. Incorporation of these dopants to the oxide can lead to the stabilization of the tetragonal and cubic phases at room temperature. In pure ZrO_2, these crystalline phases are stable only at high temperatures (above 1,200°C) [12,81,174,175]. Depending on the anodization conditions, such as the electrolyte concentration, composition, and temperature, as well as the applied current density, the incorporation of these doping species allows the partial or full stabilization of the cubic zirconia, turning the material into a good conductor and increasing its mechanical resistance. X-ray diffractometry results demonstrated the stabilization of ZrO_2 films at room temperature, with 80% in the tetragonal or cubic phases, produced by galvanostatic anodization of Zr in phosphoric acid in the presence of Ca^{2+} ions [175].

In the synthesis of TiO_2 nanotubes, the Nb^{5+} ions are widely used as dopants due to the easy insertion into the oxide crystal structure. Several studies show that doping with niobium produces materials that absorb in a wider band of the visible light spectrum [194–196]. According to Depero et al. [194], adding niobium to TiO_2 presents a series of advantages over the pure material. The formed crystallites become smaller, increasing the photoactive area of the material. In addition, the presence of the dopant promotes increased stress in the anatase phase, thus favoring the conversion to the rutile phase. Nonetheless, some authors have observed that although the mixed material absorbs a larger amount of visible light than the pure material, its photocatalytic activity is equal or lower than that of the pure oxide. This is because the formation of oxygenated Nb clusters promotes an increase in the rate of electron-hole recombination, decreasing the photoactivity.

Another way to dope the material without necessarily introducing dopant ions into the electrolyte is using the reanodization technique, in which the dopant metal is cathodically deposited over the previously anodized porous oxide. After electrodeposition of the dopant, the material is reanodized in order to introduce the metallic ions to the oxide crystal lattice [171]. Studies with copper, manganese, and chromium as dopants in TiO_2 nanotubes demonstrate that the photocurrent is greater in the doped electrodes, and the electrocatalytic efficiency increases with the dopant concentration, up to an optimum value of approximately 1%, due to the decrease in the electron-hole recombination. Above this optimum concentration value, the metal behaves as a recombination center for electron-hole pairs, decreasing the electrocatalytic efficiency of the doped material.

3.5 Conclusions

In this chapter, the use of electrochemical methods was discussed with a focus on the preparation of nanostructured materials. The approaches described present a robust method of nanomaterials fabrication, which is easy to use, produces results that are easy to interpret, is faster, and has a lower cost than other fabrication techniques, such as lithography.

Furthermore, a wide range of materials with a variety of properties can be easily obtained by electrochemical methods, given that the film properties can be controlled by the experimental conditions employed in the synthesis. Different types of materials can be obtained simply by changing voltage, current, temperature, pH, and/or the electrolyte used. In addition to obtaining electrodeposited films with different compositions, morphologies, and microstructures, it is possible to obtain highly organized structures. Examples of this are the cases of porous anodic alumina and TiO_2 nanotubes, as well as the formation of concave nanotubes, nanocolumns, and other structures. It is important to emphasize that structures with a lower order degree but large surface area and high degree of crystallinity can also be produced, and they present interesting complex phenomena during film growth, which are extensively used but yet poorly understood. An example is the case of oxide films obtained by anodization under high electric field, i.e., in conditions of dieletric breakdown of the anodic oxide.

Finally, electrochemistry also presents an interesting technique for the modification of material properties: the method of anodic doping. As discussed previously, the addition of doping ions to the electrolyte promotes significant modifications in the host material, which will depend on the nature of the inserted species. These modifications can change the material properties, such as: increased photocatalytic activity; increased biocompatibility; increased corrosion resistance; improved thermal conductivity in oxides; improved photoluminescent materials, etc. After an appropriate development time, these new materials with interesting properties can be introduced into technological niches, gaining a space in practical and commercially viable applications and becoming part of the daily life of the entire society, from the scientist to the common citizen.

List of Symbols

A Area
e Electron charge
e^- Electron
E Potential
E_0 Externally applied potential
F Faraday constant
h^+ Vacancy
I Electric current
j Electric density
l Thickness
q Qharge
R Ohmic resistance
T Temperature
t Time
z Number of moles of electrons
λ Wavelength
ρ Resistivity

References

[1] L.P. Bicelli, et al. A review of nanostructural aspects of metal electrodeposition, Int. J. Electrochem. Sci. 3 (4) (2008) 356–408.

[2] A. Mukhopadhyay, B. Basu, Consolidation–microstructure–property relationships in bulk nanoceramics and ceramic nanocomposites: a review, Int. Mater. Rev. 52 (2007) 257–288.

[3] I. Gurrappa, L. Binder, Electrodeposition of nanostructured coatings and their characterization—a review, Sci. Technol. Adv. Mater. 9 (4) (2008) 043001.

[4] G. Cao, Y. Wang, Nanostructures and Nanomaterials: Synthesis, Properties, and Applications, second ed., World Scientific Publishing Co. Pte. Ltd, Singapore, (2011) 596.

[5] J. Briggle, P.D. Nolidin, T.D. Golden, Multilayer film fabrication using flow injection coupled with electrochemical deposition, Electroanalysis 22 (19) (2010) 2157–2161.

[6] B. Jang, et al. Fabrication of Segmented Au/Co/Au nanowires: insights in the quality of Co/Au junctions, ACS Appl. Mater. Interfaces 6 (16) (2014) 14583–14589.

[7] H. Chen, et al. Electrochemical construction of porous gold nanostructures on DVD substrate and its application as nonenzymatic hydrogen peroxide sensor, Sci. China-Chem. 58 (10) (2015) p1585–1592.

[8] A. Uzunoglu, A.D. Scherbarth, L.A. Stanciu, Bimetallic PdCu/SPCE non-enzymatic hydrogen peroxide sensors, Sens. Actuat B-Chem. 220 (2015) 968–976.

[9] M. Alanyalioglu, F. Bayrakceken, U. Demir, Preparation of PbS thin films: a new electrochemical route for underpotential deposition, Electrochim. Acta 54 (26) (2009) 6554–6559.

[10] N.M.N. Rozlin, A.M. Alfantazi, Nanocrystalline cobalt–iron alloy: synthesis and characterization, Mater. Sci. Eng. A-Struct. Mater. 550 (2012) 388–394.

[11] F. Trivinho-Strixino, et al. Active waveguide effects from porous anodic alumina: an optical sensor proposition, Appl. Phys. Lett. 97 (1) (2010).

[12] J.S. Santos, F. Trivinho-Strixino, E.C. Pereira, The influence of experimental conditions on microstructure and morphology of Nb-doped ZrO_2 films prepared by spark anodization, Corros. Sci. 73 (2013) 99–105.

[13] H.A. Guerreiro, et al. Grazing angle photoluminescence of porous alumina as an analytical transducer for gaseous ethanol detection, J. Nanosci. Nanotechnol. 14 (9) (2014) 6653–6657.

[14] J.L. Zhao, et al. Crystal phase transition and properties of titanium oxide nanotube arrays prepared by anodization, J. Alloy. Compd. 434 (2007) 792–795.

[15] R.G. Freitas, E.P. Antunes, E.C. Pereira, CO and methanol electrooxidation on Pt/Ir/Pt multilayers electrodes, Electrochim. Acta 54 (7) (2009) 1999–2003.

[16] R. Nagao, et al. Oscillatory electro-oxidation of methanol on nanoarchitectured Ptpc/Rh/Pt metallic multilayer, ACS Catal. 5 (2) (2015) 1045–1052.

[17] M. Gratzel, Photoelectrochemical cells, Nature 414 (6861) (2001) 338–344.

[18] G.K. Mor, et al. A review on highly ordered, vertically oriented TiO_2 nanotube arrays: fabrication, material properties, and solar energy applications, Sol. Energy Mater. Sol. Cells 90 (14) (2006) 2011–2075.

[19] A. Emeline, et al. Spectroscopic and photoluminescence studies of a wide band gap insulating material: powdered and colloidal ZrO_2 sols, Langmuir 14 (18) (1998) 5011–5022.

[20] F. Xiao, et al. Recent progress in electrodeposition of thermoelectric thin films and nanostructures, Electrochim. Acta 53 (28) (2008) 8103–8117.

[21] P.G. Bruce, B. Scrosati, J.M. Tarascon, Nanomaterials for rechargeable lithium batteries, Angew. Chem. Int. Ed. 47 (2008) 2930–2946.

[22] G.K. Mor, et al. Fabrication of hydrogen sensors with transparent titanium oxide nanotube-array thin films as sensing elements, Thin Solid Films 496 (1) (2006) 42–48.

[23] J.J. Gooding, L.M.H. Lai, I.Y. Goon, Nanostructured electrodes with unique properties for biological and other applications, in: R.C. Alkire, D.M. Kolb, et al. (Eds.), Chemically Modified Electrodes, vol. 11, Wiley-VCH, Weinheim, 2009, pp. 1–56.

[24] A.G. Mantzila, M.I. Prodromidis, Development and study of anodic Ti/TiO_2 electrodes and their potential use as impedimetric immunosensors, Electrochim. Acta 51 (17) (2006) 3537–3542.

[25] K.I. Popov, S.S. Djokic, B.N. Grgur, Fundamentals Aspects of Electrometallurgy, Kluwer Academic Publishers, New York, (2002).

[26] F. Trivinho-Strixino, J.S. Santos, M.S. Sikora, Síntese Eletroquímica de Materiais Nanoestruturados [electrochemical synthesis of nanostructured materials], in: A.L. Da Róz, F.L. Leite, et al. (Eds.), Nanoestruturas: Princípios e Aplicações, vol. 1, Elsevier, Rio de Janeiro, 2015, pp. 63–110 Chapter 3.

[27] L. Young, Anodic Oxide Films. 1, Academic Press, New York, (1961) p. 377.

[28] J.O. Bockris, A.K.N. Reddy, Modern Electrochemistry, Plenum Press, New York, (1973) pp. 1231–1251.

[29] D.M. Kolb, An atomistic view of electrochemistry, Surf. Sci. 500 (1–3) (2002) 722–740.

[30] V.P. Santos, G. Tremiliosi, The correlation between the atomic surface structure and the reversible adsorption-desorption of hydrogen on single crystal Pt(111), Pt(100) e Pt(110), Quim. Nova 24 (2001) 856–863.

[31] C. Lebouin, et al. Electrochemically elaborated palladium nanofilms on Pt(111): characterization and hydrogen insertion study, J. Electroanal. Chem. 626 (1–2) (2009) 59–65.

[32] N.M. Markovic, H.A. Gasteiger, P.N. Ross, Copper electrodeposition on Pt(111) in the presence of chloride and (bi)sulfate—rotanting-ring Pt(111) disk electrode studies, Langmuir 11 (10) (1995) 4098–4108.

[33] I. Flis-Kabulska, Electrodeposition of cobalt on gold during voltammetric cycling, J. Appl/ Electrochem. 36 (2006) 131–137.

[34] L.H. Mendoza-Huizar, J. Robles, M. Palomar-Pardavé, Nucleation and growth of cobalt onto different substrates Part I. Underpotential deposition onto a gold electrode, J. Electroanal. Chem. 521 (2002) 95–106.

[35] M.C. Vilchenski, et al. Electrodeposition of Co and Co-Fe Films on platinum and on copper substrates, Portugaliae Electrochim. Acta 21 (2003) 33–47.

[36] B. Losiewicz, et al. The structure, morphology and electrochemical impedance study of the hydrogen evolution reaction on the modified nickel electrodes, Int. J. Hydrogen Energ. 29 (2) (2004) 145–157.

[37] L. Bonou, M. Eyraud, J. Crousier, Nucleation an growth of copper on glassy-carbon and steel, J. Appl. Electrochem. 24 (9) (1994) 906–910.

[38] C.L. Fan, D.L. Piron, P. Paradis, Hydrogen evolution on electrodeposited nickel cobalt molybdenum in alkaline water electrolysis, Electrochim. Acta 39 (18) (1994) 2715–2722.

[39] D. Stoychev, et al. Electrodeposition of platinum on metallic and nonmetallic substrates—selection of experimental conditions, Mater. Chem. Phys. 72 (3) (2001) 360–365.

[40] J.H. Marsh, S.W. Orchard, Voltammetric studies of glassy-carbon electrodes activated in air and steam, Carbon 30 (6) (1992) 895–901.

[41] S. Ranganathan, T.C. Kuo, R.L. Mccreery, Facile preparation of active glassy carbon electrodes with activated carbon and organic solvents, Anal. Chem. 71 (16) (1999) 3574–3580.

[42] Y.-J. Song, J.-Y. Kim, K.-W. Park, Synthesis of Pd dendritic nanowires by electrochemical deposition, Crys. Growth Des. 9 (1) (2009) 505–507.

[43] X. Su, C. Qiang, Influence of pH and bath composition on properties of Ni–Fe alloy films synthesized by electrodeposition, Bull. of Mater. Sci. 35 (2) (2012) 183–189.

[44] F.W.S. Lucas, A.R.F. Lima, L.H. Mascaro, Glycerol as additive in copper indium gallium diselenide electrodeposition: morphological, structural and electronic effects, RSC Adv. 5 (24) (2015) 18295–18300.

[45] A.J. Bard, L.R. Faulkner, Electrochemical Methods—Fundamentals and Applications, John Wiley & Sons, New York, (1980) p. 51.

[46] J.T. Matsushima, F. Trivinho-Strixino, E.C. Pereira, Investigation of cobalt deposition using the electrochemical quartz crystal microbalance, Electrochim. Acta 51 (2006) 1960–1966.

[47] I.L. Rakityanskaya, A.B. Shein, Anodic behavior of iron, cobalt, and nickel silicides in alkaline electrolytes, Russ. J. Electrochem. 42 (2006) 1208–1212.

[48] G.Y. Qiao, et al. High-speed jet electrodeposition and microstructure of nanocrystalline Ni–Co alloys, Electrochim. Acta 51 (1) (2005) 85–92.

[49] J.S. Santos, et al. Effect of temperature in cobalt electrodeposition in the presence of boric acid, Electrochim. Acta 53 (2007) 644–649.

[50] K.D. Song, et al. A study on effect of hydrogen reduction reaction on the initial stage of Ni electrodeposition using EQCM, Electrochem. Commun. 5 (2003) 460–466.

[51] A.R. Despic, Deposition and dissolution of metals and alloys. Part. B: mechanisms, kinetics, texture and morphology, in: B.E. Conway, J.O.M. Bockris, et al. (Eds.), Comprehensive Treatise of Electrochemistry, vol. 7, Plenum Press, New York, 1973, pp. 483.

[52] J.M. Elliott, et al. Nanostructured platinum (H-I-ePt) films: effects of electrodeposition conditions on film properties, Chem. Mater. 11 (12) (1999) 3602–3609.

[53] W.Q. Yang, et al. Nanostructured palladium-silver coated nickel foam cathode for magnesium-hydrogen peroxide fuel cells, Electrochim. Acta 52 (1) (2006) 9–14.

[54] C.M.A. Brett, A.M.O. Brett, Electrochemistry: Principles, Methods and Applications, Oxford University Press, New York, (1993).

[55] D.Y. Park, et al. Nanostructured magnetic CoNiP electrodeposits: structure-property relationships, Electrochim. Acta 47 (18) (2002) 2893–2900.

[56] C.A. Ross, Electrodeposited multilayer thin films, Annu. Rev. Mater. Sci. 24 (1994) 159–188.

[57] R.G. Freitas, E.C. Pereira, Giant multilayer electrocatalytic effect investigation on Pt/Bi/Pt nanostructured electrodes towards CO and methanol electrooxidation, Electrochim. Acta 55 (26) (2010) 7622–7627.

[58] D.S. Lashmore, R. Thomson, Cracks and dislocations in face-centered cubic metallic multilayers, J. Mater. Res. 7 (9) (1992) 2379–2386.

[59] P.N. Bartlett, et al. Non-aqueous electrodeposition of p-block metals and metalloids from halometallate salts, RSC Adv. 3 (36) (2013) 15645–15654.

[60] R. Liu, et al. Epitaxial electrodeposition of zinc oxide nanopillars on single-crystal gold, Chem. Mater. 13 (2) (2001) 508–512.

[61] M.J. Chappell, J.S.L. Leach, Passivity and breakdown of passivity of valve metals, in: R.P. Frankenthal, J. e Kruger (Eds.), The Electrochemical Society, Princeton, New Jersey, 1978.

[62] S. Ikonopisov, Theory of electrical breakdown during formation of barrier anodic films, Electrochim. Acta 22 (10) (1977) 1077–1082.

[63] V.P. Parkhutik, J.M. Albella, J.M. Martinez-Duart, Electric breakdown in anodic oxide films, in: B.E. Conway, J.O.M. Bockris, et al. (Eds.), Moderns Aspects of Electrochemistry, vol. 23, Plenum Press, New York, 1992, pp. 315–391 Chapter 5.

[64] A.L. Yerokhin, et al. Plasma electrolysis for surface engineering, Surf. Coat. Technol. 122 (2–3) (1999) 73–93.

[65] W. Lee, et al. Fast fabrication of long-range ordered porous alumina membranes by hard anodization, Nat. Mater. 5 (9) (2006) 741–747.

[66] J.S. Santos, et al. Characterization of electrical discharges during spark anodization of zirconium in different electrolytes, Electrochim. Acta 130 (2014) 477–487.

[67] M.S. Sikora, et al. Influence of the morphology and microstructure on the photocatalytic properties of titanium oxide films obtained by sparking anodization in H_3PO_4, Electrochim. Acta 56 (9) (2011) 3122–3127.

[68] B.H. Long, et al. Characteristics of electric parameters in aluminium alloy MAO coating process, J. Phys. D-Appl. Phys. 38 (18) (2005) 3491–3496.

[69] G. Sundararajan, L. Rama Krishna, Mechanisms underlying the formation of thick alumina coatings through the MAO coating technology, Surf. Coat. Technol. 167 (2–3) (2003) 269–277.

[70] A.L. Yerokhin, et al. Discharge characterization in plasma electrolytic oxidation of aluminium, J. Phys. D: Appl. Phys. 36 (17) (2003) 2110.

[71] S. Moon, Y. Jeong, Generation mechanism of microdischarges during plasma electrolytic oxidation of Al in aqueous solutions, Corros. Sci. 51 (7) (2009) 1506–1512.

[72] A.I. Maximov, A.V. Khlustova, Optical emission from plasma discharge in electrochemical systems applied for modification of material surfaces, Surf. Coat. Technol. 201 (21) (2007) 8782–8788.

[73] R.O. Hussein, et al. Spectroscopic study of electrolytic plasma and discharging behaviour during the plasma electrolytic oxidation (PEO) process, J.Phys. D: Appl. Phys. 43 (2010) 105203.

[74] F. Monfort, et al. Development of anodic coatings on aluminium under sparking conditions in sili-cate electrolyte, Corros. Sci. 49 (2) (2007) 672–693.

[75] H. Habazaki, et al. Spark anodizing of [beta]-Ti alloy for wear-resistant coating, Surf. Coat. Technol. 201 (21) (2007) 8730–8737.

[76] A. Yerokhin, et al. Spatial characteristics of discharge phenomena in plasma electrolytic oxidation of aluminium alloy, Surf. Coat. Technol. 177 (2004) 779–783.

[77] P. Gupta, et al. Electrolytic plasma technology: science and engineering—an overview, Surf. Coat. Technol. 201 (21) (2007) 8746–8760.

[78] W. Ma, et al. Preparation and in vitro biocompatibility of hybrid oxide layer on titanium surface, Surf. Coat. Technol. 205 (6) (2010) 1736–1742.

[79] X. Shi, L. Xu, Q. Wang, Porous TiO_2 film prepared by micro-arc oxidation and its electrochemical behaviors in Hank's solution, Surf. Coat. Technol. 205 (6) (2010) 1730–1735.

[80] Y. Yan, et al. Microstructure and bioactivity of Ca, P and Sr doped TiO_2 coating formed on porous titanium by micro-arc oxidation, Surf. Coat. Technol. 205 (6) (2010) 1702–1713.

[81] F. Trivinho-Strixino, et al. Tetragonal to monoclinic phase transition observed during Zr anodisation, J. Solid State Electrochem. 1–9 (2012).

[82] J.S.L. Leach, B.R. Pearson, The Effect of foreign ions upon the electrical characteristics of anodic ZrO_2 films, Electrochim. Acta 29 (9) (1984) 1271–1282.

[83] J.S.L. Leach, B.R. Pearson, Crystallization in anodic oxide-films, Corros. Sci. 28 (1988) 43.

[84] M.V. Diamanti, M.P. Pedeferri, Effect of anodic oxidation parameters on the titanium oxides forma-tion, Corros. Sci. 49 (2) (2007) 939–948.

[85] R.O. Hussein, et al. Spectroscopic study of electrolytic plasma and discharging behaviour during the plasma electrolytic oxidation (PEO) process, J. Phys. D-Appl. Phys. 43 (10) (2010).

[86] S. Stojadinović, et al. Spectroscopic and real-time imaging investigation of tantalum plasma electro-lytic oxidation (PEO), Surf. Coat. Technol. 205 (23–24) (2011) 5406–5413.

[87] C.S. Dunleavy, et al. Characterisation of discharge events during plasma electrolytic oxidation, Surf. Coat. Technol. 203 (22) (2009) 3410–3419.

[88] M.S. Sikora, et al. Theoretical calculation of the local heating effect on the crystallization of TiO_2 pre-pared by sparking anodization, Curr. Nanosci. 11 (3) (2015) 263–270.

[89] J.M. Albella, I. Montero, J.M. Martinez-Duart, Electron injection and avalanche during the anodic oxidation of tantalum, J. Electrochem. Soc.: Solid-State Sci. Technol. 131 (5) (1984) 1101–1104.

[90] E. Matykina, et al. Investigation of the growth processes of coatings formed by AC plasma electrolytic oxidation of aluminium, Electrochim. Acta 54 (27) (2009) 6767–6778.

[91] A. Yerokhin, A. Pilkington, A. Matthews, Pulse current plasma assisted electrolytic cleaning of AISI 4340 steel, J. Mater. Process. Technol. 210 (1) (2010) 54–63.

[92] W. Chanmanee, et al. Titania nanotubes from pulse anodization of titanium foils, Electrochem. Commun. 9 (8) (2007) 2145–2149.

[93] W. Lee, et al. Structural engineering of nanoporous anodic aluminium oxide by pulse anodization of aluminium, Nat. Nanotechnol. 3 (2008) 234–239.

[94] L. Sottovia, et al. Thin films produced on 5052 aluminum alloy by plasma electrolytic oxidation with red mud-containing electrolytes, Mater. Res. 17 (6) (2014) 1404–1409.

[95] D. Losic, A. Santos, Nanoporous Alumina Fabrication, Structure, Properties and Applications, Springer, New York, (2015) p. 362.

[96] F. Keller, M.S. Hunter, D.L. Robinson, Structural features of oxide coatings on aluminum, J. Electro-chem. Soc. 100 (9) (1953) 411–419.

[97] A. Despic, V.P. Parkhutik, Electrochemistry of aluminum in aqueous solution and physics of its anodic oxide, in: J.O.M. Bockris, R.E. White, et al. (Eds.), Modern Aspects of Electrochemistry, vol. 20, Plenum Press, New York and London, 1989, pp. 518 Chapter 6.

[98] J.W. Diggle, T.C. Downie, C.W. Goulding, Anodic oxide films on aluminum, Chem. Rev. 69 (3) (1969) 365–382.

[99] G.E. Thompson, G.C. Wood, Treatise on Materials Science and Technology, Academic Press, New York, (1983).

[100] H. Takahashi, K. Fujimoto, M. Nagayama, Effect of pH on the distribution of anions in anodic oxide-films formed on aluminum in phosphate solutions, J. Electrochem. Soc. 135 (6) (1988) 1349–1353.

[101] G.D. Sulka, Highly ordered anodic porous alumina formation by self-organized anodizing, in: A. Eftekhari (Ed.), Nanostructured Materials in Electrochemistry, first ed., Wiley-VCH, Berlin, 2008, pp. 1–116.

[102] G. Patermarakis, K. Masavetas, Aluminium anodising in oxalate and sulphate solutions. Comparison of chronopotentiometric and overall kinetic response of growth mechanism of porous anodic films, J. Electroanal. Chem. 588 (2) (2006) 179–189.

[103] G. Patermarakis, J. Chandrinos, K. Masavetas, Formulation of a holistic model for the kinetics of steady state growth of porous anodic alumina films, J Solid State Electrochem. 11 (9) (2007) 1191–1204.

[104] G. Patermarakis, The origin of nucleation and development of porous nanostructure of anodic alumina films, J. Electroanal. Chem. 635 (1) (2009) 39–50.

[105] G. Patermarakis, K. Moussoutzanis, Development and application of a holistic model for the steady state growth of porous anodic alumina films, Electrochim. Acta 54 (9) (2009) 2434–2443.

[106] G. Patermarakis, J. Diakonikolaou, Mechanism of aluminium and oxygen ions transport in the barrier layer of porous anodic alumina films, J. Solid State Electrochem. 16 (9) (2012) 2921–2939.

[107] M. Pashchanka, J.J. Schneider, Uniform contraction of high-aspect-ratio nanochannels in hexagonally patterned anodic alumina films by pulsed voltage oxidation, Electrochem. Commun. 34 (2013) 263–265.

[108] Y. Chen, et al. On the generation of interferometric colors in high purity and technical grade aluminum: an alternative green process for metal finishing industry, Electrochim. Acta 174 (2015) 672–681.

[109] S.-Y. Li, et al. Fabrication of one-dimensional alumina photonic crystals by anodization using a modified pulse-voltage method, Mater. Res. Bull. 68 (2015) 42–48.

[110] J.M.M. Moreno, et al. Constrained order in nanoporous alumina with high aspect ratio: smart combination of interference lithography and hard anodization, Adv. Funct. Mater. 24 (13) (2014) 1857–1863.

[111] T. Kumeria, A. Santos, Nanoporous alumina membranes for chromatography and molecular transporting, In: A. Santos, D. Losic, (Eds.), Nanoporous Alumina: Fabrication, Structure, Properties and Applications, vol. 219, Springer, 2015, p. 362 (Chapter 10), Switzerland.

[112] A. Brudzisz, A. Brzozka, G.D. Sulka, Effect of processing parameters on pore opening and mechanism of voltage pulse detachment of nanoporous anodic alumina, Electrochim. Acta 178 (2015) 374–384.

[113] W. Lee, The anodization of aluminum for nanotechnology applications, Jom 62 (6) (2010) 57–63.

[114] U.M. Garcia, et al. Comparison and construction of "mild" and "hard" anodisation reactors for the synthesis of porous alumina, Quím. Nova 38 (8) (2015) 1112–1116.

[115] Y. Li, et al. Photoluminescence and optical absorption caused by the F+ centres in anodic alumina membranes, J. Phys. Condens. Matter 13 (11) (2001) 2691–2699.

[116] G.H. Li, et al. Wavelength dependent photoluminescence of anodic alumina membranes, J. Phys. Condens. Matter 15 (49) (2003) 8663–8671.

[117] G.S. Huang, et al. Strong blue emission from anodic alumina membranes with ordered nanopore array, J. Appl. Phys. 93 (1) (2003) 582–585.

[118] G.S. Huang, et al. Dependence of blue-emitting property on nanopore geometrical structure in Al-based porous anodic alumina membranes, Appl. Phys. A: Mater. Sci. Process. 81 (2005) 1345.

[119] J.H. Chen, et al. The investigation of photoluminescence centers in porous alumina membranes, Appl. Phys. A-Mater. Sci. Process. 84 (3) (2006) 297–300.

[120] Z. Li, K. Huang, Optical properties of alumina membranes prepared by anodic oxidation process, J. Lumin. 127 (2) (2007) 435–440.

[121] T. Gao, G.M. Meng, L.D. Zhang, Blue luminescence in porous anodic alumina films: the role of the oxalic impurities, J. Phys. Conden. Matter 15 (12) (2003) 2071–2079.

[122] S. Wang, et al. The effect of nanometer size of porous anodic aluminum oxide on adsorption and fluorescence of tetrahydroxyflavanol, Spectrochim. Acta A-Mol. Biomol. Spectrosc. 59 (2003) 1139.

[123] G.S. Huang, et al. On the origin of light emission from porous anodic alumina formed in sulfuric acid, Solid State Commun. 137 (2006) 621.

[124] X. Liu, et al. Photoluminescence of poly(thiophene) nanowires confined in porous anodic alumina membrane, Polymer 49 (9) (2008) 2197–2201.

[125] C.-S. Hsiao, et al. Synthesis and luminescent properties of strong blue light-emitting Al_2O_3/ZnO nanocables, J. Electrochem. Soc. 155 (5) (2008) K96–K99.

[126] Y.Q. Cheng, et al. Photoluminescence characteristics of several fluorescent molecules on nanometer porous alumina film, Acta Chim. Sinica 62 (2) (2004) 183–187.

[127] D.F. Qi, et al. Optical emission of conjugated polymers adsorbed to nanoporous alumina, Nano Lett. 3 (9) (2003) 1265–1268.

[128] X.-J. Wu, et al. Electrochemical synthesis and applications of oriented and hierarchically quasi-1D semiconducting nanostructures, Coord. Chem. Rev. 254 (9–10) (2010) 1135–1150.

[129] Y.M. Park, et al. Artificial petal surface based on hierarchical micro- and nanostructures, Thin Solid Films 520 (1) (2011) 362–367.

[130] R.E. Sabzi, K. Kant, D. Losic, Electrochemical synthesis of nickel hexacyanoferrate nanoarrays with dots, rods and nanotubes morphology using a porous alumina template, Electrochim. Acta 55 (5) (2010) 1829–1835.

[131] A. Hoess, et al. Self-supporting nanoporous alumina membranes as substrates for hepatic cell cultures, J. Biomed. Mater. Res. A 100 (9) (2012) 2230–2238.

[132] S. Thakur, et al. Depth matters: cells grown on nano-porous anodic alumina respond to pore depth, Nanotechnology 23 (25) (2012).

[133] I. Sopyan, A. Fadli, M. Mel, Porous alumina-hydroxyapatite composites through protein foaming-consolidation method, J. Mech. Behav. Biomed. Mater. 8 (2012) 86–98.

[134] E. Bernardo, et al. Porous wollastonite hydroxyapatite bioceramics from a preceramic polymer and micro- or nano-sized fillers, J. Eur. Ceram. Soc. 32 (2) (2012) 399–408.

[135] B. Yuan, et al. Fabrication of porous alumina green bodies from suspension emulsions by gel casting, Mater. Lett. 81 (2012) 151–154.

[136] I. Sopyan, A. Fadli, Floating porous alumina from protein foaming-consolidation technique for cell culture application, Ceram. Int. 38 (6) (2012) 5287–5291.

[137] J. Martin, et al. Tailored polymer-based nanorods and nanotubes by "template synthesis": from preparation to applications, Polymer 53 (6) (2012) 1149–1166.

[138] D. Gong, et al. Titanium oxide nanotube arrays prepared by anodic oxidation, J. Mater. Res. 16 (12) (2001) 3331–3334.

[139] J. Podporska-Carroll, et al. Antimicrobial properties of highly efficient photocatalytic TiO$_2$ nanotubes, Appl. Catal. B: Environ. 176-177 (2015) 70–75.

[140] N. Shahzad, et al. Silver-copper nanoalloys-an efficient sensitizer for metal-cluster-sensitized solar cells delivering stable current and high open circuit voltage, J. Power Sources 294 (2015) 609–619.

[141] Z.P. Yang, et al. A high-performance nonenzymatic piezoelectric sensor based on molecularly imprinted transparent TiO$_2$ film for detection of urea, Biosens. Bioelectron. 74 (2015) 85–90.

[142] R. Hahn, J.M. Macak, P. Schmuki, Rapid anodic growth of TiO$_2$ and WO$_3$ nanotubes in fluoride free electrolytes, Electrochem. Commun. 9 (5) (2007) 947–952.

[143] V. Stengl, et al. Preparation, characterization and photocatalytic activity of optically transparent titanium dioxide particles, Mater. Chem. Phys. 105 (1) (2007) 38–46.

[144] A. Kar, K.S. Raja, M. Misra, Electrodeposition of hydroxyapatite onto nanotubular TiO$_2$ for implant applications, Surf. Coat. Technol. 201 (6) (2006) 3723–3731.

[145] M. Kulkarni, et al. Titanium nanostructures for biomedical applications, Nanotechnology 26 (6) (2015) 062002.

[146] K. Syrek, et al. Effect of electrolyte agitation on anodic titanium dioxide (Ato) growth and its photoelectrochemical properties, Electrochim. Acta 180 (2015) 801–810.

[147] H.-J. Song, et al. The effects of spark anodizing treatment of pure titanium metals and titanium alloys on corrosion characteristics, Surf. Coat. Technol. 201 (21) (2007) 8738–8745.

[148] S.K. Poznyak, D.V. Talapin, A.I. Kulak, Electrochemical oxidation of titanium by pulsed discharge in electrolyte, J. Electroanal. Chem. 579 (2) (2005) 299–310.

[149] S. Daothong, et al. Size-controlled growth of TiO$_2$ nanowires by oxidation of titanium substrates in the presence of ethanol vapor, Scr. Mater. 57 (7) (2007) 567–570.

[150] D. Regonini, et al. A review of growth mechanism, structure and crystallinity of anodized TiO$_2$ nanotubes, Mater. Sci. Eng. R-Rep. 74 (12) (2013) 377–406.

[151] V.M. Prida, et al. Temperature influence on the anodic growth of self-aligned Titanium dioxide nanotube arrays, J. Magn. Magn. Mater. 316 (2) (2007) 110–113.

[152] Q. Cai, L. Yang, Y. Yu, Investigations on the self-organized growth of TiO$_2$ nanotube arrays by anodic oxidization, Thin Solid Films 515 (4) (2006) 1802–1806.

[153] G.A. Crawford, N. Chawla, Porous hierarchical TiO$_2$ nanostructures: processing and microstructure relationships, Acta Mater. 57 (3) (2009) 854–867.

[154] S. Berger, et al. The origin for tubular growth of TiO$_2$ nanotubes: a fluoride rich layer between tubewalls, Surf. Sci. 605 (19–20) (2011) L57–L60.

[155] L.V. Taveira, et al. Voltage oscillations and morphology during the galvanostatic formation of self-organized TiO$_2$ nanotubes, J. Electrochem. Soc. 153 (4) (2006) B137–B143.

[156] L.V. Taveira, et al. Initiation and growth of self-organized TiO[sub 2] nanotubes anodically formed in NH$_4$F/(NH$_4$)$_2$SO$_4$ electrolytes, J. Electrochem. Soc. 152 (10) (2005) B405–B410.

[157] F. Thebault, et al. Modeling of growth and dissolution of nanotubular Titania in fluoride-containing electrolytes, Electrochem. Solid State Lett. 12 (3) (2009) C5–C9.

[158] K. Yasuda, et al. Mechanistic aspects of the self-organization process for oxide nanotube formation on valve metals, J. Electrochem. Soc. 154 (9) (2007) C472–C478.

[159] G.E. Thompson, et al. Nucleation and growth of porous anodic films on aluminum, Nature 272 (5652) (1978) 433–435.

[160] E. Matykina, et al. Morphologies of nanostructured TiO$_2$ doped with F on Ti-6Al-4V alloy, Electrochim. Acta 56 (5) (2011) 2221–2229.

[161] S.A.A. Yahia, et al. Effect of anodizing potential on the formation and EIS characteristics of TiO_2 nanotube arrays, J. Electrochem. Soc. 159 (4) (2012) K83–K92.

[162] Y. Alivov, Z.Y. Fan, D. Johnstone, Titanium nanotubes grown by titanium anodization, J. Appl. Phys. 106 (3) (2009).

[163] T. Ruff, R. Hahn, P. Schmuki, From anodic TiO_2 nanotubes to hexagonally ordered TiO_2 nanocolumns, Appl. Surf. Sci. 257 (19) (2011) 8177–8181.

[164] G.K. Mor, et al. Fabrication of tapered, conical-shaped titania nanotubes, J. Mater. Res. 18 (11) (2003) 2588–2593.

[165] J.M. Macák, H. Tsuchiya, P. Schmuki, High-aspect-ratio TiO_2 nanotubes by anodization of titanium, Angew. Chem. Int. Ed. 44 (14) (2005) 2100–2102.

[166] S. Kaneco, et al. Fabrication of uniform size titanium oxide nanotubes: impact of current density and solution conditions, Scr. Mater. 56 (5) (2007) 373–376.

[167] J.M. Macak, P. Schmuki, Anodic growth of self-organized anodic TiO_2 nanotubes in viscous electrolytes, Electrochim. Acta 52 (3) (2006) 1258–1264.

[168] A. Roguska, et al. Synthesis and characterization of ZnO and Ag nanoparticle-loaded TiO_2 nanotube composite layers intended for antibacterial coatings, Thin Solid Films 553 (2014) 173–178.

[169] B. Yuan, et al. Nitrogen doped TiO_2 nanotube arrays with high photoelectrochemical activity for photocatalytic applications, Appl. Surf. Sci. 280 (2013) 523–529.

[170] C. Das, et al. Photoelectrochemical and photocatalytic activity of tungsten doped TiO_2 nanotube layers in the near visible region, Electrochim. Acta 56 (28) (2011) 10557–10561.

[171] E.B. Gracien, et al. Photocatalytic activity of manganese, chromium and cobalt-doped anatase titanium dioxide nanoporous electrodes produced by re-anodization method, Thin Solid Films 515 (13) (2007) 5287–5297.

[172] A. Kubacka, G. Colón, M. Fernández-García, Cationic (V, Mo, Nb, W) doping of TiO_2-anatase: a real alternative for visible light-driven photocatalysts, Catal. Today 143 (3–4) (2009) 286–292.

[173] J.S.L. Leach, B.R. Pearson, The conditions for incorporation of electrolyte ions into anodic oxides, Electrochim. Acta 29 (9) (1984) 1263–1270.

[174] F. Trivinho-Strixino, F.E.G. Guimaraes, E.C. Pereira, Luminescence in anodic ZrO_2 doped with Eu(III) ions, Mol. Cryst. Liq. Cryst. 485 (2008) 766–775.

[175] E.O. Bensadon, et al. Cubic stabilized zirconium oxide anodic films prepared at room temperature, Chem. Mater. 11 (1999) 277–280.

[176] K. Smits, et al. Luminescence of Eu ion in alumina prepared by plasma electrolytic oxidation, Appl. Surf. Sci. 337 (2015) 166–171.

[177] D. Shen, et al. Effect of cerium and lanthanum additives on plasma electrolytic oxidation of AZ31 magnesium alloy, J, Rare Earth. 31 (12) (2013) 1208–1213.

[178] E. De La Rosa-Cruz, et al. Luminescence and visible upconversion in nanocrystalline ZrO_2:Er^{3+}, Appl. Phys. Lett. 83 (24) (2003) 4903–4905.

[179] R. Naccache, et al. Visible upconversion emission of Pr3+ doped gadolinium gallium garnet nanocrystals, J.Nanosci. Nanotechnol. 4 (8) (2004) 1025–1031.

[180] H. Eilers, Synthesis and characterization of nanophase yttria co-doped with erbium and ytterbium, Mater. Lett. 60 (2) (2006) 214–217.

[181] A. Boukhachem, et al. Structural, optical, vibrational and photoluminescence studies of Sn-doped MoO_3 sprayed thin films, Mater. Res. Bull. 72 (2015) 252–263.

[182] D. Chandran, et al. Structural, optical, photocatalytic, and antimicrobial activities of cobalt-doped tin oxide nanoparticles, J. Sol–Gel Sci. Technol. 76 (3) (2015) 582–591.

[183] G. Nam, B. Kim, J.-Y. Leem, Facile synthesis and an effective doping method for ZnO:In^{3+} nanorods with improved optical properties, J. Alloy. Compd. 651 (2015) 1–7.

[184] W.A. Pisarski, et al. Enhancement and quenching photoluminescence effects for rare earth—doped lead bismuth gallate glasses, J. Alloy. Compd. 651 (2015) 565–570.

[185] S. Shakir, et al. Development of copper doped titania based photoanode and its performance for dye sensitized solar cell applications, J. Alloy. Compd. 652 (2015) 331–340.

[186] R.J.D. Tilley, Defects in Solids, John Wiley & Sons Ltda, Hoboken, New Jersey, (2008).

[187] A.V. Emeline, et al. Photostimulated generation of defects and surface reactions on a series of wide band gap metal-oxide solids, J. Phys. Chem. B 103 (43) (1999) 9190–9199.

[188] E. De La Rosa, et al. Visible light emission under UV and IR excitation of rare earth doped ZrO$_2$ nanophosphor, Opt. Mater. 27 (7) (2005) 1320–1325.

[189] P.A. Arsenev, et al. X-ray and thermostimulated luminescence of 0.9ZrO$_2$-0.1Y$_2$O$_3$ single-crystals, Phys. Status Solidi A-Appl. Res. 62 (2) (1980) 395–398.

[190] E.D. Wachsman, et al. Spectroscopic investigation of oxygen vacancies in solid oxide electrolytes, Appl. Phys. A-Mater. Sci. Process. 50 (6) (1990) 545–549.

[191] R. Reisfeld, M. Zelner, A. Patra, Fluorescence study of zirconia films doped by Eu^{3+}, Tb^{3+} and Sm^{3+} and their comparison with silica films, J. Alloy. Compd. 300 (2000) 147–151.

[192] H.K. Yueh, B. Cox, Luminescence properties of zirconium oxide films, J. Nucl. Mater. 323 (1) (2003) 57–67.

[193] R. Reisfeld, et al. Rare earth ions, their spectroscopy of cryptates and related complexes in sol-gel glasses, Opt. Mater. 24 (1–2) (2003) 1–13.

[194] L.E. Depero, et al. Correlation between crystallite sizes and microstrains in TiO$_2$ nanopowders, J. Cryst. Growth 198–199 (Part I) (1999) 516–520.

[195] V. Guidi, et al. Effect of dopants on grain coalescence and oxygen mobility in nanostructured titania anatase and rutile, J. Phys. Chem. B 107 (2003) 120–124.

[196] A. Mattsson, et al. Adsorption and solar light decomposition of acetone on anatase TiO$_2$ and niobium doped TiO$_2$ thin Films, J. Phys. Chem. B 110 (3) (2006) 1210–1220.

3

Nanomedicine

J. Cancino-Bernardi, V.S. Marangoni, V. Zucolotto

GROUP OF NANOMEDICINE AND NANOTOXICOLOGY, INSTITUTE OF PHYSICS OF SÃO CARLOS, UNIVERSITY OF SÃO PAULO (UNIVERSIDADE DE SÃO PAULO—USP), SÃO CARLOS, SÃO PAULO, BRAZIL

3.1 Nanomedicine

Nanomedicine is among the numerous opportunities and advances promoted by nanotechnology. The definitions of nanomedicine accepted today were established by the National Institutes of Health of the United States and the European Science Foundation, which define nanomedicine as the "science that uses nanomaterials to the development of diagnosis, treatment and prevention of specific medical application." [1,2] The advancements of this technology have promoted innovations in different medical fields, including controlled drug delivery, biomarkers, molecular imaging, and biosensing [3–5]. Nanomedicine can make significant contributions to the health industry, including the areas of medicine, pharmaceutical sciences, and dentistry. These advances are a result of partnerships between companies and research centers, including universities, and reflect the interdisciplinarity of this area of research [6].

Many pharmaceutical companies have supported and encouraged research in drug delivery. The greatest advantage of using nanomedicine compared with traditional medicine is the use of analytical tools and treatments, such as nanoparticles. In this context, nanoparticles can overcome cellular and physiological barriers and allow for imaging and diagnostic applications at the cellular and molecular levels because of their nanoscale dimension [7–10].

After the first reports of the potential use of nanomedicine, several research centers and government agencies have contributed to the development of this area with the allocation of new investment funds for research and development or the creation of new research centers at universities. Part of the investment in nanotechnology research is devoted to nanomedicine, which includes the development of complex drug delivery systems and advanced therapeutic and imaging techniques. Many products and treatments using nanomaterials are already available on the global market, including hormone therapy, bone implants, appetite controllers, and nanodrugs for the treatment of cancer. Other products are in clinical phases of development [11].

3.2 Nanomaterials Applied to Diagnosis and Therapy

3.2.1 Use of Nanomaterials in Medicine

The rapid advancement of nanomedicine has allowed for the development of nanomaterials with novel properties, structures, and specific functions, and their application in diagnostics and therapeutics. Several nanomaterials are used to promote these changes. These nanoparticles can be metallic [12,13], magnetic [14], polymers [15], dendrimers [16,17], nanoparticles with ceramic materials [18,19], nanoparticles with silica [20], quantum dots, [21] nanotubes [22], carbon nanofibers [23], or graphene [24], all amenable to modifications [25]. Each of these nanomaterials can be modified with polymers, nanocomplexes, other nanoparticles, dendrimers, or biomolecules such as DNA, proteins, antibodies, and antigens, and we can produce a multitude of modifications to achieve higher selectivity [26–28]. For the development of nanomaterials with high specificity regardless of the application, several factors should be taken into account, including their size, shape, modifications, composition, stability, biodegradation, and biodispersion. These parameters can help identify new or improved functions of nanomaterials and consequently their capabilities. Examples of such materials are found in Fig. 3.1 [29].

Molecular imaging, drug delivery, and photodynamic and photothermal therapies using nanomaterials promise to have a major impact on diagnosis and therapy [5,30]. Several products are already available on the market, primarily in the United States and Asia, with an emphasis on the field of oncology, and the results of preclinical and clinical studies using nanostructured carrier systems, nanomaterials as contrast agents, and magnetic

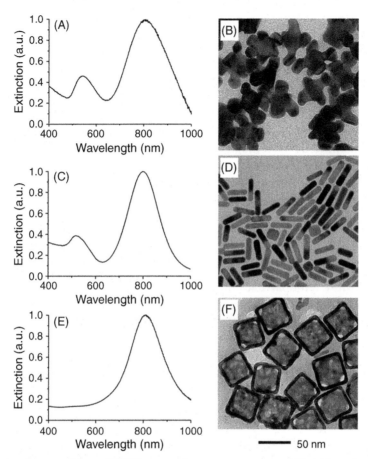

FIGURE 3.1 Examples of nanoparticles and UV-VIS spectra of aqueous suspensions of (A) gold nanohexapods, (C) gold nanorods, and (E) nanocages. Transmission electron microscopy (B, D, F) of the corresponding gold structures. *Reprinted with permission from Y. Wang et al., Comparison study of gold nanohexapods, nanorods, and nanocages for photothermal cancer treatment, ACS Nano 7 (3) (2013) 2068–2077, Copyright (2014), American Chemical Society* [29].

nanoparticles have shown improved diagnosis and treatment in terms of drug delivery, image quality, and electromagnetic functions [31–34].

The development of novel nanomaterials for use in medicine has proved viable in clinical and preclinical proof-of-concept experiments. Dendrimers have been applied in diagnostics and therapeutics for the development of a new cardiac test. This test can identify any damage to the heart muscle and significantly decrease the waiting time for blood test results [35,36]. Clinical trials with magnetic and metallic nanoparticles have yielded excellent results in the treatment of cancer via application of hyperthermia, in which the tumor is subjected to an electromagnetic field. For this application, nanoparticles are conjugated to specific antibodies or proteins to mark cancerous cells or tissues [37]. External laser irradiation in these tissues increases the temperature of the tissue to approximately 55°C and destroys tumor cells [38]. The application of nanomedicine is already a reality, and examples of this application are presented later.

3.2.2 Medical Applications of Magnetite and Core–Shells

The use of iron oxide nanoparticles in biomedical applications has increased because of their high biocompatibility, biodegradability, and superparamagnetic characteristics when applied in a magnetic field. Magnetite (Fe_3O_4) and its oxidized and stable form γ-Fe_2O_3 are by far the most commonly used magnetic nanoparticle contrast agents in magnetic resonance imaging, therapies to detect metastases, and clinical trials in nanomedicine [39,40]. Nanocrystals of magnetic iron oxide consist of magnetic metals such as iron, nickel, and cobalt. These nanomaterials are used in diagnostics and therapeutics for controlled drug delivery or imaging thanks to their nanometric scale, biocompatibility, and ability to be manipulated by an external magnetic field [41].

Many studies have focused on the encapsulation of new iron oxide nanoparticles with increasingly specific physicochemical characteristics to mark human erythrocytes by internalization of these nanoparticles through the membrane pores of these cells, without affecting cell viability [42]. The effectiveness of magnetic nanoparticles in the identification tumor metastasis is superior to that of other noninvasive methods [43,44]. Magnetic nanoparticles are versatile in the diagnostic imaging of inflammatory processes, including multiple arterioscleroses and rheumatoid arthritis in macrophages [41,45]. Magnetic nanoparticles and fluorophores are combined in preoperative treatments to diagnose glioma based on the response of these nanoparticles to fluorescence and magnetic resonance [46].

A new type of magnetic nanoparticle used in diagnostic imaging is radioluminescent core–shell nanoparticles. Their significant advantage is that a single nanoparticle allows for responses in terms of magnetic resonance, fluorescence, and radioluminescence. That is, they have magnetic and optical properties. Furthermore, magnetic luminescent nanoparticles can be used for controlled drug delivery [47]. Superparamagnetic nanoparticles with a size of 5–10 nm are used as intracellular markers for infectious or viral diseases. These nanoparticles are functionalized with antibodies, peptide ligands, or enzymes to provide greater specificity in the target cells [3,48,49]. Researchers intend to use core–shell magnetic nanoparticles coated with gold as marker probes coupled with drug delivery using an HIV antibody. These nanoparticles would trace and treat the viruses that were not eliminated with the available anti-AIDS cocktails, thereby promoting a revolution in the treatment of AIDS [49]. There are numerous applications of magnetic nanoparticles, either magnetite or core–shell, to mark tumors or lesions in tissues for improved diagnostic imaging.

3.2.3 Controlled Drug Delivery

The area that has developed the most thanks to nanomedicine is controlled drug delivery, which is the method or process of delivery of a pharmaceutical compound to achieve a therapeutic effect on an organism [50–52]. New drug delivery technologies aim to increase the release, adsorption, biodistribution, and subsequent elimination of the active ingredients, improving efficiency and the benefits to the patient. In some cases, gene-based drugs can be delivered using normal release routes, including the oral, topical, transmucosal, and inhalation routes [9,46,51,52]. In such cases, the active ingredient may be susceptible to enzymatic degradation or not be efficiently absorbed owing to its physicochemical properties [53]. Advances in nanomedicine can overcome these drawbacks. Together with pharmaceutical advances, new nanomedical technologies promise to improve the activity and delivery of medications [4].

Current advancements in drug delivery include the development of carrier systems, in which the active ingredient is released only to the target area, for example, cancerous tissues [5]. The challenge today is to develop timed-release formulations in which the active ingredient is released over a period in preset doses [50]. For the drug to reach the target site efficiently, the delivery system should avoid the defense mechanisms of the body so that the system only releases the active ingredient to the target site [54]. The greatest challenge will be to prevent physiological mechanisms from limiting the activity of delivery systems, to block drug dispersion, and to prevent immune defense responses from causing physical or biochemical changes in these systems [55].

Drug delivery systems were initially developed using polymers and later using lipid vesicles. The interest in protecting vitamins from oxidation using polymers in the 1980s marked the beginning of the development of encapsulation systems. Soon afterward, polymers such as poly(lactic-*co*-glycolic acid) (PLGA) and polylactic acid (PLA) were used in drug delivery systems owing to their biodegradability and biocompatibility [56]. However, their disadvantages were associated with their low solubility and permeability. To address these disadvantages, the development of drug carriers using polypeptide vesicles increased rapidly [51]. However, disadvantages were also associated with the use of peptide vesicles, including low bioavailability due to low stability and high degradation in biological systems, low permeability in biological membranes, and short half-life in the circulatory system. Studies in nanomedicine indicate that several nanoparticles could penetrate epithelial tissue. Therefore, nanoparticles have received more attention than vesicular systems owing to their therapeutic potential and high stability in biological fluids [50,52]. In addition, drug delivery to specific organs began to become a reality. The combination of nanoparticles with antibodies, transferases, lectins, and avidins led to the development of delivery systems with greater specificity (Fig. 3.2).

The advantage of using nanoparticles in drug delivery systems is a result of two basic properties. The first is their small size, which allows them to penetrate into cells and small capillaries, causing the accumulation of active ingredient into the delivery systems. The second characteristic is the possibility of using biodegradable materials in the preparation of nanoparticles. The primary benefits of using nanotechnology in drug delivery are the improved delivery of the active ingredient, longer shelf life of the drugs, low cost of new formulations compared with the discovery of new molecules, and particularly the reduction of the doses indicated, which reduces the costs of the drugs.

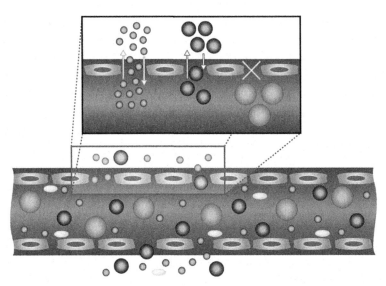

FIGURE 3.2 The efficacy of nanoparticles in drug delivery is highly dependent on their size and shape. The size of the nanoparticles affects their dispersion into and out of the vascular system, and the shape of nanoparticles affects their incorporation and penetration into the vessel walls. *Reprinted with permission from O.C. Farokhzad, R. Langer, Impact of nanotechnology on drug delivery, ACS Nano 3 (1) (2009) 16–20, Copyright (2014) American Chemical Society* [50].

The research on new biomarkers and drug delivery systems has focused on cardiovascular disease, cancer, and immunological diseases. Nanomaterials such as dendrimers, metal and silica nanoparticles, polymer nanoparticles, quantum dots, and carbon nanotubes (CNTs) have exceptional potential for direct delivery of drugs, including nitric oxide (NO) in patients with endocrine disorders [46]. In this context, the advantage of NO delivery systems using dendrimers includes the drug release control for time and site specificity and improved biocompatibility compared with conventional pharmacological applications. The primary advantage of NO release using dendrimer nanoparticles compared with other release systems is their increased capacity to store NO owing the pore structure of dendrimer molecules [57]. Furthermore, it is possible to change the external characteristics of dendrimer molecules, increasing their solubility or altering a specific function. Another example is the use of CNTs in the preferred delivery of acetylcholine in the brain for the treatment of diseases such as Alzheimer's, without damaging the mitochondria or nonspecific organelles [58].

The development of drug carriers is highly dependent on the physicochemical properties of the nanoparticles but not on the active ingredient, as was once thought. Therefore, the physicochemical properties of these carriers can be optimized to target a specific tissue or to be nonspecifically absorbed by the target cells. However, to achieve this goal, nanomedicine should be used for the correct development of these carriers by determining the best techniques for encapsulation, the solvents used, the solubility of the active ingredient and solvent, and the types of polymers and stabilizers; in addition, it allows us to control other variables, including nanoparticle size, encapsulation efficiency, temperature, pH, ionic strength, and, most importantly, the residue of the polymer or nanoparticle associated with the drug in the patient's body. The elimination of the polymer or nanoparticle from the patient's body is a major cause for concern because of the possible toxicity of these materials [59].

One challenge in developing drug delivery systems is the use of theranostic nanomaterials, which combine diagnostics with therapeutics [60]. An ideal theranostic system will combine diagnosis and treatment in a single step using preventive monitoring before disease symptoms manifest [33,46,60]. This system includes the detection and removal of fat plaques in arteries for the prevention of cardiovascular disease. The development of drug-delivery systems with multiple active ingredients is promising because it allows for the synergistically combined delivery of several drugs to biological targets, for example, an antiangiogenic compound and a DNA intercalating agent, and this approach can increase the therapeutic efficacy.

3.2.4 Nanoparticles in Photothermal and Photodynamic Therapy

The use of the plasmon absorption band of nanoparticles for phototherapy has been promising for the treatment and diagnosis of cancer, and the technique is minimally invasive. However, for photothermal treatments in vivo, it is necessary to use nanoparticles with radiation in the near-infrared (NI) region (wavelengths between 650 and 900 nm) because the absorption coefficients of hemoglobin and water are smaller in this region [20,61]. Therefore, the plasmon resonance band of the nanoparticles in the NI region has been used for therapeutic purposes, that is, the excitation of electrons from the surface of nanoparticles causes localized heating (due to the effect of localized surface plasmon resonance—LEPR) and death of cancer cells. The nanoparticles most studied for this application include rod-shaped gold nanoparticles, known as nanorods, as well as multifunctional liposomes and polymer plasmonic nanoparticles [61–63].

Standard photodynamic therapy has been applied to cancerous tumors, in which a photosensitive drug is administered to the patient, accumulates near the tumor tissue, and induces cell death after light activation. The photosensitive agents used for therapy since 1990 are porphyrin compounds, and for this reason, the photosensitive agents are classified based on the presence or absence of porphyrin [64]. Although porphyrins have a rapid adsorption and low toxicity, the disadvantages include low selectivity to smaller tumors and high hydrophobicity. Therefore, the development of highly selective photosensitive agents is the primary research topic in this area.

One of the most significant problems in photodynamic therapy is the controlled delivery of the photosensitive agent. Therefore, the development of this type of treatment is directly correlated with advances in systems for controlled drug delivery. Other limitations of photodynamic therapy are the low selectivity to tumors, low solubility, and high aggregation of these agents under physiological conditions [65]. With the advancement of research in drug delivery systems in nanomedicine, new strategies have been developed to increase the solubility of hydrophobic photosensitive agents and selectivity by modifying the surface area of nanoparticles loaded with the agent. The development of highly selective photoactive nanoparticles will ensure that this therapy targets cancer cells and preserves healthy cells [66].

Photosensitive agent formulations using nanoparticles are already available. One of the most promising formulations is the encapsulation of temoporfin as liposome (Foscan) [67], which overcome the limitations of existing commercial agents by causing a more efficient destruction of tumors and reducing damage and toxicity to healthy tissue compared with commercial agents. The encapsulation of porphyrin-based photosensitive agents in dendrimers is also possible. In this case, porphyrins and phthalocyanines are stabilized externally by dendrimers, which prevents aggregation [54]. CNTs are another possibility for delivery of photosensitive agents because they can be modified to increase specificity and adsorption in the body. However, a challenge that needs to be overcome is that CNTs absorb light in the NI region, which may cause the death of healthy cells owing to excessive local heating [68,69].

To improve the selectivity and specificity of the photosensitive agent in cancer cells, a combination of nanostructures with monoclonal antibodies and antibodies against transferrin receptors that are expressed on the surface of many solid tumors can be used [70,71]. Nanostructures can also be combined with other ligands, including vitamins, glycoproteins, peptides, aptamers, DNA, and growth factors. Another approach is the combination of different treatment modalities. For example, a single type of nanoparticle can be used in photodynamic therapy and radiotherapy. In this case, luminescent nanoparticles combined with photosensitive agents can be exposed to radiation such as X-rays, which causes the nanoparticles to emit luminescence, which in turn activates the photosensitive agents [72]; this is a major advantage for in vivo applications because the nanoparticles are luminescent and therefore no external light source is required.

3.2.5 Nanoparticles for Upconversion—Imaging of Cancer Cells

Nanoparticles for upconversion have emerged as a new class of agents for bioimaging. Upconversion is the absorption of two or more lower-energy photons and the consequent release of a higher-energy photon. This technique has been important in the development of more efficient systems in medicine [73]. Typically, these nanoparticles are excited at a wavelength in the NI region (900–1000 nm) and emit wavelengths with higher energy in the ultraviolet–visible region. This process is different from multiphoton absorption in organic molecules and quantum dots, and in general their efficiency is much higher, which allows for the use of low-cost continuous-wave diode lasers in contrast to the ultrashort pulse lasers used in the excitation of nonlinear multiphotons [73,74].

Nanoparticles used in upconversion usually consist of nanostructured systems doped with lanthanides, that is, rare-earth trivalent ions such as Yb^{3+}, Tm^{3+}, Er^{3+}, and Ho^{3+}, dispersed in nanostructures in a suitable dielectric medium [75,76]. These lanthanide dopants act as optically active centers and emit light when excited with the appropriate wavelength. These optical properties can be adjusted by selecting the lanthanide dopants to produce systems with different emission wavelengths, ranging from visible to NI or ultraviolet. Furthermore, the selection of the nanostructured host system, that is, the nanomaterials that will be doped with lanthanides, is essential for the production of highly efficient systems. Characteristics such as transparency in the spectral region of interest and chemical stability are critical in the selection of this system [73].

Nanoparticles of sodium yttrium fluoride doped with ytterbium and erbium ($NaYF_4$: Yb,Er) are efficient in converting infrared radiation into visible radiation and have shown promise in the detection of biological interactions [77]. The modification of dopants of $NaYF_4$ nanoparticles to Tm^{3+} and Yb^{3+} allows for the upconversion from NI to visible radiation, resulting in high penetration of these particles into biological tissues and acquisition of images with high optical contrast, which is supported by in vitro and in vivo experiments [78]. Several other host nanostructures have been explored in the development of these systems, including ZrO_2 [79] and LaF_3 [80].

Furthermore, the possibility to combine different materials into a single structure has made it possible to obtain systems with multiple and synergistic properties. For example, multifunctional nanoparticles can be manufactured by adsorption of Fe_3O_4 nanoparticles on the surface of upconverting nanostructures followed by the formation of a thin layer of gold on the surface [81]. Therefore, these materials have optical and magnetic properties and can be exploited as contrast agents in magnetic resonance imaging. Alternatively, the introduction of Gd^{3+} to these structures allows it to be used as a T1 contrast agent in magnetic resonance imaging [82]. The presence of gold nanoparticles on the surface of upconverting nanoparticles enables the modulation of these systems, increasing their versatility [83].

Therefore, these nanoparticles have many advantages for tumor imaging applications, including stable emission, high resistance to photodegradation, the ability to be detected in biological tissues using light in the NI region, and the possibility of biofunctionalization of their surface. Despite their great potential, rapid heating can overheat the tissue around the tumor, and lanthanides can increase the toxicity of these materials; therefore, new strategies for synthesis and functionalization are required to minimize these adverse effects [34,73,74].

3.3 Synthesis of Nanomaterials for Application in Nanomedicine

The successful use of nanomaterials in medicine is closely correlated with their properties. Characteristics such as stability, size variation, morphology, surface charge, and toxicity should be chosen carefully to achieve the desired results. In this context, increasingly effective methods have been developed to produce nanomaterials with highly controlled physicochemical parameters. The following topics represent the main methods used in the production of different types of nanomaterials explored in medicine.

3.3.1 Gold Nanoparticles

Among the metal nanoparticles, nanoparticles of noble metals such as gold have been extensively studied in recent years. These particles have a plasmon resonance band owing to collective oscillation of free electrons on their surface, whose wavelength can be changed by controlling some parameters, including size, shape, and surface coating [84].

Gold nanoparticles can be manufactured using the precipitation method, in which a reducing agent is added to a gold salt solution in the presence of a stabilizer. In the most common route of synthesis, nanoparticles are synthesized in an aqueous medium in the presence of citric acid or sodium citrate, which acts as a reducing agent and stabilizer because of its negative charge [85]. Other stabilizing molecules have been explored, including polymers. This functionalization allows us to obtain organic and inorganic hybrid nanomaterials, whose parameters, including size, shape, and surface functional groups, can be adjusted by the appropriate choice of the surface polymer [86].

Sulfur establishes a strong bond with atomic gold [87] and has been extensively explored for the preparation and stabilization of gold nanoparticles. In one method of synthesis, gold nanoparticles stabilized with sodium 3-mercaptopropionate molecules are obtained by the simultaneous addition of a citrate salt and sodium 3-mercaptopropionate to an aqueous solution of chloroauric acid ($HAuCl_4$) under boiling [88]. The particle size can be controlled by modifying the ratio between the concentration of stabilizer and the concentration of gold ions in solution [88].

3.3.2 Magnetic Nanoparticles

Magnetic nanoparticles have been extensively studied for medical applications and have proven to be highly effective as contrast agents in nuclear magnetic resonance imaging and treatment by hyperthermia [89]. These particles can be formed by different types of cubic ferrites with the general expression $M^{2+}Fe_2O_4$, where M^{2+} is a metal such as Mg, Co, Zn, or Fe [as in Fe_3O_4 or maghemite (γ-Fe_2O_3)] [90]. Fe_3O_4 can be synthesized using a simple method known as coprecipitation, which involves a controlled alkalization reaction in which a precipitator is added to an aqueous solution of iron salt at a pH between 9 and 10 at room temperature [89]. Fe_3O_4 nanoparticles synthesized from iron (II) and iron (III) salts follow the stoichiometry of the overall reaction

$$Fe^{2+} + 2Fe^{3+} + 8OH^- \rightarrow Fe_3O_4 + 4H_2O$$

Using only iron (II) salt, the reaction occurs by different mechanisms and the diameter of the obtained nanoparticles varies between 30 and 50 nm:

$$Fe^{2+} + 2OH^- \rightarrow Fe(OH)_2$$

$$3Fe(OH)_2 + \tfrac{1}{2} O_2 \rightarrow Fe(OH)_2 + 2FeOOH + H_2O$$

$$Fe(OH)_2 + 2FeOOH \rightarrow Fe_3O_4 + 2H_2O$$

This method is quite simple, but it has limitations on the control of the shape and size of the obtained particles. Methods that use organic solvents and high temperatures have been the most efficient to date. In one such method, nanoparticles (MFe_2O_4) are obtained from metal acetylacetonate (M = Fe, Co, Mn) in 1,2-hexadecanodiol in the presence of oleic acid and olamine at high temperatures [90]. After synthesis, these particles can be washed by magnetic separation and functionalized with polymeric molecules for use in an aqueous medium.

3.3.3 Core–Shell-Type Structures

Nanoparticles that incorporate multiple components and functionalities are crucial for the development of new systems for medical applications. Among them, core–shell nanostructures with different components have become increasingly important owing to the possibility of easily adjusting their new properties. For example, Oldenburg et al. [91] showed that the absorption band of gold nanoparticles can be shifted from visible to infrared by a coating of silica particles (core) with a gold layer, forming core–shell-type structures. Therefore, the absorption wavelength can be adjusted by the relative thickness of the gold layer around the silica nanoparticles [91].

Another prominent candidate for application in medicine is nanoparticles of Fe_3O_4@ Au. The advantage of this system compared with iron oxide nanoparticles is the possibility of developing systems that combine optical and magnetic properties in a single process. Furthermore, the coating of iron oxide nanoparticles protects the Fe_3O_4 core against oxidation without drastically reducing their magnetic properties [92].

Several strategies have been developed to form such structures. One example is the reduction of gold ions in the presence of Fe_3O_4 nanoparticles, 1,2-hexadecanodiol, oleic acid, and oleylamine in phenyl ether under magnetic stirring, argon atmosphere, and a temperature of 190°C [93].

Another synthesis strategy is the initial coating of iron oxide nanoparticles with silica using tetraethylorthosilicate (TEOS) and ammonium hydroxide in an alcohol solution. These particles are then functionalized with 3-aminopropyltriethoxysilane (APTES). This system is added to a suspension of gold colloids in the presence of tetrakis(hydroxymethyl) phosphonium chloride (THPC), which interacts with amino groups from the APTES. The formation of a gold nanoparticle-THPC suspension involves the reduction of chloroauric acid ($HAuCl_4$) with THPC, leading to the formation of small particles (approximately 2 nm). Finally, the nanoshell is prepared via reduction of $HAuCl_4$ with formaldehyde in a solution containing potassium carbonate (K_2CO_3), and the magnetic nanoparticles are coated with silica prepared beforehand [94,95].

3.3.4 Biofunctionalization of Nanomaterials

The possibility of incorporating various types of molecules on the surface of these nanomaterials is essential for the development of specific agents for disease diagnosis and treatment and drug delivery systems. These particles can be functionalized with biomolecules

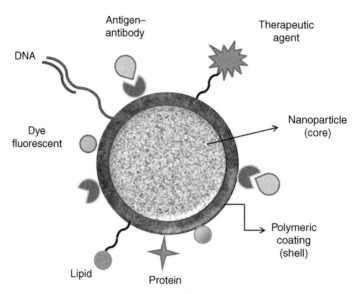

FIGURE 3.3 Nanoparticle with various ligands that confer multifunctionality using a single platform.

such as proteins [96], peptides [97], aptamers [98], and specific antibodies [99] that specifically recognize the target cells or tissues, providing additional recognition properties to the nanomaterials. Furthermore, fluorescent or antitumor molecules can be added to these nanostructures, allowing for their detection by fluorescence spectroscopy or the attenuation of the side effects caused by these drugs, respectively. The organization of these nanocomplexes is directed by ionic, covalent, and noncovalent interactions and by hydrogen bonds between the molecules to be incorporated on the surface of the nanoparticles [100] (Fig. 3.3).

For example, gold nanoparticles conjugated to oligonucleotides are of great interest because the complementarity of DNA base pairs increases the specificity of these complexes and allows their use in biosensors for the detection of specific DNA sequences associated with abnormalities or diseases [101], intracellular gene regulation [102], and supramolecular organization of nanostructures [103]. Gold nanoparticles can be functionalized with oligonucleotides by modifying the oligonucleotides with thiol groups, which form highly stable bonds with gold [104].

3.4 Nanotoxicology

Despite the rapid progress and acceptance of nanomedical technologies in diagnostics and therapeutics, their potential effect on human health due to prolonged exposure to these compounds has not been established. It is important to realize that regulatory issues in nanomedicine involve the approval of therapeutic processes, which are complex and expensive but necessary, regardless of the nanomaterial or application [105]. In the past 5 years, research in nanotoxicity has focused on cell culture systems and animal test-ing for the establishment of regulatory processes. However, data from these studies are still inaccurate, which warrants the development of new methods to elucidate these ef-fects [106,107]. In in vitro systems, the interactions between nanostructures and biologi-cal components are complex because of the diversity of proteins and substances in the cellular media, and even more complex in in vivo systems that contain biological fluids, organelles, and biomolecules [108,109].

To properly assess the potential risks of nanoparticles to health, their life cycle should be measured in all development phases, including production, storage, distribution, ap-plication, and elimination [110]. The effect on humans or the environment may vary in different stages of the life cycle. In addition, metabolic and immune responses induced by these materials are still not fully understood. Toxicological studies suggest that nano-materials can cause adverse health effects; however, the fundamental cause–effect rela-tionships are still unclear. Many nanomaterials can interact with cellular components, interrupting or changing specific functions or even causing the overproduction of reactive oxygen and nitrogen species [106].

Before nanomedicine is applied in diagnostics and therapeutics, we need to evalu-ate critically how the human body will react to different nanomaterials because they are foreign bodies. Reactions against these nanomaterials may not be identical to those pro-duced against viruses or bacteria, as in the former case there are no specific enzymes or antibodies against these agents [111]. Moreover, although these materials represent sub-stances unknown to the human body, they are immediately sensed when they reach the bloodstream (the first access route for drug delivery systems) and quickly interact with blood components through physical, chemical, and biological processes [112]. The in-teractions of nanomaterials with nuclei of mitochondrial cells have been considered the main source of cytotoxicity. Nanomaterials, such as silver nanoparticles coated with fuller-enes, micelles, copolymers, and CNTs, can interact with mitochondrial cells and cause apoptosis and DNA damage [113,114]. Owing to the complexity of the mechanisms in-volved in the interactions between nanomaterials and biological samples, the biophysical and biochemical aspects are difficult to investigate, especially in vivo and in real time.

The use of nanoparticles in nanomedicine has advantages and disadvantages. For ex-ample, the benefits of the oral, transdermal, intravenous, or respiratory administration of nanoparticles are their limited invasiveness and large surface area. The disadvantages are associated with possible interactions with mucosal membranes, local irritation, and liver toxicity. Therefore, regardless of the application of nanoparticles in medicine, their toxicity should be assessed before clinical trials. The scientific community is continually develop-ing new analytical methods to overcome the limited data on the toxicity of nanomaterials [115–117]. However, many basic standard toxicology tests are not applied to nanotoxicol-ogy because of the high reactivity of nanomaterials with chemicals used in these assays. Despite the limitations in the development of appropriate methods to assess the specific interactions between nanomaterials and biological systems, many studies using in vitro and in vivo analytical protocols are available [118–121].

3.4.1 In Vitro Assays

One type of test that is most commonly applied to evaluate the intrinsic toxicity of nano-materials is assays with different in vitro cell lines [108]. One of the most studied nano-materials to assess the degree of toxicity is CNTs [122]. CNTs have gained prominence because of their capacity to be used as carriers combined with the possibility of making modifications on their surface, which can lead to the development of novel nanomaterials with therapeutic applications, for example, in cancer [123]. However, CNTs can internal-ize into human T-cells and kidney cells and cause time- and dose-dependent cell death [124–126]. Single-wall CNTs (SWCNTs) are more toxic to alveolar macrophages compared with double-wall CNTs. The toxicity of CNTs also depends on other factors, including their concentration, shape, degree of surface functionalization, and adsorption of metals dur-ing the production process. In other cases, CNTs are not toxic, such as when they are ad-ministered in lung cells at a dose between 1.5 and 800 mg mL^{-1} [127].

Metallic and magnetic nanoparticles are biocompatible because they are made of in-ert material. However, nanoparticles made of inert material can produce morphological changes, loss of function, inflammation, and even irreversible cell damage. One study on the toxicity of gold nanoparticles in human sperm indicated changes in motility, morphol-ogy, and sperm fragmentation [128]. However, the morphology of primary endothelial cells from rat brains (rBMEC) did not change after 24 h of exposure to gold nanoparticles [129].

PLGA polymeric nanoparticles were not toxic to the A549 cell line, even at high concen-trations [130]. Silica nanoparticles administered to A549 cells in a time- and dose-depen-dent manner reduced cell viability and increased the levels of oxidative stress, indicating a cytotoxic effect [131]. Silver nanoparticles at concentrations higher than 50 µg mL^{-1} ap-plied to fibroblast cells (NIH3T3) caused apoptosis associated with ROS production [132]. The administration of silver nanoparticles at concentrations close to 2 µg mL^{-1} reduced the viability of HEK cells and induced the production of cytokines responsible for immune responses [133]. In contrast, silver nanoparticles at a concentration of 6.3 µg mL^{-1} showed no evidence of causing cell death or damage to A431 human carcinoma cells [134].

In summary, toxicological studies of nanoparticles in in vitro systems are still inconsis-tent regarding the adverse and toxic effects of nanoparticles, indicating that more research is needed to address this issue [135]. In addition, further systematic in vitro studies are necessary to regulate the medical use of nanoparticles.

3.4.2 In Vivo Studies

The in vivo characteristics of these nanoparticles are a major issue that needs to be ad-dressed [106,119,136]. It is still not known whether all nanoparticles have undesirable ef-fects on the environment and living organisms in general. At present, four routes of entry of nanoparticles into the body have been described: inhalation, ingestion, absorption through the skin, and injection during medical procedures. Once inside the body, nanoparticles are highly mobile and can remain anywhere in the body [137]. So far, neither the nanoparticles nor the products and materials that contain them are subject to regulation.

In rats, the accumulation and potential adverse effects of CNTs in several organs, including the liver, lung, and spleen, are reported [110]. Various studies have evaluated the toxicity of carbon-based nanomaterials, including multiwall CNTs, carbon nanofibers, and carbon nanoparticles, on the basis of the radius and surface chemistry of these materials. The nanotoxicity was tested in vivo using a tumor cell line culture. The results indicated that these materials are toxic in a dose-dependent manner. Furthermore, cytotoxicity was enhanced when the surface of these nanomaterials was functionalized after acid treatment [138]. Another study inoculated CNTs into primary skins and subjected this material to conjunctival irritation and sensitivity tests. The negative metagenesis results suggested that SWCNTs are not carcinogenic. In addition, the lethal dose for hamsters was 2000 mg kg^{-1} body weight, a dose higher than will ever be used in nanomedicine [139].

Changes in gene expression were observed in pregnant mice exposed to titanium oxide (TiO$_2$) nanoparticles. Exposure of mouse fetuses to TiO$_2$ nanoparticles during the prenatal period affected the expression of genes related to the development and function of the central nervous system [140]. Considering the potential toxicity of nanoparticles and nanomaterials in general and the lack of legislation governing such effects, new analytical methods and additional systematic studies are necessary to assess the nanotoxicity of these materials. It is expected that, as the field of nanomedicine progresses, new concepts in nanotoxicity will guide the creation of laws to regulate the use of nanomaterials.

References

[1] G. Oberdorster, Safety assessment for nanotechnology and nanomedicine: concepts of nanotoxicology, J. Intern. Med. 267 (1) (2010) 89–105.

[2] E.E.S. Foundation, Nanomedicine, an ESF–European Medical Research Councils (EMRC) Forward Look Report, ESF, France, 2004.

[3] E. Blanco, A. Hsiao, A.P. Mann, M.G. Landry, F. Meric-Bernstam, M. Ferrari, Nanomedicine in cancer therapy: innovative trends and prospects, Cancer Sci. 102 (7) (2011) 1247–1252.

[4] J. Wang, J.D. Byrne, M.E. Napier, J.M. DeSimone, More effective nanomedicines through particle design, Small 7 (14) (2011) 1919–1931.

[5] H. Boulaiz, P.J. Alvarez, A. Ramirez, J.A. Marchal, J. Prados, F. Rodriguez-Serrano, et al. Nanomedicine: application areas and development prospects, Int. J. Mol. Sci. 12 (5) (2011) 3303–3321.

[6] F. Caruso, T. Hyeon, V.M. Rotello, Nanomedicine 41 (7) (2012) 2537–2538.

[7] V. Labhasetwar, M. Zborowski, A.R. Abramson, J.P. Basilion, Nanoparticles for imaging, diagnosis, and therapeutics, Mol. Pharm. 6 (5) (2009) 1261–1262.

[8] M. Estanqueiro, M.H. Amaral, J. Conceicao, J.M. Sousa Lobo, Nanotechnological carriers for cancer chemotherapy: the state of the art, Colloid Surf. B 126 (2015) 631–648.

[9] A.P. Alivisatos, Less is more in medicine—sophisticated forms of nanotechnology will find some of their first real-world applications in biomedical research, disease diagnosis and, possibly, therapy, Sci. Am. 285 (3) (2001) 66–73.

[10] N. Bertrand, J. Wu, X.Y. Xu, N. Kamaly, O.C. Farokhzad, Cancer nanotechnology: the impact of passive and active targeting in the era of modern cancer biology, Adv. Drug Deliv. Rev. 66 (2014) 2–25.

[11] Nanotecnhologies TPoE. 2016 [cited 2016 03022016], Available from: http://www.nanotechproject.org/inventories/medicine/apps/

[12] R.R. Arvizo, S. Bhattacharyya, R.A. Kudgus, K. Giri, R. Bhattacharya, P. Mukherjee, Intrinsic therapeutic applications of noble metal nanoparticles: past, present and future, Chem. Soc. Rev. 41 (7) (2012) 2943–2970.

[13] E.C. Dreaden, A.M. Alkilany, X. Huang, C.J. Murphy, M.A. El-Sayed, The golden age: gold nanoparticles for biomedicine, Chem. Soc. Rev. 41 (7) (2012) 2740–2779.

[14] Y. Piao, A. Burns, J. Kim, U. Wiesner, T. Hyeon, Designed fabrication of silica-based nanostructured particle systems for nanomedicine applications, Adv. Funct. Mater. 18 (23) (2008) 3745–3758.

[15] L. Zhang, J.M. Chan, F.X. Gu, J.-W. Rhee, A.Z. Wang, A.F. Radovic-Moreno, et al. Self-assembled lipid-polymer hybrid nanoparticles: a robust drug delivery platform, ACS Nano 2 (8) (2008) 1696–1702.

[16] F. Yang, C. Jin, S. Subedi, C.L. Lee, Q. Wang, Y. Jiang, et al. Emerging inorganic nanomaterials for pancreatic cancer diagnosis and treatment, Cancer Treat. Rev. 38 (6) (2012) 566–579.

[17] M. Ciolkowski, J.F. Petersen, M. Ficker, A. Janaszewska, J.B. Christensen, B. Klajnert, et al. Surface modification of PAMAM dendrimer improves its biocompatibility, Nanomedicine 8 (6) (2012) 815–817.

[18] J.M. Rosenholm, C. Sahlgren, M. Linden, Multifunctional mesoporous silica nanoparticles for combined therapeutic, diagnostic and targeted action in cancer treatment, Curr. Drug Targets 12 (8) (2011) 1166–1186.

[19] M. Vallet-Regi, E. Ruiz-Hernandez, Bioceramics: from bone regeneration to cancer nanomedicine, Adv. Mater. 23 (44) (2011) 5177–5218.

[20] J. Croissant, J.I. Zink, Nanovalve-controlled cargo release activated by plasmonic heating, J. Am. Chem. Soc. 134 (18) (2012) 7628–7631.

[21] K.H. Lee, J.F. Galloway, J. Park, C.M. Dvoracek, M. Dallas, K. Konstantopoulos, et al. Quantitative molecular profiling of biomarkers for pancreatic cancer with functionalized quantum dots, Nanomedicine 8 (7) (2012) 1043–1051.

[22] E. Mehdipoor, M. Adeli, M. Bavadi, P. Sasanpour, B. Rashidian, A possible anticancer drug delivery system based on carbon nanotube-dendrimer hybrid nanomaterials, J. Mater. Chem. 21 (39) (2011) 15456–15463.

[23] R.J. Andrews, Nanotechnology, Neurosurgery, J. Nanosci. Nanotechnol. 9 (8) (2009) 5008–5013.

[24] Z. Liu, J.T. Robinson, S.M. Tabakman, K. Yang, H. Dai, Carbon materials for drug delivery & cancer therapy, Mater. Today 14 (7-8) (2011) 316–323.

[25] M. Swierczewska, G. Liu, S. Lee, X. Chen, High-sensitivity nanosensors for biomarker detection, Chem. Soc. Rev. 41 (7) (2012) 2641–2655.

[26] M. Perfezou, A. Turner, A. Merkoci, Cancer detection using nanoparticle-based sensors, Chem. Soc. Rev. 41 (7) (2012) 2606–2622.

[27] C.G. Palivan, O. Fischer-Onaca, M. Delcea, F. Itel, W. Meier, Protein-polymer nanoreactors for medical applications, Chem. Soc. Rev. 41 (7) (2012) 2800–2823.

[28] J.R. Siqueira Jr., M.H. Abouzar, A. Poghossian, V. Zucolotto, O.N. Oliveira Jr., M.J. Schoening, Penicillin biosensor based on a capacitive field-effect structure functionalized with a dendrimer/carbon nanotube multilayer, Biosens. Bioelectron. 25 (2) (2009) 497–501.

[29] Y. Wang, K.C.L. Black, H. Luehmann, W. Li, Y. Zhang, X. Cai, et al. Comparison study of gold nanohexapods, nanorods, and nanocages for photothermal cancer treatment, ACS Nano 7 (3) (2013) 2068–2077.

[30] C. Kranz, D.C. Eaton, B. Mizaikoff, Analytical challenges in nanomedicine, Anal. Bioanal. Chem. 399 (7) (2011) 2309–2311.

[31] J.L. Vivero-Escoto, R.C. Huxford-Phillips, W. Lin, Silica-based nanoprobes for biomedical imaging and theranostic applications, Chem. Soc. Rev. 41 (7) (2012) 2673–2685.

[32] A. Taylor, K.M. Wilson, P. Murray, D.G. Fernig, R. Levy, Long-term tracking of cells using inorganic nanoparticles as contrast agents: are we there yet?, Chem. Soc. Rev. 41 (7) (2012) 2707–2717.

[33] B. Sumer, J. Gao, Theranostic nanomedicine for cancer, Nanomedicine 3 (2) (2008) 137–140.

[34] R. Tong, D.S. Kohane, New strategies in cancer nanomedicine, in: P.A. Insel (Ed.), Annual Review of Pharmacology and Toxicology, vol. 56, Annual Reviews, Palo Alto, 2016, pp. 41–57.

[35] S.H. Bai, C. Thomas, A. Rawat, F. Ahsan, Recent progress in dendrimer-based nanocarriers, Crit. Rev. Ther. Drug Carr. Syst. 23 (6) (2006) 437–495.

[36] D.B. Buxton, S.C. Lee, S.A. Wickline, M. Ferrari, National Heart, Lung, and Blood Institute Nanotechnology Working Group. Recommendations of the National Heart, Lung, and Blood Institute Nanotechnology Working Group, Circulation 108 (22) (2003) 2737–2742.

[37] S. Nie, Understanding and overcoming major barriers in cancer nanomedicine, Nanomedicine 5 (4) (2010) 523–528.

[38] N. Lee, T. Hyeon, Designed synthesis of uniformly sized iron oxide nanoparticles for efficient magnetic resonance imaging contrast agents, Chem. Soc. Rev. 41 (7) (2012) 2575–2589.

[39] Y.-F. Li, C. Chen, Fate and toxicity of metallic and metal-containing nanoparticles for biomedical applications, Small 7 (21) (2011) 2965–2980.

[40] S.-H. Huang, R.-S. Juang, Biochemical and biomedical applications of multifunctional magnetic nanoparticles: a review, J. Nanoparticle Res. 13 (10) (2011) 4411–4430.

[41] J.R. McCarthy, R. Weissleder, Multifunctional magnetic nanoparticles for targeted imaging and therapy, Adv. Drug Deliv. Rev. 60 (11) (2008) 1241–1251.

[42] A. Antonelli, C. Sfara, E. Manuali, I.J. Bruce, M. Magnani, Encapsulation of superparamagnetic nanoparticles into red blood cells as new carriers of MRI contrast agents, Nanomedicine 6 (2) (2011) 211–223.

[43] M.W. Bourne, L. Margerun, N. Hylton, B. Campion, J.J. Lai, N. Derugin, et al. Evaluation of the effects of intravascular MR contrast media (gadolinium dendrimer) on 3D time of flight magnetic resonance angiography of the body, J. Magn. Reson. Imag. 6 (2) (1996) 305–310.

[44] G. Brown, C.J. Richards, R.G. Newcombe, N.S. Dallimore, A.G. Radcliffe, D.P. Carey, et al. Rectal carcinoma: thin-section MR imaging for staging in 28 patients, Radiology 211 (1) (1999) 215–222.

[45] C. Sun, J.S.H. Lee, M. Zhang, Magnetic nanoparticles in MR imaging and drug delivery, Adv. Drug Deliv. Rev. 60 (11) (2008) 1252–1265.

[46] V.I. Shubayev, T.R. Pisanic II, S. Jin, Magnetic nanoparticles for theragnostics, Adv. Drug Deliv. Rev. 61 (6) (2009) 467–477.

[47] H. Chen, D.C. Colvin, B. Qi, T. Moore, J. He, O.T. Mefford, et al. Magnetic and optical properties of multifunctional core-shell radioluminescence nanoparticles, J. Mater. Chem. 22 (25) (2012) 12802–12809.

[48] L. Dykman, N. Khlebtsov, Gold nanoparticles in biomedical applications: recent advances and perspectives, Chem. Soc. Rev. 41 (6) (2012) 2256–2282.

[49] P. Could, Nanoparticles probes biosystems, Mater. Today 7 (2004) 36–43.

[50] O.C. Farokhzad, R. Langer, Impact of nanotechnology on drug delivery, ACS Nano 3 (1) (2009) 16–20.

[51] S.K. Sahoo, V. Labhasetwar, Nanotech approaches to delivery and imaging drug, Drug Discov. Today 8 (24) (2003) 1112–1120.

[52] S.K. Sahoo, S. Parveen, J.J. Panda, The present and future of nanotechnology in human health care, Nanomedicine 3 (1) (2007) 20–31.

[53] S. Guo, E. Wang, Functional micro/nanostructures: simple synthesis and application in sensors, fuel cells, and gene delivery, Acc. Chem. Res. 44 (7) (2011) 491–500.

[54] B. Kateb, K. Chiu, K.L. Black, V. Yamamoto, B. Khalsa, J.Y. Ljubimova, et al. Nanoplatforms for constructing new approaches to cancer treatment, imaging, and drug delivery: what should be the policy?, Neuroimage 54 (2011) S106–S124.

[55] O. Veiseh, J.W. Gunn, M. Zhang, Design and fabrication of magnetic nanoparticles for targeted drug delivery and imaging, Adv. Drug Deliv. Rev. 62 (3) (2010) 284–304.

[56] L.D. Leserman, J. Barbet, F. Kourilsky, J.N. Weinstein, Targeting to cells of fluorescent liposomes covalently coupled with monoclonal antibody or protein-A, Nature 288 (5791) (1980) 602–624.

[57] A.B. Seabra, N. Duran, Nitric oxide-releasing vehicles for biomedical applications, J. Mater. Chem. 20 (9) (2010) 1624–1637.

[58] Z. Yang, Y. Zhang, Y. Yang, L. Sun, D. Han, H. Li, et al. Pharmacological and toxicological target organelles and safe use of single-walled carbon nanotubes as drug carriers in treating Alzheimer disease, Nanomedicine 6 (3) (2010) 427–441.

[59] A.R. Menjoge, R.M. Kannan, D.A. Tomalia, Dendrimer-based drug and imaging conjugates: design considerations for nanomedical applications, Drug Discov. Today 15 (5-6) (2010) 171–185.

[60] X. Chen, S.S. Gambhlr, J. Cheon, Theranostic nanomedicine, Acc. Chem. Res. 44 (10) (2011) 841.

[61] J.B. Song, P. Huang, H.W. Duan, X.Y. Chen, Plasmonic vesicles of amphiphilic nanocrystals: optically active multifunctional platform for cancer diagnosis and therapy, Acc. Chem. Res. 48 (9) (2015) 2506–2515.

[62] N. Frazier, H. Ghandehari, Hyperthermia approaches for enhanced delivery of nanomedicines to solid tumors, Biotechnol. Bioeng. 112 (10) (2015) 1967–1983.

[63] Y. Liu, J.R. Ashton, E.J. Moding, H.K. Yuan, J.K. Register, A.M. Fales, et al. A plasmonic gold nanostar theranostic probe for in vivo tumor imaging and photothermal therapy, Theranostics 5 (9) (2015) 946–960.

[64] T.J. Dougherty, C.J. Gomer, B.W. Henderson, G. Jori, D. Kessel, M. Korbelik, et al. Photodynamic therapy, J. Natl. Cancer Inst. 90 (12) (1998) 889–905.

[65] E. Paszko, C. Ehrhardt, M.O. Senge, D.P. Kelleher, J.V. Reynolds, Nanodrug applications in photodynamic therapy, Photodiagnosis Photodyn. Ther. 8 (1) (2011) 14–29.

[66] S. Perni, P. Prokopovich, J. Pratten, I.P. Parkin, M. Wilson, Nanoparticles: their potential use in antibacterial photodynamic therapy, Photochem. Photobiol. Sci. 10 (5) (2011) 712–720.

[67] C. Compagnin, L. Bau, M. Mognato, L. Celotti, G. Miotto, M. Arduini, et al. The cellular uptake of meta-tetra (hydroxyphenyl)chlorin entrapped in organically modified silica nanoparticles is mediated by serum proteins, Nanotechnology 20 (34) (2009) 345101–345113.

[68] Z. Zhu, Z.W. Tang, J.A. Phillips, R.H. Yang, H. Wang, W.H. Tan, Regulation of singlet oxygen generation using single-walled carbon nanotubes, J. Am. Chem. Soc. 130 (33) (2008) 10856–10857.

[69] S. Erbas, A. Gorgulu, M. Kocakusakogullari, E.U. Akkaya, Non-covalent functionalized SWNTs as delivery agents for novel Bodipy-based potential PDT sensitizers, Chem. Commun. (33) (2009) 4956–4958.

[70] L. Zhang, F.X. Gu, J.M. Chan, A.Z. Wang, R.S. Langer, O.C. Farokhzad, Nanoparticles in medicine: therapeutic applications and developments, Clin. Pharmacol. Ther. 83 (5) (2008) 761–769.

[71] V.P. Torchilin, Micellar nanocarriers: pharmaceutical perspectives, Pharm. Res. 24 (1) (2007) 1–16.

[72] B. Zhao, J.-J. Yin, P.J. Bilski, C.F. Chignell, J.E. Roberts, Y.-Y. He, Enhanced photodynamic efficacy towards melanoma cells by encapsulation of Pc4 in silica nanoparticles, Toxicol. Appl. Pharmacol. 241 (2) (2009) 163–172.

[73] G.Y. Chen, H.L. Qju, P.N. Prasad, X.Y. Chen, Upconversion nanoparticles: design, nanochemistry, and applications in theranostics, Chem. Rev. 114 (10) (2014) 5161–5214.

[74] S. Wu, H.J. Butt, Near-infrared-sensitive materials based on upconverting nanoparticles, Adv. Mater. 28 (6) (2016) 1208–1226.

[75] T. Passuello, F. Piccinelli, M. Pedroni, S. Polizzi, F. Mangiarini, F. Vetrone, et al. NIR-to-visible and NIR-to-NIR upconversion in lanthanide doped nanocrystalline GdOF with trigonal structure, Opt. Mater. 33 (10) (2011) 1500–1505.

[76] L. Tian, Z. Xu, S. Zhao, Y. Cui, Z. Liang, J. Zhang, et al. The upconversion luminescence of Er3+/Yb3+/Nd3+ triply-doped beta-NaYF4 nanocrystals under 808-nm excitation, Materials 7 (11) (2014) 7289–7303.

[77] G.S. Yi, H.C. Lu, S.Y. Zhao, G. Yue, W.J. Yang, D.P. Chen, et al. Synthesis, characterization, and biological application of size-controlled nanocrystalline NaYF4 : Yb,Er infrared-to-visible up-conversion phosphors, Nano Lett. 4 (11) (2004) 2191–2196.

[78] M. Nyk, R. Kumar, T.Y. Ohulchanskyy, E.J. Bergey, P.N. Prasad, High contrast in vitro and in vivo photoluminescence bioimaging using near infrared to near infrared up-conversion in TM3+ and Yb3+ doped fluoride nanophosphors, Nano Lett. 8 (11) (2008) 3834–3838.

[79] A. Patra, C.S. Friend, R. Kapoor, P.N. Prasad, Upconversion in Er3+: ZrO_2 nanocrystals, J. Phys. Chem. B 106 (8) (2002) 1909–1912.

[80] R. Sivakumar, F. van Veggel, M. Raudsepp, Bright white light through up-conversion of a single NIR source from sol-gel-derived thin film made with Ln(3+)-doped LaF_3 nanoparticles, J. Am. Chem. Soc. 127 (36) (2005) 12464–12465.

[81] L. Cheng, K. Yang, Y. Li, J. Chen, C. Wang, M. Shao, et al. Facile preparation of multifunctional upconversion nanoprobes for multimodal imaging and dual-targeted photothermal therapy, Angew. Chem. Int. Ed. 50 (32) (2011) 7385–7390.

[82] J. Zhou, M. Yu, Y. Sun, X. Zhang, X. Zhu, Z. Wu, et al. Fluorine-18-labeled Gd3+/Yb3+/Er3+ co-doped NaYF4 nanophosphors for multimodality PET/MR/UCL imaging, Biomaterials 32 (4) (2011) 1148–1156.

[83] H. Zhang, Y. Li, I.A. Ivanov, Y. Qu, Y. Huang, X. Duan, Plasmonic modulation of the upconversion fluorescence in NaYF4:Yb/Tm hexaplate nanocrystals using gold nanoparticles or nanoshells, Angew. Chem. Int. Ed. 49 (16) (2010) 2865–2868.

[84] P.K. Jain, I.H. El-Sayed, M.A. El-Sayed, Au nanoparticles target cancer, Nano Today 2 (1) (2007) 18–29.

[85] J. Turkevich, P.C. Stevenson, J. Hillier, A study of the nucleation and growth process in the synthesis of colloidal gold, Discuss. Faraday Soc. (11) (1951) 55.

[86] D. Tuncel, H.V. Demir, Conjugated polymer nanoparticles, Nanoscale 2 (4) (2010) 484–494.

[87] M.C. Daniel, D. Astruc, Gold nanoparticles: assembly, supramolecular chemistry, quantum-size-related properties, and applications toward biology, catalysis, and nanotechnology, Chem. Rev. 104 (1) (2004) 293–346.

[88] T. Yonezawa, T. Kunitake, Practical preparation of anionic mercapto ligand-stabilized gold nanoparticles and their immobilization, Colloid Surf. A 149 (1-3) (1999) 193–199.

[89] S. Laurent, D. Forge, M. Port, A. Roch, C. Robic, L.V. Elst, et al. Magnetic iron oxide nanoparticles: synthesis, stabilization, vectorization, physicochemical characterizations, and biological applications, Chem. Rev. 108 (6) (2008) 2064–2110.

[90] S.H. Sun, H. Zeng, D.B. Robinson, S. Raoux, P.M. Rice, S.X. Wang, et al. Monodisperse MFe2O4 (M = Fe, Co, Mn) nanoparticles, J. Am. Chem. Soc. 126 (1) (2004) 273–279.

[91] S.J. Oldenburg, R.D. Averitt, S.L. Westcott, N.J. Halas, Nanoengineering of optical resonances, Chem. Phys. Lett. 288 (2-4) (1998) 243–247.

[92] Z.H. Ban, Y.A. Barnakov, V.O. Golub, C.J. O'Connor, The synthesis of core-shell iron@gold nanoparticles and their characterization, J. Mater. Chem. 15 (43) (2005) 4660–4662.

[93] L.Y. Wang, J. Luo, M.M. Maye, Q. Fan, R.D. Qiang, M.H. Engelhard, et al. Iron oxide-gold core-shell nanoparticles and thin film assembly, J. Mater. Chem. 15 (18) (2005) 1821–1832.

[94] X.J. Ji, R.P. Shao, A.M. Elliott, R.J. Stafford, E. Esparza-Coss, J.A. Bankson, et al. Bifunctional gold nanoshells with a superparamagnetic iron oxide-silica core suitable for both MR imaging and photothermal therapy, J. Phys. Chem. C 111 (17) (2007) 6245–6251.

[95] M.P. Melancon, W. Lu, C. Li, Gold-based magneto/optical nanostructures: challenges for in vivo applications in cancer diagnostics and therapy, MRS Bull. 34 (6) (2009) 415–421.

[96] T.H.L. Nghiem, T.H. La, X.H. Vu, V.H. Chu, T.H. Nguyen, Q.H. Le, et al. Synthesis, capping and binding of colloidal gold nanoparticles to proteins, Adv. Nat. Sci. 1 (2) (2010) 025009.

[97] A.K. Oyelere, P.C. Chen, X.H. Huang, I.H. El-Sayed, M.A. El-Sayed, Peptide-conjugated gold nanorods for nuclear targeting, Bioconjug. Chem. 18 (5) (2007) 1490–1497.

[98] C.D. Medley, S. Bamrungsap, W.H. Tan, J.E. Smith, Aptamer-conjugated nanoparticles for cancer cell detection, Anal. Chem. 83 (3) (2011) 727–734.

[99] L.L. Yang, H. Mao, Y.A. Wang, Z.H. Cao, X.H. Peng, X.X. Wang, et al. Single chain epidermal growth factor receptor antibody conjugated nanoparticles for in vivo tumor targeting and imaging, Small 5 (2) (2009) 235–243.

[100] E. Katz, I. Willner, Integrated nanoparticle-biomolecule hybrid systems: synthesis, properties, and applications, Angew. Chem. Int. Ed. 43 (45) (2004) 6042–6108.

[101] G.X. Zhong, A.L. Liu, X.H. Chen, K. Wang, Z.X. Lian, Q.C. Liu, et al. Electrochemical biosensor based on nanoporous gold electrode for detection of PML/RAR alpha fusion gene, Biosens. Bioelectron. 26 (9) (2011) 3812–3817.

[102] N.L. Rosi, D.A. Giljohann, C.S. Thaxton, A.K.R. Lytton-Jean, M.S. Han, C.A. Mirkin, Oligonucleotide-modified gold nanoparticles for intracellular gene regulation, Science 312 (5776) (2006) 1027–1030.

[103] M. Pilo-Pais, S. Goldberg, E. Samano, T.H. LaBean, G. Finkelstein, Connecting the nanodots: programmable nanofabrication of fused metal shapes on DNA templates, Nano Lett. 11 (8) (2011) 3489–3492.

[104] C.A. Mirkin, R.L. Letsinger, R.C. Mucic, J.J. Storhoff, A DNA-based method for rationally assembling nanoparticles into macroscopic materials, Nature 382 (6592) (1996) 607–609.

[105] D.L. Plata, P.M. Gschwend, C.M. Reddy, Industrially synthesized single-walled carbon nanotubes: compositional data for users, environmental risk assessments, and source apportionment, Nanotechnology 19 (18) (2008) 185706–185720.

[106] H.C. Fischer, W.C.W. Chan, Nanotoxicity: the growing need for in vivo study, Curr. Opin. Biotechnol. 18 (6) (2007) 565–571.

[107] G. Oberdorster, E. Oberdorster, J. Oberdorster, Nanotoxicology: an emerging discipline evolving from studies of ultrafine particles, Environ. Health Perspect. 113 (7) (2005) 823–839.

[108] K. Donaldson, C.A. Poland, Nanotoxicity: challenging the myth of nano-specific toxicity, Curr. Opin. Biotechnol. 24 (4) (2013) 724–734.

[109] D. Docter, S. Strieth, D. Westmeier, O. Hayden, M.Y. Gao, S.K. Knauer, et al. No king without a crown—impact of the nanomaterial-protein corona on nanobiomedicine, Nanomedicine 10 (3) (2015) 503–519.

[110] S. Singh, H.S. Nalwa, Nanotechnology and health safety—toxicity and risk assessments of nanostructured materials on human health, J. Nanosci. Nanotechnol. 7 (9) (2007) 3048–3070.

[111] A.M.M. Gatti, Nanopathology: The Health Impact of Nanoparticles, Pan Stanford Publishing, Singapore, 2008.

[112] F.A. Witzmann, Effect of carbon nanotube exposure on keratinocyte protein expression, in: N.A. Monteiro-Riviere, C.L. Tran (Eds.), Nanotoxicology: Characterization, Dosing, Health, Effects, Informa Healthcare, New York, 2007.

[113] V. Stone, H. Johnston, R.P.F. Schins, Development of in vitro systems for nanotoxicology: methodological considerations, Crit. Rev. Toxicol. 39 (7) (2009) 613–626.

[114] H.J. Johnston, G. Hutchison, F.M. Christensen, S. Peters, S. Hankin, V. Stone, A review of the in vivo and in vitro toxicity of silver and gold particulates: particle attributes and biological mechanisms responsible for the observed toxicity, Crit. Rev. Toxicol. 40 (4) (2010) 328–346.

[115] J. Cancino, T.M. Nobre, O.N. Oliveira, S.A.S. Machado, V. Zucolotto, A new strategy to investigate the toxicity of nanomaterials using Langmuir monolayers as membrane models, Nanotoxicology 7 (1) (2013) 61–70.

[116] Y. Chang, S.-T. Yang, J.-H. Liu, E. Dong, Y. Wang, A. Cao, et al. In vitro toxicity evaluation of graphene oxide on A549 cells, Toxicol. Lett. 200 (3) (2011) 201–210.

[117] S. Arora, J.M. Rajwade, K.M. Paknikar, Nanotoxicology and in vitro studies: the need of the hour, Toxicol. Appl. Pharmacol. 258 (2) (2012) 151–165.

[118] K. Donaldson, A. Schinwald, F. Murphy, W.S. Cho, R. Duffin, L. Tran, et al. The biologically effective dose in inhalation nanotoxicology, Acc. Chem. Res. 46 (3) (2013) 723–732.

[119] G. Oberdoerster, Nanotoxicology: in vitro-in vivo dosimetry, Environ. Health Perspect. 120 (1) (2012) A13–A20.

[120] Y. Cao, N.R. Jacobsen, P.H. Danielsen, A.G. Lenz, T. Stoeger, S. Loft, et al. Vascular effects of multi-walled carbon nanotubes in dyslipidemic ApoE(/) mice and cultured endothelial cells, Toxicol. Sci. 138 (1) (2014) 104–116.

[121] A.B. Seabra, A.J. Paula, R. de Lima, O.L. Alves, N. Duran, Nanotoxicity of graphene and graphene oxide, Chem. Res. Toxicol. 27 (2) (2014) 159–168.

[122] H.J. Johnston, G.R. Hutchison, F.M. Christensen, S. Peters, S. Hankin, K. Aschberger, et al. A critical review of the biological mechanisms underlying the in vivo and in vitro toxicity of carbon nanotubes: the contribution of physico-chemical characteristics, Nanotoxicology 4 (2) (2010) 207–246.

[123] C.P. Firme III, P.R. Bandaru, Toxicity issues in the application of carbon nanotubes to biological systems, Nanomedicine 6 (2) (2010) 245–256.

[124] N.A. Monteiro-Riviere, R.J. Nemanich, A.O. Inman, Y.Y.Y. Wang, J.E. Riviere, Multi-walled carbon nanotube interactions with human epidermal keratinocytes, Toxicol. Lett. 155 (3) (2005) 377–384.

[125] M. Bottini, S. Bruckner, K. Nika, N. Bottini, S. Bellucci, A. Magrini, et al. Multi-walled carbon nanotubes induce T lymphocyte apoptosis, Toxicol. Lett. 160 (2) (2006) 121–126.

[126] R. Baktur, H. Patel, S. Kwon, Effect of exposure conditions on SWCNT-induced inflammatory response in human alveolar epithelial cells, Toxicol. In Vitro 25 (5) (2011) 1153–1160.

[127] M. Davoren, E. Herzog, A. Casey, B. Cottineau, G. Chambers, H.J. Byrne, et al. In vitro toxicity evaluation of single walled carbon nanotubes on human A549 lung cells, Toxicol. In Vitro 21 (3) (2007) 438–448.

[128] V. Wiwanitkit, A. Sereemaspun, R. Rojanathanes, Effect of gold nanoparticles on spermatozoa: the first world report, Fertil. Steril. 91 (1) (2009) E7–E8.

[129] W.J. Trickler, S.M. Lantz, R.C. Murdock, A.M. Schrand, B.L. Robinson, G.D. Newport, et al. Brain microvessel endothelial cells responses to gold nanoparticles: In vitro pro-inflammatory mediators and permeability, Nanotoxicology 5 (4) (2011) 479–492.

[130] K. Tahara, T. Sakai, H. Yamamoto, H. Takeuchi, N. Hirashima, Y. Kawashima, Improved cellular uptake of chitosan-modi fled PLGA nanospheres by A549 cells, Int. J. Pharm. 382 (1-2) (2009) 198–204.

[131] W. Lin, Huang Y-w, X.-D. Zhou, Y. Ma, In vitro toxicity of silica nanoparticles in human lung cancer cells, Toxicol. Appl. Pharmacol. 217 (3) (2006) 252–259.

[132] Y.-H. Hsin, C.-F. Chena, S. Huang, T.-S. Shih, P.-S. Lai, P.J. Chueh, The apoptotic effect of nanosilver is mediated by a ROS- and JNK-dependent mechanism involving the mitochondrial pathway in NIH3T3 cells, Toxicol. Lett. 179 (3) (2008) 130–139.

[133] M.E. Samberg, S.J. Oldenburg, N.A. Monteiro-Riviere, Evaluation of silver nanoparticle toxicity in skin in vivo and keratinocytes in vitro, Environ. Health Perspect. 118 (3) (2010) 407–413.

[134] S. Arora, J. Jain, J.M. Rajwade, K.M. Paknikar, Cellular responses induced by silver nanoparticles: in vitro studies, Toxicol. Lett. 179 (2) (2008) 93–100.

[135] J. Zhao, V. Castranova, Toxicology of nanomaterials used in nanomedicine, J. Toxicol. Environ. Health B 14 (8) (2011) 593–632.

[136] A. Gerber, M. Bundschuh, D. Klingelhofer, D.A. Groneberg, Gold nanoparticles: recent aspects for human toxicology, J. Occup. Med. Toxicol. 8 (2013) 6.

[137] M. Berger, Nano-Society: Pushing the Boundaries of Technology, Royal Society of Chemistry, Cambridge, UK, 2009.

[138] A. Magrez, S. Kasas, V. Salicio, N. Pasquier, J.W. Seo, M. Celio, et al. Cellular toxicity of carbon-based nanomaterials, Nano Lett. 6 (6) (2006) 1121–1125.

[139] J. Miyawaki, M. Yudasaka, T. Azami, Y. Kubo, S. Iijima, Toxicity of single-walled carbon nanohorns, ACS Nano 2 (2) (2008) 213–226.

[140] M. Shimizu, H. Tainaka, T. Oba, K. Mizuo, M. Umezawa, K. Takeda, Maternal exposure to nanoparticulate titanium dioxide during the prenatal period alters gene expression related to brain development in the mouse, Part. Fibre Toxicol. 6 (20) (2009) 1–8.

4

Nanostructured Films: Langmuir–Blodgett (LB) and Layer-by-Layer (LbL) Techniques

R.F. de Oliveira*, A. de Barros**, M. Ferreira†

*BRAZILIAN CENTER FOR RESEARCH IN ENERGY AND MATERIALS, BRAZILIAN NANOTECHNOLOGY NATIONAL LABORATORY, CAMPINAS, SÃO PAULO, BRAZIL; **STATE UNIVERSITY OF CAMPINAS, INSTITUTE OF CHEMISTRY, CAMPINAS, SÃO PAULO, BRAZIL; †FEDERAL UNIVERSITY OF SÃO CARLOS, CENTER FOR SCIENCES AND TECHNOLOGY FOR SUSTAINABILITY, SOROCABA, SÃO PAULO, BRAZIL

CHAPTER OUTLINE

105

4.1 Introduction

Ultrathin film fabrication is an important process in nanoscience and nanotechnology, as demonstrated by the large number of publications about manufacturing new thin-film materials and their applications in a wide range of fields, such as biotechnology, electronics, medicine, and so on [1–3]. Of the existing film fabrication methods, Langmuir–Blodgett (LB) and layer-by-layer (LbL) techniques are notable because they allow the film thickness and architecture to be controlled at the molecular level. For example, films with a high degree of molecular organization can be obtained, and the spatial arrangement of the film material can be investigated by the LB technique. The LbL method, in turn, is advantageous because of the simplicity of the film deposition process and the large variety of materials that can be used.

In addition to the technological importance of nanostructured films, fundamental questions about their formation are of great scientific interest because a wide variety of materials and film architectures are possible. The LB technique, which was developed in the mid-1900s, and the LbL method, which was discovered just over 20 years ago, are still attracting the interest of specialized research groups worldwide.

In this chapter, the key aspects of the LB and LbL techniques in ultrathin film fabrication, including the film formation mechanisms, main characterization techniques and applications, are addressed to provide the reader with an overview of these important nanotechnological processes.

4.2 Langmuir–Blodgett Technique

4.2.1 History

The first surface chemistry experiments that stimulated further research on ultrathin film formation were performed in the 18th century. According to Behroozi et al. [4], during one of his overseas journeys, Benjamin Franklin (1706–90) noted that the waves near ships with oil leakages appeared to be dampened. To investigate this phenomenon, he performed experiments on the damping effect of oil on water, which resulted in the first publication in the field in 1774. In one of these experiments, Franklin poured a small known amount of oil on a water surface and observed a decrease in the wind turbulence within a certain distance of the oil-coated area [5].

At that time, Franklin focused only on the wave-damping phenomenon and did not realize that this effect was related to the formation of a monomolecular layer at the surface. Dividing the oil volume by the area it occupies gives the oil layer thickness, which is on the nanometer scale [6]. This simple calculation was not performed until years later, when Lord Rayleigh (1842–1919) determined the thickness of oil layers. However, he did not know that the calculated value corresponded exactly to the length of the molecule used in the experiments, indicating that the layer was one molecule thick [6].

During the same period, Agnes Pockels (1862–1935) made an important advance in the science of monolayer formation by creating a prototype of the Langmuir trough (Fig. 4.1). This prototype was a primitive device in which barriers were used both to compress oil molecules scattered on a water surface and to remove impurities from the surface. Pockles also developed a method for measuring the surface tension of water in a container [5]. In recognition of these important contributions, Rayleigh helped Pockles publish her results in the prestigious journal *Nature*. A few years later, Irving Langmuir (1881–1957) introduced the concept of molecular conformations.

For Langmuir, molecules were asymmetrical and would therefore have identical orientations on the water surface, depending on their hydrophilic or hydrophobic character. Langmuir also estimated the size of the molecule used in his experiments, which had a large impact on the scientific community at the time. These studies motivated Langmuir to investigate monolayer formation on water surfaces, and he was awarded the Nobel Prize in Chemistry for this work in 1932. Langmuir suggested that the monolayers formed on a water surface could be transferred to a solid surface. His assistant Katherine Blodgett (1898–1979) performed the corresponding experiments, and the initial results were published in 1934 and 1935. Blodgett was awarded her own Nobel Prize for this work. During this period, the LB monolayer deposition method was developed. Several years later, Langmuir and Vincent Schaefer (1906–93) studied protein deposition on solid substrates and discovered a new approach called the Langmuir–Schaefer (LS) method for depositing Langmuir monolayers. In the LS method, a Langmuir monolayer is deposited on a surface horizontally instead of vertically, as in LB films [5,7].

The formation of Langmuir monolayers and LB and LS films was not actively studied for a long time because they had no practical application. Around 1980, research interest

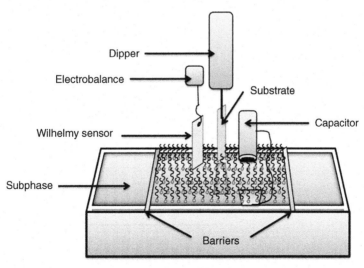

FIGURE 4.1 Schematic representation of a Langmuir trough and its accessories.

in this area was renewed by the use of LB films in organic electronics. Since then, physicists, chemists, biologists, and engineers have actively collaborated to develop this technique using various types of materials for a wide range of applications.

4.2.2 Description of the Technique

As illustrated in Fig. 4.1, the Langmuir trough is composed of an inert material, typically Teflon. The trough accessories include mobile barriers that compress the molecules spread at the air–water interface, a device (dipper) that immerses the substrate for the monolayer deposition, a Wilhelmy sensor that measures the surface tension of the subphase, and a capacitor that measures the surface potential.

Langmuir monolayers and LB films are fabricated by pouring a known volume of a given solution at the air–water interface. Ideally, the material in the solution is soluble in an organic solvent or solvent mixture. During film fabrication, the solvent is evaporated after a given amount of time, and the molecules are then compressed until maximum order is achieved.

Initially, the materials employed in the Langmuir technique were necessarily amphiphilic compounds, that is, molecules with a polar head group (hydrophilic) and a nonpolar tail group (hydrophobic) essentially insoluble in polar solvents. These compounds self-organize with their polar part in the water and nonpolar part in the air. The hydrophobic portion of these molecules generally consists of aliphatic chains, which reduce its solubility in the aqueous subphase. The hydrophilic part is responsible for spreading the molecules because it interacts strongly with the water. The molecular orientation of these molecules at the air–water interface minimizes their free energy [5,7–9].

A typical isotherm for amphiphilic molecules is illustrated in Fig. 4.2. Before compression by the barriers, the molecules are initially in the gas phase (stage A); they are fully

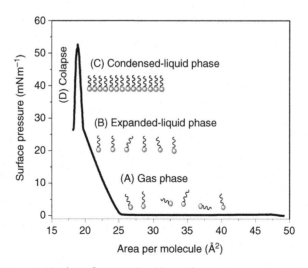

FIGURE 4.2 Typical surface pressure isotherm for a stearic acid monolayer.

dispersed and do not interact. As the molecules are compressed, their proximity leads to the formation of an expanded liquid phase (stage B). When the molecule surface density increases, the regular arrays in the film form, resulting in a compact structure called the condensed-liquid phase (stage C). However, further compression of the monolayer causes molecular disorder and a phenomenon commonly called collapse (stage D) [5,7–9].

The use of LB films composed of common amphiphilic molecules is limited by their properties. To search for new technological applications of LB films, researchers have investigated the use of various materials, including polymers [10–13], phospholipids [14–16], enzymes [17–20], and peptides [21,22]. However, the formation of monolayers of other types of molecules is not trivial because they might not be completely soluble in volatile organic solvents or might be unstable, making them difficult to spread at the surface and consequently deposit homogeneously on a substrate [8].

To minimize these difficulties, several strategies have been employed depending on the type of film material. For example, miscible solvents can be used, or the material can be spread in the subphase [10,23]. The organization, homogeneity, deposition quality, and formation of Langmuir monolayers and LB films can be evaluated experimentally by characterization methods such as Brewster angle microscopy [24,25], polarization modulation infrared reflection-absorption spectroscopy [24,26,27], and atomic force microscopy (AFM) [28,29].

4.2.3 Pressure and Surface Potential

The pressure and surface potential of LB monolayers are measured as a function of the average molecular area occupied at the air–water interface to assess molecular packing and order [5,7,9]. These characterizations are fundamental and necessary for evaluating the quality of the Langmuir film formed at the interface. To determine the surface pressure of a Langmuir monolayer, the material is spread on an aqueous subphase, and the surface pressure is measured by detecting variations in the liquid surface tension, as shown by the following equation:

$$\pi = \gamma_0 - \gamma_A \tag{4.1}$$

where γ_0 and γ_A are the surface tensions of pure water and due to the monolayer, respectively [5,7].

The surface pressure isotherms are obtained using a Wilhelmy sensor, which is suspended by a wire from an electrobalance and partially immersed in the water. The electrobalance measures the force exerted to keep the sensor stationary as the surface tension varies. The vertical force exerted by the surface tension is detected by the balance and converted into a voltage [5,7]. A typical isotherm of an amphiphilic material (stearic acid) is illustrated in Fig. 4.2.

One disadvantage of this technique is related to the contact angle (θ) formed between the liquid subphase and the balance sensor, as shown in Fig. 4.3 [5,7]. Another disadvantage is that the position of the Wilhelmy sensor relative to the barrier positions can affect the pressure isotherm during monolayer formation. These problems can be minimized by placing the sensor in the center of the trough [8].

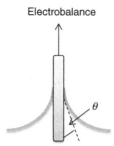

Electrobalance

θ

FIGURE 4.3 Schematic representation of the contact angle between the Wilhelmy sensor and the subphase.

The surface pressure isotherms might exhibit hysteresis. This phenomenon arises due to differences in the molecular behavior during compression and decompression. Specifically, as the barriers compress the molecules, the area that they occupy decreases, resulting in an increase in their interactions and consequently an increase in the surface pressure. When the monolayer is decompressed, the surface pressure gradually decreases, but not in the same way that it increases during the initial compression. The hysteresis in the area occupied by the molecules indicates that some of the material is lost to the aqueous subphase or to aggregation. A large hysteresis implies that the stability of the monolayer is low. Another method for characterizing monolayer stability is based on compressing the monolayer to the maximum pressure at which molecule order is observed for long periods of time. If large variations in the area are observed during this period, the monolayers are unstable and therefore unsuitable for film deposition [7–9].

Another widely used technique for characterizing Langmuir films is surface potential (ΔV) measurements. Surface potential is defined as the difference in the potentials of the aqueous subphase covered with the monolayer and the pure aqueous subphase. Surface potential can be measured using a Kelvin probe or a vibrating plate capacitor, which is more common [5,7–9,30]. In the vibrating capacitor method, surface potential is measured by a plate positioned above the water that detects the vibrations of the spread molecules. The potential of the pure water is used as a reference and is measured by a metal plate immersed in the subphase. The ΔV value is the change in the permanent electric dipoles at the air–water interface due the presence of a partially or fully ionized film. To gain insight into the effect, theoretical models were developed to relate the measured potential to the dipole moments of the film material. The most commonly used model was developed by Demchak and Fort (DF) [30]. In this model, the Langmuir monolayer is considered to be a three-layer capacitor in which each layer has a different relative permittivity, as illustrated in Fig. 4.4 [8,9,30].

According to the DF model, the surface potential is given by:

$$\Delta V = \frac{1}{A\varepsilon_0}\left(\frac{\mu_1}{\varepsilon_1}+\frac{\mu_2}{\varepsilon_2}+\frac{\mu_3}{\varepsilon_3}\right)+\Psi_0 \tag{4.2}$$

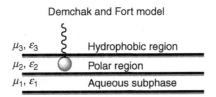

FIGURE 4.4 **Capacitor model of Demchak and Fort for a Langmuir monolayer.** *Adapted from P. Dynarowicz-Latka, A. Dhanabalan, O.N. Oliveira Jr., Modern physicochemical research on Langmuir monolayers, Adv. Colloid Interface Sci. 91 (2) (2001) 221–293 [8].*

where A is the average area per molecule; ε_0 is the vacuum permittivity; μ_1, μ_2, and μ_3 are the dipole moments attributed to the polarization and reorientation of the molecules; and ε_1, ε_2, and ε_3 are the permittivities due to changes in the dipole moments of the hydrophilic and hydrophobic groups. More specifically, the component μ_1/ε_1 corresponds to the reorientation of water molecules that is induced by the presence of the monolayer, μ_2/ε_2 is the contribution of the hydrophilic portion of the molecule, μ_3/ε_3 is the contribution of the hydrophobic portion, and Ψ_0 is attributed to the electrical double layer that forms when the film is partially or fully ionized [5,7–9]. In polymer films, for example, it is impossible to estimate values for $\mu_1/\varepsilon_1 + \mu_2/\varepsilon_2 + \mu_3/\varepsilon_3$ and Ψ_0, mainly because the polymer is often processed under different conditions, which induce structural changes in the polymer chain. These changes make it difficult to explain the experimental results using a theoretical surface potential model.

Molecular behavior during the surface potential measurement can be described as follows. Initially, the molecules spread on the subphase cover an area so large that their interactions are too weak to induce a detectable change in the surface potential of the aqueous subphase. During compression, the area covered by the molecules reaches a critical value at which the potential is no longer zero, and it increases sharply as the area per molecule is further decreased. Due to this critical area, surface potential measurements are more sensitive to the film organization than surface pressure measurements [5,7–9]. Fig. 4.5 shows a typical surface potential curve for stearic acid based on the DF model, which considers the changes in the dielectric constants of the three layers as functions of the area per molecule.

4.2.4 Langmuir Monolayer Deposition: Transfer Ratio

Langmuir monolayers can be deposited by two methods, the vertical (LB) and horizontal (LS) methods, which are as illustrated in Fig. 4.6. The vertical method is more commonly used because the amount of material deposited on the substrate cannot be effectively controlled during horizontal deposition, that is, it is unknown if the entire surface is actually covered by the monolayer [5,31]. The effectiveness of the monolayer deposition process is described by the transfer ratio (TR). A TR of close to 1.0 implies that the material exhibits good adhesion to the substrate when it is immersed in and then removed from the subphase. However, a low TR value indicates that the material does not adhere to the

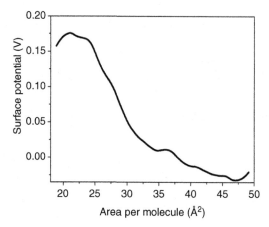

FIGURE 4.5 Typical surface potential isotherm for a stearic acid monolayer.

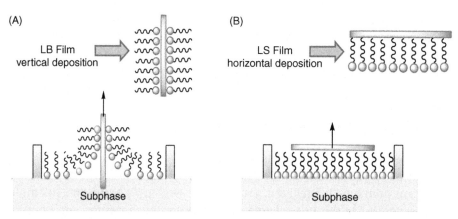

FIGURE 4.6 Schematic representation of the different Langmuir monolayer deposition methods. (A) Vertical deposition (LB) and (B) horizontal deposition (LS).

substrate well, and consequently, the second monolayer deposited on the film might be easily removed, resulting in a low-quality film [7]. The TR value is calculated using the following equation:

$$\tau = \frac{A_L}{A_S} \tag{4.3}$$

where A_L is the decrease in the area occupied by the molecules at the air/water interface (at constant pressure) and A_S is the area of the substrate covered with the monolayer [7].

Different LB film architectures can be obtained by varying the deposition parameters, substrate characteristics, and film material. The possible architectures are called X-, Y-, and Z-type films and are illustrated in Fig. 4.7. Y-type LB films are generally obtained when substrates with hydrophilic surfaces are used. For the X- and Z-type films, deposition

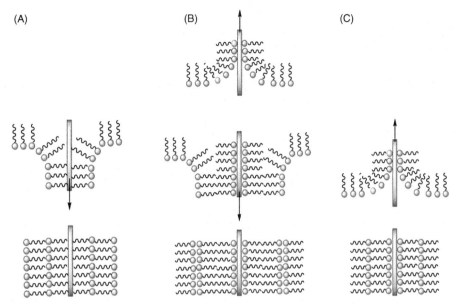

FIGURE 4.7 Different LB film types: X (A), Y (B), and Z (C).

occurs preferentially during substrate immersion and removal from the subphase, which favors the formation of films in which the polar and nonpolar groups in adjacent monolayers interact. It is important to note that Z-type films form on hydrophilic substrates, whereas X-type films form onto hydrophobic substrates [5,7,32].

4.2.5 Applications

Numerous applications of LB and LS films, such as nonlinear optical and piezoelectric devices [33,34], chemical sensors and biosensors [10,12,13,17–20], cell membrane applications [35–37], and so on [1,38–40] have been reported in the literature.

LB and LS films are widely used in sensors, biosensors, and cell membrane model studies because their molecular organization can be highly controlled, which results in satisfactory experimental results. In one study, enzymes were immobilized on lipid monolayers of LS films, and the catalytic activity of the biomolecule-containing films in sucrose hydrolysis was evaluated. The experimental results indicated that 78% of the invertase enzymatic activity was maintained in this system [17]. In another study, the interactions between mucin and chitosan were studied using dimyristoyl phosphatidic acid (DMPA) Langmuir monolayers as cell membrane models. The experimental results for the DMPA–chitosan–mucin Langmuir and LB films indicated that a mucin–chitosan complex forms via electrostatic interactions that are crucial for the mucoadhesion mechanism [35].

De Barros et al. developed electrochemical sensors composed of polyaniline and montmorillonite clay using LB technique for the detection of copper, lead, and cadmium metal ions. The Langmuir isotherms and ultraviolet-visible (UV-vis) absorption spectroscopy, Fourier transform infrared spectroscopy and Raman spectroscopy results indicated the existence of synergistic molecular-level interactions between the film materials, which were favorable for the sensor performance, that is, the sensors had high sensitivity and a low detection limit on the order of $\mu g\ L^{-1}$ [10]. Thin films of polythiophene derivatives were fabricated using LS and spin-coating techniques to compare the effect of the deposition method on the electrical properties of the films. The experimental results revealed an electrochromic effect during analysis, and the electrical and electrochemical measurements revealed the effects of the film organization on its electrical properties [11].

For further reading on the theoretical concepts and applicability of this technique, the reader is referred to the work of M. Petty titled "Langmuir–Blodgett Films: An Introduction" and other texts in this field.

4.3 Layer-by-Layer Technique

4.3.1 History

The LbL method became widely spread during the 1990s due to several publications by G. Decher [41–46], including a paper in the prestigious journal *Science* [46] in 1997. This technique, however, was first described in 1966 by R.K. Iler as a study titled "Multilayers of Colloidal Particles" [47]. This study showed that multilayer films could be obtained by alternately immobilizing oppositely charged colloidal particles. Iler observed that the thicknesses of alternating cationic (alumina) and anionic (silica) layers on glass could be controlled and observed by the change in color of light reflected on the substrate surface [47].

The LbL technique was further developed in several subsequent studies that described, for example, the successive deposition of inorganic ionic compounds [48] and polyanion adsorption on cellulosic fibers containing preadsorbed polycations [49], among other systems [50]. Decher contributed not only to the rediscovery of this technique but also to its use as an alternative to the LB method for fabricating nanostructured films. He also demonstrated that it could be used to deposit films of various types of materials, including bipolar amphiphilic molecules [42], polyelectrolytes [42,44], and proteins and DNA [43,44]. According to Decher and Schlenoff [50], the field of nanostructure fabrication via the LbL method is still popular and growing more than 20 years after the technique was reintroduced.

4.3.2 Description of the Technique

As the terminology itself suggests, the LbL technique is a method for obtaining nanostructured films by successively depositing layers of different materials, such as polymers, nanoparticles, enzymes, cells, and so on, with highly controlled thicknesses at the molecular level. The ultrathin films fabricated using LbL method have nanostructures that can be used in optical devices, electronics, sensors, and biotechnological applications [2,3,51,52].

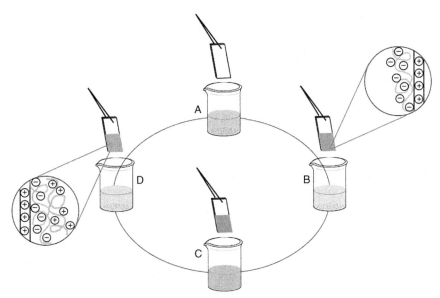

FIGURE 4.8 Schematic representation of the traditional LbL deposition process.

Unlike the sophisticated LB method, the main advantages of the LbL technique are its simplicity and low cost. Fig. 4.8 illustrates the LbL deposition process and film formation.

Initially, a substrate is immersed in a solution containing charged species, for example, polyanions (step A). During this step, the anionic polyelectrolyte adsorbs on the substrate surface via simple electrostatic interactions, resulting in the formation of a negative charge network on the surface. Then, the substrate is immersed in a rinsing solution (step B) to remove weakly adsorbed material to prevent contamination of the next solution. The substrate is subsequently immersed in a polycation solution to generate a new network of positive charges on the surface (step C). Finally, the substrate is again immersed in a rinsing solution (step D). Thus, a bilayer of the ionic materials, in this case polyelectrolytes, is obtained. This procedure can be repeated as many times as desired to obtain multilayer films with controlled structures and thicknesses.

4.3.3 Mechanisms of LbL Film Formation

The multilayer formation process can be divided into two stages: (1) the initial adsorption of the material on the surface (fast process) and (2) the relaxation of the adsorbed layer (slower process) [53]. In the first stage, the polyelectrolyte segments diffuse through the solution to adsorb on the surface until a repulsive electrical potential is generated, preventing the adsorption of more chains [54]. At this point, the polyelectrolyte layer has reached saturation, which is only possible when the adsorption rate is not limited, that is, when the species concentration in solution is much larger than the saturation concentration of the adsorbed species [46,53,54]. In this stage, the polyelectrolyte segments adsorb on the surface because of favorable electrostatic interactions, and the surface becomes oppositely charged, which enables the adsorption of new polyions with opposite charges to the surface [46].

When the initial substrate charge density is small, multiplication of surface functionality occurs. In this process, the charge density of the first adsorbed layer is larger than that of the substrate, which favors the subsequent adsorption of oppositely charged species [46]. It should be noted that the surface charge and roughness of the substrate (glass, quartz, silicon, paper, cloth, etc.) might affect the deposition of the first few layers [53]. The influence of the substrate on the deposited layers becomes negligible after a few deposition cycles. Then, the film growth and structure are fundamentally governed by the film materials. This important feature allows a film to be fabricated on a different surface for a given application than on the substrate that was used to study the film properties [50].

During the second stage of the layer formation process, the conformations of the adsorbed polyelectrolyte chains change as they relax [54]. The relaxation process explains the interpenetration of adjacent polyelectrolyte chains observed in experimental studies. However, it is assumed that the adsorbed segments do not diffuse along the surface during LbL film formation because this process would involve the breaking of many electrostatic bonds simultaneously, which requires more energy than that gained by thermodynamic contributions to the system and is therefore highly improbable [54].

Although electrostatic interactions are involved in the formation of polyelectrolyte films (Fig. 4.8), they are not the only interactions that can be exploited in LbL film formation. Depending on the characteristics of the selected materials, covalent bonds, van der Waals forces, and hydrogen bonds can also govern the film formation mechanism, further enhancing the versatility of the LbL technique [55].

Hydrogen bonds, for example, are sensitive to the medium and can be broken and reformed by changing the pH. This phenomenon is observed in the production of some porous films, such as polyacid acrylic (PAA)/polyvinyl pyridine (PVP) multilayers. In a basic solution, the PAA layers dissolve, allowing the PVP layers to subsequently rearrange to yield the porous structure [56]. The film characteristics, such as the pore size and pore size distribution, can also be tuned by varying the pH and temperature of the solution and the immersion time and nature of the substrate during treatment in a basic solution [55,57]. Hydrogen bonds also enable the production of multilayers of a single material; for example, films can be fabricated from a dendrimer with carboxylic acid groups that can act as both hydrogen bond donors and acceptors [56]. Other examples of different multilayer film formation mechanisms include charge transfer reactions (electron donor and acceptor materials are alternately deposited) [58] and covalent bonding [59].

In addition to these different interactions, other strategies, such as using avidin–biotin biospecificity [60], DNA hybridization [61], and other interactions [2,55,62,63], have been employed to obtain LbL films. In LbL films, different interactions not only drive the film formation mechanism but also participate to a minor degree in films formed via electrostatic interactions, influencing their properties, such as their stability, morphology, and thickness [53].

Of all the characteristics of a film, its thickness might be one of the most important [50]. The thickness depends on the deposition conditions of each layer, which in turn influence the subsequently deposited layer and, consequently, the final film properties, including the roughness, uniformity, chemical resistance, and so on. The intrinsic film material properties, such as the nature and density of the charged groups; the solution characteristics, such as the concentration, pH, and ionic strength; the fabrication operating parameters, such as the deposition and rinse times; and many other factors affect the film formation. For example, studies indicate that in general, increasing the ionic strength of a polyelectrolyte solution causes the polymer chains to contract, which leads to an increase in the surface density of the adsorbed segments and thus to an increase in film thickness [54]. However, exposing a polyelectrolyte film to a concentrated salt solution after fabrication leads to a decrease in film roughness due to increased chain mobility [50]. Moreover, changing the solution pH can alter the degree of dissociation and the conformation of some polyelectrolytes and also the enzymatic activity of biological films. Although many factors in the deposition process affect the film properties, understanding their effects are crucial to optimizing the film deposition process.

4.3.4 Spray- and Spin-LbL Methods

In addition to the conventional immersion method, other LbL film fabrication procedures have been reported in the literature. In the spin-assisted LbL method, the film materials are alternately added to a substrate rotating at a given speed, whereas in the spray-LbL method, the materials are sprayed in small liquid particles on the substrate surface. These methods are depicted in Fig. 4.9.

These techniques can be advantageous over the conventional method and therefore more appropriate, depending on the desired film characteristics and application. The

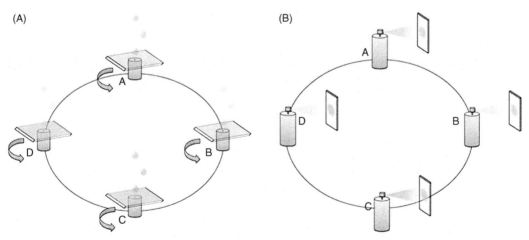

FIGURE 4.9 Novel LbL deposition strategies. (A) Spin-assisted LbL and (B) spray-LbL.

spin-assisted LbL and spray-LbL methods are faster than the conventional immersion method, particularly in the production of thick (micrometer) films because they do not depend on the diffusion kinetics of the species in the liquid medium. They also consume smaller amounts of material. However, the spin-assisted LbL method can only be used with flat substrates and produces nonuniform films on large-area substrates. The spray technique is more appropriate for these types of substrates [64]. The characteristics of films produced via the spray and conventional methods are very similar, whereas films produced by the spin-assisted LbL method are generally smoother due to decreased chain interpenetration, for example, in the case of polyelectrolytes [64]. It should be noted that all three deposition methods can be automated, which increases the reproducibility of the fabricated films [65].

4.3.5 Characterization Methods

Regardless of the deposition procedure, materials, and desired application, the characterization of LbL films is frequently necessary. In particular, UV-vis spectroscopy, zeta potential determination, and quartz crystal microbalances are commonly used to monitor the film during the deposition steps. For example, the amount of material adsorbed in each step can be determined by UV-vis monitoring in which the absorption band intensities of compounds with chromophore groups are measured as a function of the number of deposited layers. However, this and some other techniques require the film to be dried after rinsing, which might not be desirable depending on the application. Alternatively, film growth can be monitored in situ by a quartz crystal microbalance. In this method, the frequency of a resonant quartz crystal changes in response to changes in the material mass, enabling the amount of material adsorbed during each step to be measured. This technique allows film growth to be monitored when UV-vis spectroscopy is not possible, for example, when the material of interest absorbs only at frequencies outside the ultraviolet and visible regions [50].

Several other techniques, such as infrared spectroscopy, Raman spectroscopy, AFM, scanning electron microscopy (SEM), and transmission electron microscopy (TEM), are often used to characterize LbL films. Raman and infrared spectroscopies are used to determine the functional groups present in the film materials by recording their vibrational modes. Techniques such as AFM, SEM, and TEM are used to analyze film morphology, including its surface roughness, uniformity, and so on. The reader is referred to other chapters in this book for more information on various nanostructure characterization techniques.

4.3.6 Applications

Although the LbL technique has been widely used to fabricate polyelectrolyte films, it is currently employed for many different materials and applications. A few of the commonly used synthetic and natural materials include polypeptides, proteins,

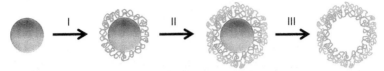

FIGURE 4.10 Schematic illustration of the fabrication of spherical shells by the LbL method.

enzymes, polysaccharides, metallic nanoparticles, oxides and clays, DNA, biopolymers, dendrimers, and carbon nanotubes [2,50,66–71]. LbL films are employed in biotechnological applications, such as antibacterial and antiadhesive biopolymer surface coatings [72], sensors [66,71] and electrochemical biosensors [73–75], and also in biomedical applications, such as self-supporting polymeric shells for controlled drug delivery [53]. In the latter application, polystyrene, polylactic acid and silica particles are coated with polyelectrolyte or biopolymer multilayers (steps I and II, Fig. 4.10) and then decomposed in an appropriate solvent to obtain spherical shells (step III, Fig. 4.10). The drug of interest is introduced into the shells via a stimulus, such as a change in the temperature, magnetic field, or pH, and then released at the target using the same stimulus [53]. Fig. 4.10 illustrates the formation process of these spherical shells.

LbL films are also used in electronic devices; they serve as charge injection layers in light-emitting diodes [76], are employed in solar cells [77], constitute the active layer in field-effect transistors [78], and are used in memory applications [79]. In addition, these films are used as ion-exchange membranes in fuel cells [80] and as antireflective surface coatings in optical applications [81], among other applications [2,3,9].

For a more comprehensive study of the concepts and applicability of the LbL technique, the reader is referred to the publication of G. Decher titled "Multilayer Thin Films: Sequential Assembly of Nanocomposite Materials, 2nd Edition" and other texts in this field.

4.4 Final Considerations

The unique properties of nanostructured films constructed from various materials and the versatility of the LB and LbL methods have contributed considerably to the development of nanotechnology. LB and LbL methods have enabled the use of various materials with diverse architectures in many technological applications, from electronics to medicine. It should be noted that these methods have limitations and are thus complementary. Although ultrathin film fabrication has been extensively studied, it is a growing and popular field with many possibilities yet to be explored.

Acknowledgment

The authors thank *Bianca Martins Estevão* for producing some of the figures.

References

[1] K. Ariga, et al. 25th anniversary article: what can be done with the Langmuir–Blodgett method? Recent developments and its critical role in materials science, Adv. Mater. 25 (45) (2013) 6477–6512.

[2] K. Ariga, et al. Layer-by-layer nanoarchitectonics: invention, innovation and evolution, Chem. Lett. 43 (1) (2014) 36–68.

[3] K. Ariga, et al. Thin film nanoarchitectonics, J. Inorg. Organomet. Polym. Mater. 25 (3) (2015) 466–479.

[4] P. Behroozi, et al. The calming effect of oil on water, Am. J. Phys. 75 (2007) 407–414.

[5] A. Ulman, Langmuir–Blodgett films, in: An Introduction to Ultrathin Organic Films from Langmuir–Blodgett to Self-Assembly, [s.I.] Academic Press, USA, 1991, pp. 101–219.

[6] J.N. Israelachvili, Historical perspective, in: Intermolecular and Surface Forces, third ed., [s.I] Academic Press, USA, 2011, pp. 15–16.

[7] M.C. Petty, Langmuir–Blodgett films: an introduction [s.I.], Cambridge University Press, Cambridge, England, (1996).

[8] P. Dynarowicz-Latka, A. Dhanabalan, O.N. Oliveira Jr., Modern physicochemical research on Langmuir monolayers, Adv. Colloid. Interface Sci. 91 (2) (2001) 221–293.

[9] O.N. Oliveira Jr., F.J. Pavinatto, D.T. Balogh, Fundamentals and applications of organised molecular films, in: Nanomaterials and Nanoarchitectures. first ed., [s.I] Springer, 2015, pp. 301–343.

[10] A. De Barros, et al. Synergy between polyaniline and OMt clay mineral in Langmuir–Blodgett films for the simultaneous detection of traces of metal ions, ACS Appl. Mater. Interfaces 7 (12) (2015) 6828–6834.

[11] M.L. Braunger, et al. Electrical and electrochemical measurements in nanostructures films of polythiophenes derivates, Electrochim. Acta 164 (1–6) (2015).

[12] X. Chen, et al. Enhanced of polymer solar cells with a monolayer of assembled gold nanoparticle films fabricated Langmuir–Blodgett technique, Mater. Sci. Eng. B 178 (1) (2013) 53–59.

[13] S. Jayaraman, L.T. Yu, M.P. Srinivasan, Polythiphene-gold nanoparticle hybrid systems: Langmuir–Blodgett assembly of nanostructure films, Nanoscale 5 (2013) 2974–2982.

[14] E. Guzmán, et al. DPPC-DOPC Langmuir monolayers modified by hydrophilic silica nanoparticles: phase behaviour, structure and rheology, Colloids Surf. A Physicochem. Eng. Asp. 413 (2012) 174–183.

[15] H. Nakahara, et al. Interfacial properties in Langmuir monolayers and LB films of DPPC with partially fluorinated alcohol (F$_8$H$_7$OH), J. Oleo Sci. 62 (12) (2013) 1017–1027.

[16] M. Saha, S.A. Hussain, D. Bhattacharjee, Interaction of nano-clay platelets with a phospholipid in presence of a fluorescence probe, Mol. Cryst. Liq. Cryst. 608 (1) (2015) 198–210.

[17] J. Cabaj, et al. Biosensing invertase-based Langmuir–Schaefer films: preparation and characteristic, Sens. Actuators. B 166/167 (2012) 75–82.

[18] L. Caseli, J.R. Siqueira, High enzymatic activity preservation with carbon nanotubes incorporated in urease-lipid hybrid Langmuir–Blodgett films, Langmuir 28 (12) (2012) 5398–5403.

[19] J.M. Rocha, L. Caseli, Adsorption and enzyme activity of sucrose phosphorylase on lipid Langmuir and Langmuir–Blodgett films, Colloids Surf. B 116 (2014) 497–501.

[20] F.A. Scholl, L. Caseli, Langmuir and Langmuir–Blodgett films of lipids and penicillinase: studies on adsorption and enzymatic activity, Colloids Surf. B 126 (2015) 232–236.

[21] K. Togashi, et al. Fabrication of Langmuir–Blodgett films using amphiphilic peptides, J Nanosci. Nanotechnol. 12 (1) (2012) 568–572.

[22] N. Kato, T. Sasaki, Y. Mukai, Partially induced transition from horizontal to vertical orientation of helical peptides at the air-water interface and the structure of their monolayers transferred on the solid substrates, Biochim. Byophys. Acta 1848 (4) (2015) 967–975.

[23] S.A. Hussain, et al. Incorporation of nano-clay saponite layers in the organo-clay hybrid films using anionic amphiphile stearic acid by Langmuir–Blodgett technique, Thin Solid Films 536 (2013) 261–268.

[24] L. Ballesteros, et al. Directionally oriented LB films of and OPE derivate: assembly, characterization and electrical properties, Langmuir 27 (7) (2011) 3600–3610.

[25] A.C. Kucuk, J. Matsui, T. Miyashita, Langmuir–Blodgett films composed of amphiphilic double-decker shaped polyhedral oligomeric silsesquioxanes, J. Colloid Interface Sci. 355 (1) (mar 2011) 106–114.

[26] M.J. Shultz, et al. Aqueous solution/air interfaces probed with sum frequency generation spectroscopy, J. Phys. Chem. B 106 (2002) 5313–5324.

[27] D. Blaudez, et al. Polarization-modulated FT-IR spectroscopy of a spread monolayer at the air/water interface, Appl. Spectrosc. 47 (1993) 869–874.

[28] S. Talu, N. Patra, N. Salerno, Micromorphological characterization of polymer-oxide nanocomposite thin films by atomic force microscopy and fractal geometry analysis, Prog. Org. Coat. 89 (2015) 50–56.

[29] J. Makowiecki, et al. Molecular organization of perylene derivatives in Langmuir–Blodgett multilayers, Opt. Mater. 46 (2015) 555–560.

[30] R.J. Demchak, T Fort Jr., Surface dipole moments of close-packed un-ionized monolayers at the air-water interface, J. Colloid Interface Sci. 46 (2) (1974) 191–202.

[31] Y.S. Lee, Nanostructures films, in: Self Assembly and Nanotechnology, [s.I.] John Wiley & Sons, Inc., Hoboken, New Jersey, 2008, pp. 259–262.

[32] G. Cao, Two-dimensional nanostructures: thin films, in: Nanostructures & Nanomaterials: Synthesis, Properties & Applications, [s.I.] Imperial College Press, United States, 2004, pp. 173–224.

[33] G. Giancane, et al. Investigations and application in piezoelectric phenol sensor of Langmuir–Schaefer films of a copper phthalocyanine derivate functionalized with bulky substituents, J. Colloid. Interface Sci. 377 (2012) 176–183.

[34] R. Fernández, et al. Optical storage in azobenzene –containing epoxy polymers process as Langmuir–Blodgett films, Mater. Sci. Eng. C 33 (3) (2013) 1403–1408.

[35] C.A. Silva, et al. Interaction of chitosan and mucin in a biomembrane model environment, J. Colloid Interface Sci. 376 (1) (2012) 289–295.

[36] MT. Nobre, et al. Interaction of bioactive molecules and nanomaterials with Langmuir monolayers as cell membrane models, Thin Solid Films 593 (2015) 158–178.

[37] A. Mangiarotti, B. Caruso, N. Wilke, Phase coexistence in films composed of DLPC and DPPC: a comparison between different model membrane system, Biochim. Biophys. Acta 1838 (7) (2014) 1823–1831.

[38] V. Arslanov, et al. Designed and evaluation of sensory system based on amphiphilic anthraquinones molecular receptors, Colloids Surf. A 483 (2015) 193–203.

[39] Z. Yin, et al. Real-time DNA detection using Pt nanoparticle-decorated reduced graphene oxide field-effect transistors, Nanoscale 4 (2012) 293–297.

[40] L. Wang, et al. A new strategy for enhancing electrochemical sensing from MWCNTs modified electrode with Langmuir–Blodgett film and used in determination of methylparaben, Sens. and Actuators B 211 (2015) 332–338.

[41] G. Decher, J. Hong, Buildup of ultrathin multilayer films by a self-assembly process: II. Consecutive adsorption of anionic and cationic bipolar amphiphiles and polyelectrolytes on charged surfaces, Phys. Chem. Chem. Phys. 95 (11) (1991) 1430–1434.

[42] G. Decher, J.D. Hong, J. Schimitt, Buildup of ultrathin multilayer films by self-assembly process: III. Consecutively alternating adsorption of anionic and cationic polyelectrolytes on charged surfaces, Thin Solid Films 210 (1992) 831–835.

[43] Y. Lvov, G. Decher, G. Sukhorukov, Assembly of thin films by means of successive deposition of alternate layers of DNA and poly(allylamine), Macromolecules 26 (20) (1993) 5396–5399.

[44] G. Decher, Y. Lvov, J. Shimitt, Proof of multilayer structural organization in self-assembled polycation-polyanion molecular films, Thin Solid Films 244 (1) (1994) 772–777.

[45] G. Decher, et al. New nanocomposite films for biosensors: layer-by-layer adsorbed films of polyelectrolytes, proteins or DNA, Biosens. Bioelectron. 9 (9) (1994) 677–684.

[46] G. Decher, Fuzzy nanoassemblies: toward layered polymeric multicomposites, Science 277 (5330) (1997) 1232–1237.

[47] R.K. Iler, Multilayers of colloidal particles, J. Colloid Interface Sci. 21 (6) (1996) 569–594.

[48] Y.F. Nicolau, Solution deposition of thin solid compound films by a successive ionic-layer adsorption and reaction process, Appl. Surf. Sci. 22 (1985) 1061–1074.

[49] R. Aksberg, L. Ödberg, Adsorption of anionic polyacrylamide on cellulosic fibers with pre-adsorbed polyelectrolytes, Nord. Pulp. Pap. Res. J. 5 (1990) 168–171.

[50] G. Decher, F.F. Schlenoff, Multilayer Thin Films: Sequential Assembly of Nanocomposite Materials, second ed., John Wiley & Sons, Germany, 2012.

[51] G. Decher, Layered nanoarchitectures via directed assembly of anionic and cationic molecules, in: J.P. Sauvage, M.W. Hosseini (Eds.), Comprehensive supramolecular chemistry [s.I.], 9, Pergamon Press, Oxford, 1996, pp. 507–528.

[52] R.J. El-Khouri, et al. Multifunctional layer-by-layer architectures for biological applications, in: W. Knoll, R.C. Advincula (Eds.), Functional polymer films [s.I.], Wiley-VCH, Germany, 2011, pp. 11–72.

[53] M.M. De Villiers, et al. Introduction to nanocoatings produced by layer-by-layer (LbL) self-assembly, Adv. Drug Deliv. Rev. 63 (9) (2011) 701–715.

[54] M. Schönhoff, Layered polyelectrolyte complexes: physics of formation and molecular properties, J. Phys. Condens. Matter (2003) 1781–1808.

[55] J. Borges, j.F. Mano, Molecular interactions driving the layer-by-layer assembly of multilayers, Chem. Rev. 114 (18) (2014) 8883–8942.

[56] S. Bai, et al. Hydrogen-bonding-directed layer-by-layer polymer films: substrate effect on the microporous morphology variation, Eur. Polym. J. 42 (4) (2006) 900–907.

[57] X. Zhang, H. Chen, H. Zhang, Layer-by-layer assembly: from conventional to unconventional methods, Chem. Commun. 14 (2007) 1395.

[58] J. Zhang, et al. Layer-by-layer assembly of azulene-based supra-amphiphiles: reversible encapsulation of organic molecules in water by charge-transfer interaction, Langmuir 29 (2013) 6348–6353.

[59] A.H. Broderick, U. Manna, D.M. Lynn, Covalent layer-by-layer assembly of water-permeable and water-impermeable polymer multilayers on highly water-soluble and water-sensitive substrates, Chem. Mater. 24 (2012) 1786–1795.

[60] S. Takahashi, K. Sato, J.-I. Anzai, Layer-by-layer construction of protein architectures through avidin–biotin and lectin–sugar interactions for biosensor applications, Anal. Bioanal. Chem. 402 (2012) 1749–1758.

[61] L. Lee, et al. Influence of salt concentration on the assembly of DNA multilayer films, Langmuir 26 (5) (2010) 3415–3422.

[62] M. Matsusaki, et al. Layer-by-layer assembly through weak interactions and their biomedical applications, Adv. Mater. 24 (4) (2012) 454–474.

[63] H. Xu, M. Schönhoff, X. Zhang, Unconventional layer-by-layer assembly: surface molecular imprinting and its applications, Small 8 (4) (2012) 517–523.

[64] Y. Li, X. Wang, J. Sun, Layer-by-layer assembly for rapid fabrication of thick polymeric films, Chem. Soc. Rev. 41 (18) (2012) 5998.

[65] J.J. Richardson, M. Bjornmalm, F. Caruso, Technology-driven layer-by-layer assembly of nanofilms, Science 348 (2015) 411–423.

[66] Y. Kim, et al. Stretchable nanoparticle conductors with self-organized conductive pathways, Nat. Lett. 500 (2013) 59–64.

[67] Q. Xi, et al. Gold nanoparticle-embedded porous graphene thin film fabricated via layer-by-layer self-assembly and subsequent thermal annealing for electrochemical sensing, Langmuir 28 (2012) 9885–9892.

[68] A. De Barros, et al. Nanocomposites based on LbL films of polyaniline and sodium montmorillonite clay, Synth. Met. 197 (2014) 119–125.

[69] A.Y.W. Sham, S.M. Notley, Graphene polyelectrolyte multilayer film formation driven by hydrogen bonding, J. Colloid Interface Sci. 456 (2015) 32–41.

[70] X. Xu, et al. Multifunctional drug carriers comprised of mesoporous silica nanoparticles and polyamidoamine dendrimers based on layer-by-layer assembly, Mater. Des. 88 (2015) 1127–1133.

[71] J.S. Silva, et al. Layer-by-layer films based on carbon nanotubes and polyaniline for detecting 2-chlorophenol, J. Nanosci. Nanotechnol. 14 (9) (2014) 6586–6592.

[72] H.D.M. Follmann, et al. Anti-adhesive and antibacterial multilayer films via layer-by-layer assembly of TMC/heparin complexes, Biomacromolecules 13 (2012) 3711–3722.

[73] R.F. De Oliveira, et al. Exploiting cascade reactions in bienzyme layer-by-layer films, J. Phys. Chem. C 115 (39) (2011) 19136–19140.

[74] P.P. Campos, et al. Amperometric detection of lactose using β-galactosidase immobilized in layer-by-layer films, ACS Appl. Mater. Interfaces 6 (14) (2015) 11657–11664.

[75] J.S. Graça, et al. Amperometric glucose biosensor based on layer-by-layer films of microperoxidase-11 and liposome-encapsulated glucose oxidase, Bioelectrochemistry 96 (2014) 37–42.

[76] P.K.H. HO, et al. Molecular-scale interface engineering for polymer light-emitting diodes, Nature 404 (mar 2000) 481–484.

[77] S.K. Saha, A. Guchhait, A.J. Pal, Organic/inorganic hybrid pn-junction between copper phthalocyanine and CdSe quantum dot layers as solar cells, J. Appl. Phys. 112 (4) (2012) 044507.

[78] H. Hwang, et al. Highly tunable charge transportation in layer-by-layer assembled graphene transistor, ACS Nano 6 (3) (2012) 2432–2440.

[79] B. Koo, H. Baek, J. Cho, Control over memory performance of layer-by-layer assembled metal phthalocyanine multilayers via molecular-level manipulation, Chem. Mater. 24 (6) (2012) 1091–1099.

[80] L. Zhao, et al. Fabrication of ultrahigh hydrogen barrier polyethyleneimine/graphene oxide films by LbL assembly fine-tuned with electric field application, Compos. Part A Appl. S. 78 (2015) 60–69.

[81] K. Katagiri, et al. Anti-reflective coatings prepared via layer-by-layer assembly of mesoporous silica nanoparticles and polyelectrolytes, Polym. J. 47 (2015) 190–194.

4

Nanoneurobiophysics

R.P. Camargo, C.C. Bueno, D.K. Deda, F. de Lima Leite

*NANONEUROBIOPHYSICS RESEARCH GROUP, DEPARTMENT OF PHYSICS,
CHEMISTRY & MATHEMATICS, FEDERAL UNIVERSITY OF SÃO CARLOS,
SOROCABA, SÃO PAULO, BRAZIL*

CHAPTER OUTLINE

4.1 Introduction

Nanoneurobiophysics is a new field of study created in 2011 and established globally in 2015 [1]. The main objective of this field is to investigate autoimmune and neurodegenerative diseases, such as multiple sclerosis (MS), neuromyelitis optica (NMO), Parkinson's disease, and Alzheimer's disease, using advanced nanocharacterization techniques, such as scanning electron microscopy (SEM) and atomic force microscopy (AFM) combined with computer simulation methods. These techniques and methods enable the understanding of neurobiological disorders at a molecular level and contribute to the development of new diagnostic methods and treatment regimens.

Nanoneurobiophysics allows the contextualization of different fields that intersect with the main axis formed by physics, biology, neuroscience, and nanoscience. Nanoneurobiophysics allows collaboration and integration of these four basic sciences into a transdisciplinary whole, avoiding the fragmentation of knowledge into distinct fields. The interactions between the basic fields create a synergy and a networked research process, allowing for interaction between other areas of research, such as nanopharmacology [2,3], nanoneuroscience [4–6], and nanoneuropharmacology [7].

The application of nanotechnology concepts in research fields, such as pharmacology, biophysics, and medicine, associated with contributions from basic sciences, such as chemistry, physics, biology, and even informatics, could promote important breakthroughs in diagnostics and treatment. Naturally, there are intrinsic synergies between the cited disciplines due to their general similarities, as well as other aspects associated with the complexity of the biologic systems. Despite these similarities, such synergies are not completely understood and used together in research. Hence, new lines of research are born. These include nanoneurobiophysics, which, although not yet established, has the primary objective of applying the theories and methods of physics in conjunction with nanotechnology tools to promote breakthroughs in the field of neuroscience. This is a new area of study that will contribute to the search for solutions to diseases of the nervous systems that currently lack a cure or have limited forms of treatment and diagnostics.

4.2 Nanopharmacology

Nanopharmacology is the study of nanoscale interactions between medications with proteins, DNA (deoxyribonucleic acid), and RNA (ribonucleic acid), cells, tissues, or organs [8]. Furthermore, it involves the interactions between traditional medications and physiological systems, using nanoscale investigative techniques [3]. This new line of research has many peculiar characteristics that distinguish it from traditional pharmacology, for example, the use of nanotechnology tools, such as AFM probes. It is important to note that although it is a relatively new technique, AFM has been used to define structural targets for pharmaceutical action in a specific and desired location, as well as for the design of pharmaceuticals and delivery systems and the monitoring, mapping, and analysis of target-pharmaceutical interaction at a molecular level. Nanopharmacology is also intimately related to the study of nanoparticulate pharmaceuticals and therefore contributes to the design of more efficient pharmaceuticals atom by atom (bottom-up assembly) [9].

In this scenario, nanopharmacology has emerged as an interdisciplinary area, joining the integrated fields of nanotechnology, nanomedicine, biophysics, and others. Hence, it is converging on a single point that is capable of revolutionizing pharmacology as we currently understand it and even the clinical protocols that are currently used in diagnostic and treatment methods.

4.3 Nanoneuroscience and Nanoneuropharmacology

The high specificity of nanomaterials and nanodevices may be used to direct a selective and controlled drug activity in biologic systems, reaching target cells, and specific tissues. Therefore, these materials could contribute to treatment and diagnostic breakthroughs, improving the physiological response to and reducing the side effects of many therapies [10]. In this context, nanotechnology applied to diagnostic and treatment methods and to monitoring and control of the biological systems is called nanomedicine [11]. In the last few years, many nanomedicine studies were dedicated to the application of nanotechnology to nanoscience, originating the term nanoneuroscience [12]. This new subject is divided into two broad areas of study according to the type of application [10]:

- *Basic* nanoneuroscience: dedicated to the investigation of molecules, cells, physiological processes, and the mechanism of action of pharmaceuticals and other molecules acting on or in the central nervous system (CNS).
- *Clinical* nanoneuroscience: works toward limiting and reversing neuropathological statuses, while aiming for the regeneration of cells of the nervous system and the development of strategies for neuroprotection through resources provided by nanotechnology, such as controlled drug delivery.

4.3.1 Basic Nanoneuroscience

There are many nanotechnology tools for the study of cell mechanisms that could be used to develop and improve therapy and diagnostic methodologies [12]. Particularly in the case of nanoscience, special attention should be drawn to the contributions of AFM and the use of quantum dots.

4.3.1.1 AFM

In general terms, AFM is an imaging technique that consists of scanning the surface of a sample with a nanometric probe (cantilever) to obtain its topographic image with atomic resolution, in addition to mapping certain mechanical and physicochemical properties of the composing materials [13,14]. Further details on the function of AFM as well as on imaging techniques and interaction mechanisms between the probe and sample are presented in Chapter 2 (Atomic Force Microscopy: A Powerful Tool for Electrical Characterization) of volume 3 in this work. Additionally, some aspects related to the use of AFM as a nanoparticle characterization tool will be discussed in Chapter 5 (*Low Dimensionality Systems: Nanoparticles*) of volume 1. One of the applications of AFM in nanoneurobiophysics is based on the calibration of nerve cell responses, through their stimulus using microelectrodes. This tool also allows the analysis of topographical variations in these cells [15].

Furthermore, the functionalization of the cantilever/probe (i.e., the chemical modification of its surface) allows the observation of specific interactions (antibody–antigen or enzyme–substrate, for instance). This feature makes the probe extremely selective as a powerful nanosensor [16,17]. Nanosensors based on functionalized cantilevers and some examples of their applications are discussed in Chapter 5 (Nanosensors) of volume 2 of this work. Particularly in the case of neuroscience, the use of cantilever-based sensors remains poorly explored, although it is very promising. For example, there are specific ion channels in the brain that are responsible for neural excitability and are therefore very important for the physiological equilibrium of the CNS. Maciaszek et al. [18] performed mapping of SK potassium channels by functionalizing the end of a cantilever with apamin, a bee venom toxin. This toxin bonds specifically to SK channels, and their interaction was verified by mapping living rat neurons, which in turn made it possible to identify that these channels are concentrated in dendrites.

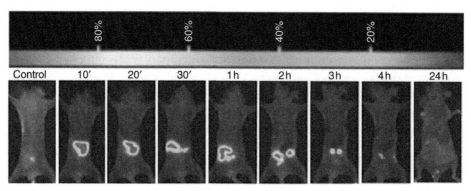

FIGURE 4.1 Quantum dots administered to rats and visualization of their fluorescence, demonstrating possible biological and biomedical applications in the gastrointestinal tract. The quantum dots used were synthesized by a combination of polythiol ligands with a silica coating for stabilization in acid media. The images correspond to analyses at different times. *Reproduced from Y.F. Loginova, S.V. Dezhurov, V.V. Zherdeva, N.I. Kazachkina, M.S. Wakstein, A.P. Savitsky. Biodistribution and stability of CdSe core quantum dots in mouse digestive tract following per os administration: advantages of double polymer/silica coated nanocrystals, Biochem. Biophys. Res. Commun. 419 (1) (2014) 54–59, Copyright (2014) [19]. With the kind permission of Elsevier Publisher.*

Together, the aforementioned studies demonstrate that previously unknown information about CNS cells may be discovered using AFM, aiding the identification of new therapeutic targets. Additionally, this ability to measure intermolecular interactions could be employed in the study of pharmaceutical mechanisms of action in the CNS. With all of these data, it is possible to design pharmaceuticals and more efficient drug delivery systems, which could represent important breakthroughs in the treatment of CNS diseases [9].

4.3.1.2 Quantum Dots

Quantum dots are semiconductor crystals of nanometric size, constituted of a heavy metal nucleus, such as cadmium selenide telluride or cadmium telluride, and an exterior coating of bioactive molecules selected for a specific application [19]. Quantum dots are extremely small, approximately 10–50 atoms in diameter and 1–5 nm in radius, and best known for their high fluorescence [19,20].

Quantum dots emit an intense fluorescent signal for up to 48 h when subjected to a luminous excitation with energy higher than the gap between the valence and conduction bands (Fig. 4.1). In general, the largest quantum dots emit fluorescence in the orange, red, or gray regions of the spectrum, while the smallest emit in the blue, green, or yellow wavelengths. When incorporated in cells, the fluorescence of these nanocrystals can be detected through at least four cell generations, and some cells remain marked for up to 2 weeks. This characteristic allows the identification of very small quantities of biomolecules, making it possible to investigate and follow cellular processes, such as mitosis or even a virus [21]. The intensity and persistence of the fluorescent signal of quantum dots is one of the great advantages of these nanosystems when compared to conventional fluorophores. Commonly used fluorescent markers are generally constituted of organic molecules, whose fluorescence is rapidly extinguished after exposure to a source of radiation. Importantly, each fluorophore is excited by a specific wavelength, which is in contrast to quantum dots, which can be excited by a single radiation source. Additionally, while fluorophores can be used for marking and diagnostic purposes, they can present considerable toxicity [22].

Although they do not have toxic effects in vitro, the use of fluorophores in vivo leads to challenges due to potential local and systemic toxicity, which is one of the main targets of existing studies [10,19,20]. Some reports indicate that quantum dots are rapidly removed from the bloodstream, accumulating in organs, such as the liver, kidney, and spleen, where they can stay for up to several months [23].

The toxicity and biological behavior of quantum dots are generally regulated by three factors: the material that constitutes the quantum dot nucleus, the molecules/biomolecules present on the surface, and the external conditions. According to Hoshino et al. [23], the primary factor responsible for toxic effects and for modulating the interaction of quantum dots with biological systems is modification of the particle surface and not the heavy metal nucleus itself. Additionally, interactions with the environment, such as the oxidative, photolytic, and mechanical conditions, could lead to destabilization of quantum dots and result in compounds that are potentially toxic to the organism [24]. Due to this, Prasad et al. [25] used a gelatin coating on cadmium telluride quantum dots to act as a barrier against the escape of toxic ions from the nucleus. This coating proved to be effective, efficiently reducing the toxicity of the dots in PC12 cells (a line derived from rat pheochromocytoma, which are rare tumors located in the medullar region of the suprarenal glands) for a period of up to 17 days.

Quantum dots, especially those made from CdSe or ZnS [24], can be used for many purposes depending on the characteristics of the surface. They can be used as vascular markers [26], for lymph node mapping [27], for detection of cancer cells [28], and as cell markers for the study of cell biology and the development of diagnostic methods for different diseases [29].

Marking cells in culture with quantum dots is well established and has been used for a wide variety of cell strains [10], especially for the detection of cancer [20,28]. Additionally, this technique has been used for studies of the CNS, such as the conjugation of MMP-9-siRNA on the surface of quantum dots for its release into nervous tissue. In that experiment, the group found a decrease in the expression of metalloproteinase MMP-9, preventing neuroinflammation and preserving the integrity of the hematoencephalic barrier (HEB), thereby preventing aggravation of diseases, such as HIV-associated dementia and cerebral ischemia [30].

4.3.2 Clinical Nanoneuroscience

The applications of clinical nanoneuroscience are directed at limiting and reversing neuropathological statuses. Thus, alternative approaches seeking breakthroughs in the mechanisms of neuroprotection and regeneration have been investigated, applying nanosystems, such as carbon and fullerene nanotubes [10,31–34]. In addition, the field of

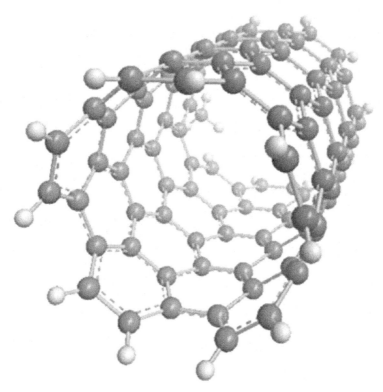

FIGURE 4.2 Structure of a carbon nanotube.

nanoneuropharmacology has progressed with the study and development of new pharmaceuticals and treatments. This has led to promising new controlled release systems.

4.3.2.1 Neuroprotection and Neuroregeneration
4.3.2.1.1 CARBON NANOTUBES
As illustrated in Fig. 4.2, carbon nanotubes are cylindrical and hollow structures with closed extremities whose diameter and length can vary from 0.4 to 2.0 nm and from 1 to 100 nm, respectively. They are formed by the organization of carbon atoms in sheets. Each sheet is then rolled, forming a cylinder (tube), and can be classified in relation to the number of sheets (layers) as single or multilayer nanotubes [33,35].

Nanotubes may permit the adsorption of molecules and/or biomolecules at their wall or extremities, a process known as functionalization, which is responsible for making them biocompatible and facilitates their interaction with organic molecules, biological structures, pharmaceuticals, or DNA [36]. Additionally, functionalization reduces their toxicity, increasing their potential for application in the biological and medical fields [37]. The functionalization of carbon nanotubes is also discussed in Chapter 3 (Nanomedicine) of this volume, and its application in other fields, such as nanoelectronics, is described in Chapter 2 of this volume.

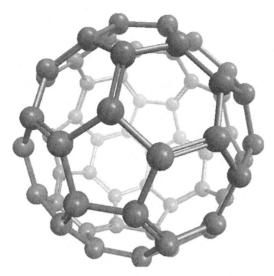

FIGURE 4.3 Fullerene structure.

In nanomedicine, nanotubes can be used for the regeneration of neural tissues [38] and the controlled administration of pharmaceuticals, as biological markers [39] and DNA vectors in gene therapy [40], and in microelectrode implants for drug delivery [41]. Although it is at an early stage, the potential application of carbon nanotubes for the therapy of neuro-degenerative diseases has also been demonstrated [33,34]. Tosi et al. [42] reported the ability of these nanosystems to pass through the HEB of rats when functionalized with sialic acid. Many studies also describe the potential application of nanotubes for the regeneration of neuronal tissue due to their favorable size, form, flexibility, and electrical conductivity [43–45]. According to these studies, hippocampal neurons in culture are able to survive and continue to grow on substrates constituted by carbon nanotubes. Additionally, studies have reported that altering surface characteristics of the nanostructure, such as charge and functionaliza-tion with growth factors or biomolecules responsible for increasing the adhesion of cells to the substrate, are efficient alternatives for stimulating neuronal growth [43,46,47].

4.3.2.1.2 FULLERENES

Fullerenes constitute a class of spheroid, stable nanomolecules formed exclusively by car-bon atoms (Fig. 4.3). The best-known and most stable representative of the fullerene family is C_{60}, formed by 60 carbon atoms to give a diameter of approximately 1 nm. In contrast to the other allotropic forms of carbon, such as graphite, diamonds, and nanotubes, fullerenes are a molecular form of carbon. Since fullerene is an excellent acceptor of electrons in its funda-mental state, it acts as a sponge that absorbs free radicals in a biological medium, preventing damage to cell structures, such as lipids, proteins, DNA, and macromolecules [32,48].

Studies show that fullerenes and their derivatives inhibit apoptosis, limit excitotoxicity [32,49,50], delay motor degeneration in rats with lateral amyotrophic sclerosis [10], pro-mote neuroprotection and reversion of axonal loss in experimental models of MS [31], and improve neuronal function in Alzheimer's disease [32]. It is believed that the mechanism of neuroprotective action of fullerenes is through inhibition of glutamate receptors [10]. Bar-Shir et al. [51] described the incorporation of amino adamantyl (already employed in the treatment of some illnesses, including diseases of the CNS) on the surface of fullerene, with the goal of treating Parkinson's and other neurologic diseases.

In addition to studies on their protective action against free radicals, some researchers have investigated the radioprotective properties of fullerenes [52–54]. This application is of great interest in situations of exposure to radiotherapies, such as in the treatment of cancer, where the intention is to protect healthy cells from the systemic effects of radiation. However, the desired specificity for this application has not yet been achieved, with effects on healthy as well as sick cells [52].

4.3.2.2 Nanoneuropharmacology: Controlled Delivery Systems Applied to CNS Diseases

Approximately 40% of discovered therapeutic compounds do not reach clinical studies because of pharmaceutical limitations, such as the impossibility of increasing the dosage, poor rates of retention or degradation of the therapeutic agent, low solubility, side effects associated with high concentrations, and absorption difficulties [55,56]. In this context, nanotechnology has contributed extensively in the past few years to enabling the development of appropriate formulations, decreasing the toxicity of the pharmaceutical, improving its solubility, absorption and stability, and promoting the extension of its action in the organism [8]. According to Zhang [3], the main driver of nanopharmacology is related to the possibility of directing the pharmaceutical to a specific target and allowing its slow and controlled delivery. All of this is possible using nanosystems capable of carrying and delivering the active agent in an efficient and intelligent manner, based on the therapeutic needs of the local microenvironment [56].

With the use of these systems at a nanoscale, the objective is also to penetrate biological barriers, such as the HEB, to treat diseases by reaching locations where traditional pharmaceuticals cannot arrive [57]. This is the case for the neurological diseases, where the HEB represents one of the limiting factors in diagnosis and treatment [58]. The HEB is only permeable to lipophilic molecules or molecules with a molecular mass smaller than 400–600 Da. Its function is to protect the CNS from pathogens, such as viruses, parasites, and toxins [58]. However, the efficacy of the HEB in protecting the CNS also stops many pharmaceuticals from crossing it. Therefore, nanopharmacology has been working toward the development of systems capable of crossing the HEB and has potential application for the treatment and early diagnosis of neurological diseases [59].

Nanomedicine describes the controlled delivery of pharmaceuticals through systems composed of particles smaller than 500 nm, with development focusing on therapeutic properties [56]. The conjugation of biomolecules (such as albumin and transferrin) to the surface of these nanosystems improves their biocompatibility, helps their passage through the HEB, and improves their pharmacokinetics. In this sense, it should be noted that other characteristics such as the solubility, size, and surface charge of nanosystems have a fundamental role in the pharmacokinetic alterations.

Pharmacokinetic regulation allows configuration of the moment when and where a pharmaceutical will be delivered without altering its pharmacodynamics. This control can be performed with the development of nanosystems sensitive to temperature, pH variations, or other biological stimuli [60].

Among the benefits of controlling pharmacokinetics is the reduction of side effects due to specificity of the site of action; reduction in the number of administrations because it is possible to develop a system that maintains a constant release during a given period; greater comfort for the patient and, consequently, better compliance; and pharmacotechnical benefits, such as improving dissolution, better stability, and the possibility of carrying insoluble compounds [61]. Controlled drug delivery is performed using systems such as nanocapsules, nanospheres, liposomes [62,63], and other more recently studied systems such as nanosponges [64]. In general, the materials used in these systems are biocompatible and biodegradable because they are expected to not interfere with the action of the medication or cause other reactions or toxic effects.

Generally speaking, nanotechnology applied to pharmaceuticals enables the development of carriers capable of directing and delivering medication exclusively to the affected tissue, maximizing the therapeutic effect and minimizing side and adverse effects. Due to this characteristic, which allows the carriers to reach and accumulate in the target tissue, controlled drug delivery systems are considered the "Trojan Horses of Nanotechnology" [34]. In addition to the previously cited advantage, there is a range of pharmacotechnical possibilities, such as the production of new delivery systems for a specific drug (carried in capsules, solutions, lotions, etc.) and new administration routes (oral, transdermic, etc.).

There are many nanosystems used in nanopharmacology; however, those that will be briefly discussed in this chapter are the most common to nanoscience, such as the nanogels, liposomes, nanoparticles, micelles, dendrimers, and nanosponges. Each system differs in how it is produced (type of material and process), in the type of substance that can be incorporated, and in its physicochemical characteristics.

4.3.2.2.1 NANOGELS

Nanogels are obtained by the emulsification of ionic and nonionic polymers (Fig. 4.4) and solvent evaporation. They expand in water and are able to incorporate molecules, such as oligonucleotides, siRNA, DNA, proteins, and drugs with low molecular weight. The capacity for drug incorporation is approximately 40–60% [65].

Nanogels are able to markedly decrease the degradation of biomolecules and drugs during their delivery in the organism [59] as well as decrease the eventual toxic effects of pharmaceuticals [48]. Oishi et al. [66] reported the incorporation of doxorubicin, a drug commonly used in the treatment of cancer, in a pH-sensitive nanogel. Their results indicated higher activity of the drug in HuH-7 tumor cells when incorporated in a nanogel relative to its free form (traditional method). Studies have also been performed in vitro and reported the delivery of oligonucleotides to the brain, considerably reducing their absorption by the liver and spleen, making nanogels strong candidates for the delivery of pharmaceuticals to the CNS [35].

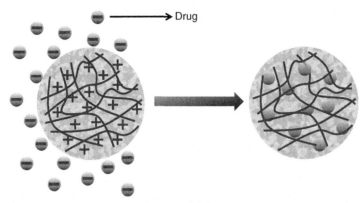

FIGURE 4.4 Schematic of a nanogel. Polymeric structure containing a positive charge and incorporation of an active ingredient with a negative charge.

4.3.2.2.2 LIPOSOMES

Liposomes are aqueous vesicles surrounded by concentric phospholipid bilayers and can carry lipophilic, hydrophilic, and amphiphilic pharmaceuticals. Since they are primarily constituted by lipids, liposomes are biodegradable, nontoxic, and nonimmunogenic compounds [62]. Fig. 4.5 illustrates the structure of a liposome and the appropriate locations for incorporation of pharmaceuticals with the different characteristics cited.

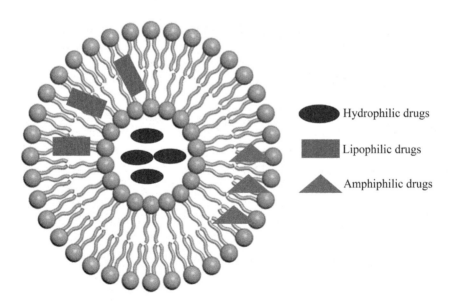

FIGURE 4.5 Structure of a single-layer liposome indicating the regions where pharmaceuticals with different characteristics can be incorporated.

Since they are constituted by a double lipid layer, liposomes are very similar to biological membranes. This characteristic promotes the permeation and transportation of pharmaceuticals through cell membranes and protects the drug from degradation by the reticuloendothelial system [63].

Liposomes are formed in water by the aggregation of amphiphilic molecules (usually phospholipids) that contain a polar group and an aliphatic chain. This process forms bilayers that close, forming an internal aqueous compartment. Volume 1 of this work, Chapter 2 (Supramolecular Systems), describes the primary methods of liposome preparation.

Liposomes are considered excellent delivery systems because of their structural flexibility, and they represent the most clinically established system for delivery of cytotoxic pharmaceuticals, genes, and vaccines [57]. There are countless medications on the market that use this delivery system [55,56,67], and many researchers have studied the possibility of using liposomes to carry pharmaceuticals for the treatment of Alzheimer's disease [32,59]. However, improvements are still needed before liposomes can be used in the treatment of neurodegenerative diseases because their short half-life and low stability represent disadvantages yet to be overcome [58,59].

4.3.2.2.3 BIODEGRADABLE NANOPARTICLES
Biodegradable particles and capsules at the nano- and micrometric scale are considered an alternative to the use of liposomes, as they allow a larger dose of pharmaceutical to be encapsulated and greater control of its delivery process. These advantages are provided by the large variety of materials available and the different methods for their preparation [68].

Microparticles are small, solid, and spherical particles with sizes that range from 1 to 1000 μm, while nanoparticles are drug carrier systems with diameters smaller than 100 nm. The term "nanoparticle" includes nanocapsules and nanospheres, which differ according to their composition and structural organization. The capsules are formed by a polymeric layer around an oily nucleus. The pharmaceutical can be dissolved in this nucleus and/or adsorbed in the polymeric wall. In turn, the spheres, which did not contain oil, are formed by a polymer matrix where the pharmaceutical of interest can be retained or absorbed [69]. Fig. 4.6 shows an illustration of these two systems, containing molecules of an incorporated medication.

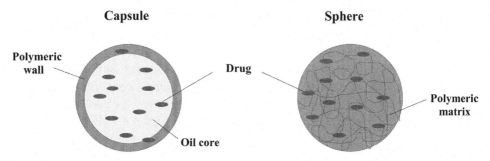

FIGURE 4.6 Schematic illustration of nanospheres and nanocapsules with incorporated pharmaceuticals.

Once these nanosystems reach the target tissue, the drug starts to be released by processes, such as desorption, diffusion through the polymeric matrix, or erosion of the particle [56]. Nanoencapsulated drugs are more protected against degradation and phagocytosis by the reticuloendothelial system [58].

Countless studies have been performed to develop formulations of encapsulated pharmaceuticals, including for the treatment of CNS disease. For example, *Rivastigmine*, a drug used in the treatment of Alzheimer's disease, was incorporated in poly(*n*-butyl cyanoacrylate) nanocapsules coated with polysorbate-80. According to the results of that study, when the encapsulated drug was employed, there was an increase in the availability of the pharmaceutical in the brain relative to the free form. This caused a decrease in the side effects of the drug, demonstrating the efficacy of the nanosystem [70].

In other studies, *Tacrine*, a medication also indicated for the treatment of Alzheimer's disease, and a *Bromocriptine*, commonly used in the treatment of Parkinson's disease, were nanoencapsulated in poly(*n*-butyl cyanoacrylate) coated with polysorbate-80 and poloxamer-188, respectively [58,71]. Similar to the previously cited study [70], a considerable increase was observed in the availability of the pharmaceuticals in the brain relative to the administration of their nonencapsulated free forms. Also according to these reports, *Ritonavir* (a medication used to treat HIV infections and AIDS) was carried by conjugated nanoparticles toward specific peptides for the treatment of neurological pathologies associated with HIV. These results indicated that nanosystems described previously are able to cross the HEB, providing an accumulation of the pharmaceutical in the CNS that is approximately 800 times higher than the conventional form of administration. Additionally, the use of this nanoencapsulated form increased the bioavailability of the pharmaceutical, maintaining the therapeutic concentration of the drug for up to 4 weeks [72].

Thus, nanoparticles represent an excellent alternative for the encapsulation of pharmaceuticals, especially for the diagnosis and treatment of neurological diseases, where crossing the HEB is the limiting factor in ensuring the appropriate concentration of the pharmaceutical in the CNS.

4.3.2.2.4 MICELLES

Micelles are colloidal dispersions composed of particles with sizes between 5 and 100 nm [56]. These nanosystems have drawn much attention in drug delivery studies, largely due to their capacity to carry hydrophobic molecules and reach specific tissues. They are structures formed by an aggregate of molecules (generally surfactants or polymers) dispersed in a liquid [73]. Micelle preparation methods are discussed in Chapter 2 (Supramolecular Systems) in volume 1 of this work.

In regards to their geometry, micelles can be globular, ellipsoid, or layered, depending on their constitution and the conditions of the solution, such as concentration, temperature, pH, and ionic force. Micelles have specific characteristics, such as thermodynamic stability, and their "shell-core" structure imitates certain natural delivery systems, which promotes the absorption and distribution of the encapsulated medication. In comparison to liposomes, micelles are much smaller and can provide a more efficient method for guiding the pharmaceutical to the target site [56,58].

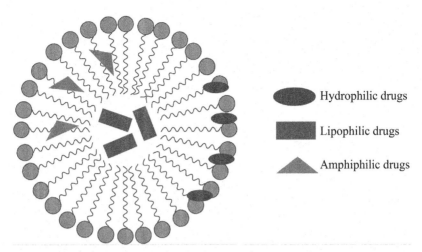

Hydrophilic drugs

Lipophilic drugs

Amphiphilic drugs

FIGURE 4.7 Structure of a micelle and substances with different polarities incorporated or adsorbed on its surface.

Due to their hydrophilic layer, micelles are not easily captured by macrophages from the reticuloendothelial defense system, instead protecting the incorporated drug from rapid degradation [59]. Micelles can solubilize lipophilic substances in their nucleus, hydrophilic substances adsorbed on their surface, and substances with neutral polarity (amphiphilic) along the molecules that form their structure [73]. Fig. 4.7 illustrates the structure of a micelle and the sites for the solubilization of substances with different polarities.

Although less extensive than some of the methodologies discussed previously, the use of micelles for the incorporation of drugs used in the treatment of CNS diseases has also been under study. For example, the effects of *Haloperidol* (a neuroleptic drug) in both its free form and when incorporated in Pluronic P85 micelles were compared, revealing greater effectiveness of the form incorporated into the micellar system [74].

Doxorubicin is an important drug used for the therapy of different types of tumors. However, it is not able to cross the HEB, so it cannot be used for the treatment of brain tumors. Nevertheless, a study has demonstrated that when this pharmaceutical is carried by polysorbate-80 micelles, considerable concentrations of the drug were detected in the brain [75].

Therefore, exploration of the use of micelles to release pharmaceuticals or diagnostic agents in the CNS could be a promising alternative for the development of efficient and versatile formulations.

4.3.2.2.5 DENDRIMERS

The word dendrimer derives from the Greek *dendron*, which means tree. Dendrimers are structures on the order of 5–20 nm, formed by radial growth in layers from a polyfunctionalized core, where the number of monomer units incorporated in each layer is successively doubled or tripled in relation to the previous cycle (Fig. 4.8). The resulting structure has a large number of ramifications and functional agglomerations on its surface [76].

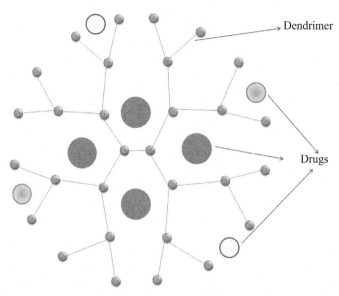

FIGURE 4.8 Illustration of a dendrimer containing molecules of a pharmaceutical. *Adapted from P. Kesharwani, K. Jain, N.K. Jain, Dendrimer as nanocarrier for drug delivery, Prog. Polym. Sci. 39 (2) (2014) 268–307, Copyright (2014) [77].With kind permission from Elsevier Publisher.*

Encapsulation of a medication in dendrimers is performed through Van der Wals interactions and/or hydrogen bonds [56]. The advantage relative to other synthetic polymers is high molecular uniformity and high predictability of the molecular mass, size, and functional groups, providing more safety in relation to the composition and quantity of the incorporated pharmaceutical [78].

Some dendrimers are being used in biomedicine as cardiac markers for rapid diagnosis of heart attack, as a tool to promote in vitro gene transfection and as strategic biological sensors of anthrax or botulinum toxin [78]. There are also studies for the application of this nanosystem in the treatment of cancer [34,35].

In the specific case of neurological diseases, Kannan et al. [76] treated rabbits suffering from induced cerebral palsy using *N-acetyl cysteine* dendrimers. The drug was able to cross the HEB and reach the cells responsible for the inflammatory process involved in this disease, improving motor conditions and decreasing inflammation. Therefore, dendrimers also represent a nanosystem that could be used to carry drugs with neurological applications.

4.3.2.2.6 NANOSPONGES

Nanosponges represent a new class of controlled release system. These particles can carry hydrosoluble substances and promote the dispersion of certain lipophilic substances in the water. They are tiny structures with dimensions comparable to a virus and take the form of a mesh that is often produced from a type of biodegradable polyester (alginate, for instance), with segments that form spheres or cavities where the medication is incorporated [79] (Fig. 4.9).

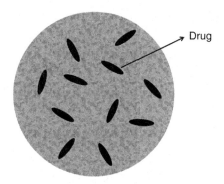

FIGURE 4.9 Illustration of a nanosponge with incorporated medication.

Compared to other nanosystems, nanosponges are more soluble in water as well as in organic solvents. They have high porosity, are atoxic, and are stable at temperatures of up to 300°C. The only disadvantage of this system is that only small molecules can be incorporated [64].

Nanosponge-based formulations can be administered through oral, parenteral, topical, and inhaled routes. Studies performed with medications that are already commercialized in their free forms are being performed with the same compounds incorporated in nanosponges. These include *Paclitaxel* [80] and *Tamoxifen* [81] (both used for the treatment of cancer), *Dexamethasone* (antiinflammatory/antiallergy), and *Itraconazole* (antifungal) [82] as well as vaccines, antibodies, proteins, and enzymes [64]. To date, there are no reports of studies on the application of this system in neurological diseases, but they certainly have great potential to be explored.

Thus far, this chapter has briefly described the contributions of nanotechnology to the field of pharmacology that are promoting important breakthroughs in different areas of medicine, such as neurology. However, given the complexity of biological systems, the development of new pharmaceuticals and the study of their mechanisms of action still depend on the contribution of other fields of study, including informatics.

4.4 Computational Resources in Nanomedicine

From a broad perspective, informatics is itself the application of information and computational science methods to collect, analyze, and apply data in a specific situation. The use of prefixes in informatics, such as *bio*informatics, *eco*informatics, and *neuro*informatics, became the standard descriptor for the application of these methods to study problems within a specific field or subject. Likewise, computational science is the use of advanced computational techniques and sophisticated algorithms to position and solve problems. Here, the use of descriptive words in n-computational, such as in computational biophysics, refers to the use of computational methods within the field or subject of biophysics.

In the last two decades, the exploration of large amounts of data began to combine oriented experimental data with computer science and methods that use massive computational networks and state-of-the-art scientific tools as well as social network system technologies [83–85]. Pioneer projects to sequence and analyze the human genome [86] and the *Xylella fastidiosa* genome [87] are excellent examples of this combination of sciences and can demonstrate how computational sophistication and coordination of domain expertise can unite to overcome great scientific challenges (Fig. 4.10).

4.4.1 Computational Neuroscience, Neuroinformatics, and Neurobiophysics

Computational neuroscience is an interdisciplinary science that unites different areas of neuroscience, computer science, applied physics, and mathematics. It is the primary theoretical method for investigating the function and mechanism of the nervous system. Computational neuroscience differs from the psychological connectionism and from the theories from disciplines such as machines learning, neural networks, and statistical learning theory, which emphasize the realist, functional, and biological description of neurons as well as their physiology and dynamics [83]. These models capture the essential characteristics of biological systems in many space–time scales, from membrane barriers and protein to chemical coupling in network oscillations, learning and memory. These computational resources are used to test hypotheses that could be directly verified through biological experiments. Currently, this field is rapidly expanding and leading to a revolution in neuroscience. Many pieces of software allow the realistic, rapid, and systematic in silico modeling of neurons, such as *NEURON* (http://www.neuron.yale.edu/). Furthermore, there are also methodologies of systems grouped by patterns of molecular expression and of neuronal population information, such as the "Brain Architecture Knowledge Management System" (*BAMS*—http://brancusi.usc.edu/bkms/), an interface created to integrate the results of different experiences and forms of neurological data [88].

Neuroinformatics combines neuroscience and informatics research, aiming to develop and apply advanced tools and methods essential for breakthroughs in the understanding of brain structure and function [89,90]. Using these methods, the principles of brain signal processing are extracted from biological systems and applied to the construction of artificial intelligent systems [83]. The construction of artificially intelligent systems generates valuable feedback on the sufficiency of neurobiological concepts. Therefore, neuroinformatics varies between the modeling of neurophysiological processes and the biologically inspired development of artificial information processing systems, such as many approaches to artificial neural networks [83,91].

Klopper (2005) [92] also defines neuroinformatics as an emerging area of informatics where microscopic implants are performed, allowing *nanocomputers* to improve the sensory input to the brain. This may enable the blind to see again, the deaf to hear again, and the paralyzed to recover their motion. Yamaji et al. [93] added that neuroinformatics is a new research tool that promotes the organization of data from neuroscience and databases using Internet technology, such as open access databases.

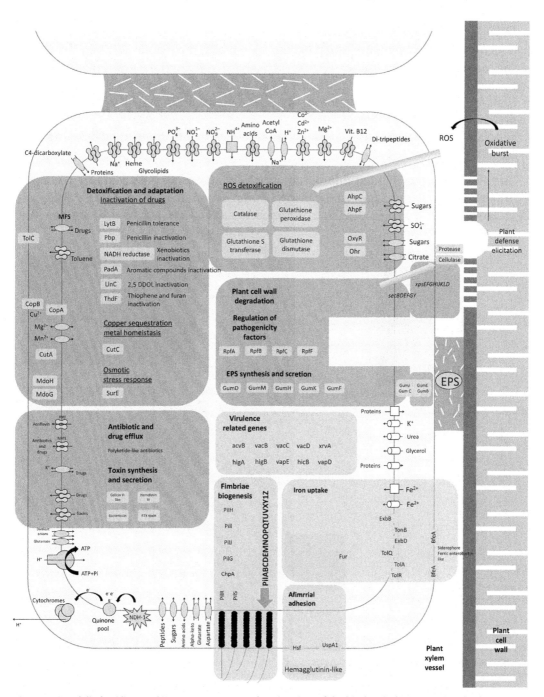

FIGURE 4.10 *Xylella fastidiosa* and its genome: a comprehensive view of the biochemical processes involved in its pathogenicity and survival in the xylem host. The main functional categories are shown in bold, and the bacterial genes and gene products related to that function are organized in sections [87]. *Reproduced from A.J.G. Simpson, F.C. Reinach, P. Arruda, F.A. Abreu, M. Acencio, et al., The genome sequence of the plant pathogen* Xylella fastidiosa, *406 (6792) (2014) 151–157, Copyright (2014) [87]. With the permission of Nature Publishing Group.*

Generally, this subject comprises three main areas [84]:

1. data related to neuroscience and databases, increasingly capable of dealing with the complexity and organization of the nervous system, from the molecular to the behavioral level;
2. tools for acquisition, visualization, analysis, and distribution of data from the nervous system; and
3. simulation of theoretical and computational environments for the modeling and understanding of the brain.

The main objective of neuroinformatics is to help neuroscientists deal with the analysis, modeling, simulation, and management of information resources before, during, and after research (Fig. 4.11). As such these scientists need to have access to a common and

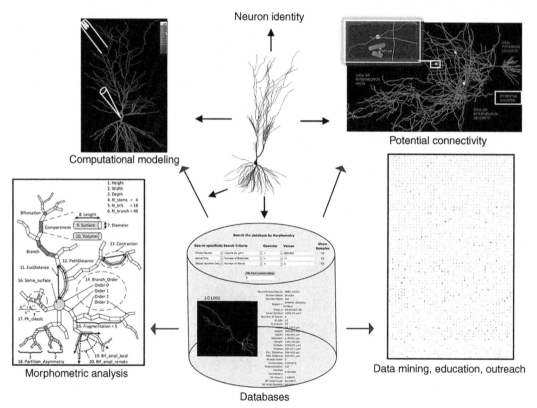

FIGURE 4.11 Scientific applications of three-dimensional reconstruction: the tracking of axonal and dendritic morphology is often carried out for many purposes, such as the creation of neuronal identity, anatomical implementation, and realistic biophysical simulations of neuronal electrophysiology, as well as for performing morphometric and stereological analyses to determine the potential for connectivity. Data deposition in central databases makes reconstructions easily accessible for reuse in any of these applications as well as in data mining, education, and propagation [95]. *Reproduced from R. Parekh, G.A. Ascoli, Neuronal morphology goes digital: a research hub for cellular and system neuroscience, Neuron 77 (6) (2014) 1017–1038, Copyright (2014) [95]. With permission from Elsevier Publisher.*

remote environment able to provide tools for the organization of data and the storage of results. One example of this environment is the *Visiome* (http://platform.visiome.neuro-inf.jp) Neuroinformatics platform, which approaches questions related to neuroscience and provides different portals for different fields of brain research [89,90,94].

Gillies and Willshaw [96] showed in their study that neuroinformatics techniques, particularly computational modeling, can provide a reliable method to unite and develop new pharmacological concepts and phenomena, such as the loss of dopamine, with electrophysiological characteristics. These precise models of brain dynamics can directly connect diseases, such as Parkinson's disease, to effective treatments by simulating pharmaceutical input through the membrane channels of the brain as well as critical physiology and mechanisms of brain stimulation.

Complementing the previously cited concepts, neurobiophysics is the application of the basic principles of physics to the function of the nervous system. The methods of neurobiophysics are identical to those of other quantitative scientific fields and include the observation of phenomena under controlled conditions and the subsequent replication of those phenomena as well as the elaboration of models where the laws of physics can be applied and provide quantitative explanations of relevant observations [97].

A basic assumption in neurobiophysics is that all neuronal activity has a possible explanation based on the application of known laws of physics [98]. In this view, the morphological complexity of the neuron and the structural complexity of the neuronal interconnections are practical barriers for understanding the nervous system, and thus, neurobiophysics uses the computational sciences to investigate areas ranging from cell biology to complex neural network systems [97]. Table 4.1 shows the primary tools and resources for digital reconstructions that are currently available for the study of neural networks.

4.4.2 Nano(bio)informatics

Nanotechnology is a means of integrating the biological and the artificial world [99] given that, in fact, nanoparticles and nanodevices can interact with biological systems. Therefore, from the perspective of computer science, nanotechnology can catalyze a paradigm change in the domain of Informatics. That is, the interaction of computational processes with biological, natural information from complex systems and materials designed at the nanoscale.

The application of nanotechnology to healthcare is known as nanomedicine [99]. Thus, the application of nanomedicine to the field of Informatics is called nanoinformatics. Nanoinformatics can be used with biomolecular approaches for diagnosis, preventive medicine, and monitoring and treatment of illnesses [99], in addition to promoting the creation of medical tools under the umbrella of nanobioinformatics. It is important to emphasize that while this discussion will only include its biomedical applications, nanoinformatics can also be related to other areas of nanotechnology.

Nanoinformatics has the purpose of building a bridge between nanomedicine and information technologies by applying computational methods to manage the information

Table 4.1 Three-Dimensional Reconstruction Resources and Tools of Neuron Morphology [95]

Tool	Platform	Support	Web Address
Design			
Neurolucida[a]	W	EFGTDA	mbfbioscience.com/neurolucida
Filament Tracer[a]	WM	ETDA	bitplane.com/go/products/filamenttracer
Amira[a]	WML	EGDA	www.vsg3d.com/amira/skeletonization
Neuron 3DMA[a]	L	DA	www.ams.sunysb.edu/~lindquis/3dma/3dma_neuron/3dma_neuron.html
NTS/Eutectic[b]	W	G	bellsouthpwp.net/c/a/capowski/NTSPublic.html
NeuronJ[b]	WML	GNDA	imagescience.org/meijering/software/neuronj
NeuRA[b]	M	G	neura.org
NeuronStudio[b]	W	EGA	research.mssm.edu/cnic/tools-ns.html
ORION[b]	WL	FGDA	cbl.uh.edu/ORION
Neuron_Morpho	WM (I)	GN	www.personal.soton.ac.uk/dales/morpho
Farsight	WML (C)	LTDA	farsight-toolkit.org/wiki/Open_Snake_Tracing_System
Trees Toolbox	WML (O)	EGTDA	treestoolbox.org
Vaa3D	WML (C)	FGTDA	Vaa3D.org
Simple Neurite Tracer	WML (I)	EGLTDA	pacific.mpi-cbg.de/wiki/index.php/Simple_Neurite_Tracer
Neuromantic	W	ETDA	www.reading.ac.uk/neuromantic
Analysis			
Neurolucida Explorer[a]	W	EFGTDA	mbfbioscience.com/neurolucida
Neuronland[b]	WM	GLDA	neuronland.org/NL.html
Cvapp	WM (J)	G	www.compneuro.org/CDROM/docs/cvapp.html
L-Measure	WML (CJ)	FGDA	cng.gmu.edu:8080/Lm
Computational modeling			
L-Neuron[c]	WL (C)	GNA	krasnow1.gmu.edu/cn3/L-Neuron
NeuGen[c]	WML (CJ)	GNDA	neugen.org
NetMorph[c]	WML (C)	GNDA	netmorph.org
CX3D[c]	WM (J)	FT	www.ini.uzh.ch/~amw/seco/cx3d
NEURON	WML (C)	FGLTDA	www.neuron.yale.edu/neuron
GENESIS	WML (C)	FGLTDA	genesis-sim.org
NeuronC	ML (C)	GNDA	retina.anatomy.upenn.edu/~rob/neuronc.html
Surf-Hippo	L (O)	GL	www.neurophys.biomedicale.univ-paris5.fr/~graham/surf-hippo.html
Catacomb[b]	WML	GNDA	catacomb.org
neuroConstruct	WML (J)	GTNDA	neuroconstruct.org
Moose	WML (O)	FGDA	moose.ncbs.res.in
PSICS	WML (JO)	EFGDA	psics.org
Database			
NeuroMorpho.Org	WML	EGDA	neuromorpho.org
Cercal DB	WML	NDA	apps.montana.edu/cercaldb
CCDB	WML	GLDA	ccdb.ucsd.edu
Fly Circuit	WML	EGT	flycircuit.tw

Platform: W, Windows; M, Mac; L, Linux; (C), C/C++; (J), Java; (I), plugin ImageJ; (O) others, that is, Matlab, Python, Lisp, or Fortran.

Support: E, email; F, forum; G, guide (manual); L, discussion list; N, newsletter; T, tutorial; D, continuous development; A, actively maintained.

[a]Tools: Commercial.

[b]Tools: Open source, it is not open, all of the others, free.

[c]Tools: Morphological modeling (as opposed to electrophysiological).

Source: Reproduced from R. Parekh, G.A. Ascoli, Neuronal morphology goes digital: a research hub for cellular and system neuroscience, Neuron 77 (6) (2014) 1017–1038, Copyright (2014) [95]. With permission from Elsevier Publisher.

created within the domain of nanomedicine. Hence, nanoinformatics is simply the management of information, an approach for efficiently coordinating the resources available in scientific study and making predictions through molecular and biological modeling [100]. While bioinformatics is usually applied in the context of the analysis of DNA sequence data and the prediction of relationships between genetic mutations and diseases [101], Nanoinformatics is applied to organizing, interpreting, and making predictions of structures, physicochemical properties, environments, and applications associated with nanoparticles and nanomaterials, thereby promoting the evolution of medical fields in general [100,102,103]. These new applications are being born at a time when Genomic Medicine and customized treatment are still being developed and therefore hold great promise for Modern Medicine.

Currently, some fields are largely focused on the study of nanoinformatics [101]: nanoparticle characterization, nanotoxicology, modeling and simulation, imaging, terminology, ontologies and standardization, data integration and exchange, interoperability of systems, data and text mining for nanomedical research, connection of nanoinformation to computerized medical records, basic and translational research, creation of international networks of researchers, projects and labs, nanoinformatics education, and ethical matters.

Nanoinformatics education is still in its early phases. The subject of nanoinformatics was created with the support by the *US National Science Foundation* in 2007 and is therefore a very recent field of science. In Europe, this term was launched in 2008 in a project called ACTION-Grid and supported by the European Commission. Its purpose was to analyze the challenges surrounding and create a schedule for the development of this discipline, which is intrinsically connected to nanotechnology, biomedicine, and informatics [100,104].

In regards to nanotoxicology, nanoinformatics is present in studies related to the investigation of the toxicity of nanoscale products in natural systems (in vivo or in vitro). This type of information is fundamental for guaranteeing the safe use of nanomaterials for medical and pharmacological products [100] and is used in new clinical trials and nanoscale simulations generated by informatics. When speaking of toxicology at the nanoscale, it is important to emphasize that it refers to the nanometric size of particles as well as their unprecedented electrical, optical, chemical, and biological properties. A common methodology in nanoinformatics, with regard to toxicity, is the quantitative structure–activity relationship (QSAR), a process through which the chemical structure is quantitatively correlated to a well-defined process, such as the biological activity or chemical reactivity of a nanotechnology material. The development of QSAR generally requires information on well-characterized nanomaterial structures, experimental data that establish the activity of these materials within the environment where they are used, and a theoretical base founded on the verification of specific literature [105].

There are databases with information on toxicological effects, such as those from the National Institute for Occupational Safety and Health (NIOSH—https://www.cdc.gov/niosh/) and the oregon nanoscience and Microtechnologies Institute (ONAMI—http://www.

onami.us/). In the near future, researchers will use nanoinformatics to model and simulate toxicity processes, relating them to the information of real patients from computerized medical reports to predict the human response to products, such as nanopharmaceuticals [102]. In Europe and the United States, similar initiatives, such as the Advancing Clinical Genomic Trials on Cancer (ACGT) and the Cancer Biomedical Informatics Grid (caBIG), have developed new approaches for data sharing, modeling, and simulation of drug delivery [102,106]. These initiatives are designed to accelerate the trials necessary to guarantee the efficacy and safety of pharmaceuticals. By using the advanced computational methods offered by nanoinformatics, doctors and researchers will be able to reduce the time taken by the laboratory–clinical practice to make nanopharmaceuticals available to patients [102].

4.4.3 Application of Informatics Methods and Tools to Nanomedical Data

In computer science, ontologies are used in artificial intelligence, web semantics, software engineering, and information architecture as a form to represent knowledge of the world in whole or part [107]. In turn, nanoinformatics ontologies can be considered models of data that represent a set of concepts within a domain, represented by nanomedical structures. These models contains countless concepts, nanomedical categories, and relationships that could be used as a controlled terminology for the management of data and analysis, allowing different investigation tasks, such as the registration of experimental data or the integration of heterogeneous data sources [108]. Considering the extensive medical, biological, and nanotechnological vocabulary, the ontologies provide a mechanism to standardize the representation of the nanomedical domain, which could be very useful to propagate information and knowledge, as shown in Fig. 4.12.

Currently, there are relevant initiatives for the development of nanomedical ontologies, such as nanoparticle ontology (NPO) [109,110], developed by researchers at the National Cancer Institute. In addition to the representation of domain knowledge, nanoinformatics could lead to another relevant challenge for nanomedicine: the management of large volumes of data and information generated by nanomedical research. This challenge demands methods to improve efficiency in the search and recovery of information, such as that provided by mining data in the medical literature or the location of information in large databases to access new discoveries and other information on reactions, chemical properties of nanoparticles, nanomaterials, nanotoxicity, and physical interactions of nanomaterials.

Based on the history of research and interdisciplinary cooperation, there is a clear potential for synergy among all areas to which computational nanoscience could contribute. In the not too distant future, this junction of disciplines could improve our current knowledge of medicine and, especially, could lead science to find currently unknown cures for many diseases.

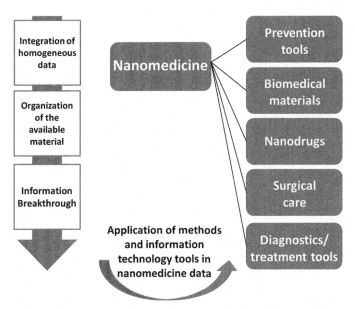

FIGURE 4.12 Medical nanoinformatics.

Abbreviations

AFM	Atomic force microscopy
CNS	Central nervous system
DNA	Deoxyribonucleic acid
HEB	Hematoencephalic barrier
MS	Multiple sclerosis
NMO	Neuromyelitis optica
NPO	Nanoparticle ontology
QSAR	Quantitative structure-activity relationship
RNA	Ribonucleic acid
SEM	Scanning electron microscopy

Glossary

Biocompatible The presence of a material in an organism does not cause harmful reactions in the organism. For example, when the body adapts to an implant by forming a collagen capsule. In this case, the organism recognizes the implant as an "intruder", but its presence does not cause harmful effects.

Biodegradable A material with biocompatible characteristics and that creates minimal organic interference because it is degraded by biological processes.

Excitotoxicity Toxicity due to excessive and nonnatural excitation of a nervous cell by a neurotransmitter. This is common in pathologies such as Parkinson's disease and often causes cell death.

Pharmacodynamics The physiological effects of pharmaceuticals in an organism, the mechanism of action, and the relationship between the pharmaceutical concentration and effect.

Pharmacokinetics The "path" taken by a medication through an organism. It includes absorption, delivery, biotransformation, and excretion.

References

[1] F.L. Leite, et al. Nanoneurobiophysics: new challenges for diagnosis and therapy of neurologic disorders, Nanomedicine 10 (23) (2015) 3417–3419.

[2] P. O'malley, Nanopharmacology for the future-think small, Clin. Nurse Spec. 24 (3) (2010) 123–124.

[3] Y. Zhang, Nanopharmacology, Acta. Pharmacologica. Sinica. 27 (2006) 444–1444.

[4] H.S. Sharma, A. Sharma, Neurotoxicity of engineered nanoparticles from metals, CNS Neurol. Disord. Drug Targets 11 (1) (2012) 65–80.

[5] H.S. Sharma, A. Sharma, Nanowired drug delivery for neuroprotection in central nervous system injuries: modulation by environmental temperature, intoxication of nanoparticles, and comorbidity factors, Wiley Interdiscip. Rev. Nanomed. Nanobiotechnol. 4 (2) (2012) 184–203.

[6] H.S. Sharmaet al., Nanoparticles influence pathophysiology of spinal cord injury and repair, in: H.S. Sharma, (Ed.). Nanoneuroscience and Nanoneuropharmacology, vol.180, 2009, pp. 155–180 (Progress in brain research).

[7] H.S. Sharma, Nanoneuroscience and nanoneuropharmacology preface, Nanoneurosci. Nanoneuropharmacol. 180 (2009) VII–VII10.

[8] C.E. Mora-Huertas, H. Fessi, A. Elaissari, Polymer-based nanocapsules for drug delivery, Int. J. Pharm. 385 (1–2) (2010) 113–142.

[9] R. Lal, S. Ramachandran, M.F. Arnsdorf, Multidimensional atomic force microscopy: a versatile novel technology for nanopharmacology research, AAPS J. 12 (4) (2010) 716–728.

[10] G.A. Silva, Neuroscience nanotechnology: progress, opportunities and challenges, Nat. Rev. Neurosci. 7 (1) (2006) 65–74.

[11] Y. Barenholz, Nanomedicine: shake up the drug containers, Nat. Nanotechnol. 7 (8) (2012) 483–484.

[12] H.S. Sharma, Nano neuro science: emerging concepts on nanoneurotoxicity and nanoneuroprotection, Nanomedicine 2 (6) (2007) 753–758.

[13] G. Binnig, C.F. Quate, C. Gerber, Atomic force microscope, Phys. Rev. Lett. 56 (9) (1986) 930–933.

[14] R. Garcia, R. Perez, Dynamic atomic force microscopy methods, Surf. Sci. Rep. 47 (6–8) (2002) 197–301.

[15] M.B. Shenai, et al. A novel MEA/AFM platform for measurement of real-time, nanometric morphological alterations of electrically stimulated neuroblastoma cells, IEEE Trans. Nanobiosci. 3 (2) (2004) 111–117.

[16] A. Noy, D.V. Vezenov, C.M. Lieber, Chemical force microscopy, Annu. Rev. Mater. Sci. 27 (1997) 381–421.

[17] T. Ito, I. Grabowska, S. Ibrahim, Chemical-force microscopy for materials characterization, TrAC Trends Anal. Chem. 29 (3) (2010) 225–233.

[18] J.L. Maciaszek, et al. Topography of native SK channels revealed by force nanoscopy in living neurons, J. Neurosci. 32 (33) (2012) 11435–11440.

[19] Y.F. Loginova, et al. Biodistribution and stability of CdSe core quantum dots in mouse digestive tract following per os administration: advantages of double polymer/silica coated nanocrystals, Biochem. Biophys. Res. Commun. 419 (1) (2012) 54–59.

[20] J. Zhou, Y. Yang, C.Y. Zhang, Toward biocompatible semiconductor quantum dots: from biosynthesis and bioconjugation to biomedical application, Chem. Rev. 115 (21) (2015) 11669–11717.

[21] K. Donaldson, et al. Nanotoxicology, Occup. Environ. Med. 61 (9) (2004) 727–728.

[22] J.N. Marsh, et al. Molecular imaging with targeted perfluorocarbon nanoparticles: quantification of the concentration dependence of contrast enhancement for binding to sparse cellular epitopes, Ultrasound Med. Biol. 33 (6) (2007) 950–958.

[23] A. Hoshino, S. Hanada, K. Yamamoto, Toxicity of nanocrystal quantum dots: the relevance of surface modifications, Arch. Toxicol. 85 (7) (2011) 707–720.

[24] R. Hardman, A toxicologic review of quantum dots: toxicity depends on physicochemical and environmental factors, Environ. Health Perspect. 114 (2) (2006) 165–172.

[25] B.R. Prasad, et al. Effects of long-term exposure of gelatinated and non-gelatinated cadmium telluride quantum dots on differentiated PC12 cells, J. Nanobiotechnol. 10 (2012) 4.

[26] J.D. Smith, et al. The use of quantum dots for analysis of chick CAM vasculature, Microvas. Res. 73 (2) (2007) 75–83.

[27] S. Kim, et al. Near-infrared fluorescent type II quantum dots for sentinel lymph node mapping, Nat. Biotechnol. 22 (1) (2004) 93–97.

[28] D. Onoshima, H. Yukawa, Y. Baba, Multifunctional quantum dots-based cancer diagnostics and stem cell therapeutics for regenerative medicine, Adv. Drug Deliv. Rev. 95 (2015) 2–14.

[29] D.R. Larson, et al. Water-soluble quantum dots for multiphoton fluorescence imaging in vivo, Science 300 (5624) (2003) 1434–1436.

[30] A. Bonoiu, et al. MMP-9 gene silencing by a quantum dot-siRNA nanoplex delivery to maintain the integrity of the blood brain barrier, Brain Res. 1282 (2009) 142–155.

[31] A.S. Basso, et al. Reversal of axonal loss and disability in a mouse model of progressive multiple sclerosis, J. Clin. Invest. 118 (4) (2008) 1532–1543.

[32] F. Re, M. Gregori, M. Masserini, Nanotechnology for neurodegenerative disorders, Maturitas 73 (1) (2012) 45–51.

[33] Z. Yang, et al. Pharmacological and toxicological target organelles and safe use of single-walled carbon nanotubes as drug carriers in treating Alzheimer disease, Nanomed. Nanotechnol. Biol. Med. 6 (3) (2010) 427–441.

[34] E.D. Melita, G. Purcel, A.M. Grumezescu, Carbon nanotubes for cancer therapy and neurodegenerative diseases, Rom. J. Morphol. Embryol. 56 (2) (2015) 349–356.

[35] A.V. Kabanov, H.E. Gendelman, Nanomedicine in the diagnosis and therapy of neuro degenerative disorders, Prog. Polym. Sci. 32 (8–9) (2007) 1054–1082.

[36] L.W. Zhang, et al. Biological interactions of functionalized single-wall carbon nanotubes in human epidermal keratinocytes, Int. J. Toxicol. 26 (2) (2007) 103–113.

[37] C.M. Sayes, et al. Functionalization density dependence of single-walled carbon nanotubes cytotoxicity in vitro, Toxicol. Lett. 161 (2) (2006) 135–142.

[38] J. Hwang, et al. Biofunctionalized carbon nanotubes as substrates for drug delivery and neural regeneration, Tissue Eng. Part A 21 (2015) S366–S1366.

[39] A.V. Vergoni, et al. Nanoparticles as drug delivery agents specific for CNS: in vivo biodistribution, Nanomed. Nanotechnol. Biol. Med. 5 (4) (2009) 369–377.

[40] W. Cheung, et al. DNA and carbon nanotubes as medicine, Adv. Drug Deliv. Rev. 62 (6) (2010) 633–649.

[41] E.W. Keefer, et al. Carbon nanotube coating improves neuronal recordings, Nat. Nanotechnol. 3 (7) (2008) 434–439.

[42] G. Tosi, et al. Targeting the central nervous system: in vivo experiments with peptide-derivatized nanoparticles loaded with Loperamide and Rhodamine-123, J. Controll. Release 122 (1) (2007) 1–9.

[43] H. Hu, et al. Chemically functionalized carbon nanotubes as substrates for neuronal growth, Nano Lett. 4 (3) (2004) 507–511.

[44] M.P. Mattson, R.C. Haddon, A.M. Rao, Molecular functionalization of carbon nanotubes and use as substrates for neuronal growth, J. Mol. Neurosci. 14 (3) (2000) 175–182.

[45] E. Jan, N.A. Kotov, Successful differentiation of mouse neural stem cells on layer-by-layer assembled single-walled carbon nanotube composite, Nano Lett. 7 (5) (2007) 1123–1128.

[46] K. Matsumoto, et al. Neurite outgrowths of neurons with neurotrophin-coated carbon nanotubes, J. Biosci. Bioeng. 103 (3) (2007) 216–220.

[47] S.K. Powell, et al. Neural cell response to multiple novel sites on laminin-1, J. Neurosci. Res. 61 (3) (2000) 302–312.

[48] D. Brambilla, et al. Nanotechnologies for Alzheimer's disease: diagnosis, therapy, and safety issues, Nanomed. Nanotechnol. Biol. Med. 7 (5) (2011) 521–540.

[49] L.L. Dugan, et al. Carboxyfullerenes as neuroprotective agents, Proc. Natl. Acad. Sci. USA 94 (17) (1997) 9434–9439.

[50] G.A. Silva, Nanotechnology approaches for the regeneration and neuroprotection of the central nervous system, Surg. Neurol. 63 (4) (2005) 301–306.

[51] A. Bar-Shir, Y. Engel, M. Gozin, Synthesis and water solubility of adamantyl-OEG-fullerene hybrids, J. Org. Chem. 70 (7) (2005) 2660–2666.

[52] A.P. Brown, et al. Evaluation of the fullerene compound DF-1 as a radiation protector, Radiat. Oncol. 5 (2010) 1–9.

[53] B. Daroczi, et al. In vivo radioprotection by the fullerene nanoparticle DF-1 as assessed in a zebrafish model, Clin. Cancer Res. 12 (23) (2006) 7086–7091.

[54] H.S. Lin, et al. Fullerenes as a new class of radioprotectors, Int. J. Radiat. Biol. 77 (2) (2001) 235–239.

[55] M. Patlak, Nanotechnology takes a new look at old drugs, J. Natl. Cancer Inst. 102 (23) (2010) 1753–1755.

[56] D.K. Deda, K. Araki, Nanotechnology, light and chemical action: an effective combination to kill cancer cells, J. Braz. Chem. Soc. 26 (12) (2015) 2448–2470.

[57] E.A. Nance, et al. A dense poly(ethylene glycol) coating improves penetration of large polymeric nanoparticles within brain tissue, Sci. Transl. Med. 4 (2012) 149.

[58] J.R. Kanwar, et al. Nanoparticles in the treatment and diagnosis of neurological disorders: untamed dragon with fire power to heal, Nanomed. Nanotechnol. Biol. Med. 8 (4) (2012) 399–414.

[59] G. Modi, et al. Nanotechnological applications for the treatment of neurodegenerative disorders, Prog. Neurobiol. 88 (4) (2009) 272–285.

[60] W. Abdelwahed, et al. Freeze-drying of nanoparticles: formulation, process and storage considerations, Adv. Drug Deliv. Rev. 58 (15) (2006) 1688–1713.

[61] G. Cevc, Lipid vesicles and other colloids as drug carriers on the skin, Adv. Drug Deliv. Rev. 56 (5) (2004) 675–711.

[62] V.P. Torchilin, Multifunctional nanocarriers, Adv. Drug Deliv. Rev. 58 (14) (2006) 1532–1555.

[63] T.X. Xiang, B.D. Anderson, Liposomal drug transport: a molecular perspective from molecular dynamics simulations in lipid bilayers, Adv. Drug Deliv. Rev. 58 (12–13) (2006) 1357–1378.

[64] S. Subramanian, et al. Nanosponges: a novel class of drug delivery system—review, J. Pharm. Pharm. Sci. 15 (1) (2012) 103–111.

[65] S.V. Vinogradov, et al. Polyplex nanogel formulations for drug delivery of cytotoxic nucleoside analogs, J. Controll. Release 107 (1) (2005) 143–157.

[66] M. Oishi, et al. Endosomal release and intracellular delivery of anticancer drugs using pH-sensitive PEGylated nanogels, J. Mater. Chem. 17 (35) (2007) 3720–3725.

[67] M.K. Teli, S. Mutalik, G.K. Rajanikant, Nanotechnology and nanomedicine: going small means aiming big, Curr. Pharm. Des. 16 (16) (2010) 1882–1892.

[68] Y.N. Konan, R. Gurny, E. Allemann, State of the art in the delivery of photosensitizers for photodynamic therapy, J. Photochem. Photobiol. B Biol. 66 (2) (2002) 89–106.

[69] S.R. Schaffazick, et al. Physicochemical characterization and stability of the polymeric nanoparticle systems for drug administration, Quim. Nov. 26 (5) (2003) 726–737.

[70] B. Wilson, et al. Poly(n-butylcyanoacrylate) nanoparticles coated with polysorbate 80 for the targeted delivery of rivastigmine into the brain to treat Alzheimer's disease, Brain Res. 1200 (2008) 159–168.

[71] B. Wilson, et al. Targeted delivery of tacrine into the brain with polysorbate 80-coated poly(n-butyl-cyanoacrylate) nanoparticles, Eur. J. Pharm. Biopharm. 70 (1) (2008) 75–84.

[72] K.S. Rao, et al. TAT-conjugated nanoparticles for the CNS delivery of anti-HIV drugs, Biomaterials 29 (33) (2008) 4429–4438.

[73] S.H. Voon, et al. Vivo studies of nanostructure-based photosensitizers for photodynamic cancer therapy, Small 10 (24) (2014) 4993–5013.

[74] E.V. Batrakova, et al. Effects of pluronic P85 unimers and micelles on drug permeability in polarized BBMEC and Caco-2 cells, Pharm. Res. 15 (10) (1998) 1525–1532.

[75] J. Kreuter. Application of nanoparticles for the delivery of drugs to the brain, In: A.G. Deboer (Ed.), Drug Transport(ers) and the Diseased Brain, vol. 1277, Elsevier Science Bv, Amsterdam, 2005, pp.85–94 (International Congress Series).

[76] S. Kannan, et al. Dendrimer-based postnatal therapy for neuroinflammation and cerebral palsy in a rabbit model, Sci. Transl. Med. 4 (130) (2012).

[77] P. Kesharwani, K. Jain, N.K. Jain, Dendrimer as nanocarrier for drug delivery, Prog. Polym. Sci. 39 (2) (2014) 268–307.

[78] P. Forstner, et al. Cationic PAMAM dendrimers as pore-blocking binary toxin inhibitors, Biomacromolecules 15 (7) (2014) 2461–2474.

[79] F. Trotta, et al. The application of nanosponges to cancer drug delivery, Expert Opin. Drug Deliv. 11 (6) (2014) 931–941.

[80] S.J. Torne, et al. Enhanced oral paclitaxel bioavailability after administration of paclitaxel-loaded nanosponges, Drug Deliv. 17 (6) (2010) 419–425.

[81] J. Alongi, et al. Role of beta-cyclodextrin nanosponges in polypropylene photooxidation, Carbohydr. Polym. 86 (1) (2011) 127–135.

[82] S. Swaminathan, et al. Formulation of betacyclodextrin based nanosponges of itraconazole, J. Incl. Phenom. Macrocycl. Chem. 57 (1–4) (2007) 89–94.

[83] J. Wiemer, et al. Informatics united–exemplary studies combining medical informatics, neuroinformatics, and bioinformatics, Methods Inf. Med. 42 (2) (2003).

[84] S. Usui, Neuroinformatics research in vision, in: L. Rutkowski, J. Kacprzyk (Eds.), Neural Networks and Soft Computing, 2003 (Advances in Soft Computing).

[85] S. Usui, Neuroinformatics research for vision science: NRV project, Biosystems 71 (1–2) (2003) 189–193.

[86] P. De Jong, et al. Initial sequencing and analysis of the human genome (vol 409, pg 860, 2001), Nature 412 (6846) (2001).

[87] A.J.G. Simpson, et al. The genome sequence of the plant pathogen *Xylella fastidiosa*, Nature 406 (6792) (2000) 151–159.

[88] M. Bota, O. Sporns, L.W. Swanson, Neuroinformatics analysis of molecular expression patterns and neuron populations in gray matter regions: the rat BST as a rich exemplar, Brain Res. 1450 (2012) 174–193.

[89] S. Usui, et al. Visiome environment: enterprise solution for neuroinformatics in vision science, Neurocomputing 58 (2004) 1097-L1101.

[90] S. Usui, Visiome: neuroinformatics research in vision project, Neural Netw. 16 (9) (2003) 1293–1300.

[91] A. Wrobel, The need of neuroinformatic approach in functional neurophysiology, Acta Neurobiol Exp 65 (4) (2005) 421–423.

[92] R. Klopper, Future communications: mobile communications, cybernetics, neuroinformatics and beyond, Alternation 12 (1a) (2005) 121–144.

[93] K. Yamaji, et al. Customizable neuroinformatics database system: XooNIps and its application to the pupil platform, Comput. Biol. Med. 37 (7) (2007) 1036–1041.

[94] S. Usui, et al. Keyword extraction, ranking, and organization for the neuroinformatics platform, Bio-systems 88 (3) (2007) 334–342.

[95] R. Parekh, G.A. Ascoli, Neuronal morphology goes digital: a research hub for cellular and system neuroscience, Neuron 77 (6) (2013) 1017–1038.

[96] A. Gillies, D. Willshow, Neuroinformatics and modeling of the basal ganglia: bridging pharmacology and physiology, Expert Rev. Med. Dev. 4 (5) (2007) 663–672.

[97] K. Franze, J. Guck, The biophysics of neuronal growth, Rep. Prog. Phys. 73 (9) (2010).

[98] I.G. Iliev, Neurobiophysics, Encycl. Appl. Phys 11 (1994) 297–322.

[99] S. Chiesa et al., Building an index of nanomedical resources: an automatic approach based on text mining, in: I. Lovrek, R.J. Howlett, et al. (Eds.), Knowledge-Based Intelligent Information and Engineering Systems, Pt 2, Proceedings, vol. 5178, 2008 (Lecture Notes in Artificial Intelligence).

[100] D. De la Iglesia, et al. , Nanoinformatics: new challenges for biomedical informatics at the nano level: Studies in health technology and informatics, IOS Press BV 150, Amsterdam, 2009, pp. 987–991.

[101] V. Maojo, et al. Nanoinformatics: developing new computing applications for nanomedicine, Computing 94 (6) (2012) 521–539.

[102] V. Maojo, et al. Nanoinformatics and DNA-based computing: catalyzing nanomedicine, Pediatr. Res. 67 (5) (2010) 481–489.

[103] F. Gonzalez-Nilo, et al. Nanoinformatics: an emerging area of information technology at the intersection of bioinformatics, computational chemistry and nanobiotechnology, Biol. Res. 44 (1) (2011) 43–51.

[104] V. L.-A., et al. Action GRID: assessing the impact of nanotechnology on biomedical informatics, AMIA Annu. Symp. Proc. 6 (2008) 1046.

[105] F. Martin-Sanchez et al., A primer in knowledge management for nanoinformatics in medicine, in: I. Lovrek, R.J. Howlett et al. (Eds.), Knowledge-Based Intelligent Information and Engineering Systems, Pt 2, Proceedings, vol. 5178, 2008 (Lecture notes in artificial intelligence).

[106] E. Jakobsson, M.D. Wang, L. Molnar, Bio-nano-info integration for personalized medicine, 2007.

[107] T.R. Gruber, Toward principles for the design of ontologies used for knowledge sharing, Int. J. Hum. Comput. Stud. 43 (5–6) (1995) 907–928.

[108] D. De La Iglesia, et al. International efforts in nanoinformatics research applied to nanomedicine, Methods Inf. Med. 50 (1) (2011) 84–95.

[109] D.G. Thomas et al., Ontologies for cancer nanotechnology research, Embc: 2009 Annual International Conference of the Ieee Engineering in Medicine and Biology Society, vols. 1–20, 2009.

[110] D.G. Thomas, R.V. Pappu, N.A. Baker, NanoParticle Ontology for cancer nanotechnology research, J. Biomed. Inform. 44 (1) (2011) 59–74.

5

Nanosensors

C.C. Bueno*, P.S. Garcia*, C. Steffens**, D.K. Deda*, F. de Lima Leite*

*NANONEUROBIOPHYSICS RESEARCH GROUP, DEPARTMENT OF PHYSICS, CHEMISTRY & MATHEMATICS, FEDERAL UNIVERSITY OF SÃO CARLOS, SOROCABA, SÃO PAULO, BRAZIL; **REGIONAL INTEGRATED UNIVERSITY OF UPPER URUGUAI AND MISSIONS, ERECHIM, RIO GRANDE DO SUL, BRAZIL*

5.1 Introduction

One of the main attractions of nanomaterial engineering in the context of sensing is the great sensitivity and selectivity/specificity potential it can offer in addition to the possibility of detecting multiple analytes and controlling their presence in real time [1] and in situ [2]. Such characteristics are obtained at the nanoscale because of the large number of sites available for molecular interactions originating from the high area/surface ratio of nanomaterials [3]. For these reasons, nanostructured sensors have been gaining more marketing attention because of the need to detect and measure chemical and physical properties and properties of complex biological and industrial systems [4].

According to Business Communications Company Inc. (BCCI), the tips and (micro) cantilevers used in atomic force microscopy (AFM) and scanning probe microscopy (SPM) currently represent the largest market share in the nanosensors market [3]. The next largest market share are the interdigitated sensors, *lab-on-a-chip*, because of their wide use in sensing applications in various areas such as industrial, medical, and environmental. Thus, this chapter will focus mainly on the discussion of aspects related to nanosensors with AFM (micro)cantilevers and tips in addition to introducing some concepts related to other types of nanosensors that are currently being researched.

5.2 Sensors and Nanosensors: New Detection Tools

The technologies related to sensors and nanosensors are at the forefront of analytical instrumentation research [2]; to better understand this technology, one must realize that a sensor receives the prefix *nano* if one or more of its spatial properties are in the scale between 10 and 100 nm. Therefore, nanosensors can be defined as any sensitive material used to transmit chemical, physical, or biological information about nanomaterials and molecules from recognition events. They can also be defined as molecular devices for analyses that incorporate a recognition element associated with a transduction system that allows the processing of the signal produced by the interaction between the sensing element and the analyte [5]. Therefore, these sensors can detect physical, chemical, biological, and quantum phenomena at the nanoscale.

Nanosensors can be classified according to four main criteria: (1) type of interaction established between the sensing element and the analyte; (2) method used to detect this interaction; (3) nature of the recognition element (e.g., biological or inorganic); and (4) transduction system. Therefore, there are different types of nanosensors that are defined and categorized by their different detection targets, their constituent materials, and the signals they use to transmit information, as shown in Fig. 5.1.

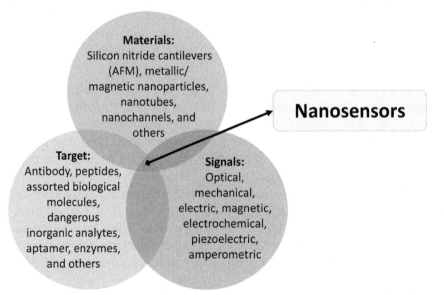

FIGURE 5.1 Fundamental components of a nanosensor. *Adapted from M. Swierczewska, et al., High-sensitivity nanosensors for biomarker detection, Chem. Soc. Rev. 41 (7) (2012) 2641–2655 [6]. With permission of The European Society for Photobiology, the European Photochemistry Association, and The Royal Society of Chemistry.*

Table 5.1 shows a brief summary of the main types of nanosensors that are currently being researched, as well as their main applications.

The development of high performance sensors, such as AFM tip and cantilever sensors, and others cited in Table 5.1 make use of nanofabrication technologies to maximize the advantages offered by miniaturized devices, to integrate the low cost and efficiency of these systems [52]. These structures are being researched for the purpose of finding innovative tools that can be applied in various areas including medical diagnosis, for example, disease markers, antibodies, deoxyribonucleic acid sequencing, DNA, hormones, and antibiotics; monitoring of gases in general; biochemical assays and investigating biomolecular interactions (drug carriers, antigen–antibodies, nucleic acids, enzyme assays, and proteins); environmental monitoring (atmosphere pollutants); agribusiness (pesticides, pheromones, and heavy metal ions); quality control of food (microorganism detection, water quality, and fruit ripening); and safety (drugs and explosives), among others [15,53–55].

5.3 Atomic Force Spectroscopy (Force Curve)

Atomic force microscopy, in addition to being useful for imaging various surfaces and in supplying information about their mechanical properties [56] (a topic previously discussed in Volume 1 of the Nanoscience and Nanotechnology collection, in the chapter titled "Low-Dimensional Systems: Nanoparticles"), can also be used as a tool for specific recognition at the nanoscale. The same device used for surface scan analysis in AFM, the cantilever with a tip (Fig. 5.2), can be used for point measurements for the interaction and specific recognition of several analytes. Sensors based on this device are called AFM tip or probe nanosensors. They are based on the interaction response between the tip of the cantilever and a specific substrate, or even between molecules attached to their surfaces, as a function of the relative distance between them. Data from this recognition are supplied by force curves versus distance, or simply by force curves, that consist of data describing the vertical position of the tip and of the deflection of the cantilever, registered and converted into numerical information that is then translated into curves by the microscopy software. The force curves result from a technique known as atomic force spectroscopy (AFS), a dynamic and analytic technique that allows the study of the chemical and physical bonds between molecules of interest [14]. By means of AFS, it is possible to obtain data regarding the behavior of molecules present on surfaces and being subjected to elongation and torsion forces, as well as their response to indentation [14,57].

Through these principles, AFM tips can be used as sensing elements and in this context are referred to as AFM tip nanosensors. These sensors can detect both inorganic and organic analytes and can be constructed by means of chemical functionalization (chemical modification of the surface) of the AFM tips with biological elements (biomolecules) [58] or biomimetics [5], constituting nanobiosensors.

5.3.1 Theoretical Considerations Regarding Force Curves

The optimization of nanosensor recognition, as well as their construction, requires a thorough understanding of the components and variables that constitute the specific

Table 5.1   Types of Current Nanosensors and Their Main Applications

Types of Nanosensors	Main Applications
Gas nanosensors	Nonspecific sensors (global selectivity), in which the variation of electrical conductivity, voltage difference, or frequency as a function of its exposure to the monitored analyte is verified. For example, electronic nose systems to monitor the degree of ripeness of fruits and vegetables for harvest, as well as for quality control of food in general [7]. They are often used for recognizing specific chemical compounds in a complex mixture of fumes [8–12]. They may also serve as a portable breathalyzer for diagnosis of diseases [13].
Enzyme nanobiosensors and nanosensors	In general, based on interdigitated sensors, AFM *cantilevers* or tips that are chemically functionalized to bond covalently to an enzyme or antibody [14–16]. Generally, their specific interactions are reversible, which permits the continuous monitoring of samples [17]. They are tools that are very useful for the diagnosis of diseases (in situ or not) and for other analytical purposes, such as the detection of toxic analytes in complex environments, because they are highly selective and sensitive [11,18].
Fluorescent nanosensors (quantum dots)	The fluorescent properties of quantum dots are applied in procedures to detect tumors and chemical changes (intracellular) inside the body and to monitor important diseases in vivo, such as hyponatremia and diabetes [19–22]. They have potential applications in the fields of catalysis [20] and in the monitoring of the environmental quality of toxic substances [23].
Chemical nanosensors	Sensors with biometric sensory receptors, usually constructed with molecularly imprinted polymers (MIP). They are useful for applications in industrial processes for control and automation, to monitor specific analytes and toxic substances in biomedical, pharmaceutical, and environmental analyses, and in the area of defense and security [24–26].
Electric/ionic nanosensors	Sensors with electric/ionic detection, primarily field-effect transistors (FETs), offering simple and direct measurements in real time, as well as the characteristic of being portable [27,28]. Typically, these sensors use nanowires, nanoribbons, and nanotubes to measure the change in induced resistivity of a target molecule. Sensors of this nature using carbon nanotubes exhibit weaker detection limits compared to silicon nanowires; however, they are highly sensitive to the detection of deoxyribonucleic acid (DNA). The main limitation of electric/ionic nanosensors may be related to the inability to detect molecules of interest in physiological solutions, especially in those that have elevated concentrations of sodium, because these solutions may filter electric signals, thus reducing the detection sensitivity [6]. They have advantages such as nondestructive measurements of cells, allowing the constant monitoring of these analyses of various liquids and the detection of proteins, metal ions, various chemical species, and viral particles [6,29]. Recently, these sensors have also been used in research for the biomimicry of muscles and artificial heart valves [27].
Magnetic nanosensors	These are sensors that use magnetic materials and the principles of magnetism to compose sensitive sensing systems. Many types use magnetic nanoparticles (diameter 5–300 nm) or magnetic particles (diameter 300–5000 nm), usually chemically functionalized to recognize specific molecular targets [30]. There are basically three types of magnetic biosensors that use different biosensing principles, magnetic materials, and instrumentation: magnetic relaxation switches, magnetic particle relaxation sensors, and magnetoresistive sensors [30,31]. In general, magnetic nanosensors are related to the detection of analytes such as nucleotides, proteins, viruses, cancerous cells, antibodies, and bacteria [30,32,33]. They also enable several industrial applications, such as the monitoring of processes, robotics, testing of magnetic fields for machines and motors, speed detection for transmissions, and position sensors for ferromagnetic structures and in the silent motor controls in hard drives. One of the major disadvantages of these sensors in technological applications is the need to operate at relatively low temperatures.

Table 5.1   Types of Current Nanosensors and Their Main Applications (*cont.*)

Types of Nanosensors	Main Applications
Optical nanosensors/pH nanosensors/Raman and hyper-Raman nanosensors	Optical nanosensors can be defined as sensing devices for chemical products or biological events using optical signals [34,35]. The sensors known as probes encapsulated by biologically localized embedding (PEBBLE) are also in this class. These nanosensors consist of an inert matrix in which a detection component and/or an optical correlator is entrapped [34]. Fluorescence is used as a transduction method because it is highly sensitive and relatively easy to measure. Optical nanosensors present advantages such as intracellular measurements with low or no physical disturbance [34], allowing the precise determination of oxygen changes in plant cells [36] as well as optical memory devices with high data density [37]. The pH nanosensors are based on optical fluorescence, making them excellent alternatives to methods involving glass electrodes. Their applications range from marine research to toxicological tests. They have advantages such as not requiring reference elements and being immune to electrical interference [38–41]. There are also optical nanosensors based on surface enhanced Raman scattering (SERS). These nanosensors based on SERS have been used in research involving various types of cells that present strong and specific Raman spectrum signals, such as macrophages, fibroblasts, and epithelial cells, and molecules in general [42–45].
Optical fiber nanosensors	Optical fiber nanosensors are noted as an alternative to electronic nanosensors. They are of passive nature, presenting immunity to electromagnetic noise and having high durability [46]. Most optical fiber nanosensors are based on the modulation of the intensity of the light produced by the substance to be detected. The detection element can be any optical fiber or an external component attached to the fiber. In any case, one of the optical properties, such as color, index of refraction, or fluorescence must be modified as a function of the substance monitored in the environment [46,47].
Twin-action nanosensors	This type of nanosensor is usually based on polymers whose physical–chemical surface properties respond to two different stimuli at the same time. As an example, there are sensors that can identify metal ions and their temperature via colorimetric response [48]. They have the advantage that their synthesis process is cheaper, compared to the synthesis of nanoparticles and quantum dots, in addition to the fact that they are extremely sensitive to external stimuli.
Graphene nanosensors/ injectable	These are sensors of active or passive approach that usually transmit sensing data by wireless systems or by smartphone apps via fluorescence. The graphene is imprinted in tissue soluble in water (as adhesives) to be compatible with biological tissue, thus being able to offer an in situ monitoring of the level of bacterial infection and contamination [49]. Injectable nanosensors, in contrast, are more often used in the in situ monitoring of diseases such as diabetes and arterial hypertension. With them, it is possible to constantly measure glucose [50] and sodium [51] levels in blood, for example. Injectable nanosensors are similar to a tattoo and are implanted in patients in a minimally invasive form. These sensors have advantages such as being highly sensitive and allowing medical monitoring, even at a distance, thus ensuring the quality of life of the patient.

recognition system at the nanoscale as well as the theoretical principles involved. Understanding AFS principles and how functionalization affects specific interaction events of a nanosensor is of fundamental importance for the development of a recognition system with high sensitivity, selectivity, and reproducibility. Therefore, important considerations for a better understanding of the principle of operation of these nanosensors will be introduced next.

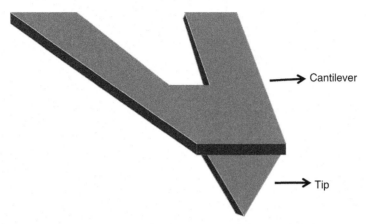

FIGURE 5.2 Device used for AFM measurements and nanoscale sensing (tip) composed of a cantilever and tip.

5.3.1.1. Basic AFS Principles

AFS is one of the most promising tools for obtaining information from (bio)recognition processes with molecular resolution. For biological studies, the force curve allows measurements of the specific interactions between molecules in physiological conditions, as well as the capacity to supply detailed information about the kinetics and the thermodynamics of a pair of interacting molecules [59]. Force curves can also complement traditional biochemical approaches, offering the possibility of elucidating nonconventional aspects related to specific recognition processes, conformational changes, and heterogeneity of the molecular population of a sample [60]. Due to these capabilities, this technique is particularly useful for investigating biological systems.

In Chapter 5 (Low-Dimensional Systems: Nanoparticles) of Book 1 from this collection, in the Biosensors section, the authors emphasize molecular recognition as the fundamental point of biosensors because of the possibility of promoting high selectivity and specificity when performing these molecular interactions. Cited examples include antigen–antibody interactions (immunosensors) and ligand–receptor interactions, which exhibit all of the requirements to be well performed with AFS technique.

In addition, the force curve is a high resolution tool based on force measurements [60]. In this technique, the tip of the cantilever is moved only in the vertical direction, z, (perpendicular to the sample plane) downward, until it is in contact with the surface of the sample. Once the interaction is detected, the tip is forced to release the sample (upward movement) until it returns to its initial position. This induced movement is reflected in the force curves and is a function of piezoelectric displacement (Fig. 5.3). From these curves it is possible to investigate how biomolecules interact, how ligand–receptor complexes are formed, and how the molecules are oriented both on the tip and on the substrate; therefore, a configuration favorable for specific recognition is used [15,61,62]. Furthermore, other information such as the length of the bonds (decoupling length) [61], the tendency and behavior of the decoupling process of the molecules, and whether the bonding

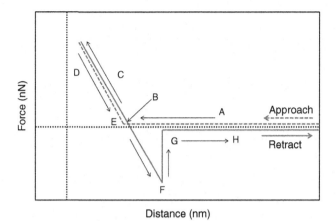

Distance (nm)

FIGURE 5.3 Typical force curve. (A–B) The cantilever tip is induced to come into contact with the sample surface. (B–C) The tip reaches the C region because of attractive forces near the sample surface. (C–D) The cantilever bends up as the tip is pushed against the sample surface. (D–E) The cantilever comes into equilibrium with the sample surface forces. (E–F) The cantilever bends from attractive forces. (F–G) Rupture moment (F) bounded by tip interaction and sample surface. (G) The cantilever is restored to the initial position. Vertical distance (F–G) represents tip-sample force adhesion (F_{adh}), and in (G–H) the tip and sample are not in contact. *Adapted from D.A. Smith, et al., Chemical force microscopy: applications in surface characterization of natural hydroxyapatite, Anal. Chim. Acta 479 (1) (2003) 39–57 [63]. Copyright (2014), with permission of Elsevier Publishing House.*

process is reversible or not can also be obtained. These investigations can mimic what occurs in biological systems, opening wide horizons for applications for new medications and environmental bioremediation systems, for example.

Force curves can be statistically significant when carried out in series and repeated n times, where $n \geq 50$ (preferably). Thus, the validation of force curves is performed by statistical approaches, evaluating the obtained data using normality tests, for example. As shown in Fig. 5.4, for an analysis to represent the sample as a whole, it is useful to construct whiskbroom scanning lines of the adhesion forces (F_{adh}), and these can be specified by the AFM topographic image previously supplied during the analysis and/or by the functions provided by the equipment software.

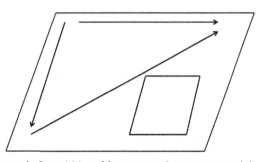

FIGURE 5.4 Arrangements (patterns) of acquisition of force curves that represent real characteristics of the sample: horizontal, vertical, and diagonal lines and assorted geometric shapes.

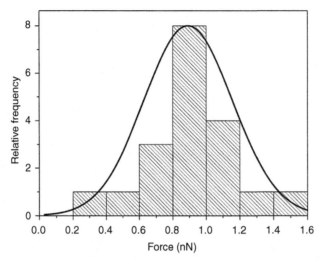

FIGURE 5.5 Histogram pattern associated with the mean relative frequency of adhesion force values measured by force curves.

After obtaining a sample that statistically represents the data, a relative frequency histogram of the adhesion force values is obtained (Fig. 5.5), allowing one to obtain a range of standard values of F_{adh}. Thus, these histograms display the statistical distribution of the adhesion forces, given by the vertical distance found in the force curve. These results are extremely useful for commercial applications using the nanobiosensor and can evaluate if an actual sample satisfies the statistical conditions previously standardized and, therefore, verify if the actual sample contains the molecule of interest (analyte).

To augment the analyses introduced, one can produce an adhesion map (Fig. 5.6). In this type of analysis, both topographic and force data are used to form the adhesion image, pixel by pixel. The adhesion map indicates whether the molecule of interest is uniformly distributed over the surface of the sample. In addition, these maps provide the incidence

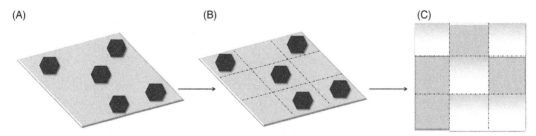

FIGURE 5.6 **Schematic representation of an adhesion map.** (A) Surface covered with target analytes; (B) surface divided in pixels; and (C) adhesion map image generated by the adhesion force calculation of punctual force curves generated. The brightest pixels *(white)* represent the specific interaction, and the darkest pixels *(gray)* are those in which the adhesion force was low and there was therefore no specific interaction.

of specific interaction events per square centimeter or square milimeter [14]. Adhesion maps have another notable application: the pattern of shining pixels for specific recognition events can establish a fingerprint for each molecule, substance, receptor, and complexation to recognize and locate specific substances on a surface [64]. Therefore, this type of analysis can be a useful tool that allows the differentiation of surfaces at the nanoscale, via the sensing of systems with AFM tips [55].

5.3.1.2 Theoretical Models for Analysis of Force Curves

Molecular recognition processes can be framed in several theories, such as "key-lock" or "induced-fit " [65,66], with the latter currently preferred. The induced-fit theory assumes that the associated molecules perform an important role in the final determination of each other's geometric forms; that is, a molecule that acts as a receptor is flexible and adaptable.

These theories influence the way in which adhesion forces are interpreted and therefore how they act in the formation of specific complexes. Thus, the detection of the force that binds inorganic or biological molecules to other surfaces, and also to each other, is an important prerequisite for the characterization of adhesion processes [67]. There are several theoretical models that accurately explain and quantify the several types of forces present in sensing systems and, therefore, in the operation of AFS [68,69]. In addition to providing data from the process of adhesion between two surfaces, these theoretical models provide information about material properties such as hardness and elasticity. In a study by Leite et al. [70], one can find a detailed description of these theoretical bases for several systems studied with AFS regarding intermolecular interactions and surface forces in vacuum, in air, and in solution. As we are discussing specific recognition in nanosensors with AFM tips, the main theories related to the contact region of the force curves only will be presented next, followed by the introduction of adhesion models for biological materials.

To understand the theories related to the contact region, it is necessary to know that the contact lines in the force curves describe the elastic/plastic behavior of the materials and its resilience. Considering first an ideal elastic material, after contact is established between the AFM tip and the sample (indentation), the latter will recover its initial shape. After its shape is recovered, the sample will exert on the tip the same force that is applied to its surface. When this occurs, the retraction curve perfectly overlaps the approximation curve (Fig. 5.7A) [70,71]. In a second scenario, when the sample exhibits ideal plastic behavior, it neither recovers its initial form nor can it exert a force on the tip. In this case, the approximation and retraction curves of the force curves do not overlap (Fig. 5.7B). Most samples exhibit mixed behavior and therefore usually present force curves without the curves overlapping [71–73]. This difference in the curve shift is called hysteresis, in which the force used in the retraction curve is smaller than the force used in the approximation curve (Fig. 5.7C).

In addition to interpreting elastoplastic behavior, it is necessary to be familiar with the geometry of the tip because it is related to the contact region with the sample (Fig. 5.8).

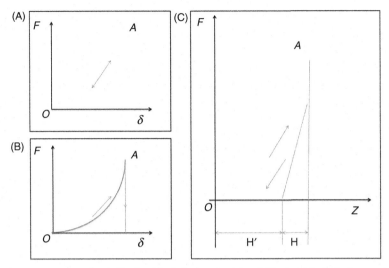

FIGURE 5.7 (A) Force curves versus depth (δ) of an ideal elastic material and (B) of an ideal plastic material. (C) Force–depth curve of an elastoplastic material. *H'* is the indentation depth at which the force equals zero, and *H* is the distance over which the sample recovers. *Adapted from B. Cappella, G. Dietler, Force–distance curves by atomic force microscopy, Surf. Sci. Rep. 34 (1–3) (1999) 1–104 [68]. Copyright (2014), with permission of Elsevier Publishing House.*

The determination of the spring constant of the sample (k_s) depends on this contact area, Young's modulus (E), and Poisson's ratio (ν), as shown in Eq. 5.1 [68]:

$$k_s = 2a\frac{E}{1-\nu^2} \tag{5.1}$$

where a is the contact radius.

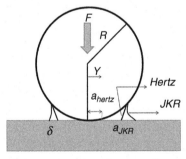

FIGURE 5.8 Elastic sphere deformation scheme over a plain surface according to the Hertz and Johnson–Kendall–Roberts (JKR) theories. The profile of a spherical tip assumed using the Derjaguin–Muller–Toporov (DMT) theory is equal to the Hertz theory, in which *F* is the force applied, *R* is the sphere radius, *y* is the distance from the center of the contact area, δ is indentation depth, and a_{hertz} and a_{JKR} are the contact radius from the Hertz and JKR theories. *Adapted from B. Cappella, G. Dietler, Force–distance curves by atomic force microscopy, Surf. Sci. Rep. 34 (1–3) (1999) 1–104 [68]. Copyright (2014), with permission of Elsevier Publishing House.*

5.3.1.2.1 HERTZ AND SNEDDON THEORY

To quantify the mechanical properties of biological samples, the most widely used theory is the Hertz theory. This model describes the elastic deformation of two plane, rigid, and homogeneous surfaces arising from contact with each other but does not take into account superficial and adhesion forces. Regarding force curves, the forces measured are dominated by the elastic properties because viscosity presents a minimal contribution over time. Therefore, given a sphere of radius R that comes into contact with a surface by a force F, the adhesion force F_{adh}, the contact radius a, the contact radius with zero load a_0, the deformation of the spherical tip δ, and the pressure P are given, respectively, by Eqs. 5.2–5.6:

$$F_{adh} = 0 \tag{5.2}$$

$$a = \sqrt[3]{\frac{RF}{K}} \tag{5.3}$$

$$a_0 = 0 \tag{5.4}$$

$$\delta = \frac{a^2}{R} = \frac{F}{Ka} \tag{5.5}$$

$$P(x) = \frac{3Ka\sqrt{1-x^2}}{2\pi R} = \frac{3F\sqrt{1-x^2}}{2\pi a^2} \tag{5.6}$$

where $x = y/a$, y is the distance from the contact center, and the reduced Young's modulus K is given by Eq. 5.7 [68]:

$$\frac{1}{K} = \frac{3}{4}\left(\frac{1-v^2}{E} + \frac{1-v_i^2}{E_i}\right) \tag{5.7}$$

However, the Hertz theory assumes that the characteristics of the materials in contact are isotropic and homogeneous and also that the contact does not create friction because its geometry has axial and continuous symmetry (zones with high and low exerted superficial forces) [68]. However, in a large portion of the AFM/AFS experiments, the tip has a higher hardness than the surface of the sample, and the deformations resulting from this fact, both on the tip and on the sample, do not exhibit axial or continuous geometry. Therefore, one can conclude that the Hertz theory is experimentally valid only for contact events that create small deformations. In situations where large deformations occur, friction and adhesion (indentation), for example, must be taken into account. In these cases, the Hertz theory is not valid, and a modification of it must be used: the Sneddon model [74].

In the Sneddon model, the force F exerted on the surface of the sample and the deformation δ resulting from the indentation are given by [75,76]:

$$F = \frac{3}{8}K\left[a^2 + R^2 \ln\left(\frac{R+a}{R-a}\right) - 2aR\right] \tag{5.8}$$

$$\delta = \frac{1}{2}a\ln\left(\frac{R+a}{R-a}\right) \tag{5.9}$$

The deformation and the contact force are calculated for heterogeneous and asymmetric indentation profiles (depth) [68,77]:

$$\delta = \int_0^1 \frac{f'(x)}{\sqrt{1-x^2}}\,dx \tag{5.10}$$

$$F = \frac{3}{4}Ka\int_0^1 \frac{x^2 f'(x)}{\sqrt{1-x^2}}\,dx \tag{5.11}$$

where $f(x)$ is the function that describes the indentation profile.

Both models are used mainly to estimate the value of Young's modulus for biological cells via force curves. This type of analysis has become increasingly important in the medical and biological fields for the nanomechanical analysis of different groups of cells, especially to differentiate between healthy and sick cells.

5.3.1.2.2. BRADLEY, DERJAGUIN–MULLER–TOPOROV, AND JOHNSON–KENDALL–ROBERTS THEORIES

This section presents theories that take into account the effects of surface energy on contact deformation: Bradley, DMT, and JKR [68].

Bradley Theory: In 1932, the author theoretically modeled the interaction of two rigid spheres, for which the total force between these spheres is given by Eq. 5.12 [78] (z_0, separation equilibrium; R, reduced radius of the spheres $(1/R_1 + 1/R_2)^{-1}$; and W, work of adhesion on contact) [68]:

$$F(z) = \frac{8\pi WR}{3}\left[\frac{1}{4}\left(\frac{z}{z_0}\right)^{-8} - \left(\frac{z}{z_0}\right)^{-3}\right] \tag{5.12}$$

DMT Theory: In addition to the deformation of the elastic sphere (Hertz), there is also the action of an external force (F) as well as of the forces between the two bodies outside the contact region; such forces produce a finite area of contact. There may be an increase or a decrease of the contact area when forces are applied: an external force implies an increase in contact area, whereas a negative force implies a decrease of this area until zero is reached, when, to leave the contact state, it reaches its maximum value [68]. This theory can be applied to systems with low adhesion that have tips with a small radius [68].

Eqs. 5.13–5.17 correspond to Hertz and Sneddon equations, however, with minimization of the sum of the elastic and surface energies [70]:

$$F_{\text{adh}} = 2\pi RW \tag{5.13}$$

$$a = \sqrt[3]{(F + 2\pi RW)\frac{R}{K'}} \tag{5.14}$$

$$a = \sqrt[3]{\frac{2\pi W}{K}R^2} \tag{5.15}$$

$$\delta = \frac{a^2}{R} \tag{5.16}$$

$$P(x) = \frac{3Ka\sqrt{1-x^2}}{2\pi R} = \frac{3F\sqrt{1-x^2}}{2\pi a^{\text{W}}} \tag{5.17}$$

JKR Theory: Only short-range forces inside the contact region are considered [70]. Its behavior incorporates hysteresis, and it is appropriate for highly adhesive systems with tips of low rigidity and large radius [79]. Thus, the presented JKR equations, 5.18–5.22, are equivalent to Eqs. 5.2–5.5 from Hertz and Sneddon:

$$F_{\text{adh}} = \frac{3}{4}\pi RW \tag{5.18}$$

$$a = \sqrt[3]{\frac{R}{K}\left[F + 3\pi RW + \sqrt{6\pi RWF + (3\pi RW)^2}\right]} \tag{5.19}$$

$$a_0 = \sqrt[3]{\frac{6\pi R^2 W}{K}} \tag{5.20}$$

$$\delta = \frac{a^2}{R} - \frac{2}{3}\sqrt{\frac{6\pi Wa}{K}} \tag{5.21}$$

$$P(x) = \frac{3Ka}{2\pi R}\sqrt{1-x^2} - \sqrt{\frac{3KW}{2\pi a}}\frac{1}{\sqrt{1-x^2}} \tag{5.22}$$

It is worth mentioning that there are several theoretical models that are more appropriate for biological systems; they feature adjustments related to the concentration of, kinetics of, and the complexes formed by organic molecules such as, for example, the Bell and Bell–Evans models [80–82] and the extended Bell model [83].

5.3.2 Chemical Force Microscopy

5.3.2.1 Introduction
Chemical force microscopy (CFM) is a technique used to chemically map the surface of a sample at the nanoscale (chemical nanocharacterization) by means of chemically

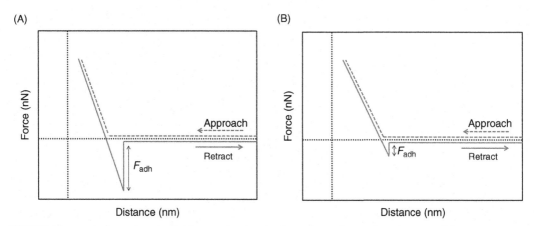

FIGURE 5.9 Interaction among molecules is strongly related to its conformation on the substrates. If the correct conformation is reached, the molecules will have spatial freedom to move; thus, they will reach the expected specific sensing active sites and will therefore achieve specific binding, and the adhesion force (F_{adh}) will be high (A). Otherwise, binding will not be specific, and the adhesion force will be low (B).

modifying the AFM tip, making it sensitive to specific interactions. This technique not only allows the modulation of the interaction between functional groups by using chemically modified cantilevers, but it also allows the detection and quantification of these interactions [84]. This chemical modification, termed functionalization, ensures that the cantilever exhibits selectivity and sensitivity for interactions at the molecular level, thereby allowing detailed mapping of the chemical surface at the nanoscale [85]. The functionalization of the cantilevers has been extensively used in the development of biosensors (nanobiosensors) sufficiently powerful to detect a single molecule. CFM is an appealing tool because it presents fast response and selectivity [85] and is used both for characterization and for applications involving the detection of various types of analytes [86–90].

5.3.2.2 Methodology for the Functionalization and Characterization of AFM Probes
Several strategies have been developed to immobilize molecules on the tip of a cantilever and on the substrate for the purpose of favoring the correct interaction between them (molecule architecture/geometrical *design*), to reduce multiple events, and to differentiate specific (Fig. 5.9A) and unspecific (Fig. 5.9B) recognition events within the studied system.

Regarding nanosensors, it is well known that the devices used in AFS (cantilevers, AFM tips) are based on silicon technology (e.g., SiO_2, Si, and Si_3N_4) [91] and that they do not have, on their own, specific recognition characteristics. To enable such materials to be highly selective and sensitive sensors, their surfaces are usually chemically modified (functionalized) with organosilanes (such as 3-aminopropyltriethoxysilane, APTES), which have chemically active groups to provide an appropriate interface between the silicon transducer and the immobilized molecules. An important element of chemical functionalization of

AFM tips is the *linker*, that is, a flexible molecule between the modified surface and the biomolecule that will be immobilized at its end to perform detection. Some examples of linkers are glutaraldehyde, alkanethiols, and polyethylene glycol (PEG). The functions of these linkers in a nanoscale sensing system are to give mobility and structural orientation freedom to the molecules; to help differentiate between specific and unspecific recognitions; to prevent molecule distortions and denaturation; to contribute to the uniform orientation of surface molecules; and to act as a bridge between the organic surfaces and the biomolecules [60]. All of these functions significantly contribute to prioritizing the specific bonds.

The immobilizations of these chemical groups to surfaces such as AFM tips are similar to a group of assorted molecular pieces in which the fitting of blocks (molecules) can occur via cross-linking or physical adsorption, in which the environment and the surfaces are controlled to immobilize molecules through the natural forces present, or in which the fitting can occur through the formation of covalent bonds, which is the methodology most often used, in which surface modification is reached by chemical reactions [50,61,92]. Concerning the chemical modification of surfaces for sensing systems based on AFM tips, it is necessary to know the geometry of the tip used, as previously noted, as well as its elastic constant [58]. Additionally, it is necessary to adapt the choice of molecule that will be immobilized to the tip so that it can specifically recognize the analyte [15,61,93]. For example, if the developed sensor must detect molecules of a specific antigen, the biomolecule bonded to the AFM tip must be its specific antibody. The use of molecular modeling together with recognition systems at the nanoscale is fundamental because this tool allows one to predict the types of interactions and structural conformations, binding energies, and electrical properties, among other parameters, aiding the understanding of the molecular behavior of the system [58].

Fig. 5.10A illustrates the tip of a cantilever coated with an organic monolayer whose terminal groups consist of methyl. Fig. 5.10B, in contrast, shows the tip of a cantilever functionalized with biotin, which is connected to the silicon nitride surface by a layer of bovine serum albumin (BSA). This functionalization renders the cantilever an extremely selective biosensor for detecting streptavidin molecules.

The characterization of the functionalized tip is usually necessary to ensure that the molecules/biomolecules on the surface are oriented appropriately to interact with the substrate, in addition to validating the functionalization process. For this purpose, several techniques can be used: imaging of the tip using scanning electron microscopy (SEM) [94,95] and AFM [96–98], infrared reflection-absorption spectroscopy [94,97], Fourier transform infrared spectroscopy (FTIR) [99,100], X-ray photoelectron spectroscopy (XPS) [96,98], fluorescence spectroscopy [101], and surface plasmon resonance [96,102] and contact angle measurements [97,98].

Additionally, other less common techniques can also supply information about the functionalization. For example, Volcke et al. [98] used techniques such as ellipsometry, chemical force titration, and dual-polarization interferometry.

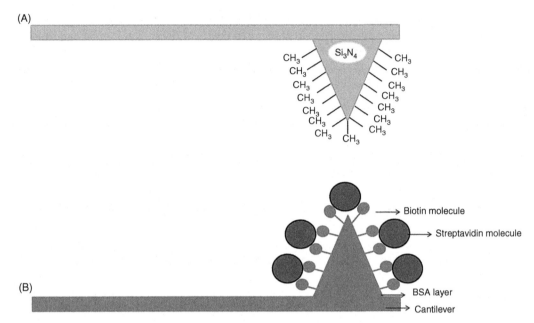

FIGURE 5.10 (A) Schematic representation of a silicon nitride tip covered by a thin organic methyl-terminated layer. (B) Schematic representation of a silicon nitride tip covered by biotin, which converts it to a nanobiosensor to detect streptavidin. *Part A, Adapted from J.E. Headrick, C.L. Berrie, Alternative method for fabricating chemically functionalized AFM tips: silane modification of HF-treated Si3N4 probes, Langmuir 20 (10) (2004) 4124–4131 [94]; part B, adapted from M. Micic, et al., Scanning electron microscopy studies of protein-functionalized atomic force microscopy cantilever tips, Scanning 21 (6) (1999) 394–397 [95].*

5.4 Applications for AFM Tip Sensors

AFM tip nanosensors are frequently used to study various types of biological structures and processes, such as antigen–antibody bonds, ligand–receptor interactions [55], protein unfolding, molecular stretching [61], and the detection of molecules of interest for environmental [93,103], industrial [104], medical [105], military, and safety monitoring [26].

These devices can be used, for example, to detect agrochemical molecules at low concentration levels for environmental and food safety [93]. This technique is a viable and more sensitive and selective alternative when compared to the analytical methods used recently to detect agrochemicals, such as mass spectroscopy [14].

Silva et al. [93] describe the development of an AFM tip nanosensor devised for the detection of the herbicide metsulfuron-methyl via the formation of a specific bond with the acetolactate-synthase (ALS) enzyme. This sensor is based on the mechanism of inhibition of the action of the enzyme by the agrochemical and relies on a specific bond that can be mimicked by AFM tip sensors. In this same study, the authors demonstrated that the process of molecule immobilization (functionalization) was essential for the successful operation of the sensor. In addition, Franca et al. [58] proposed a theoretical sensing system based on the interaction of the acetyl-coenzyme-A-carboxylase (ACCase) enzyme for

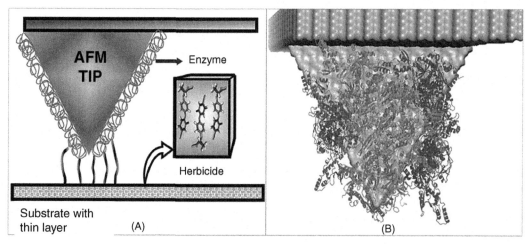

FIGURE 5.11 Schematic representation of an AFM tip enzymatic nanobiosensor. (A) Schematic of the detection principle: enzyme molecules were immobilized on the AFM tip in solution as a result of electrostatic attraction. Enzyme-herbicide interaction was obtained through force curve measurements of the system. (B) Immobilized, oriented enzyme molecules over a functionalized AFM tip surface [58]. *Reproduced from Franca et al., Designing an enzyme-based nanobiosensor using molecular modeling techniques, Phys. Chem. Chem. Phys. 13 (19) (2011) 8894–8899 [58]. With permission of the European Society for Photobiology, the European Photochemistry Association, and The Royal Society of Chemistry.*

the detection of the agrochemical diclofop-methyl, based on the same specific inhibition principle presented by Silva et al. [93] (Fig. 5.11). These techniques have also been used in research regarding autoimmune and autoneurodegenerative diseases such as multiple sclerosis and neuromyelitis optica [106]. The group has recently created the research field termed nanoneurobiophysics and has pioneered the development of AFM tip and microcantilever nanosensors for the detection of enzyme-inhibiting herbicides and demyelinating diseases [106–112].

5.5 Microcantilever Sensors

Recent advances in microelectromechanical systems (MEMS) have facilitated the development of sensors involving mechanical transduction or other mechanical phenomena [113]. One of the advantages of MEMS systems, including AFM microcantilevers, is the capacity to adapt the size and the structure of the device, in addition to its low cost. The microcantilever used in AFM consists of a thin and flexible beam with a thin pyramidal needle at the bottom, whose radius of the vertex at the end of the tip is a few nanometers. The microcantilevers can be fabricated with one or more silicon or silicon nitride beams with an average length between 100 and 500 μm and thickness between 0.5 and 5 μm. They may have a "V" (triangular) or a "T" (rectangular) shape [114].

To act as high quality sensors, the microcantilevers must be covered (i.e., functionalized) with a sensitive layer specific to the molecules of interest. The adsorption of these molecules to the beam creates excess surface tension, resulting in the curvature of the

FIGURE 5.12 Schematic representation of the microcantilever operation modes [118], (A) resonance frequency change (f_{res}) resulting from an effective mass change and (B) surface tension resulting from molecular adsorption [119]. *Reproduced from A chemical sensor based on a microfabricated cantilever array with simultaneous resonance-frequency and bending readout, 77/1-2, M. Battiston, J.P. Ramseyer, H.P. Lang, M.K. Baller, Ch Gerber, J.K Gimzewski, E Meyer, H.-J Güntherodt, 122–131, Copyright (2014), with permission of Elsevier Publishing House.*

microcantilever [115] if it occurs preferentially on one of the sides of the surface. Selective adsorption is usually controlled by functionalizing the top surface of the microcantilever only [116,117].

5.5.1 Operation Modes of Microcantilever Sensors

There are two basic operation modes for microcantilever sensors, termed the static and dynamic modes. The static mode uses the change of the physical deflection of the microcantilever, whereas the dynamic mode uses the change in resonance frequency of the microcantilever resulting from the increase in mass adsorbed on the surface [115]. These modes differ because of the principles of the transduction, functionalization, and method of detection [116,117]. Fig. 5.12 shows the schematic representation of both operation modes for the microcantilevers.

In the static deflection mode, the difference between the surface functionalized on only one side of the microcantilever (active side) and the surface that is not functionalized (passive side) causes a change in surface tension, causing the deflection of the microcantilever [116]. The deflection is a function of the spring constant (k), the properties of the material, and the geometry of the microcantilever. Stoney's equation allows one to obtain the theoretical ratio of the response of the microcantilever (Z_{max}) to the tension difference between the active and the passive surfaces [115].

$$Z_{max} = \frac{3l^2(1-\upsilon)}{Et^2}\Delta\sigma \tag{5.30}$$

where υ is the *Poisson* ratio ($\upsilon = 0.24$ for a rectangular microcantilever), $\Delta\sigma$ is the surface tension induced by the analyte ($\Delta\sigma_{active\ side} - \Delta\sigma_{passive\ side}$ (N·m^{-1})), E is Young's modulus, t is the thickness, and l is the length of the cantilever. This relation is valid only for thin films (20% of the thickness of the microcantilever) [64].

According to Lang and Gerber [116], the adsorption of molecules on the surface (functionalized on one side only) will result in a deflection upward or downward because of the change in surface tension between the functionalized and nonfunctionalized sides. If the opposite situation occurs, that is, when both sides of the microcantilever (top and bottom) are functionalized, both sides will be exposed to compressive stress predominantly on the bottom surface of the microcantilever and also to tensile stress on the top surface. For this reason, it is extremely important to functionalize the surface appropriately so that the nonfunctionalized side does not respond when exposed to an analyte or in deflection measurements for the microcantilever sensor in the static mode.

In the dynamic mode, the resonance frequency (f_{res}) of an oscillating microcantilever is constant as long as the elastic properties remain unaltered during the adsorption or desorption of the molecules and the damping effects are insignificant [120]. These conditions can be attained as long as the surface molecules are in dynamic equilibrium with the molecules in the environment. The change in mass during the molecular adsorption can be related to the change of f_{res} using Eq. 5.31 [121]:

$$\Delta m = \frac{k}{4\pi^2 v}\left(\frac{1}{f_{res(0)}^2} - \frac{1}{f_{res(1)}^2}\right)$$ (5.31)

where $f_{res(0)}$ is the initial resonance frequency and $f_{res(1)}$ is the resonance frequency after mass is added.

The deflection and f_{res} can be accurately monitored by using a variety of methods, of which optical deflection, piezoresistivity, capacitance, and electron tunneling are especially notable [121,122]. The simplest way of evaluating the deflection is the optical method, in which a laser diode is focused at the extremity of the microcantilever and the reflection of the laser is monitored by using a photodetector. The minimum deflection detected by this method is of approximately 0.1 nm. Its advantages are simplicity of operation, linear response, and the lack of requirement for electrical connections [123]. However, the calibration at each measurement is indispensable to register the signal in terms of the real deflection of the microcantilever.

In this context, the development of these devices based on the microcantilevers used in AFM has been very attractive in various application fields because recent advances in design and in the development of these sensors make them able to detect very small mechanical deflections. This high sensitivity is reached because the spring constant of a microcantilever is between the order of 10^{-3} and 10^1 N·m^{-1}, thus allowing the detection of very small forces (10^{-12}–10^{-9} N) [124]. These factors result in a fast response time, lower manufacturing costs, possibility of constructing a sensor array with small dimensions, and the ability to explore microenvironments [125].

The difficulty of comparison between the sensitivities of microcantilever sensors is compensated by the high level of detection, on the order of parts per trillion (ppt) [126–128], rarely reached by other types of sensors. It is emphasized that the sensitivity of microcantilever sensors is directly related to the sensitive layer.

Some aspects must be considered when choosing the microcantilever to be used as a sensor [122]; for example, the material must contain low internal damping (Young's modulus), and the geometry must provide an elevated quality factor (*Q*); the surface of the microcantilever must reflect the optical reading with high quality, and so it cannot be rough

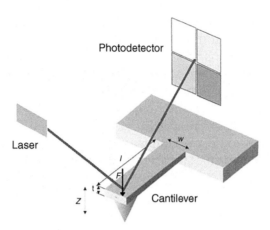

FIGURE 5.13 Schematic illustration of a microcantilever deflection (*Z*), in which *F* is the applied force and *t*, *w*, and *l* are the dimensions of the microcantilever [55]. *Reproduced from Atomic Force Microscopy as a Tool Applied to Nano/Biosensors, Sensors, 12/6, Clarice Steffens, Fabio L. Leite, Carolina C. Bueno, Alexandra Manzoli and Paulo Sergio De Paula Herrmann, 8278–8300.*

because it would disperse the light in all directions; and the coating (functionalization) of the surface must be specific for the desired application.

5.5.2 Theoretical Considerations

One of the simplest mechanical structures is a microcantilever, in which one end is free and the other end is fixed to a support. When a force F (N) perpendicular to the laser beam is applied to its extremity, a deflection Z (nm) will occur in the material with Young's modulus E (N·m^{-2}) [115] (Fig. 5.13).

The theoretical mass-spring mechanism can be used to describe the resonance frequency (f_{res}) for a rectangular microcantilever with spring constant (k) [121]:

$$f_{res} = \frac{1}{2\pi}\sqrt{\frac{k}{m}} \tag{5.32}$$

where m is the mass of a microcantilever (kg) and can be expressed as:

$$m = \rho.t.l.w \tag{5.33}$$

where ρ is the density of the material (kg·m^{-3}), and t, w, and l denote the thickness, width, and length of the microcantilever, respectively. Microcantilevers with lengths in the micrometer (μm) scale and with width and thickness dimensions in the nanometer (nm) scale vibrate at resonance frequencies (Hz) that are measurable because of thermal and environmental noise [121].

The expression of the spring constant (k) is given as a function of the geometric dimensions and parameters of the material [129]:

$$k = \frac{Ewt^3}{4l^3} \tag{5.34}$$

where E is Young's modulus ($E = 1.3 \times 10^{11}$ N·m^{-2} for silicon in the <100> plane of its crystalline structure) [121]. The spring constant of a microcantilever is related to Young's modulus, which depends on the properties of the material [130], thus defining the sensitivity of the microcantilever [131].

Microcantilever sensors can be functionalized with different types of materials, such as the inorganic and organic types, which enables the creation of bimaterial structures (silicon cantilever plus the sensitive layer) [132]. Among the inorganic sensitive layers most often used are metals such as gold, which is often used for the detection of mercury vapor [124].

To increase the functionality of these sensors, several types of organic layers have been researched for functionalization, such as biomolecules (DNA, RNA, specific antibodies, polypeptides, and nucleotides) [133], lipid layers [134], polymers [117], nanocomposites [135], and hydrogels [136], among others, for the detection of various physical, chemical, and biological stimuli [64]. Polymers are also often used to functionalize sensors for the detection of volatile compounds because the adsorption is completely reversible [137].

5.5.3 Applications for Microcantilever Sensors

In a study using microcantilever sensors functionalized with a conducting polymer, Steffens [137] observed that these devices present a detection limit for relative humidity (RH) of 1 ppm$_\text{v}$. In addition, functionalized microcantilever sensors presented a fast response time, were repetitive during several RH cycles, and exhibited durability longer than 6 months. However, nonfunctionalized sensors were found to not interact with RH (%), that is, they did not provide a deflection response. In this same report [137], the deflection response of the microcantilever was essentially linear regarding the relative water vapor content at 20°C. The sensitive layer deposited on the surface of the microcantilever caused changes in the surface tension corresponding to the changes in the water vapor. A schematic demonstration of this research is illustrated in Fig. 5.14, in which one can observe the deflection of the silicon microcantilever sensors functionalized with polyaniline compared to a nonfunctionalized microcantilever when exposed to water vapor.

Tests related to the use of the cantilever as a humidity sensor were carried out by Singamaneni et al. [64], whose research evaluated the use of flexible silicon structures covered with methacrylonitrile (bimaterial). The same authors correlated the deflection of the bimaterial with humidity and observed that the sensitivity change obtained with the cantilever was of the order of ±10 ppb, with response intervals in the range of milliseconds. Fig. 5.15 shows the deflection of the silicon cantilever functionalized with a layer of polyimide; according to another report [138], the observed upward deflection is due to the shrinkage of the polymer after curing. During the operation, when the cantilever comes into contact with the humidity in the environment, the polyimide layer swells, thus inducing a compressive stress so that the cantilever tends to return to its horizontal position. These examples show the great possibility of using cantilever sensors for measuring vapors and volatile organic compounds in food, with low detection limits and high selectivity.

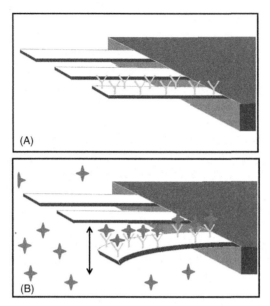

FIGURE 5.14 Schematic representation of nanomechanical microcantilever responses in two distinct moments. (A) System consisting of microcantilevers (one functionalized microcantilever while the others are without any covering) in PBS solution and (B) when a specific antigen is added in the solution, binding to the functionalized microcantilever and leading to deflection.

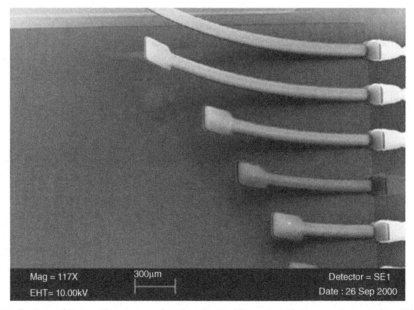

FIGURE 5.15 Deflections of the cantilever sensor functionalized with a polyimide layer and exposed to different relative humidity values [138]. *Reproduced from Simulation of capacitive type bimorph humidity sensors, 10, J Fragakis, S Chatzandroulis, D Papadimitriou and C Tsamis, 305–308, Copyright (2014), with permission of © IOP Publishing. Reproduced with permission. All rights reserved.*

The first studies involving microcantilever chemical sensors are related to the evaluation of physical and environmental parameters such as temperature, pressure, and humidity. Thundat et al. [139] reported the deflection of silicon nitride microcantilevers functionalized with 4 nm of chromium and 40 nm of gold, functionalized with 5–13 nm of nonfunctionalized aluminum, and exposed to different values of relative humidity. The deflection was essentially linear and reversible for the microcantilevers coated with gold, negligible for those without coating, and extremely sensitive for those functionalized with aluminum.

The absorption of water vapor by the sensitive layer was investigated by Thundat et al. [124], in which the authors functionalized one of the sides of two microcantilevers, one with a thin gelatin film and the other with phosphoric acid, both hygroscopic materials. The results indicated that the resonance frequency decreased for the microcantilever functionalized with phosphoric acid and increased for the microcantilever functionalized with thin gelatin film, when exposed to different values of relative humidity.

Ma et al. [140] described the use of silicon microcantilevers covered with a layer of polyimide for the measurement of humidity and gas flow rate. The change in humidity induced the microcantilever to suffer a deflection, creating a change in the measured resistance. This occurs because when the air flow passes over the microcantilever, a small deformation occurs, and the change in air flow rate can be determined by the change in resistance caused by the deflection of the laser beam.

Tests related to the use of microcantilevers as humidity sensors were also carried out by Singamaneni et al. [134], who evaluated the use of silicon flexible structures covered with methacrylonitrile (bimaterial). The results indicate that the change in sensitivity was on the order of ±10 ppb, with response intervals in the range of milliseconds.

The surfaces of the microcantilevers can also be covered with different polymers to analyze a mixture of volatile organic compounds [119,141,142]. The exposure of the polymers to the gases causes the polymeric matrix to swell because of the analyte adsorption process, which changes the density and the mass of the polymer. Lahav et al. [143] reported the development of an electro-oriented device with the deposition of polyaniline by redox activation on one of the sides of the microcantilever. The degree of deflection was controlled by repulsive electrostatic interaction by applying a voltage from −0.5 to 0.8 V. The oxidation of the polymer on top of the microcantilever caused electrostatic repulsion and the swelling of the polymer chains, thus resulting in deflection. The shrinkage of the polymer released the tension applied on it; as a result, the microcantilever returned to its original position.

Wu et al. [144] evaluated the use of microcantilevers functionalized with specific antibodies of prostate-specific antigen (PSA) for the diagnosis of prostate cancer via the optical detection method. When the functionalized microcantilevers were placed in contact with blood samples from patients with prostate cancer, an antigen–antibody complex was formed, deflecting the cantilever because of the adsorbed mass related to the antigen molecules. According to the authors, these results were more sensitive than conventional biochemical techniques for the detection of PSA.

Arrays of silicon nitride nanocantilevers were used to detect pieces of DNA with 1587 base pairs via resonance frequency change that resulted from the mass increase of the device because of analyte binding [145]. The reading of the frequency for the cantilevers was performed by means of laser beam scanning. To calibrate the response of the nanosensors with high sensitivity and precision (to count a small number of bonded molecules), researchers inserted gold nanodots on the cantilevers, which captured the modified sulfide agents from the double-stranded DNA. In this way, molecules were identified with a mass of 0.23 ag (1 ag = 10^{-18} g), indicating the detection of a simple DNA molecule, with 1587 base pairs.

Microcantilever sensors have also been applied in research involving bacterial resistance in infections, which involve several forms of the disease, ranging from inflammation to food poisoning [146]. Conventional methods for tests of resistance to antibiotics, such as the agar dilution method described by the National Committee for Clinical Laboratory Standards, require analysis times from 16 h up to 2 days [147]. In a study by Gfeller et al. [148], the authors evaluated the use of a cantilever array sensor functionalized with agarose as a nutrient medium for the detection of *Escherichia coli*. The sensors were exposed to suspensions of *E. coli* with growth in Luria-Bertani broth containing 10 µg·mL^{-1} tetracycline or kanamycin for 10 min. The resonance frequency was monitored to evaluate the increase in mass of the cantilever arrays, with an observed mass equal to 50 pg·Hz^{-1} (this mass corresponds to a sensitivity of approximately 100 *E. coli* cells). The sensors were able to detect growth during 1 h. Thus, this specific system of cantilever array sensors presents a large potential for applications in methods for fast detection of bacterial growth.

Microscopic fungi are human pathogens and serious contaminants in food production and industry [149]. Therefore, the fast detection of these species of yeasts and molds is of great importance for medical diagnosis and food quality control, among other applications. Nugaeva et al. [150] evaluated the use of cantilever arrays functionalized with concanavalin A, fibronectin, and immunoglobulin G for the selective immobilization and fast quantitative detection of fungal spores. Two morphological forms were evaluated, the fungus *Aspergillus niger* and the yeast *Saccharomyces cerevisiae*, and it was possible to detect with a sensor array a range of 10^3–10^6 colony forming units (CFU)·L^{-1}. The measurement was proportional to the mass of individual fungal spores and can be used to evaluate the levels of spore contamination, with a response time of approximately 4 h, that is, 10 times faster than conventional analysis techniques.

Therefore, cantilever sensors and nanosensors have countless applications in various areas, thus presenting large potential to be explored as devices that are extremely sensitive, selective, and simple, have low cost, are able to detect a low quantity of analyte (in µL), and have fast analysis response [64]. Compared to other techniques for the detection of traces of substances in ppb and in ppt, such as high performance liquid chromatography (HPLC), thin layer chromatography (TLC), gas chromatography (GC), among others, analysis by these microdevices is much simpler and requires less time and instrumentation [133].

Cantilever sensors can consist of a single sensor element or an array of sensor elements. Simple cantilevers present some disadvantages because they are susceptible to changes in deflection because of temperature changes or chemical interactions with the environment (if operating in a liquid medium). Therefore, it is important to use a reference cantilever to exclude influences on deflection [115].

By using microcantilever arrays, it is possible to use a reference cantilever, whereas the other cantilevers are functionalized with the molecules of interest (sensitive layer with affinity), for the detection of the target molecules, thus evaluating multiple signals [115]. This type of array can also be expanded to integrated multisensors, supplying a viable platform for the development of high performance electronic noses [129]. The electronic nose was developed in the early 1980s [151], and the most accepted definition is given by Gardner et al. [152], who defined it as an "instrument which comprises an array of electronic chemical sensors with partial specificity and appropriate pattern recognition system, capable of recognizing simple or complex odors". Since then, the analysis of volatile compounds has grown in prominence, and several studies have been dedicated to the improvement of odor measurement. Electronic noses can be applied to several areas, such as in agriculture for the detection of diseases in cultures, for the identification of insect infestations, and for the monitoring of food quality [7,153]. The responses obtained by electronic noses are typically formed by comparison of the profiles of the volatile organic compounds (VOCs) in the sample and in a reference profile, such as VOCs emitted by healthy and sick plants/fruits [154]. Several studies are being performed using electronic noses for the detection of the ripening degree of fruits (banana) [153], for monitoring changes of the odor from apples during storage [155], and to evaluate the post-harvest of fruits and vegetables [156].

Polymers have often been used to functionalize these different cantilevers with distinct sensitive layers because they present high sensitivity [132]. This sensitivity is obtained through the surface tension of the sensor and can be a result of the swelling of polymer chains because of the absorption of analytes [157]. Research performed with sensitive polymer layers has addressed the detection of humidity, pH changes, alcohols, alkanes, ethers, aromatics, esters, ketones, perfume essences, beverage flavors [115,117,158,159], and also as the inert layer in reference cantilevers [160,161].

5.6 Challenges and Tendencies

From a marketing viewpoint, both academia and industry must face the challenge of reducing the costs of the materials and devices used to build a sensor/nanosensor, with the purpose of making the products that use this technology compatible with most of the technologies currently used. Additionally, there is a need to develop a process of mass production that is efficient as well as to identify key elements that lead to strategic options for technological connection with current and future market conditions. There is also the need to invest in Research & Development, from basic to specialized research, to create more knowledge, information and, consequently, sensing products that are broadly characterized, reliable, useful, and able to perform multiple analyses. In addition, nanosensors must satisfy certain requirements, such as early and safe detection with high specificity and selectivity, simplicity, and the ability to be connected to a minimally invasive or even noninvasive procedure of detection (for the nanobiosensor case) [6].

It is worth noting that there is also a social and ethical challenge related to sensors built at the nanoscale. With these sensors, the capacity to measure extremely small quantities of atmospheric pollutants or toxic materials in diverse environments brings questions and dilemmas of risk thresholds. Additionally, nanosensors for medical applications will not only aid diagnosis and treatment of patients but also enable prediction of a healthy or nonhealthy patient profile. This capability will have a direct impact on how health insurance companies will act because they may use this information to grant or deny coverage of a service. In the food industry, the use of nanosensors can lead to high thresholds of food safety and quality. Finally, when applied to information technology, nanosensors may significantly increase the efficiency of devices for surveillance and monitoring, which are directly coupled to issues related to privacy and security.

Given that the advances in the detection and sensing of different biological and chemical species (organic and inorganic) can be achieved with greater capacity and accuracy through the use of nanosensors, the social and ethical mechanisms involved can be positively designed with accurate and reliable information, thus heading to a future without negative implications but offering great discoveries and innovations.

List of Symbols

$\Delta\sigma$	Surface tension induced by the analyte
Δm	Change in mass
δ	Deformation of the spherical tip
a	Contact radius
a_0	Contact radius with zero load
E	Young's modulus
F	Force
f_{res}	Resonance frequency
$f_{res(0)}$	Initial resonance frequency
$f_{res(1)}$	Resonance frequency after mass is added
K	Reduced Young's modulus
K	Spring constant
k_s	Spring constant of the sample
P	Material's density
P	Pressure
R	Sphere radius
ν	Poisson's ratio
W	Work of adhesion upon contact

References

[1] R. Bogue, Nanosensors: a review of recent research, Sens. Rev. 29 (4) (2009) 310–315.

[2] B.M. Cullum, G.D. Griffin, T. Vo-Dinh, Nanosensors: design and application to site-specific cellular analyses, Fourth Conference on Biomedical Diagnostic, Guidance, and Surgical-Assist Systems, January, San Jose, CA, 2002, pp. 20–21.

[3] R. Bogue, Nanosensors: a review of recent progress, Sens. Rev. 28 (1) (2008) 12–17.

[4] T.-C. Lim, S. Ramakrishna, A conceptual review of nanosensors, Zeitschrift Fur Naturforschung A 61 (7–8) (2006) 402–412.

[5] M.N. Velasco-Garcia, T. Mottram, Biosensor technology addressing agricultural problems, Biosyst. Eng. 84 (1) (2003) 1–12.

[6] M. Swierczewska, et al. High-sensitivity nanosensors for biomarker detection, Chem. Soc. Rev. 41 (7) (2012) 2641–2655.

[7] A. Manzoli, et al. Low-cost gas sensors produced by the graphite line-patterning technique applied to monitoring banana ripeness, Sensors 11 (6) (2011) 6425–6434.

[8] O. Lupan, et al. Selective hydrogen gas nanosensor using individual ZnO nanowire with fast response at room temperature, Sens. Actuators B 144 (1) (2010) 56–66.

[9] J. Tamaki, et al. Ultrahigh-sensitive WO_3 nanosensor with interdigitated Au nano-electrode for NO_2 detection, Sens. Actuators B 132 (1) (2008) 234–238.

[10] T. Zhang, et al. A gas nanosensor unaffected by humidity, Nanotechnology 20 (25) (2009) 255501.

[11] Y. Cui, et al. Biomimetic peptide nanosensors, Acc. Chem. Res. 45 (5) (2012) 696–704.

[12] C. Steffens, et al. Development of gas sensors coatings by polyaniline using pressurized fluid, Sens. Actuators B 171–172 (0) (2012) 627–633.

[13] P. Gouma, M. Stanacevic, Selective nanosensor array microsystem for exhaled breath analysis, in: G. Kaltsas,C.e. Tsamis (Eds.). Eurosensors Xxv, vol. 25, Procedia Engineering, 2011.

[14] D.K. Deda, et al. Atomic force microscopy-based molecular recognition: a promising alternative to environmental contaminants detection, in: A. Méndez-Vilas (Ed.), Current Microscopy Contributions to Advances in Science and Technology, vol. 2, Formatex Research Center, Badajoz, 2012, pp. 1337–1348.

[15] A.C.N. da Silva, et al. Nanobiosensors exploiting specific interactions between an enzyme and herbicides in atomic force spectroscopy, J. Nanosci. Nanotechnol. 14 (2014) 6678–6684.

[16] X. Sisquella, et al. A single-molecule force spectroscopy nanosensor for the identification of new antibiotics and antimalarials, FASEB J. 24 (11) (2010) 4203–4217.

[17] R.M. Ferraz, et al. High-throughput, functional screening of the anti-HIV-1 humoral response by an enzymatic nanosensor, Mol. Immunol. 43 (13) (2006) 2119–2123.

[18] T. Vo-Dinh, IEEE, Nanosensors and Nanoprobes for Environmental Health Sensing and Biomedical Screening, 2008 Digest of the Leos Summer Topical Meetings, 2008, pp. 59–60.

[19] M.K. Balaconis, H.A. Cark, Biodegradable optode-based nanosensors for in vivo monitoring, Anal. Chem. 84 (13) (2012) 5787–5793.

[20] Z. Wu, et al. Well-defined nanoclusters as fluorescent nanosensors: a case study on Au25(SG)18, Small 8 (13) (2012) 2028–2035.

[21] M. Brasuel, et al. Fluorescent nanosensors for intracellular chemical analysis: decyl methacrylate liquid polymer matrix and ion exchange-based potassium PEBBLE sensors with real-time application to viable rat C6 glioma cells, Anal. Chem. 73 (10) (2001) 2221–2228.

[22] J.M. Dubach, D.I. Harjes, H.A. Clark, Fluorescent ion-selective nanosensors for intracellular analysis with improved lifetime and size, Nano Lett. 7 (6) (2007) 1827–1831.

[23] J.P. Sumner, et al. Cu^+- and Cu^{2+}-sensitive PEBBLE fluorescent nanosensors using DsRed as the recognition element, Sens. Actuators B 113 (2) (2006) 760–767.

[24] M. Bompart, Y. De Wilde, K. Haupt, Chemical nanosensors based on composite molecularly imprinted polymer particles and surface-enhanced Raman scattering, Adv. Mater. 22 (21) (2010) 2343–2348.

[25] G. Barbillon, et al. Biological and chemical gold nanosensors based on localized surface plasmon resonance, Gold Bull. 40 (3) (2007) 240–244.

[26] S.N. Shtykov, T.Y. Rusanova, Nanomaterials and nanotechnologies in chemical and biochemical sensors: capabilities and applications, Russ. J. Gen. Chem. 78 (12) (2008) 2521–2531.

[27] M. Shahinpoor, Ionic polymeric conductor nanocomposites (IPCNCs) as distributed nanosensors and nanoactuators, Bioinspir. Biomim. 3 (3) (2008) 035003.

[28] J. Gao, B. Xu, Applications of nanomaterials inside cells, Nano Today 4 (1) (2009) 37–51.

[29] M. Gebinoga, et al. Nanosensors for label-free measurement of sodium ion fluxes of neuronal cells, Mater. Sci. Eng. B 169 (1–3) (2010) 182–185.

[30] I. Koh, L. Josephson, Magnetic nanoparticle sensors, Sensors 9 (10) (2009) 8130–8145.

[31] C. Kaittanis, et al. The assembly state between magnetic nanosensors and their targets orchestrates their magnetic relaxation response, J. Am. Chem. Soc. 133 (10) (2011) 3668–3676.

[32] M. Colombo, et al. Femtomolar detection of autoantibodies by magnetic relaxation nanosensors, Anal. Biochem. 392 (1) (2009) 96–102.

[33] G. Liang, et al. Magnetic nanosensors for highly sensitive and selective detection of bacillus Calmette-Guerin, Analyst 137 (3) (2012) 675–679.

[34] J.W. Aylott, Optical nanosensors—an enabling technology for intracellular measurements, Analyst 128 (4) (2003) 309–312.

[35] B.M. Cullum, T. Vo-Dinh, The development of optical nanosensors for biological measurements, Trends Biotechnol. 18 (9) (2000) 388–393.

[36] C. Ast, et al. Optical oxygen micro- and nanosensors for plant applications, Sensors 12 (6) (2012) 7015–7032.

[37] C. Connolly, Nanosensor developments in some European universities, Sens. Rev. 28 (1) (2008) 18–21.

[38] L. Xue, et al. Carboxylate-modified squaraine dye doped silica fluorescent pH nanosensors, Nanotechnology 21 (21) (2010) 215502.

[39] R.V. Benjaminsen, et al. Evaluating nanoparticle sensor design for intracellular pH measurements, ACS Nano 5 (7) (2011) 5864–5873.

[40] X. He, et al. Research of the relationship of intracellular acidification and apoptosis in Hela cells based on pH nanosensors, Sci. China Ser. B 50 (2) (2007) 258–265.

[41] T. Doussineau, S. Trupp, G.J. Mohr, Ratiometric pH-nanosensors based on rhodamine-doped silica nanoparticles functionalized with a naphthalimide derivative, J. Colloid Interface Sci. 339 (1) (2009) 266–270.

[42] A.K.M. Jamil, et al. Molecular recognition of 2,4,6-trinitrotoluene by 6-aminohexanethiol and surface-enhanced Raman scattering sensor, Sens. Actuators B 221 (2015) 273–280.

[43] J. Kneipp, et al. Novel optical nanosensors for probing and imaging live cells, Nanomedicine 6 (2) (2010) 214–226.

[44] N.H. Ly, S.W. Joo, Silver nanoparticle-enhanced resonance Raman sensor of chromium(III) in seawater samples, Sensors 15 (5) (2015) 10088–10099.

[45] K.L. Nowak-Lovato, B.S. Wilson, K.D. Rector, Sers nanosensors that report pH of endocytic compartments during Fc epsilon RI transit, Anal. Bioanal. Chem. 398 (5) (2010) 2019–2029.

[46] C. Elostua, et al. Volatile alcoholic compounds fibre optic nanosensor, Sens. Actuators B 115 (1) (2006) 444–449.

[47] C.Y. Liu, et al. Preparation of surface-enhanced Raman Scattering(SERS)-active optical fiber sensor by laser-induced Ag deposition and its application in bioidentification of biotin/avidin, Chem. Res. Chin. Univ. 31 (1) (2015) 25–30.

[48] Q. Yan, et al. Dual-sensing porphyrin-containing copolymer nanosensor as full-spectrum colorimeter and ultra-sensitive thermometer, Chem. Commun. 46 (16) (2010) 2781–2783.

[49] M.S. Mannoor, et al. Graphene-based wireless bacteria detection on tooth enamel, Nat. Commun. 3 (2012) 763.

[50] K. Billingsley, et al. Fluorescent nano-optodes for glucose detection, Anal. Chem. 82 (9) (2010) 3707–3713.

[51] J.M. Dubach, et al. In vivo sodium concentration continuously monitored with fluorescent sensors, Integr. Biol. 3 (2) (2011) 142–148.

[52] R. Bashir, BioMEMS: state-of-the-art in detection, opportunities and prospects, Adv. Drug Deliv. Rev. 56 (11) (2004) 1565–1586.

[53] M. Arroyo-Hernández, et al. Micro and Nanomechanical Biosensors. Nanomedicine and Nanorobotics: Handbook of Nanophysics, Boca Raton: CRC Press, 2010, pp. 1–16.

[54] G. Zhao, H. Wang, G. Liu, Advances in biosensor-based instruments for pesticide residues rapid detection, Int. J. Electrochem. Sci. 10 (12) (2015) 9790–9807.

[55] C. Steffens, et al. Atomic force microscopy as a tool applied to nano/biosensors, Sensors 12 (6) (2012) 8278–8300.

[56] N.A. Burnham, R.J. Colton, H.M. Pollock, Interpretation of force curves in force microscopy, Nanotechnology 4 (1993) 64–80.

[57] R. Garcia, R. Perez, Dynamic atomic force microscopy methods, Surf. Sci. Rep. 47 (6–8) (2002) 197–301.

[58] E.F. Franca, et al. Designing an enzyme-based nanobiosensor using molecular modeling techniques, Phys. Chem. Chem. Phys. 13 (19) (2011) 8894–8899.

[59] A.R. Bizzarri, S. Cannistraro, The application of atomic force spectroscopy to the study of biological complexes undergoing a biorecognition process, Chem. Soc. Rev. 39 (2) (2010) 734–749.

[60] A.R. Bizzarri, S. Cannistraro, The application of atomic force spectroscopy to the study of biological complexes undergoing a biorecognition process, Chem. Soc. Rev. 39 (2) (2010) 734–749.

[61] P. Hinterdorfer, et al. A mechanistic study of the dissociation of individual antibody–antigen pairs by atomic force microscopy, Nanobiology 4 (1998) 177–188.

[62] A.D. Moraes, et al. Evidences of detection of atrazine herbicide by atomic force spectroscopy: a promising tool for environmental sensing, Acta Microsc. 24 (1) (2015) 53–63.

[63] D.A. Smith, et al. Chemical force microscopy: applications in surface characterization of natural hydroxyapatite, Anal. Chim. Acta 479 (1) (2003) 39–57.

[64] S. Singamaneni, et al. Bimaterial microcantilevers as a hybrid sensing platform, Adv. Mater. 20 (4) (2008) 653–680.

[65] D.E. Koshland, The key-lock theory and the induced fit theory, Angew. Chem. Int. Ed. Engl. 33 (23–24) (1995) 2375–2378.

[66] B.L. Stoddard, D.E. Koshland, Prediction of the structure of a receptor protein complex using a binary docking method, Nature 358 (6389) (1992) 774–776.

[67] T. Maver, et al. Functionalization of AFM tips for use in force spectroscopy between polymers and model surfaces, Mater. Tehnol. 45 (3) (2011) 205–211.

[68] B. Cappella, G. Dietler, Force-distance curves by atomic force microscopy, Surf. Sci. Rep. 34 (1–3) (1999) 1–104.

[69] H.J. Butt, B. Cappella, M. Kappl, Force measurements with the atomic force microscope: technique, interpretation and applications, Surf. Sci. Rep. 59 (1–6) (2005) 1–152.

[70] F.L. Leite, et al. Theoretical models for surface forces and adhesion and their measurement using atomic force microscopy, Int. J. Mol. Sci. 13 (10) (2012) 12773–12856.

[71] A.L. Weisenhorn, et al. Measuring adhesion, attraction, and repulsion between surfaces in liquids with an atomic-force microscope, Phys. Rev. B 45 (19) (1992) 11226–11232.

[72] A.L. Weisenhorn, et al. Forces in atomic force microscopy in air and water, Appl. Phys. Lett. 54 (26) (1989) 2651–2653.

[73] Y. Mo, K.T. Turner, I. Szlufarska, Friction laws at the nanoscale, Nature 457 (7233) (2009) 1116–1119.

[74] M. Heuberger, G. Dietler, L. Schlapbach, Elastic deformations of tip and sample during atomic force microscope measurements, J. Vacuum Sci. Technol. B 14 (2) (1996) 1250–1254.

[75] B. Poon, D. Rittel, G. Ravichandran, An analysis of nanoindentation in linearly elastic solids, Int. J. Solids Struct. 45 (24) (2008) 6018–6033.

[76] D. Maugis, M. Barquins, Adhesive contact of a conical punch on an elastic half-space, J. Phys. Lett. 42 (5) (1981) L95–L97.

[77] D. Vallet, M. Barquins, Adhesive contact and kinetics of adherence of a rigid conical punch on an elastic half-space (natural rubber), Comput. Methods Contact Mech. V 5 (2001) 41–53.

[78] X.-F. Wu, Y.A. Dzenis, Adhesive contact in filaments, J. Phys. D 40 (14) (2007) 4276–4280.

[79] B. Capella, et al. Force-distance curves by AFM—a powerful technique for studying surface interactions, IEEE Eng. Med. Biol. Mag. 16 (2) (1997) 58–65.

[80] G.I. Bell, Theoretical-models for the specific adhesion of cells to cells or to surfaces, Adv. Appl. Probab. 12 (3) (1980) 566–567.

[81] E. Evans, Probing the relation between force—lifetime—and chemistry in single molecular bonds, Ann. Rev. Biophys. Biomol. Struct. 30 (2001) 105–128.

[82] E. Evans, K. Ritchie, Dynamic strength of molecular adhesion bonds, Biophys. J. 72 (4) (1997) 1541–1555.

[83] S.S.M. Konda, et al. Chemical reactions modulated by mechanical stress: Extended Bell theory, J. Chem. Phys. 135 (16) (2011) 164103.

[84] A. Beaussart, et al. Chemical force microscopy of stimuli-responsive adhesive copolymers, Nanoscale 6 (1) (2014) 565–571.

[85] A. Noy, D.V. Vezenov, C.M. Lieber, Chemical force microscopy, Ann. Rev. Mater. Sci. 27 (1997) 381–421.

[86] M. Alvarez, et al. Development of nanomechanical biosensors for detection of the pesticide DDT, Biosens. Bioelectron. 18 (5–6) (2003) 649–653.

[87] H. Kim, et al. Characterization of mixed self-assembled monolayers for immobilization of streptavidin using chemical force microscopy, Ultramicroscopy 108 (10) (2008) 1140–1143.

[88] H. Kim, et al. Selective immobilization of proteins on gold dot arrays and characterization using chemical force microscopy, J. Colloid Interface Sci. 334 (2) (2009) 161–166.

[89] C.R. Suri, et al. Label-free ultra-sensitive detection of atrazine based on nanomechanics, Nanotechnology 19 (23) (2008) 235502.

[90] P.S. Waggoner, H.G. Craighead, Micro- and nanomechanical sensors for environmental, chemical, and biological detection, Lab on a Chip 7 (10) (2007) 1238–1255.

[91] Y. Seo, W. Jhe, Atomic force microscopy and spectroscopy, Rep. Prog. Phys. 71 (1) (2008) 016101.

[92] K. Awsiuk, et al. Spectroscopic and microscopic characterization of biosensor surfaces with protein/amino-organosilane/silicon structure, Colloids Surf. B 90 (2012) 159–168.

[93] A.C.N. da Silva, et al. Nanobiosensors based on chemically modified AFM probes: a useful tool for metsulfuron-methyl detection, Sensors 13 (2) (2013) 1477–1489.

[94] J.E. Headrick, C.L. Berrie, Alternative method for fabricating chemically functionalized AFM tips: silane modification of HF-treated Si_3N_4 probes, Langmuir 20 (10) (2004) 4124–4131.

[95] M. Micic, et al. Scanning electron microscopy studies of protein-functionalized atomic force microscopy cantilever tips, Scanning 21 (6) (1999) 394–397.

[96] A. Berquand, et al. Antigen binding forces of single antilysozyme Fv fragments explored by atomic force microscopy, Langmuir 21 (12) (2005) 5517–5523.

[97] M. Fiorini, et al. Chemical force microscopy with active enzymes, Biophys. J. 80 (5) (2001) 2471–2476.

[98] C. Volcke, et al. Plasma functionalization of AFM tips for measurement of chemical interactions, J. Colloid Interface Sci. 348 (2) (2010) 322–328.

[99] Z. Long, K. Hill, M.J. Sepaniak, Aluminum oxide nanostructured microcantilever arrays for nanomechanical-based sensing, Anal. Chem. 82 (10) (2010) 4114–4121.

[100] M.E. Drew, et al. Nanocrystalline diamond AFM tips for chemical force spectroscopy: fabrication and photochemical functionalization, J. Mater. Chem. 22 (25) (2012) 12682–12688.

[101] R. Janissen, L. Oberbarnscheidt, F. Oesterhelt, Optimized straight forward procedure for covalent surface immobilization of different biomolecules for single molecule applications, Colloids Surf. B 71 (2) (2009) 200–207.

[102] R.P. Lu, et al. Scanning spreading resistance microscopy current transport studies on doped III-V semiconductors, J. Vacuum Sci. Technol. B 20 (4) (2002) 1682–1689.

[103] S.R. Nalage, et al. Development of Fe_2O_3 sensor for NO_2 detection, in: A.K. Chauhan, C. Murli, et al. (Eds.), Solid State Physics, vol. 57, 1512, American Institute of Physics, Melville, 2013, pp. 504–505 (AIP Conference Proceedings).

[104] S. Fatikow, et al. Automated handling of bio-nanowires for nanopackaging, IEEE/RSJ 2010 International Conference on Intelligent Robots and Systems, New York, 2010, pp. 3999–4004.

[105] J.J. Lin, et al. Characteristics of polysilicon wire glucose sensors with a surface modified by silica nanoparticles/gamma–APTES nanocomposite, Sensors 11 (3) (2011) 2796–2808.

[106] F.L. Leite, et al. Nanoneurobiophysics: new challenges for diagnosis and therapy of neurologic disorders, Nanomedicine 10 (2015) 3417–3419.

[107] P.S. Garcia, et al. A nanobiosensor based on 4-hydroxyphenylpyruvate dioxygenase enzyme for mesotrione detection, IEEE Sens. J. 15 (2015) 2106–2113.

[108] C.C. Bueno, et al. Nanobiosensor for diclofop detection based on chemically modified AFM probes, IEEE Sens. J. 14 (2014) 1467–1475.

[109] A.M. Amarante, et al. Modeling the coverage of an AFM tip by enzymes and its application in nanobiosensors, J. Mol. Graph. Model. 53 (2014) 100–104.

[110] C. Steffens, et al. Bio-inspired sensor for insect pheromone analysis based on polyaniline functionalized AFM cantilever sensor, Sens. Actuators B 191 (2014) 643–649.

[111] D.K. Deda, et al. The use of functionalized AFM tips as molecular sensors in the detection of pesticides, Mater. Res. 16 (2013) 683–687.

[112] G.S. de Oliveira, et al. Molecular modeling of enzyme attachment on AFM probes, J. Mol. Graph. Model. 45 (2013) 128–136.

[113] A.D. Romig, M.T. Dugger, P.J. Mcwhorter, Materials issues in microelectromechanical devices: science, engineering, manufacturability and reliability, Acta Mater. 51 (19) (2003) 5837–5866.

[114] L.G. Carrascosa, et al. Nanomechanical biosensors: a new sensing tool, Trends Anal. Chem. 25 (3) (2006) 196–206.

[115] H. Lang, M. Hegner, C. Gerber, Nanomechanical cantilever array sensors, in: B. Bhushan (Ed.), Springer Handbook of Nanotechnology, Springer, Berlin, Heidelberg, 2010, pp. 427–452 (Chapter 15).

[116] H. Lang, C. Gerber, Microcantilever sensors, in: P. Samorì (Ed.), STM and AFM Studies on (Bio)molecular Systems: Unravelling the Nanoworld, vol. 285, Springer, Berlin, Heidelberg, 2008, pp. 1–27 (Chapter 28, Topics in Current Chemistry).

[117] C. Steffens, et al. Atomic force microscope microcantilevers used as sensors for monitoring humidity, Microelectron. Eng. 113 (2014) 80–85.

[118] V. Tabard-Cossa, Microcantilever Actuation Generated by Redox-Induced Surface Stress, 2005, 169 (Thesis Doctor of Philosophy).

[119] F.M. Battiston, et al. A chemical sensor based on a microfabricated cantilever array with simultaneous resonance-frequency and bending readout, Sens. Actuators B 77 (1–2) (2001) 122–131.

[120] J. Wang, X. Qu, Recent progress in nanosensors for sensitive detection of biomolecules, Nanoscale 5 (9) (2013) 3589–3600.

[121] C. Wang, et al. Ultrasensitive biochemical sensors based on microcantilevers of atomic force microscope, Anal. Biochem. 363 (1) (2007) 1–11.

[122] A. Boisen, et al. Cantilever-like micromechanical sensors, Rep. Prog. Phys. 74 (3) (2011) 036101.

[123] G. Binning, et al. Atomic resolution with atomic force microscope, Europhys. Lett. 3 (1987) 1281–1286.

[124] T. Thundat, et al. Detection of mercury-vapor using resonating microcantilevers, Appl. Phys. Lett. 66 (13) (1995) 1695–1697.

[125] B.C. Fagan, et al. Modification of micro-cantilever sensors with sol–gels to enhance performance and immobilize chemically selective phases, Talanta 53 (3) (2000) 599–608.

[126] H.F. Ji, et al. A novel self-assembled monolayer (SAM) coated microcantilever for low level caesium detection, Chem. Commun. 6 (2000) 457–458.

[127] L.A. Pinnaduwage, et al. Detection of 2,4-dinitrotoluene using microcantilever sensors, Sens. Actuators B 99 (2–3) (2004) 223–229.

[128] Z. Long, et al. Landfill siloxane gas sensing using differentiating, responsive phase coated microcantilever arrays, Anal. Chem. 81 (7) (2009).

[129] H.P. Lang, et al. Nanomechanics from atomic resolution to molecular recognition based on atomic force microscopy technology, Nanotechnology 13 (5) (2002) R29–R36.

[130] G.Y. Chen, et al. Adsorption-induced surface stress and its effects on resonance frequency of microcantilevers, J. Appl. Phys. 77 (8) (1995) 3618–3622.

[131] T.A. Betts, et al. Selectivity of chemical sensors based on micro-cantilevers coated with thin polymer films, Anal. Chim. Acta 422 (1) (2000) 89–99.

[132] C. Steffens, et al. Microcantilever sensors coated with a sensitive polyaniline layer for detecting volatile organic compounds, J. Nanosci. Nanotechnol. 14 (9) (2014) 6718–6722.

[133] G.H. Wu, et al. Origin of nanomechanical cantilever motion generated from biomolecular interactions, Proc. Natl. Acad. Sci. USA 98 (4) (2001) 1560–1564.

[134] K.-W. Liu, S.L. Biswal, Probing insertion and solubilization effects of lysolipids on supported lipid bilayers using microcantilevers, Anal. Chem. 83 (12) (2011) 4794–4801.

[135] V. Seena, et al. Polymer nanocomposite nanomechanical cantilever sensors: material characterization, device development and application in explosive vapour detection, Nanotechnology 22 (29) (2011) 295501.

[136] Y.F. Zhang, et al. Detection of CrO_{42}—using a hydrogel swelling microcantilever sensor, Anal. Chem. 75 (18) (2003) 4773–4777.

[137] C. Steffens, et al. Microcantilever sensors coated with doped polyaniline for the detection of water vapor, Scanning 35 (4) (2013) 1–6.

[138] J. Fragakis, et al. Simulation of capacitive type bimorph humidity sensors, in: A.G. Nassiopoulou, N. Papanikolaou, et al. (Eds.), Second Conference on Microelectronics, Microsystems and Nanotechnology, vol. 10, IOP Publishing Ltd, Bristol, 2005, pp. 305–308 (Journal of Physics Conference Series).

[139] T. Thundat, et al. Thermal and ambient-induced deflections of scanning force microscope cantilevers, Appl. Phys. Lett. 64 (21) (1994) 2894–2896.

[140] R.-H. Ma, et al. Microcantilever-based weather station for temperature, humidity and flow rate measurement, Microsyst. Technol. 14 (7) (2008) 971–977.

[141] S. Singamaneni, et al. Polymer-silicon flexible structures for fast chemical vapor detection, Adv. Mater. 19 (23) (2007) 4248–4255.

[142] B.H. Kim, et al. Multicomponent analysis and prediction with a cantilever array based gas sensor, Sens. Actuators B 78 (1–3) (2001) 12–18.

[143] M. Lahav, et al. Redox activation of a polyaniline-coated cantilever: an electro-driven microdevice, Angew. Chem. Int. Ed. 40 (21) (2001) 4095–4097.

[144] G.H. Wu, et al. Bioassay of prostate-specific antigen (PSA) using microcantilevers, Nat. Biotechnol. 19 (9) (2001) 856–860.

[145] B. Ilic, et al. Enumeration of DNA molecules bound to a nanomechanical oscillator, Nano Lett. 5 (5) (2005) 925–929.

[146] S.M. Lee, et al. Protocol for the use of a silica nanoparticle-enhanced microcantilever sensor-based method to detect HBV at femtomolar concentrations, Methods Mol. Biol. 903 (2012) 283–293.

[147] Standards, N.C.F.C.L. Methods for dilution antimicrobial susceptibility tests for bacteria that grow aerobically, vol. 4, National Committee for Clinical Laboratory Standards, Wayne, PA, 1997, M7-A4.

[148] K.Y. Gfeller, N. Nugaeva, M. Hegner, Rapid biosensor for detection of antibiotic-selective growth of Escherichia coli, Appl. Environ. Microbiol. 71 (5) (2005) 2626–2631.

[149] S. Moises, M. Schäferling, Toxin immunosensors and sensor arrays for food quality control, Bioanal. Rev. 1 (1) (2009) 73–104.

[150] N. Nugaeva, et al. Micromechanical cantilever array sensors for selective fungal immobilization and fast growth detection, Biosens. Bioelectron. 21 (6) (2005) 849–856.

[151] K. Persaud, G. Dodd, Analysis of discrimination mechanisms in the mammalian olfactory system using a model nose, Nature 299 (5881) (1982) 352–355.

[152] J.W. Gardner, P.N. Bartlett, A brief history of electronic noses, Sens. Actuat. B: Chem. 18 (1994) 211–220.

[153] C. Steffens, et al. Gas sensors development using supercritical fluid technology to detect the ripeness of bananas, J. Food Eng. 101 (4) (2010) 365–369.

[154] S. Sankaran, et al. A review of advanced techniques for detecting plant diseases, Comput. Electron. Agric. 72 (1) (2010) 1–13.

[155] J. Brezmes, et al. Fruit ripeness monitoring using an Electronic Nose, Sens. Actuators B 69 (3) (2000) 223–229.

[156] L.Q. Pan, et al. Early detection and classification of pathogenic fungal disease in post-harvest strawberry fruit by electronic nose and gas chromatography-mass spectrometry, Food Res. Int. 62 (2014) 162–168.

[157] N.L. Privorotskaya, W.P. King, The mechanics of polymer swelling on microcantilever sensors, Microsyst. Technol. 15 (2) (2009) 333–340.

[158] R. Bashir, et al. Micromechanical cantilever as an ultrasensitive pH microsensor, Appl. Phys. Lett. 81 (16) (2002) 3091–3093.

[159] C.P. Cheney, et al. In vivo wireless ethanol vapor detection in the Wistar rat, Sens Actuators B 138 (1) (2009) 264–269.

[160] W. Shu, et al. Highly specific labe-free protein detection from lysed cells using internally referenced microcantilever sensors, Biosens. Bioelectron. 24 (2008) 233–237.

[161] D.A. Raorane, et al. Quantitative and label-free technique for measuring protease activity and inhibition using a microfluidic cantilever array, Nano Lett. 8 (9) (2008) 2968–2974.

5

Low-Dimensional Systems: Nanoparticles

C.M. Miyazaki*, A. Riul Jr**

*FEDERAL UNIVERSITY OF SÃO CARLOS, CENTER OF SCIENCES AND TECHNOLOGY FOR SUSTAINABILITY, SOROCABA, SÃO PAULO, BRAZIL; **"GLEB WATAGHIN" INSTITUTE OF PHYSICS, STATE UNIVERSITY OF CAMPINAS, CAMPINAS, SÃO PAULO, BRAZIL

CHAPTER OUTLINE

5.1 Introduction

Nanoparticles have been extensively explored because of their unique properties, which are dependent on their size and shape and enable their application in an extensive variety of fields, such as, cancer treatment and diagnosis [1–3]. Nanoparticles also enable the exploitation and production of existing materials with novel properties that, combined with advances in the synthesis and characterization processes, facilitate the development of new products, such as fabrics, dyes, cosmetics, and sports products.

In the 4th century, the Romans employed metal nanoparticles (silver and gold) to produce colorful glasses. The Lycurgus cup is a classic example, with its variable coloration according to the incident light, which ranges from green to red. In the mid-18th century, photography was discovered based on the production of light-sensitive silver nanoparticles. In 1857, Michael Faraday synthesized and performed experiments with a colloidal suspension of gold via the reduction of $[AuCl_4]^-$ salt using phosphorus as a reducing agent [4]. In 1951, Turkevitch et al. described a detailed syntheses study, using transmission electron microscopy (TEM), of the preparation of gold nanoparticles using different reducing agents to elucidate the processes of nucleation, growth and agglomeration of nanoparticles, and the synthesis of monodisperse and reproducible suspensions [4]. Richard Feynman, who was awarded the 1965 Nobel Prize in Physics, predicted the technological potential of nanostructures, which suggests that the manipulation of individual atoms can provide a new material with new properties, and their application in electronic circuits on a nanometric scale for more powerful computers, as well as the manipulation of biologic systems. Currently, many examples and applications show that Feynman's vision was ahead of his time [5].

Nanoparticles are defined as materials with sizes that range from 1 to 100 nm in at least one of their dimensions [6]; by this definition, nanotubes, fullerene, and nanothreads are also considered to be nanoparticles. However, this chapter focuses on zero-dimensional nanostructures. The term cluster (or nanocluster) will be extensively applied to define colloidal particle aggregates and may sometimes refer to nanoparticles.

The synthesis of nanoparticles involves two different approaches: the top-down method and the bottom-up method; these approaches will be detailed in Section 5.2. The important requirements for the practical application of nanoparticles are as follows: nanometric dimension, uniform size distribution, morphology, chemical composition, and identical crystal structure [7]. In addition to advancements in the synthesis of nanoparticles, the need for characterization techniques that can demonstrate the physico-morphological characteristics of these nanometric entities is evident. Currently, a range of tools are available to investigate these characteristics, including electronic microscopy and X-ray diffraction (XRD) techniques and spectroscopy.

Given the extent of the content that involves synthesis, properties, and applications of nanoparticles, this chapter will focus on the main synthesis processes, with a simplified approach to the nanoparticle properties, describe the main characterization methods, and indicate some applications in the literature.

5.2 Synthesis Methods

As previously mentioned, the unique properties of a nanoparticle are dependent on its shape and size, which are dependent on the synthesis process that is employed for the fabrication of nanoparticles. The process in which nanoparticles derive from atomic precursors that aggregate to form a cluster and subsequently form a nanostructure is referred to as a bottom-up process. The process in which nanoparticles are obtained by the physical wear of a larger volume is referred to as a top-down process. The main top-down (mechanical attrition and lithography) and bottom-up (chemical synthesis via sol–gel and reduction of metal salts) techniques are described in the next section.

5.2.1 Top-Down Methods

The mechanical attrition and lithography are the most well-known top-down methods for the production of nanometric structures, in which the former is commonly applied in the industry for large-scale production and the latter is a more sophisticated technique for the production of electronic and optical devices.

5.2.1.1 Mechanical Friction

Mechanical friction is extensively applied by the metallurgical industry for the production of new alloys and mixtures with different properties due to the incorporation of defects in the crystal lattice of a metal [8,9]. This technique enables syntheses that are impossible via traditional fusion routes, such as, the production of uniform dispersions of ceramic particles in a metal matrix or metal dispersions with different melting points [8,10]. Mechanical friction also allows the solubility of immiscible binary systems due to the segregation of solute in the grain boundaries [9].

In the high-energy grinding process, particles with diameters near 50 μm are placed together with steel balls or tungsten binary composites inside sealed chambers and subjected to intense agitation (Fig. 5.1). The high energy of the grinding process can be obtained by applying high-frequency and low-amplitude vibration [8]. The grain size decreases with the grinding time until a constant value is obtained depending on the melting point of the material. The high-energy grinding process enables the creation of mono- or multicomponent nanoparticles at the industrial scale; however, contamination by the grinding medium is a major disadvantage [9]. Lam et al. produced Si nanoparticles via the milling process with stainless steel balls. The reaction on the solid phases between graphite and silicon oxide ($C + SiO_2 \rightarrow Si + CO_2$) produces nanoparticles with a 5-nm Si nucleus and 1-nm external layer of amorphous silicon oxide [11]. Details about the process of nanoparticle production by mechanical friction are provided by Koch (1993).

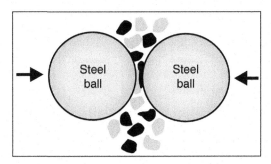

FIGURE 5.1 Schematic of the ball-milling process using steel balls.

5.2.1.2 Lithography

The lithographic process consists of the transfer of patterns to the desired substrate and is extensively applied for the production of electronic and optical devices due to the high resolution obtained [12]. Briefly, this process consists of the deposition of a photoresist, exposure of its specific areas using a mask, the exposure of the resist and consequently, the transfer of the desired pattern.

The first step includes the deposition of the resist on a substrate, using the controlled rotation of the substrate by a technique denominated as spin-coating [13]. The second step includes the irradiation of the photoresist using a polymeric material that undergoes chemical structural alterations when irradiated, such as, the rupture of the polymeric chain (positive resist) or the formation of cross-links (negative resist). In the case of the positive resist, the irradiated part will be dissolved during the exposure process due to the breakdown of the macromolecule into smaller parts; for the negative resist, the formation of cross-links render the irradiated parts insoluble [12].

Depending on the type of radiation that is used to sensitize the resist, such as ultraviolet, X-ray, electron or ion beam, different resolutions can be obtained. Patterns of up to 5 nm can be obtained with an electron beam, whereas 30-nm patterns are obtained using X-ray [12]. After the desired model is created, the transfer can be performed via the corrosion or the lift-off metallization process. The chemical corrosion involves the protection of certain areas by the pattern created in the resist and the corrosion of the other areas by specific chemical solution or plasma with reactive radicals [12]. Metal structures can be created by metallization followed by lift-off, that is, the substrate receives the resist with the desired pattern, and the metal is evaporated on top of it. The metal is fixed at the substrate-free zone, and the resist is removed using an appropriate solvent, leaving only the desired metal pattern, as illustrated in Fig. 5.2.

Lithography is an extensive process, including various details. For a better understanding of the technique, refer to Dobisz et al. (1996) and Hilleringmann (2005).

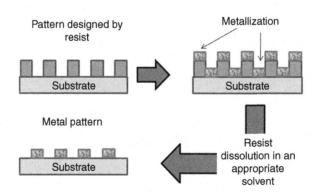

FIGURE 5.2 Schematic of the lift-off process for the production of metal patterns.

5.2.2 Bottom-Up Methods

In the bottom-up process, the starting points are the atomic or ionic precursors that unite to form larger particles. Initially, the discussion will cover the physicochemical aspects of the process of nanoparticle production in solution, the nanoparticle stabilization in suspensions, the synthesis of metal nanoparticles via metal reduction and the synthesis of oxide nanoparticles via the sol–gel method. Although we will only discuss the techniques based on chemical synthesis, other bottom-up methods, such as molecular beam epitaxy, chemical vapor deposition, pulse laser deposition, laser ablation and sputtering deposition, are extremely important in the semiconductor industry. However, these sophisticated techniques are dependent on specific equipment for temperature and pressure control; details are provided in Refs. [14–16]. We also emphasize the high control and quality of nanoparticles that are obtained by physical methods in relation to size and stoichiometric composition [17,18]. The metal nanoparticles should be protected against the oxidative effects of the environment for their subsequent technological applications. We also emphasize the complexity that is required by these experimental setups, which justify the selected descriptions of the chemical methods discussed in this paper.

For each application, the synthesis conditions should be well controlled to ensure that the product has well-defined characteristics, such as equivalent dimension (uniform size distribution), shape and morphology, and similar composition and crystal structure between particles to prevent the formation of agglomerates [7].

5.2.2.1 Physicochemical Aspects of Nanoparticle Formation

The mechanism of nanoparticle formation in a liquid phase can be explained by nucleation and growth phenomena [7,19]. According to the theory developed by Lamer [20], particle formation consists of the following steps:

1. increase in monomer concentration in the solution until supersaturation and the start of nuclei formation;
2. continuous aggregation of monomers in the nucleus, which causes a gradual decrease in the monomer concentration in the solution; and
3. stabilization of the surface of the resulting particles by surfactants.

If the solute concentration increases to the limit of solubility or the temperature decreases to a point that enables phase transformation, the system will have a high Gibbs free energy. To minimize the energy of the system, the solute aggregates and the variation of the Gibbs free energy per unit volume (ΔG_v) of the solid phase is given by Eq. 5.1, which is dependent on the solute concentration [7].

$$\Delta G_v = \frac{-KT}{\Omega} \ln\left(\frac{C}{C_0}\right) \tag{5.1}$$

where C is the solute concentration, C_0 is the solubility, K is the Boltzmann constant, T is the temperature and Ω is the atomic volume. Consider the supersaturation $\sigma = (C - C_0)/C_0$:

$$\Delta G_v = \frac{-KT}{\Omega}\ln(1+\sigma) \tag{5.2}$$

From Eq. 5.2, when $\sigma = 0$, the Gibbs free energy will be zero and supersaturation will not occur. If $C > C_0$, the free energy will be negative and the nucleation process will be spontaneous.

A spherical nucleus of radius r can be assumed, with the Gibbs free energy given in Eq. 5.3 [7]:

$$\Delta\mu_v = \frac{4}{3}\pi r^3 \Delta G_v \tag{5.3}$$

In addition to the decrease in free energy, the total energy of the system will also be influenced by the appearance of the surface energy $\Delta\mu_s$. Thus, Eq. 5.4 indicates that the reduction in the free energy of the system is influenced by the appearance of the surface energy, which is expressed as $4\pi r^2 \gamma$ [7].

$$\Delta G = \Delta\mu_v + \Delta\mu_s = \frac{4}{3}\pi r^3 \Delta G_v + 4\pi r^2 \gamma \tag{5.4}$$

where γ is the surface energy per unit area. The formed nuclei should have a minimum radius to be stable; otherwise, they dissolve and return to their volume [7,19]. This minimum radius is referred to as the critical radius r^*, and the nuclei with $r \geq r^*$ will act as nucleation centers for the formation of particles. Fig. 5.3 illustrates the nucleation and growth processes.

After nucleation starts, the concentration of the species in solution and the free energy of the system decreases [7,21]. The concentration decreases to values below a specific

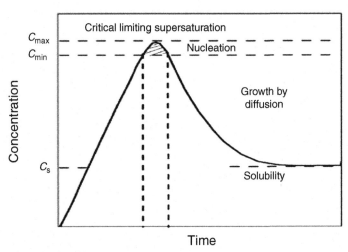

FIGURE 5.3 Schematic of the nucleation and growth process by LaMer. *Modified from V.K. LaMer, R.H. Dinegar, Theory, production and mechanism of formation of monodispersed hydrosols, J. Am. Chem. Soc. 72 (1950) 4847–4854 [20].*

concentration when the nucleation stops, and the growth process continues until an equilibrium concentration C_s is attained. When a nucleus is formed, the growth process immediately starts, with larger particles that increase at the expense of the smaller particles by a process named Ostwald ripening, which reduces the surface energy. Above a minimum concentration, the nucleation and growth process simultaneously occur but at different rates.

5.2.2.2 Nanoparticle Stabilization

Nanometric particles have a wider surface area and tend to form aggregates to minimize their surface energy. The agglomeration process can initiate during synthesis; thus, the use of different types of surfactants and stabilizers has been explored [22]. To minimize the surface energy of the particles and prevent the agglomeration process, two paths can be followed: (1) electrostatic repulsion between covered particles, which is effective in diluted aqueous or polar organic systems, and (2) steric effects, which are active in aqueous and nonaqueous systems, and dispersions at high concentration; however, they are less sensitive to impurities or additives when compared with electrostatic repulsion [4,22].

For example, when suspended, gold nanoparticles that are synthesized via sodium citrate are surrounded by an electron double layer, which is formed by the adsorbed citrate, chloride ions and cations that are attracted to the surface. In this case, any disturbance in the system, such as an increase in ionic force, may cause the formation of agglomerates. The protection by steric effects includes the adsorption of molecules, such as polymers, surfactants and other types of ligands, on the surface of nanoparticles. Polymers are commonly applied and selected according to the solubility of the precursor in the polymeric solution and the ability to stabilize the product [4]. Some natural polymers, such as chitosan [23,24] and cyclodextrins [25,26], as well as synthetic polymers, such as PVP (polyvinylpyrrolidone) [27–29] and PVA (polyvinyl alcohol) [30,31], have been employed.

5.2.2.3 Sol–Gel Synthesis

Sol–gel synthesis is extensively employed for the production of colloidal dispersions of oxides and the production of core-shell nanostructures, which is also useful for the production of inorganic materials and organic–inorganic hybrids, particularly oxides [7,32]. The precursor can be an inorganic metal salt (acetate, chloride, nitrate, and sulfate) or a metal organic species, such as a metal alkoxide [33]. The precursor can be dissolved in an aqueous or organic solvent, and a catalyst is added to promote the hydrolysis and condensation reactions [7].

A typical sol–gel production process consists of the use of metal alkoxide as a precursor in an aqueous system that undergoes hydrolysis and condensation. The alkoxides are precursors for silica, alumina, titanium, and zirconium [32]. With hydrolysis, the metal alkoxide is transformed into sol (colloidal dispersion of particles in a liquid) and a gel after condensation [33]. The hydrolysis and condensation steps occur sequentially and in parallel with the condensation step, which frequently causes the formation of aggregates of oxides or metal hydroxides with incorporated or linked organic groups. These organic groups may derive from incomplete hydrolysis or may be introduced as organic nonhy-drolyzable ligands [7].

The synthesis in an aqueous medium has the disadvantage of different reactivities of metal alkoxides, which hinder the control of the composition and its homogeneity in the synthesis of multicomponent oxides [33,34]. The nonaqueous synthesis can be divided into synthesis by surfactant control or by synthesis controlled by solvents. In the first case, the precursor is commonly injected into a heated solvent that contains surfactants, with the surfactant molecules avoiding agglomeration, which produces reasonable colloidal stability in organic solvents [34]. Zeng et al. produced $MnFe_2O_4$ magnetic nanoparticles with different sizes by controlling the stabilizers/Fe ratio. Depending on the proportion, the produced particles presented a spherical, polyhedron or cubic shaped [35].

In the case in which the presence of surfactants is undesirable, for example, when it affects the catalytic activity, the synthesis is performed via the solvent-control path. Ba et al. produced monodisperse tin oxide nanoparticles with a diameter of 3.5 nm. For the synthesis, a tin chloride solution was dripped into a benzyl alcohol solvent under agitation at 100°C for 24 h. The precipitate was collected, centrifuged, and suspended in tetrahydro-furan (THF) to form a suspension of transparent nanoparticles that are stable even in the absence of surfactants without the formation of agglomerates [36]. Additional information about the sol–gel hot-injection method is provided in Ref. [37].

5.2.2.4 Metal Reduction

The reduction of salts to obtain colloidal suspensions of metal nanoparticles is the most common synthesis technique [4], with the use of an extensive variety of metal salts, sur-factants, and reducing agents. The following materials can be applied as metal precursors: oxides, nitrates, chlorides, acetates, and acetyl ketones [38]. The following materials can be employed as reducing agents: hydrides (such as sodium borohydride), dehydrogenat-ing gases, citrates, ascorbic acid, hydrazine, and ethylene glycol. Depending on the reduc-tion potential of the reducing agents, a reaction may occur at room temperature or at high temperatures [19].

The reduction of gold by citrate is one of the most common examples of metal reduc-tion. In 1951, Turkevitch demonstrated the synthesis of nanoparticles in the scale of 20 nm via the reduction of $HAuCl_4$ in an aqueous medium with sodium citrate [39]. Yonezawa et al. proposed a modification of the synthesis via citrate using 3-mercaptopropionate as a stabilizer. The nanoparticle size can be controlled by varying the stabilizer/precursor ratio [40]. Zhu et al. produced functionalized Au nanoparticles with carboxylic groups via reduction with citrate using 2-mercaptosuccinic acid [41], in which the product was more stable to pH changes.

Nanoreactors have been applied to improve the control of the synthesis processes. The dendrimers are extensively employed because they are macromolecules of tree-like mor-phology with regularly spaced ramifications, three-dimensional structures and an abun-dance in surface groups [42]. The advantages of their application are as follows: (1) uniform structure and composition; (2) the stabilization and encapsulation of nanoparticles inside

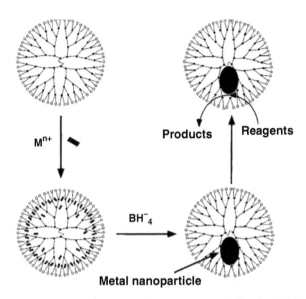

FIGURE 5.4 Schematic of the synthesis of metal nanoparticles via reduction by borohydride inside a dendrimer. *R.M. Crooks, M. Zhao, L. Sun, V. Chechik, L.K. Yeung, Dendrimer-encapsulated metal nanoparticles: synthesis, characterization, and applications to catalysis, Acc. Chem. Res. 34 (2001) 181–190 [43].*

the dendrimers without the formation of agglomerates; (3) the encapsulation of nanoparticles by steric effects; thus, a large part of the particle surface is available to participate in the reactions; and (4) the peripheral groups of dendrimers can be adapted for the control of solubility, which helps in surface adsorption [43]. Fig. 5.4 illustrates the synthesis of metal nanoparticles inside the structure of a dendrimer via reduction by borohydride.

Polyamidoamine (PAMAM) is one of the most employed nanoreactors because it produces very small nanoparticles with diameters less than 4 nm. This type of method has been applied to the fields of biosensors and catalysis and the fabrication of electronic devices. Crespilho et al. developed an enzymatic biosensor for the detection of glucose. Initially, PAMAM-Au was synthesized by reduction of $KAuCl_4$ solution with formic acid in the presence of G4 PAMAM. PAMAM-Au were alternated with poly(vinylsulfonic acid) (PVS) on ITO electrodes using the self-assembly technique [layer-by-layer (LbL)] to form bilayers. The hexacyanoferrate was electrodeposited, which modified the gold nanoparticles, and the glucose oxidase enzyme was subsequently immobilized on the self-assembled film to form the biosensor [44].

5.3 Properties

As predicted by Feynman, new phenomena govern the characteristics of materials that affect their properties at the nanoscale, which can be properly explored in many applications. When the size of a nanoparticle or nanocrystal is smaller than the wavelength of the incident radiation, its de Broglie wavelength will be comparable to its diameter.

Consequently, the conduction electrons can become trapped in the nanoparticles; this effect is known as quantum trapping, which induces alterations in the bandgap and in the electron energy levels of the material. This effect is more pronounced in semiconductor nanoparticles due to an increase in the bandgap that is caused by a decrease in their size, which generates interband transitions that cause electron displacements to higher frequencies. In a simplistic manner, different electronic configurations exist compared with the configurations observed in the same materials at the macroscopic scale. The control of the nanostructure size generates control of the wavelength of the scattered light and, consequently, of the color of the observed sample [6,7,45].

A nanoparticle with a diameter of 10 nm, for example, contains ~10% of its atoms at the surface, which become more active than the atoms in the volume due to the number of bonds that the volumetric atoms make with the closest neighbors. Consequently, the surface atoms in the nanoparticles have more electron energy levels due to the imperfections or active sites caused by the bonds that are not completed. When an electromagnetic wave is incident on the nanoparticle, these surface conduction electrons interact with the electric field of the incident wave, which induces the polarization of the free electrons in relation to the ions of the crystal structure of the material. These free electrons start to oscillate together from one side to the other side at the surfaces of the nanoparticles. When the incident radiation frequency approaches the oscillation frequency of the conduction electrons (dipolar resonance), strong energy absorption is induced at the surfaces of the nanoparticles (region in which the oscillation occurs), which is known as surface plasmon absorption (which is dependent on the nanoparticle size and shape). To illustrate this phenomenon, Fig. 5.5 (top panels) presents the micrographs of the seeds (a) that originate the rod-shaped gold nanoparticles (b–e) with different aspect ratios (length divided by width). The bottom panels show the optical spectra that correspond to each suspension [46].

The macroscopic metal materials have a continuous conduction band, and the nanoparticles have discrete electronic states that are dependent on their size. In nanometric scale, a significant percentage of the material surface is exposed, which intensifies the surface and interface effects, which, consequently, also produce a significant variation in their magnetic properties [45].

A magnetic field can be created due to the electron movement inside the material, whereas a magnetic model can arise due to the spin orientations inside the sample. Because small magnetic particles tend to form monodomains in the range of 20–2000 nm, the magnetic behavior of the majority of experimental systems can be attributed to the contributions of the effect of size and the interactions among the nanoparticles. The correlation between the nanostructure and its magnetic properties has isolated nanometric particles that generally have small interactions with magnetism due to the spin polarized tunneling effect and oscillatory coupling between magnetic/nonmagnetic multilayers and magnetoresistance. Conversely, volumetric materials with nanoscale structures (nanostructured films), whose magnetic properties are predominantly caused by the interactions involved in these systems, are observed. However, when the nanoparticle size is reduced to a size smaller than the critical diameter, the formation of domains is energetically unfavorable,

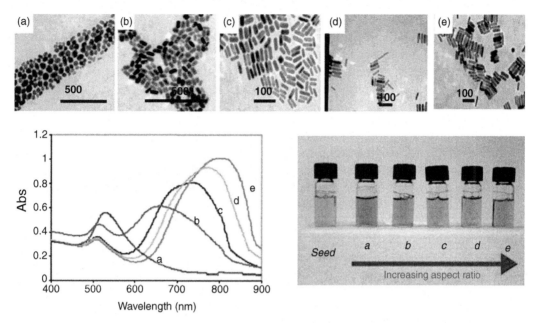

FIGURE 5.5 Transmission electron micrographs (top), optical spectra (left) and photographs (right) of gold nanorod aqueous solutions of different aspect ratios (length divided by width). Scale: 500 nm for (a) and (b); 100 nm for (c–e). *C.J. Murphy, T.K. Sau, A.M. Gole, C.J. Orendorff, J. Gao, L Gou, et al., Anisotropic metal nanoparticles: synthesis, assembly, and optical applications, J. Phys. Chem. B. 109 (2005) 13857–13870 [46].*

and the variations in the magnetization are not caused by the movement of domains but by the coherent rotation of spins, which is frequently affected by thermal fluctuations (the system starts to present paramagnetic behaviors) [45]. An example that is often cited in studies is the possibility of the application of magnetic nanoparticles for cancer treatment. The nanoparticles are encapsulated with specific chemical agents that bond in tumor regions; with the control of magnetization, they agitate, which heats the affected region by a hyperthermia process and selectively destroys cancer cells [19].

5.4 Characterization Methods

Various experimental techniques can be employed for the characterization of nanoparticles, such as spectroscopic methods (UV-vis and infrared), XRD, TEM, scanning electron microscopy, and atomic force microscopy (AFM). Some methods and examples of nanoparticle characterization by XRD, TEM, and AFM are presented in the next section.

5.4.1 X-Ray Diffraction

The XRD technique enables the investigation of the crystallinity degree and the size of the crystallites of different materials, including nanoparticles [47], according to the spaces between adjacent atomic planes [48]. With the incidence of X-rays of wavelength λ, with the incidence angle θ on a crystal, they can pass with no interruptions or interact with the atoms of the crystal lattice of the material. Bragg's law (Eq. 5.5) indicates that diffracted X-rays of various intensities, which represent a specific interplanar distance d in the lattice, can be detected when X-rays of known wavelength and incidence angle are directed to a crystal lattice.

$$n\lambda = 2\,d\,sen\theta \tag{5.5}$$

Scanning at different angles and a constant wavelength generates a diffraction pattern [48]. The determination of the unit cell size and interplanar atomic distances can be evaluated by the angle position of the peaks in the diffractogram, and the atomic arrangement inside each unit cell can be associated with the relative intensities of these peaks [49].

XRD has also been explored in the estimation of nanoparticle size with Scherrer's equation (Eq. 5.6), which relates the full width at half maximum (FWHM) of the diffraction peak with the crystallite size D [7].

$$D = \frac{k\lambda}{w\cos\theta} \tag{5.6}$$

where λ is the wavelength of the incidence photons (nm), θ is Bragg's reflection angle, w is the FWHM and k is a constant that is dependent on geometrical factors [50]. The widening of the peaks represents crystallites of increasingly smaller sizes [51]; thus, nanoparticles present wide diffraction peaks that differ from the materials at the macroscopic scale, which present fine lines in the diffractograms. Note that Scherrer's equation can be applied to crystallites of average size in the range of approximately 100–200 nm; above these dimensions, the diffraction peak width can be affected by other factors than the crystallite size [52].

Borchert et al. investigated the nanoparticle size of $CoPt_3$ via three different methods: TEM, small-angle X-ray scattering (SAXS) and XRD. The correctly adapted Scherrer's equation for the near-spherical geometry of $CoPt_3$ nanoparticles provides diameters that are highly consistent with the diameters obtained by the other methods: 8.5 nm by TEM, 8.11 by SAXS and 8.4 with the adapted Scherrer's equation (Eq. 5.7), which yields a difference less than 5%.

$$D = \frac{4}{3}\frac{0.9\lambda}{w\cos\theta} \tag{5.7}$$

The method for the determination of the average crystallite size by XRD is indirect, and its use is preferably combined with microscopic techniques. With the appropriate care in the experiments and analyses, the results have been consistent with the values determined by direct methodologies. Details about Scherrer's equation in the investigation of nanoparticle size are provided in Refs. [50,53].

5.4.2 Transmission Electron Microscopy

TEM is one of the most prevalent techniques for the nanoparticle size and shape analysis. Electrons are generally accelerated at 100 KeV or higher energy values, displaced on a thin

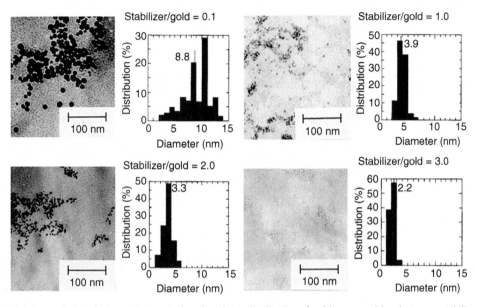

FIGURE 5.6 Transmission electron microscopy (TEM) and size distribution of gold nanoparticles that were stabilized in 3-mercaptopropionate reduced via citrate, with different stabilizer/gold ratios. *T. Yonezawa, T. Kunitake, Practical preparation of anionic mercapto ligand-stabilized gold nanoparticles and their immobilization, Colloids Surf. A 149 (1999) 193–199 [40].*

layer of the sample, and elastically or inelastically scattered. The inelastic scattering occurs in heterogeneous regions, such as intergrain boundaries, defects, and density variations, and cause scattering effects that produce differences in the intensities of the transferred electrons [7]. The wavelength of the electron beam limits the image resolution, which may attain the Angstrom scale depending on the equipment.

Yonezawa and Kunitake, for example, produced gold nanoparticles that were stabilized in 3-mercaptopropionate reduced via citrate. The TEM investigation provided a detailed analysis of the form and size distribution (Fig. 5.6). Particles presented spherical shape and smaller dimensions when the stabilizer/gold ratio was higher [40]. From the analysis of the histograms, a narrow size distribution was observed in greater stabilizer/gold ratio, and a broad variation in size was observed when the ratio was 0.1.

5.4.3 Atomic Force Microscopy

AFM is an important tool for the investigation of the surface topography and roughness at the atomic level [54]. A flexible probe that is sensitive to an interaction force is displaced with a tracking pattern on the surface of a solid sample. The force that acts between the probe and the sample surface causes small deflections in the cantilever, which are detected by optical systems [54]. Surface scanning with a continuous pattern *xy*, which is performed with a cantilever with a tiny tip that moves up and down along the *z*-axis according to the topographic change of the surface can be translated by a computer into a topography image [54].

Three modes can be employed for AFM image acquisition: contact, noncontact and in-termittent. The contact mode is not adequate for soft surfaces, such as biological samples and polymers, because the tip will be in constant contact with it. In some cases, the damage can be avoided by the intermittent mode, in which the tip is periodically in contact with the sample surface for a short period of time. In the noncontact mode, the tip fluctuates over the surface at a few nanometers of distance [54].

Sruanganurak et al. investigated the modification of the surface of natural latex rubber to minimize the friction effect, which is the main problem of latex application in gloves. Poly(methyl methacrylate) (PMMA) nanoparticles were deposited by the self-assembly technique (LbL) on the natural rubber substrates that were treated with polyacrylamide (natural rubber grafted with polyacrylamide, NR-g-PAAm). The AFM analysis that was employed in the intermittent mode (tapping mode) enabled the analysis of the surface morphology and roughness. In Fig. 5.7, C_s represents the surface coverage, which is expressed by C_s (%) = $(N/N_{max})*100$, where N is the number of adsorbed particles per unit area and N_{max} is the maximum number of adsorbed particles in the same area, assuming hexagonal close packing [55].

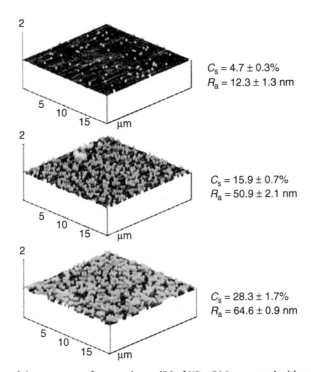

$C_s = 4.7 \pm 0.3\%$
$R_a = 12.3 \pm 1.3$ nm

$C_s = 15.9 \pm 0.7\%$
$R_a = 50.9 \pm 2.1$ nm

$C_s = 28.3 \pm 1.7\%$
$R_a = 64.6 \pm 0.9$ nm

FIGURE 5.7 Investigation of the average surface roughness (R_a) of NR-g-PAAm covered with poly(methyl methacrylate) (PMMA) particles, with different C_s values. A. Sruanganurak, K. Sanguansap, P. Tangboriboonrat, Layer-by-layer assembled nanoparticles: a novel method for surface modification of natural rubber latex film, Colloids Surf. A 289 (2006) 110–117 [55].

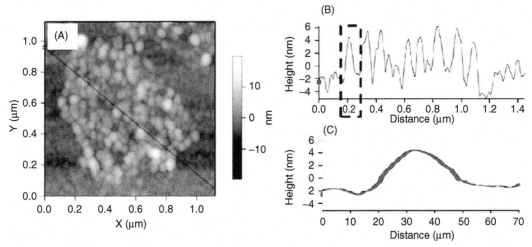

FIGURE 5.8 Atomic force microscopy (AFM) image of (A) cast film of graphene–platinum colloidal suspension, (B) topology of the cross-section at the line showed in the image in (A), and (C) expanded image of the topology of the cross-section of a cluster [represented by the dashed rectangle in (B)]. *Adapted from W. Chartarrayawadee, S.E. Moulton, D. Li, C.O. Too, G.G. Wallace, Novel composite graphene/platinum electro-catalytic electrodes prepared by electrophoretic deposition from colloidal solutions, Electrochim. Acta 60 (2012) 213–223 [56].*

Chartarrayawadee et al. employed the AFM technique to confirm the anchorage of platinum nanoparticles on graphene sheets, as illustrated in Fig. 5.8. The composite was successfully applied for the catalysis in the reactions of solar cells, which are sensitized by dyes, and the generation of hydrogen from acid. According to the topological analysis of the cross-section, the heights of the clusters in the graphene-platinum composite varies from 3 to 6 nm, whereas the diameter varies from 40 to 100 nm. With the amplification of the topological image of a cluster [dashed rectangle in (B)], its dimension and shape can be observed [56].

5.5 Applications

5.5.1 Biosensors

Nanoparticles can exert many functions in applications such as biosensors, which favor the immobilization of biomolecules and the catalysis of electrochemical reactions, in addition to an increase in the charge transfer and biomolecule labeling [57]. The immobilization of biomolecules in the volume of a material can cause the denaturation or loss of bioactivity; however, the use of nanoparticles enables the immobilization and the preservation of the biocompatibility. The use of nanoparticles with catalytic properties produces high-sensitivity sensors, which attain limits of detection lower than 2 fM [57].

Molecular recognition is one of the most important points of biosensor selectivity because some biological entities can be recognized and linked to one another with high selectivity and specificity [7]. These biological entities include antibodies, oligonucleotides, and enzymes. The antibodies—proteins of the immune system—recognize a virus as an intruder and bond to viruses to destroy them. For example, antibodies and oligonucleotides can easily bond to the surfaces of nanoparticles via thiol–Au bonds in the case of gold nanoparticles or covalent bonds at silanized surfaces via biotin–avidin bonds, in which avidin is bonded to the material surface [7].

Gold nanoparticles have been applied for the amplification of the analytical signal due to its ease of synthesis, narrow size distribution, efficient surface modification by thiols and other ligands and biocompatibility [58]. The metal nanoparticles can be guided to specific regions using the antigen-antibody recognition in the ligand–receptor interactions, for example, by bonding to cancer cells [19].

Biosensors that are based on molecular recognition, such as the antigen–antibody interaction, are referred to as immunosensors [59]. Li et al. developed a fast-detection immunosensor for *Escherichia coli*. The gold electrode was treated to receive -NH$_2$ functional groups and after, alternated nanoparticle layers of gold and the chitosan–MWNT–thionine composite were produced via the LbL technique. Voltammetric readings guarantee the modification of electrodes, and the sensitivity and stability of the sensor are related to the amount of thionine mediator [60]. Subsequently, anti-*E. coli* O157:H7 was immobilized, followed by a period of incubation for 60 min in bovine serum albumin. The sensor proved to be efficient for the detection of *E. coli* in milk and water samples.

5.5.2 Catalysis

Since the surface atoms of nanoparticles are more reactive than the volumetric atoms of nanoparticles, they have sites with missing bonds, which increases the activity of nanoparticles and provides additional electronic states and a higher reactivity [6].

Fig. 5.9 illustrates the increase in surface area with the decrease in the volume size of the nanoparticles and the increase in surface area, which are proportional to its reactivity and catalytic activity.

The versatility of the nanoparticles enables their use as homogeneous and heterogeneous catalysts. In the first case, only the nanoparticle is employed; in the second case, it

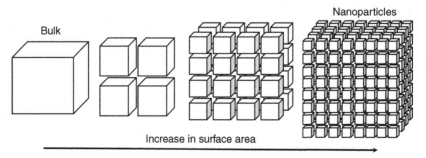

FIGURE 5.9 Increase in surface area, which is proportional to its reactivity and catalytic activity.

can be anchored or deposited on a substrate. Nanoparticles exhibit excellent performance as catalysts in many areas, including dehydrogenation, halogenation, oxidation, reduction, decomposition, and electron transfer reactions, and the catalytic efficiency is dependent on their shape, composition and size [6].

For example, the catalysis of oxidoreduction in fuel cells, in which a suitable catalyst is required for efficient redox reactions, can be cited. Metal colloidal particles, especially platinum, are very interesting due to their catalytic action in methanol oxidation reactions and oxygen reduction, which are responsible for the energy generation in direct methanol fuel cells (DMFC) [61]. Platinum nanoparticles have been applied with graphene as LbL films, and their catalytic activities were analyzed by cyclic voltammetry [62]. Graphene was modified with an ionic liquid to obtain a positively charged suspension, whereas the platinum nanoparticles were stabilized in citrate with negative charges. The LbL film that electrostatically formed presented high electrocatalytic activity in the reduction of oxygen [62]. The production of composite catalysts with more than one metal (generally Pt alloys) is interesting for applications in methanol oxidation in DMFCs. Yola et al. synthesized catalysts composed of Pt nanoparticles and bimetal particles (contain gold and silver with platinum) that were anchored in functionalized graphene oxide for the oxidation of methanol. The formed Au-Pt/GO was superior than other tested structures, which indicates a wide active surface, high electrocatalytic activity, and high tolerability to contamination by carbon monoxide (product of the methanol oxidation reaction) [63]. To reduce the mass of Pt that is required by the electrode and reduce the cost of energy generation, Zhao et al. synthesized core-shell-structured nanoparticles, an Au nucleus and a shell composed of Pt and Cu alloy to obtain a larger electrochemically active area that is superior to other structures (PtCu/C, AuPtCu/C and commercial Pt/C) due to the synergic effects between the nanostructured core and the shell with uniform dispersion [64].

The synthesis process by the Rampino and Nord method, for example, is extensively applied for the production of an aqueous suspension of platinum nanoparticles. This method involves K_2PtCl_4 and polyacrylic acid solution that receives argon gas and is reduced by hydrogen gas [65]. Hao-Lin et al. proposed the modified synthesis of platinum nanoparticles with Nafion. The particles presented a size of approximately 4 nm, were adsorbed by the surfaces of the carbon nanotubes and employed as catalysts in proton exchange membrane fuel cells, which demonstrate a superior performance [66]. Other researchers applied in situ polymerization with platinum [67,68] for the modification of Nafion membranes to target the application in DMFCs. The in situ polymerization of pyrrole was performed on the Nafion membrane, followed by the reduction of platinum salt to form nanoparticles. The methanol permeability (one of the main problems in DMFCs) decreased with the time of pyrrole monomer impregnation [67].

5.5.3 Magnetic Nanoparticles in Biomedicine

The quantum effect and the large surface area of the magnetic nanoparticles produce a material with altered magnetic properties, which presents a superparamagnetic phenomenon because each particle is considered to be a single magnetic domain [1]. Factors such as the biocompatibility and possible functionalization of the surfaces of magnetic nanoparticles elevate the potential for their use in biomedical applications.

The magnetic nanoparticles can be synthesized and encapsulated by polymers, which enables the functionalization of the surface according to the objective. Thus, different biological molecules, such as antibodies, proteins, and target ligands, can bond to the surfaces of these nanoparticles by amide or ester chemical bonds [1]. Among the most investigated applications are the controlled release of pharmaceuticals, applications in hyperthermia, imaging by nuclear magnetic resonance and the separation and selection of molecules [3,69].

The iron oxides are the most investigated materials, with an emphasis on maghemite (γ-Fe_2O_3) and magnetite (Fe_2O_3). The possibility of nanoparticles to be controlled by an external magnetic field enables their conduction through specific body parts and favors the release of medication in certain regions. The use of nanometric systems enables the direct delivery of the drug to the cells or specific tumor regions surrounded by a healthy tissue [70]. Peptides and proteins can bond to nanoparticles, permeate membranes and enable intracellular drug delivery [1]. When subjected to an external magnetic field of alternated frequency, the magnetic nanoparticles cause the heating of the location where they are located (magneto hyperthermia, as previously mentioned). Since tumor cells are more sensitive to a significant temperature increase than normal cells, they are destroyed at temperatures between 41 and 47°C [1,69].

Tissue regeneration is another well-explored area because stem cells have significant potential for substituting degenerated cells or repairing damaged tissue [1]. Superparamagnetic nanoparticles can be associated with stem cells to be transported to the location of interest. Many proteins and growth factors can be associated with these nanoparticles and, therefore, are distributed in the damaged tissue, where they can act to regenerate the affected area.

5.6 Final Considerations

The unique properties of nanoparticles have been extensively explored by the most diverse fields, which corroborates the prediction by Feynman regarding the potential of nanostructures. The changes in the physical and chemical properties of the nanoparticles are related to the increase in the number of surface atoms in relation to the volume. Similarly, the electronic properties are altered by the formation of discrete energy levels instead of the conventional band structure for volumetric materials. This chapter discussed typical synthesis processes for the production of metal and oxide nanoparticles according to the need of application. Additionally, the improvements in the processes of material characterization during the last decade significantly contributed to the investigation of nanoparticles. Although various studies have been performed, additional exploration is needed.

List of Symbols

C	Solute concentration
C_0	Solubility
C_s	Surface coverage
d	Interplanar distance
D	Particle diameter
σ	Concentration of supersaturation
ΔGv	Gibbs free energy of volume
$\Delta\mu_v$	Surface energy
γ	Surface energy per unit area
k	Constant dependent on geometric factors
K	Boltzmann constant
λ	Wavelength
N	Number of particles adsorbed per unit area
N_{max}	Maximum number of particles adsorbed per unit area
nm	Nanometers
Ω	Atomic volume
r	Radius of spherical nucleus
r^*	Critical radius
R_a	Roughness average
T	Temperature
θ	Incidence angle in relation to the considered plane
w	Full width at half maximum

References

[1] A.K. Gupta, M. Gupta, Synthesis and surface engineering of iron oxide nanoparticles for biomedical applications, Biomaterials 26 (2005) 3995–4021.

[2] J. Lee, D.K. Chatterjee, M.H. Lee, S. Krishnan, Gold nanoparticles in breast cancer treatment: promise and potential pitfalls, Cancer Lett. 347 (2014) 46–53.

[3] Q.A. Pankhurst, J. Connolly, S.K. Jones, J. Dobson, Applications of magnetic nanoparticles in biomedicine, J. Phys. D: Appl. Phys. 36 (2003) R167.

[4] J.S. Bradley, G. Schmid, Noble Metal Nanoparticles, In: Nanoparticles: From Theory to Application, Wiley-VCH, Weinheim, 2004, pp. 186–198.

[5] E. Katz, A.N. Shipway, I. Willner, Biomaterial-nanoparticle hybrid systems: synthesis, properties, and applications, In: Nanoparticles: From Theory to Application, Wiley-VCH, Germany, 2004, pp. 368–421.

[6] Y.S. Lee, Self-assembly and nanotechnology: a force balance approach, Jonh Wiley & Sons, New Jersey, (2008).

[7] G. Cao, Nanostructures & Nanomaterials: Synthesis, Properties & Applications, Imperial College Press, London, (2004).

[8] H.J. Fecht, Formation of nanostructures by mechanical attrition, In: Nanomaterials: Synthesis, Properties and Applications, Taylor & Francis, New York, 1996, pp. 89–110.

[9] C.C. Koch, The synthesis and structure of nanocrystalline materials produced by mechanical attrition: a review, Nanostruct. Mater. 2 (1993) 109–129.

[10] H.J. Fecht, Nanostructured materials and composites prepared by solid state processing, In: Nano-structured Materials: Processing, Properties and Applications, Noyes Publication, New York, 2002, pp. 73–114.

[11] C. Lam, Y.F. Zhang, Y.H. Tang, C.S. Lee, I. Bello, S.T. Lee, Large-scale synthesis of ultra"ne Si nanopar-ticles by ball milling, J. Cryst. Growth 220 (2000) 466–470.

[12] E.A. Dobisz, F.A. Buot, C.R.K. Marrian, Nanofabrication and Nanoelectronics, In: Nanomaterials: Syn-thesis, Properties and Applications, Taylor & Francis, London, 1996, pp. 495–540.

[13] U. Hilleringmann, Lithography procedures, In: Nanotechnology and Nanoelectronis: Materials, Devices, Measurement Techniques, Springer-Verlag, Berlin, 2005, pp. 154–171.

[14] H. Frey, H.R. Khan, Handbook of Thin Film Technology, Springer Science & Business Media, Berlin, (2015).

[15] Muttaqin, T. Nakamura, S. Sato, Synthesis of gold nanoparticle colloids by highly intense laser irradia-tion of aqueous solution by flow system, Appl. Phys. A. 120 (2015) 881–888.

[16] M. Vinod, K.G. Gopchandran, Ag@Au core–shell nanoparticles synthesized by pulsed laser ablation in water: effect of plasmon coupling and their SERS performance, Spectrochim. Acta A Mol. Biomol. Spectrosc. 149 (2015) 913–919.

[17] M. Hillenkamp, G. di Domenicantonio, C. Félix, Monodispersed metal clusters in solid matrices: a new experimental setup, Rev. Sci. Instrum. 77 (2006) 025104.

[18] A.D.T. de Sá, V.T.A. Oiko, G. di Domenicantonio, V. Rodrigues, New experimental setup for metallic clusters production based on hollow cylindrical magnetron sputtering, J. Vac. Sci. Technol. B 32 (2014) 061804.

[19] T.-H. Tran, T.-D. Nguyen, Controlled growth of uniform noble metal nanocrystals: aqueous-based synthesis and some applications in biomedicine, Colloids Surf. B Biointerfaces 88 (2011) 1–22.

[20] V.K. LaMer, R.H. Dinegar, Theory, production and mechanism of formation of monodispersed hydro-sols, J. Am. Chem. Soc. 72 (1950) 4847–4854.

[21] R. Viswanatha, D.D. Sarma, Growth of nanocrystals in solution, In: Nanomaterials Chemistry: Recent Developments and New Directions, Wiley-VCH, Weinheim, 2007, pp. 139–170.

[22] G.M. Chow, Gonsalves, K.E., Particle synthesis by chemical routes, In: Nanomaterials: Synthesis, Properties and Applications, Taylor & Francis, New York, 1996, pp. 55–71.

[23] H. Huang, X. Yang, Synthesis of chitosan-stabilized gold nanoparticles in the absence/presence of tripolyphosphate, Biomacromolecules 5 (2004) 2340–2346.

[24] A. Murugadoss, H. Sakurai, Chitosan-stabilized gold, gold–palladium, and gold–platinum nano-clusters as efficient catalysts for aerobic oxidation of alcohols, J. Mol. Catal. A: Chem. 341 (2011) 1–6.

[25] A. Abou-Okeil, A. Amr, F.A. Abdel-Mohdy, Investigation of silver nanoparticles synthesis using ami-nated β-cyclodextrin, Carbohydr. Polym. 89 (2012) 1–6.

[26] Y. Huang, D. Li, J. Li, β-Cyclodextrin controlled assembling nanostructures from gold nanoparticles to gold nanowires, Chem. Phys. Lett. 389 (2004) 14–18.

[27] D. Chen, X. Qiao, X. Qiu, J. Chen, Synthesis and electrical properties of uniform silver nanoparticles for electronic applications, J. Mater. Sci. 44 (2009) 1076–1081.

[28] L. Guo, S. Yang, C. Yang, P. Yu, J. Wang, W. Ge, et al. Synthesis and characterization of poly(vinylpyrrolidone)-modified zinc oxide nanoparticles, Chem. Mater. 12 (2000) 2268–2274.

[29] M. Habib Ullah, T. Hossain, C.-S. Ha, Kinetic studies on water-soluble gold nanoparticles coordinated to poly(vinylpyrrolidone): isotropic to anisotropic transformation and morphology, J. Mater. Sci. 46 (2011) 6988–6997.

[30] A. Gautam, G.P. Singh, S. Ram, A simple polyol synthesis of silver metal nanopowder of uniform par-ticles, Synth. Met. 157 (2007) 5–10.

[31] P.S. Roy, J. Bagchi, S.K. Bhattacharya, Size-controlled synthesis and characterization of polyvinyl alcohol coated palladium nanoparticles, Transit. Metal Chem. 34 (2009) 447–453.

[32] L.C. Klein, Processing of nanostructured sol–gel materials, In: Nanomaterials: Synthesis, Properties and Applications, Taylor & Francis, New York, 1996, pp. 147–164.

[33] M. Niederberger, M. Antonietti, Nonaqueous sol--gel routes to nanocrystalline metal oxides, In: Nanomaterials Chemistry: Recent Developments and New Directions, Wiley-VCH, Weinheim, 2007, pp. 119–138.

[34] M. Niederberger, Nonaqueous Sol–Gel routes to metal oxide nanoparticles, Acc. Chem. Res. 40 (2007) 793–800.

[35] H. Zeng, P.M. Rice, S.X. Wang, S. Sun, Shape-controlled synthesis and shape-induced texture of Mn-Fe$_2$O$_4$ nanoparticles, J. Am. Chem. Soc. 126 (2004) 11458–11459.

[36] J. Ba, J. Polleux, M. Antonietti, M. Niederberger, Non-aqueous synthesis of tin oxide nanocrystals and their assembly into ordered porous mesostructures, Adv. Mater. 17 (2005) 2509–2512.

[37] C. de Mello Donegá, P. Liljeroth, D. Vanmaekelbergh, Physicochemical evaluation of the hot-injection method, a synthesis route for monodisperse nanocrystals, Small 1 (2005) 1152–1162.

[38] N.A. Frey, S. Peng, K. Cheng, S. Sun, Magnetic nanoparticles: synthesis, functionalization, and applications in bioimaging and magnetic energy storage, Chem. Soc. Rev. 38 (2009) 2532.

[39] J. Turkevich, P.C. Stevenson, J. Hillier, A study of the nucleation and growth processes in the synthesis of colloidal gold, Discuss. Faraday Soc. 11 (1951) 55–75.

[40] T. Yonezawa, T. Kunitake, Practical preparation of anionic mercapto ligand-stabilized gold nanoparticles and their immobilization, Colloids Surf. A 149 (1999) 193–199.

[41] T. Zhu, K. Vasilev, M. Kreiter, S. Mittler, W. Knoll, Surface modification of citrate-reduced colloidal gold nanoparticles with 2-mercaptosuccinic acid, Langmuir 19 (2003) 9518–9525.

[42] F.G. Zhao, W.S. Li, Dendrimer/inorganic nanomaterial composites: tailoring preparation, properties, functions, and applications of inorganic nanomaterials with dendritic architectures, Sci. China Chem. 54 (2011) 286–301.

[43] R.M. Crooks, M. Zhao, L. Sun, V. Chechik, L.K. Yeung, Dendrimer-encapsulated metal nanoparticles: synthesis, characterization, and applications to catalysis, Acc. Chem. Res. 34 (2001) 181–190.

[44] F.N. Crespilho, M. Emilia Ghica, M. Florescu, F.C. Nart, O.N. Oliveira Jr., C.M.A. Brett, A strategy for enzyme immobilization on layer-by-layer dendrimer–gold nanoparticle electrocatalytic membrane incorporating redox mediator, Electrochem. Commun. 8 (2006) 1665–1670.

[45] G.L. Hornyak, H.F. Tibbals, J. Dutta, J.J. Moore, Introduction to Nanoscience and Nanotechnology, CRC Press, Boca Raton, (2009).

[46] C.J. Murphy, T.K. Sau, A.M. Gole, C.J. Orendorff, J. Gao, L. Gou, et al. Anisotropic metal nanoparticles: synthesis, assembly, and optical applications, J. Phys. Chem. B. 109 (2005) 13857–13870.

[47] C.P. Poole Jr., F.J. Owens, Introduction to Nanotechnology, Wiley-Interscience, New Jersey, (2003).

[48] B.S. Mitchell, The structure of materials, in: B.S. Mitchell (Ed.), An Introduction to Materials Engineering and Science: For Chemical and Materials Engineers, John Wiley & Sons, New Jersey, 2004, pp. 1–135.

[49] W.D. Callister, Ciência e engenharia de materiais [Materials sciences and engineering], seventh ed., LTC, Rio de Janeiro, (2008).

[50] H. Borchert, E.V. Shevchenko, A. Robert, I. Mekis, A. Kornowski, G. Grübel, et al. Determination of nanocrystal sizes: a comparison of TEM, SAXS, and XRD studies of highly monodisperse CoPt$_3$ particles, Langmuir 21 (2005) 1931–1936.

[51] V.S. Zaitsev, D.S. Filimonov, I.A. Presnyakov, R.J. Gambino, B. Chu, Physical and chemical properties of magnetite and magnetite-polymer nanoparticles and their colloidal dispersions, J. Colloid Interface Sci. 212 (1999) 49–57.

[52] U. Holzwarth, N. Gibson, The Scherrer equation versus the 'Debye-Scherrer equation', Nat. Nanotechnol. 6 (2011) 534–1534.

[53] A.L. Patterson, The Scherrer formula for X-ray particle size determination, Phys. Rev. 56 (1939) 978.

[54] D.A. Skoog, F.J. Holler, S.R. Crouch, Instrumental Analysis, sixth ed., Cengage Learning, Belmont, CA, (2007).

[55] A. Sruanganurak, K. Sanguansap, P. Tangboriboonrat, Layer-by-layer assembled nanoparticles: a novel method for surface modification of natural rubber latex film, Colloids Surf. A 289 (2006) 110–117.

[56] W. Chartarrayawadee, S.E. Moulton, D. Li, C.O. Too, G.G. Wallace, Novel composite graphene/platinum electro-catalytic electrodes prepared by electrophoretic deposition from colloidal solutions, Electrochim. Acta 60 (2012) 213–223.

[57] X. Luo, A. Morrin, A.J. Killard, M.R. Smyth, Application of nanoparticles in electrochemical sensors and biosensors, Electroanalysis 18 (2006) 319–326.

[58] X. Cao, Y. Ye, S. Liu, Gold nanoparticle-based signal amplification for biosensing, Anal. Biochem. 417 (2011) 1–16.

[59] M.C. Daniel, D. Astruc, Gold nanoparticles: assembly, supramolecular chemistry, quantum-size-related properties, and applications toward biology, catalysis, and nanotechnology, Chem. Rev. 104 (2004) 293–346.

[60] Y. Li, P. Cheng, J. Gong, L. Fang, J. Deng, W. Liang, et al. Amperometric immunosensor for the detection of *Escherichia coli* O157:H7 in food specimens, Anal. Biochem. 421 (2012) 227–233.

[61] G.S. Devi, V. Rao, Room temperature synthesis of colloidal platinum nanoparticles, Bull. Mater. Sci, 23 (2000) 467–470.

[62] C. Zhu, S. Guo, Y. Zhai, S. Dong, Layer-by-layer self-assembly for constructing a graphene/platinum nanoparticle three-dimensional hybrid nanostructure using ionic liquid as a linker, Langmuir 26 (2010) 7614–7618.

[63] M.L Yola, T. Eren, N. Atar, H. Saral, İ. Ermiş, Direct-methanol fuel cell based on functionalized graphene oxide with mono-metallic and Bi-metallic nanoparticles: electrochemical performances of nanomaterials for methanol oxidation, Electroanalysis 28 (2015) 570–579.

[64] Z.L. Zhao, L.Y. Zhang, S.J. Bao, C.M. Li, One-pot synthesis of small and uniform Au@PtCu core–alloy shell nanoparticles as an efficient electrocatalyst for direct methanol fuel cells, Appl. Catal. B: Environ. 174–175 (2015) 361–366.

[65] L.D. Rampino, F. Nord, Preparation of palladium and platinum synthetic high polymer catalysts and the relationship between particle size and rate of hydrogenation, J. Am. Chem. Soc. 63 (1941) 2745–2749.

[66] T. Hao-lin, P. Mu, M. Shi-chun, Y. Run-zhang, Synthesis of platinum nanoparticles modified with nafion and the application in PEM fuel cell, J. Wuhan Univ. Technol.–Mater. Sci. Ed. 19 (2004) 7–9.

[67] L. Li, Y. Zhang, J.-F. Drillet, R. Dittmeyer, K.-M. Jüttner, Preparation and characterization of Pt direct deposition on polypyrrole modified Nafion composite membranes for direct methanol fuel cell applications, Chem. Eng. J. 133 (2007) 113–119.

[68] M.A. Smit, A.L. Ocampo, M.A. Espinosa-Medina, P.J. Sebastián, A modified Nafion membrane with in situ polymerized polypyrrole for the direct methanol fuel cell, J. Power Sources 124 (2003) 59–64.

[69] A. Singh, S.K. Sahoo, Magnetic nanoparticles: a novel platform for cancer theranostics, Drug Discov. Today 19 (2014) 474–481.

[70] G.A. Hughes, Nanostructure-mediated drug delivery, Nanomedicine: Nanotechnology, Biology and Medicine. 1 (2005) 22–30.

6

Magnetic Nanomaterials

M.A.G. Soler*, L.G. Paterno**

*INSTITUTE OF PHYSICS, UNIVERSITY OF BRASÍLIA, UNIVERSITY CAMPUS
DARCY RIBEIRO, BRASILIA, BRAZIL; **INSTITUTE OF CHEMISTRY, UNIVERSITY
OF BRASILIA, UNIVERSITY CAMPUS DARCY RIBEIRO, BRASILIA, BRAZIL

CHAPTER OUTLINE

6.1 Introduction

Magnetic materials of nanometric dimensions are found in nature, including the magnetite nanoparticles (NPs) present in many bacteria, insects, and larger animals. Many migratory animals also have magnetic NPs in their body and use them as biomagnetic compasses. For instance, salmon use the magnetite NPs present in their nasal fossa for orientation during their migratory travels [1]. Artificial magnetic nanomaterials have been developed since the 1980s, especially because of the availability of proper instrumentation to characterize structures and properties at the nanoscale and to develop many synthesis routes and elaborate surface treatments. Magnetic nanostructures can be simply produced in the form of NPs of a specific magnetic material (metals, magnetic alloys, or oxides), molecular magnets or even as one-, two-, or three-dimensional arrangements such as nanothreads, mono- and multilayer films, NP agglomerates (clusters), dispersions in nonmagnetic lattices (nanocomposites), and others [2]. NPs can be prepared in different shapes such as spheres, rods, fibers, and polyhedrons in general, from cubes to multifaceted prisms [3,4].

Among the magnetic materials prepared at a nanometric dimension, special attention is drawn to those produced with the transition metals such as, Fe, Co, Ni, and their alloys; pure ferrites such as magnetite (Fe_3O_4) and maghemite (γ-Fe_2O_3); mixed ferrites such as, cobalt ($CoFe_2O_4$), nickel ($NiFe_2O_4$), manganese ($MnFe_2O_4$), zinc ($ZnFe_2O_4$), and copper ($CuFe_2O_4$) ferrite; $BaFe_{12}O_{19}$; $SmCo_5$; manganese compounds; or core–shell structure [3]. Magnetic NPs in their colloidal state, with sizes on the order of 10 nm, have also been created, especially because of their relatively easy preparation and stability in aqueous or organic medium suspensions. In addition, the colloidal suspensions of magnetic nanomaterials known as ferrofluids, nanofluids (NFs), or magnetic fluids (MFs) respond to the action of a magnetic field gradient as though they were a single liquid and magnetic phase, making them interesting materials for different purposes [5].

Superparamagnetic cubic ferrites (e.g., maghemite, magnetite, and cobalt ferrite) present a spinel-like structure and form monodomains with diameters ranging from 5 to 20 nm. These NPs can be synthesized via wet chemical routes with a reasonable control of shape, size, composition, crystallinity, and physical properties [6–9]. Because of their superparamagnetic behavior [5,10], these nanomaterials are known as superparamagnetic iron oxides (SPIOs), and they consist essentially of an iron oxide core with a diameter of a few nanometers and a subnanometric surface layer composed of iron oxyhydroxide groups [11,12]. In the colloidal form, the surface must be functionalized to achieve stability in suspension. Functionalization can be obtained by a simple acid–base reaction that introduces surface charges to the particles or by coating with molecular species of small molecules (e.g., citric acid) as well as surfactants or polymers. This coating also promotes the anchorage of other chemical species that are able to perform diagnosis and therapy tasks in vivo [13]. Because of their biocompatibility, SPIO particles have great potential for biomedical applications (e.g., drug delivery systems, markers, magnetic hyperthermia, and improving magnetic resonance imaging contrast). In addition, these systems might have a large effect on new industrial technologies such as insulating oils for transformers, spintronics structures, bioelectrochemistry, catalysis, and chemical sensors [14–17]. Therefore, the design and control of the properties of low-dimensional structures are a challenge for the fields of fundamental and applied magnetism.

This chapter presents the principles of magnetism for low-dimensional systems in a general way, emphasizing the properties of the magnetic nanostructures formed by iron oxide-based colloids. The discussion also includes details about the most used chemical synthesis methods for the production of iron oxide NPs as well as for surface functionalization and the preparation of films. Lastly, their major physico-chemical properties and some of their biomedical applications are discussed.

6.2 Basic Concepts of Magnetism

When a solid material is subjected to the action of an external magnetic field H, a magnetic field B is induced inside the material. The relationship between B and H is a material property, in SI units [18]:

$$\vec{B} = \mu_0 \left(\vec{H} + \vec{M} \right) \tag{6.1}$$

where $\vec{M}$ is the magnetization of the medium given by the magnetic moment per volume unit ($\vec{M} = \dfrac{\vec{m}}{V}$, where $\vec{m}$ is the magnetic moment); μ_0 is the magnetic permeability of vacuum, equal to $4\pi \times 10^{-7}$ Henry m^{-1} (SI). The units of M and H are (A m^{-1}), and those for μ_0 are weber (A m)$^{-1}$, also known as Henry m^{-1}; therefore, the units of B are weber m^{-2}, or tesla (T), where 1 Gauss = 10^{-4} T. The magnitude of M is proportional to the applied magnetic field H,

$$M = \chi H \tag{6.2}$$

where χ is the magnetic susceptibility of the material. In SI units, χ is dimensionless.

The response of solid materials to the action of an external magnetic field depends on their atomic structure, electromagnetic excitation, pressure, and temperature. Even in the classical approximation, the electrons and nuclei of solids produce magnetic fields because they are moving charges that have intrinsic magnetic dipole moments. The factors that contribute to the atomic magnetic moment are the orbital angular momentum, the electron spin, and their interaction. Hence, magnetism is essentially a quantum mechanical phenomenon. The solution of Schrödinger's equation provides information about the energy shells or states occupied by the electrons and (with this solution) magnetic moments. The state of one electron is characterized by four quantum numbers [19,20]:

1. The principal quantum number (n), with values of 1, 2, 3, ... corresponding to shells K, L, M, ...
2. The orbital angular momentum (l), which describes the angular momentum of the orbital movement. For a given value of l, the angular momentum of an electron due to its orbital movement is equal to $\hbar\sqrt{l(l+1)}$. The number l can have integer numbers of 0, 1, 2, 3, ... $n - 1$. The electrons with $l = 1, 2, 3, 4, ...$ refer to the electrons s, p, d, f, ...
3. The magnetic quantum number (m_l), which describes the component of the orbital angular momentum along a particular direction. For most cases, this direction is selected along an applied field. For a given value of l, $m_l = l, l-1, ..., 0, ..., -l+1, -l$. For instance, the values allowed for the angular moment along a direction are $2\hbar$, $\hbar$, 0, $-\hbar$, and $-2\hbar$ for a d electron.
4. The spin quantum number (m_s), which describes the electron spin component along a particular direction, generally in the direction of the applied field. The electron spin is the intrinsic angular momentum that corresponds to the rotation of each electron around its internal axis. The possible values for m_s are $\pm 1/2$, and the corresponding spin angular momentum components are $\pm\hbar/2$.

According to the Pauli exclusion principle, it is not possible for two electrons to occupy the same state (i.e., each electron should have a unique set of quantum numbers n, l, m_l, and m_s). The maximum number of electrons that occupy a given shell is given by [20]

$$2\sum_{l=0}^{n-1}(2l+1)=2n^2 \tag{6.3}$$

The movement of an electron can be considered as a stream flowing through a conducting wire. An electron with orbital angular momentum $\hbar l$ has an associated magnetic moment $\vec{\mu}_l$ given by

$$\vec{\mu}_l =-\frac{|e|}{2m}\hbar\vec{l}=-\mu_B\vec{l} \tag{6.4}$$

where μ_B is the Bohr magneton. The absolute value of the magnetic moment is given by

$$|\vec{\mu}_l|=\mu_B\sqrt{l(l+1)} \tag{6.5}$$

and its projection along the direction of the applied field is

$$\mu_{lz}=-m_l\mu_B \tag{6.6}$$

In the case of the spin moment, the associated magnetic moment is expressed as

$$\vec{\mu}_s=-g_e\frac{|e|}{2m}\hbar\vec{s}=-g_e\mu_B\vec{s} \tag{6.7}$$

where g_e (=2.002290716) is the spectroscopic splitting factor (or the g-factor for the free electron). The Bohr magneton expresses the magnetic dipole moment of the electron (i.e., the most fundamental magnetic moment), whose magnitude in SI is $9.27 \times 10^{-24}\,\mathrm{J\,T^{-1}}$. For each electron in an atom, the spin magnetic moment is $\pm\mu_B$.

The net magnetic moment of an atom is equal to the sum of the magnetic moments of each one of its electrons, including the contributions of the orbitals and their spin. For an atom with completely filled electron shells and subshells, there is a mutual of the moments (orbital and spin). Therefore, the materials formed by atoms with completely filled electron shells are unable to be permanently magnetized. These materials are known as diamagnetic materials. The eventual formation of a dipole moment in these materials only occurs when they are subjected to an external field that causes a temporary variation of the orbital movement. In terms of susceptibility, the diamagnetic materials have negative susceptibility; thus, the magnitude of the field induced internally by the action of an external field is lower than that in vacuum. In addition, the magnetic moment induced by the external field in these materials is extremely low and has a direction opposite to the applied field. In turn, the paramagnetic materials are formed by atoms with half-filled electron shells; therefore, they present a dipole moment. Nevertheless, dipoles are randomly oriented and are only aligned by the action of an external field, creating a small but

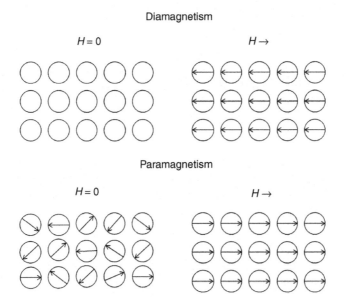

FIGURE 6.1 An illustration of the magnetic dipole arrangement for diamagnetism and paramagnetism.

positive susceptibility. A scheme of these behaviors is presented in Fig. 6.1. Both diamagnetic and paramagnetic materials are considered as nonmagnetic because they only present magnetization in the presence of an external field [19].

The transition metals, for instance, iron, cobalt, nickel, and some rare-earth metals such as, gadolinium, are ferromagnetic, with positive susceptibility on the order of 10^6. The antiferromagnetic materials present an antiparallel arrangement of spins that cancel each other and result in a net magnetic moment of zero. From a macroscopic point of view, ferrimagnetic materials are similar to ferromagnetic materials. The difference between them is in the origin of their net magnetic moments. Because these compounds are essentially ionic compounds such as, dual iron oxides (or ferrites), the magnetism of the ferrimagnetic materials is due to the spin moments of their unpaired metal ions in the crystalline lattice.

Magnetic materials have magnetic moments with a spontaneous long-range order. This order is due to a quantum electrostatic interaction known as exchange or interchange interaction. The interaction responsible for the magnetic order might be short-range (i.e., direct exchange interaction), long-range, or an indirect interaction [21]. Magnetic ordering (e.g., observed in the parallelism of atomic magnetic moments in ferromagnetic materials) is due to the exchange interaction. The presence of other interactions results in the formation of magnetic domains, regions of the sample where (to a first approximation) perfectly ordered moments can be considered. When an external magnetic field is applied, the borders of these domains (the domain walls) are displaced, causing changes in the ferromagnet magnetization [21]. Only materials with ferromagnetic or ferrimagnetic behaviors are considered as magnetic materials.

The formulation of the exchange interaction can be obtained from Schrödinger's equation for a two-electron system with spatial coordinates $\vec{r}_1$ and $\vec{r}_2$, spin coordinates σ, and nondegenerate energy states. Conventionally, σ can be +1 or −1, corresponding to the z projection of the spins equal to +1/2 and −1/2, respectively. Assuming the existence of an interaction between the electrons, a Coulombic potential term is introduced into the Hamiltonian, which (in the CGS system) is given by [21]

$$V_{1,2}(\vec{r}_1,\vec{r}_2) = \frac{e^2}{r_{1,2}} \tag{6.8}$$

The obtained eigenvalues have a term that refers to the Coulombic energy and another term known as the exchange integral (J) [21]. Hence, the introduction of the interaction term $V_{1,2}$ between spins leads to a new energy term that depends on the relative orientation of these spins. In addition, it is necessary to include the atom pair interaction (i,j), each one with an associated spin moment $\vec{S}$ that considers the contribution of the spin moments of all of their electrons. The Heisenberg Hamiltonian equation describes the interactions for the neighbor atom pairs in the lattice, and it is given by [22]

$$\mathcal{H} = -2 \sum_{i,j} \mathcal{J}_{ij} \vec{S}_i \cdot \vec{S}_j \tag{6.9}$$

where, i and j represent the positions of the atoms in the lattice, and J is the exchange constant. If $J > 0$, then the ferromagnetic interaction occurs, which couples the spins in a parallel manner (↑↑). In turn, if $J < 0$, then the interaction is antiferromagnetic, and the spins are coupled in an antiparallel manner (↑↓). The first type of interaction occurs in ferromagnets, whereas the second type occurs in antiferromagnetic and ferrimagnetic materials. Fig. 6.2 illustrates the arrangement of the dipole moments in the ferromagnetic, antiferromagnetic, and ferrimagnetic materials with their respective exchange constants.

As discussed at the beginning of this chapter, any description of magnetism (ferromagnetism and ferrimagnetism) should consider the total moments (orbital and spin) of the magnetic atom [20,21]. The total orbital angular momentum $(\vec{L})$ of an atom is given by the sum of the orbital momenta of all of its electrons [20]:

$$\vec{L} = \sum_i \vec{l}_i \tag{6.10}$$

Naturally, the sum for a complete electron shell is zero. The same logic can be applied to the total spin moment $(\vec{S})$, defined by [20]

$$\vec{S} = \sum_i \vec{s}_i \tag{6.11}$$

The result of the spin–orbit interaction, the total angular momentum $\left(\vec{J}\right)$, is given by [21]

$$\vec{J} = \vec{L} + \vec{S} \tag{6.12}$$

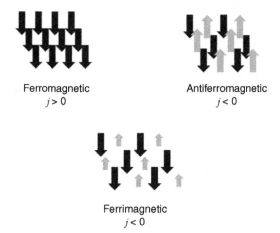

Ferromagnetic
$j > 0$

Antiferromagnetic
$j < 0$

Ferrimagnetic
$j < 0$

FIGURE 6.2 **An illustration of the dipole arrangements in ferromagnetic, antiferromagnetic, and ferrimagnetic systems, and the values of their exchange constant.** *Modified from D. Ortega, Structure and Magnetism in magnetic nanoparticles, in: N.T.K Thanh (Ed.), Magnetic Nanoparticles From Fabrication to Clinical Applications, Taylor & Francis Group, Boca Raton, FL, 2012 [22].*

The value of $\vec{J}$, also known as the spin–orbit or Russell–Saunders coupling, expresses the interaction between the spin and angular momenta of an electron orbit (note that $\vec{J}$ differs from $\mathcal{J}$). Its magnitude and effect on the energy shells of the atom depend on the relative orientation of the magnetic moments of spin and orbit. The value of J can range from $J = (L - S), (L - S + 1)...(L + S - 1), (L + S)$. The spin–orbit interaction energy is given by $\lambda \vec{L}.\vec{S}$, where λ is the coupling constant in cm^{-1}. When the coupling constant is positive (which occurs when the electron shell is less than half-filled), the minimum energy configuration (i.e., fundamental state) is obtained for antiparallel L and S (i.e., $J = L - S$). For a more than half-filled shell, the minimum energy configuration is given by $J = L + S$. For instance, consider the cobalt(II) ion: The 3d subshell of Co^{2+} contains 7 of the 10 electrons that it can accommodate; the subshell is more than half-filled, so the fundamental state is characterized by the quantum number $J = L + S = 9/2$ [22]. The elements that have incomplete shells are known as "transition elements" and belong to groups 3d (iron group), 4d (palladium group), 5d (platinum group), 4f (lanthanides), and 5f (actinides). Unlike what occurs with closed shells where the sum of the projections of the angular momenta $\vec{m}_l$ and $\vec{m}_s$ is zero, the angular momenta of the incomplete shells are not zero; consequently, their magnetic moment is not zero. Thus, transition elements are the important elements for magnetism. However, a more external and incomplete shell (e.g., 4s) does not result in magnetic effects because the unpaired electron participates in the chemical bond.

Ferromagnetic materials consist of magnetic domains and present spontaneous magnetization at room temperature as well as the phenomena of saturation magnetization, and hysteresis. Spontaneous magnetization disappears above the critical temperature (T_c) known as the Curie temperature, above which the material becomes paramagnetic. The graph of M versus H is known as the magnetization curve or $M \times H$ curve. A magnetization

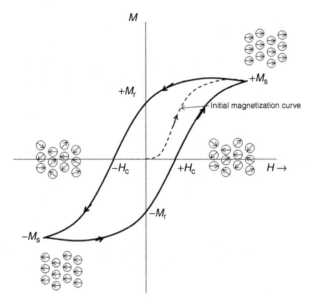

FIGURE 6.3 A magnetic hysteresis curve of a ferromagnetic (or ferrimagnetic) material. The major magnetic magnitudes are indicated in the curve. The evolution of the magnetic ordering in a system as a function of the field is illustrated.

curve for the ferrimagnetic and ferromagnetic materials is presented in Fig. 6.3. The curve presents a saturation magnetization M_S above a determined value of the applied field. As the field decreases to zero after saturation, the magnetization is not reduced to zero, a phenomenon known as hysteresis, that is, important for technological applications. If the field is applied to a magnetic material at its demagnetized state, then the magnetization will have a value that ranges from zero to M_S when the field increases in the positive direction. When the field is reduced after saturation, the magnetization decreases to M_r (the remanent magnetization). The field applied in the opposite direction needed to reduce the magnetization to zero is known as the coercive field or coercivity, H_c. Magnetic materials are classified as hard or soft depending on their coercivity value. A hard magnet requires the application of a large field to reduce the remanence to zero or to saturate the magnetization. A soft magnet reaches saturation with small fields and is easily demagnetized. When the reverse field is increased further, magnetization saturation is reached with the opposite orientation. When the ferromagnet is demagnetized, the magnetization vectors in the different domains have random orientations and the total magnetization is equal to zero. The magnetization process causes the orientation of all domains to be the same.

6.2.1 Ferrimagnetism

The most important ferrimagnetic materials are the dual oxides composed of iron and another metal, generally known as ferrites. These ferrites can take a cubic or hexagonal shape. Cubic ferrites have a molecular formula of the type MFe_2O_4 or $MO \cdot Fe_2O_3$, where M is a divalent transition metal such as Fe, Co, Ni, Mn, Cu, or Zn. Cobalt ferrite is considered as magnetically hard, whereas the other cubic ferrites are considered as soft. The most important hexagonal ferrites are barium hexaferrite ($BaO \cdot 6Fe_2O_3$) and strontium hexaferrite ($SrO \cdot 6Fe_2O_3$) [18].

Ferrites are ionic compounds, and their magnetic ions are responsible for their magnetic properties. The spins, whose action in ferrimagnetism and related phenomena result in spontaneous magnetic moments, occur in the ionic lattice. Only the contribution of the spins is considered because the orbital contribution can be negated in ferrites. Oxygen ions do not have a resultant moment. In the transition elements, the d shell can contain five spin-up electrons ($+1/2$) and five spin-down electrons ($-1/2$). The first electrons spin up to maximize the moment. According to the Pauli exclusion principle, the sixth electron must have a spin down. An ion with four electrons in 3d, such as Fe^{2+}, should have a moment of $4\mu_B$, considering only the spin. In this case, the ionic compounds that are insulators are considered. In these materials, the electron energy levels do not superpose as they do in metals. Therefore, an integer number of electrons can be associated with each ion of the solid, such as, in the case of the free ion. The interaction between the atomic spins responsible for the magnetic order is the exchange interaction, an interaction of electronic origin. Therefore, the ferrimagnetic state is the result of the exchange interaction between the resultant spins of the different ions. Ferrimagnetism implies the existence of two non-identical sublattices, A and B, and usually at least three different exchange interactions, J_{AA}, J_{AB}, and J_{BB} [23]. The ions of sublattice A are magnetized in one direction, whereas the ions of sublattice B are magnetized in the opposite direction. The opposite moments do not cancel each other out, resulting in spontaneous magnetization.

6.2.2 Cubic Ferrites

Cubic ferrites have a complex structure known as a spinel because it is similar to that of the spinel mineral ($MgO \cdot Al_2O_3$). This structure has 56 ions per unit cell, which can be divided into eight subunits. The oxygen ions are larger (with an ionic radius on the order of 0.13 nm), and they are arranged in a face-centered cubic configuration with interstitials half-filled with much smaller metal ions (with an ionic radius ranging between 0.07 and 0.08 nm). The two types of metal ion coordination result in two magnetic sublattices in cubic ferrites: one tetrahedral (site A), in which the ions are located at the center of a tetrahedron and the vertices are occupied by oxygen ions, and another octahedral (site B), in which the metal ions are coordinated with six oxygen ions. The crystalline symmetry is cubic and corresponds to the O_h^7 ($Fd3m$) spatial group. Fig. 6.4 shows the types of tetrahedral (T_d) and octahedral (O_h) coordination symmetries as well as the ion spatial geometries for a spinel-like cubic ferrite.

Direct spinel (or normal) and inverse structures might exist depending on the distribution of divalent and trivalent ions at sites A and B. In the direct spinel structure, all of the divalent metal ions occupy only the tetrahedral sites (A), whereas the trivalent metal ions occupy the octahedral sites (B). In the inverse spinel structure, all of the divalent ions

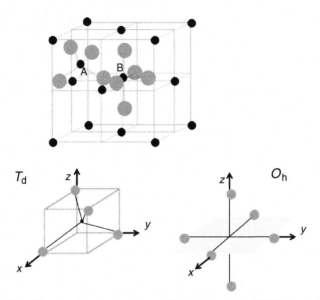

FIGURE 6.4 A representation of the atomic arrangement as well as the tetrahedral (T_d) and octahedral (O_h) coordination geometries for a spinel-like cubic ferrite. The spheres in gray represent the O^{2-} ions, and the spheres in black represent the iron ions (Fe^{2+} or Fe^{3+}). *Adapted from B.D. Cullity, C.D. Graham, Introduction to Magnetic Materials, second ed., John Wiley & Sons, New Jersey, 2009 [18].*

occupy the octahedral sites, whereas the trivalent ions are equally distributed between the two sites (i.e., half occupy the octahedral sites, and the other half occupy the tetrahedral sites). Fig. 6.5 shows the spin distribution in the inverse spinel structure of magnetite (Fe_3O_4) [24]. As shown, the spin moment due to the Fe^{3+} ions is zero because their spin moments cancel each other out between sites A and B. In turn, the spin moments of the Fe^{2+} ions are aligned in parallel and in the same direction (sites B), which results in an effective magnetic moment responsible for the magnetization of the material. Moreover, in both the direct and inverse spinel structures, not all of the available sites are occupied by metal ions; only 1/8 of sites A and half of sites B are occupied. In practice, the synthesis conditions influence the arrangement of metal ions in ferrites of spinel-like structure, and it is not rare to obtain mixed ferrites. The mixed ferrites represent an intermediary arrangement between the direct and inverse structures. The degree of inversion is the parameter that describes this type of ferrite [23].

Magnetite is a typical inverse ferrite. In turn, maghemite is considered as a form of magnetite with an iron deficiency, and it has the chemical formula of $\square_{1/3}Fe_{8/3}O_4$, where $\square$ represents a vacant site in the crystal. Given the crystallographic and magnetic representation of the magnetite structure in the two neighboring sites A and B, a given iron atom is antiferromagnetically coupled via superexchange to another iron atom of the same valence and ferromagnetically coupled via double exchange with an iron atom of a different valence. The two exchange interactions occur with regard to the oxygen ion.

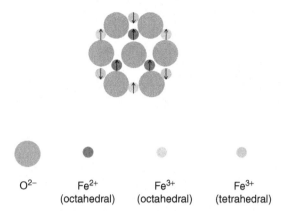

FIGURE 6.5 **Scheme of the spin moment distribution for magnetite (Fe$_3$O$_4$) with an inverse spinel structure.** *Adapted from W.D. Callister, Materials Science and Engineering: an Introduction, seventh ed., John Wiley & Sons New York 2007 [24].*

6.2.3 Superparamagnetism

Superparamagnetic behavior is associated with the dependency of the structure of magnetic domains on the size of the material, and it occurs when its dimension reaches the nanometric scale. Any ferromagnetic or ferrimagnetic material below its Curie temperature contains small regions where mutual alignment of all magnetic dipole moments exists in the same direction. This region is known as the magnetic domain (Fig. 6.6). Within each domain, the magnetization has its own saturation value. The domain structure is organized to reduce magnetostatic energy. The size, shape, and orientation of the domains

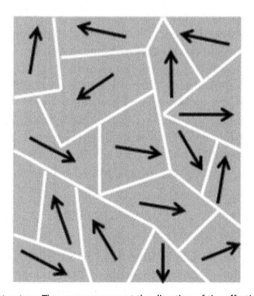

FIGURE 6.6 **Magnetic domain structure.** The *arrows* represent the direction of the effective magnetic moment in each domain. The white separating lines represent the domain walls.

depend on the interaction between the exchange, magnetostatic, and anisotropic energies of the system. In turn, the domains inside a material are separated by boundaries known as domain walls, through which the direction of the magnetization varies gradually. In a nonmagnetized material, the weighted sum of the magnetization vectors of all of the domains is zero, and the material does not have net magnetization.

When the volume of the magnetic material decreases to the nanometric scale, the typical domain size and the width of the wall-domain interfaces are reduced, thereby modifying its internal structure. The energy for the creation of a wall-domain interface in particles with volumes smaller than a given critical value is higher than the corresponding reduction in magnetostatic energy. In this boundary, no division into smaller domains occurs, and the magnetic structure of a single domain is maintained. For spherical particles, the critical diameter (D_c), above which the material could be considered as a monodomain, was determined by Kittel, and it is on the order of 10–20 nm [25].

The magnetic behavior of a monodomain system is characterized by the mutual alignment of the atomic moments along the direction of the easy axis of magnetization, creating a large magnetic moment. In this type of magnetism, the magnetic moments of the particles behave similarly to a paramagnetic system, albeit with a total moment of a much larger magnitude (10^2–10^4 μ_B) than those of the individual atoms ($\mu_e \sim \mu_B$). This fact originated the name "superparamagnetism."

In the superparamagnetic regimen, the orientation of the magnetic moments of particles floats around the easy axis of magnetization. The simplest model that describes the dynamics of the particle magnetic moments is the one-dimensional double-well potential [26], which is schematically illustrated in Fig. 6.7. The primary contribution to the double-well potential boundary is the magnetocrystalline anisotropy that aligns the magnetic moments in a given crystallographic direction (e.g., in the [111] direction for magnetite). The magnetization of a monodomain particle is oriented along the easy axis of magnetization as determined by the shape and magnetocrystalline anisotropy of the system. Depending on the domain size, the thermal energy kT can be sufficiently high, high enough ($U = KV$, the effective magnetocrystalline energy) to promote the reorientation of the magnetic moments against the magnetocrystalline energy boundary, thereby resulting in the fluctuation of the magnetic moments on a time scale of nanoseconds.

The typical relaxation time, which corresponds to the mean time taken by the magnetic moment to reorient itself in relation to the easy axis for a system of monodomain particles that do not interact among themselves, was calculated and is known as the Néel relaxation time (τ_N) (13) [27]

$$\tau_N = \tau_0 \exp^{(KV/k_B T)} \tag{6.13}$$

where the value of τ_0 é typically ranges from 10^{-11} to 10^{-9} s, k is the Boltzmann constant (8.7×10^{-5} eV K^{-1}), K is the effective anisotropy constant, and V is the particle volume. KV represents the effective magnetocrystalline energy barrier. Below a certain temperature, that is, the blocking temperature (T_B), the changes in direction due to the thermal

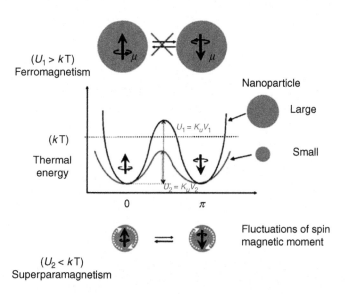

$(U_1 > kT)$
Ferromagnetism

(kT)

Thermal
energy

Nanoparticle

Large

$U_1 = K_u V_1$

Small

$U_2 = K_u V_2$

0   π

Fluctuations of spin
magnetic moment

$(U_2 < kT)$
Superparamagnetism

FIGURE 6.7 **The one-dimensional double-well potential representing the dynamics of magnetic moments in monodomain particles and the difference between the ferromagnetic and superparamagnetic behavior.** *Reproduced with authorization from Y.-W. Jun, et al., Nanoscaling laws of magnetic nanoparticles and their applicabilities in biomedical sciences, Acc. Chem. Res. 41 (2008) 179–189 [26].*

activation occur at longer time scales than the observation time, causing the moments to appear as frozen. For a given observation time (τ_{obs}), typical for the experimental technique, the blocking temperature is defined as (6.14) [28]

$$T_B = \frac{KV}{k_B \ln(\tau_{obs}/\tau_0)} \tag{6.14}$$

At the blocking temperature, the reverse magnetization of a set of identical monodomain particles changes from blocked (with a hysteresis curve) to unblocked (without a hysteresis curve); the latter reveals the superparamagnetic behavior. Except for a scale factor that depends on the characteristic measurement time, the blocking temperature represents the point at which the thermal energy ($k_B T_B$) is comparable to the effective magnetocrystalline energy barrier (KV).

For the magnetization measurements with a typical measurement time on the order of 100 s, the blocking temperature of monodomain magnetic particles is $T_B \cong KV/25 k_B$. The dependency of magnetization on the temperature $M(T)$ at the zero-field-cooled (ZFC) condition has a maximum around the blocking temperature. On the time scale of the magnetization measurement (and above T_B), the magnetic moment of the particle is free to align itself with the applied magnetic field. The magnetic behavior for a monodomain system as a function of temperature can be described using the first-order Langevin function (6.15) [29]

$$M(T) = M_S \left[coth\left(\frac{\mu H}{k_B T}\right) - \frac{k_B T}{\mu H} \right] \tag{6.15}$$

where M_S is the saturation magnetization and μ is the magnetic moment of the monodomain particle.

Magnetic NPs dispersed in an FM can have two magnetic moment relaxation mechanisms: the previously discussed Néel mechanism and the Brownian mechanism. In the Brownian mechanism, the NPs change the orientation of their intrinsic magnetic moments via the rotation of the particle itself inside the carrier liquid due to thermal activation. Hence, in the absence of a magnetic field, the particles are randomly oriented in the FM sample, and the net magnetization is zero. The magnetization decay time of an FM sample due to its rotational diffusion is described as the Brownian relaxation time (τ_B) [30]

$$\tau_B = \frac{3\eta V_{hidr}}{k_B T_B} \tag{6.16}$$

where V_{hidr} is the hydrodynamic volume of the particle and η is the viscosity of the carrier fluid.

If both the Néel and Brownian relaxation mechanisms are present, then the mechanism with the briefer relaxation time will determine the behavior of the system, and the effective relaxation time (τ_{eff}) will be given by (6.17) [31]

$$\tau_{eff} = \frac{\tau_N \tau_B}{\tau_N + \tau_B} \tag{6.17}$$

6.3 SPIOs

SPIOs, as well as magnetite, maghemite, and cobalt ferrite NPs with diameters less than 20 nm, are currently receiving intense study because they have applications in biomedicine, electronics, data storage, catalysis, and other fields. Their easy production using chemical methods is clearly one of their major advantages.

Of the chemical methods used to synthesize iron oxide NPs, the most commonly cited are the coprecipitation of salts in aqueous medium and the thermal decomposition of organometallic precursors or coordination compounds [32]. Both methods essentially result in colloidal systems. Each presents advantages and disadvantages; however, they share the advantage of producing considerable amounts of nanomaterial at a lower cost than physical methods such as powder grinding or more sophisticated techniques such as epitaxy or plasma.

The production of stable colloidal particles requires the initial preparation of a core with high magnetization and the proper surface treatment. In addition, the majority of the aforementioned applications require a material with particles smaller than 20 nm on average, with a narrow size distribution and a homogeneous structure to guarantee uniform physical and chemical properties. Hence, the greatest challenge in the synthesis of SPIO particles is the precise control of the following parameters: size, size distribution, shape, core phase, and surface physico-chemical properties. The chemical methods regarding the synthesis and functionalization of SPIO systems are described in detail later.

6.3.1 Synthesis of Iron Oxides via Coprecipitation

The coprecipitation of Fe^{2+} and Fe^{3+} ions with a molar ratio of 1:2 in an alkaline aqueous medium (pH > 9) at room temperature results in the immediate formation of a black precipitate of magnetite NPs according to the reaction [7]

$$Fe^{2+}(aq) + 2Fe^{3+}(aq) + 8OH^-(aq) \rightarrow Fe_3O_4(s) + 4H_2O$$

Eventually, the Fe(II) ions can be replaced by the ions of other divalent metals, such as, Co(II), Ni(II), Cu(II), and Zn(II), when the objective is to obtain mixed ferrites of the type MFe_2O_4, where M represents the divalent metal [32–35]. The precipitation reaction of the mixed ferrites is generally conducted at temperatures between 95 and 99°C, followed by a surface passivation step (i.e., Fe enrichment) to prevent dissolution and oxidation [36].

In principle, any base can be used for precipitation; however, NaOH and NH_4OH are generally preferred because of their cost, availability, and high water solubility. The aqueous medium can also contain surfactants, neutral polymers, or polyelectrolytes that act as NP functionalization agents [37]. Regardless of these variants, the precipitate is already magnetic and immediately responds to the application of an external magnetic field. In fact, magnets are usually applied to help in the decantation and separation of the supernatant NPs.

The proposed mechanism [12,38] for this reaction considers that the hydrolysis in the alkaline aqueous medium is responsible for the initial formation of hydroxylated iron aquo complexes. These complexes undergo condensation reactions that form solid iron oxide lattices. The condensation reactions depend on the stoichiometric composition of the medium, the pH, and the ionic force. In particular, the pH and the ionic force of the medium have preponderant effects on the synthesis thermodynamics because they act in the protonation–deprotonation equilibrium of the hydroxylated groups at the surface of the new NPs. Its effect is the decrease of the particle/reaction medium interfacial tension and the increase of the surface area, which contributes to the coupling of aquo complexes in solution and the formation of the solid iron oxide lattice. The average particle size decreases with the increase of the pH [12,38].

Maghemite NPs are not directly obtained via coprecipitation; rather, they are obtained via a subsequent step with the oxidation of the magnetite particles according to the reaction [7]

$$4Fe_3O_4(s) + O_2(g) \rightarrow 6\gamma\text{-}Fe_2O_3(s)$$

The oxidation can be performed by bubbling oxygen or adding $Fe(NO_3)_3$ in an aqueous magnetite NP suspension kept under agitation and heat [14], according to the scheme in Fig. 6.8. The magnetite→maghemite conversion can be observed with the naked eye; the initial black solid becomes brown as oxidation occurs.

NPs prepared via coprecipitation have an approximately spherical shape, with a diameter smaller than 10 nm. This method invariably produces polydisperse samples (i.e., the NPs produced by a single process have different diameters). In some cases, the polydispersion can be greater than 20%. Fig. 6.9A shows a transmission electron microscopy (TEM) image of a maghemite NP sample obtained via coprecipitation, followed by

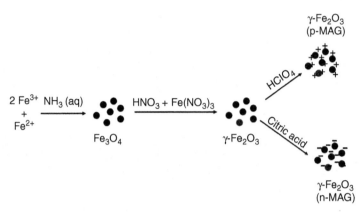

FIGURE 6.8 An illustration of the synthesis process via the coprecipitation of magnetite (Fe_3O_4), subsequent oxidation, and surface treatments to obtain maghemite NPs with a positively (p-MAG) or negatively (n-MAG) charged surface.

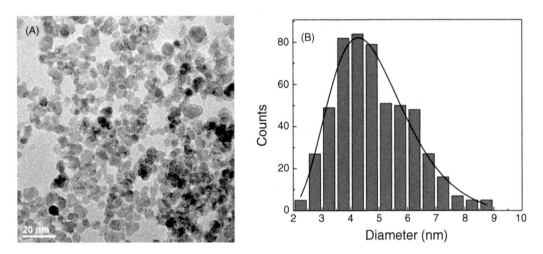

FIGURE 6.9 TEM image obtained for maghemite NPs (A) and the corresponding particle size distribution (B).

functionalization with citric acid (obtained according to the scheme in Fig. 6.8). Fig. 6.9B shows the particle size histograms obtained from TEM images, and the solid line represents the data fit using a log-normal distribution. The images indicate that the particles have approximately a spherical shape with a mean diameter (D_m) and standard deviation of 4.6 ± 0.1 nm and 0.28 ± 0.04, which were obtained by fitting the data presented in Fig. 6.9B.

6.3.2 Synthesis of Iron Oxides via Thermal Decomposition

Another synthesis method used to produce SPIO particles is based on the thermal decomposition of metal precursors in the presence of high boiling point solvents [9,39–42]. The precursors can be organometallic compounds such as, iron pentacarbonyl ($Fe[CO]_5$)

[40] or complexes such as iron acetylacetonate (Fe[acac]$_3$) [41,42]. Alternately, the synthesis can be performed from iron oleate (Fe[olea]$_3$), which is of low cost and lower toxicity [42]. Because the reactivity of the iron aquo complexes is high, it is difficult to control the condensation reactions that produce the solid; therefore, synthesis via coprecipitation is limited to the production of particles with a broader size distribution [12,38]. In turn, the decomposition method takes advantage of the strong interaction between an iron complex and a high boiling-point-coordinating solvent to control the nucleation and particle growth processes [39]. The elimination of the first ligand molecule is responsible for the nucleation of the particles, which occurs at a temperature approximately 50°C lower than that for the elimination of the other two. The particle growth at this temperature range is slow, guaranteed by the remaining ligand molecules that are only eliminated at higher temperatures. This dynamic allows the temporary separation of the nucleation and growth processes, resulting in samples with more uniform particles. Another important factor is the use of high boiling point solvents that enable the complete decomposition temperature of the precursor to be reached and that act as surfactants to promote the colloidal stability of the synthesized particles. Fig. 6.10 presents a TEM image of the monodisperse maghemite crystals obtained by Hyeon et al. [9].

Nevertheless, the decomposition method has two disadvantages. The first is the presence of an organic phase that, in the case of biomedical applications, requires one additional ligand exchange step for the particles to disperse in the aqueous medium. The second is the difficulty to produce equivalent amounts of NP materials in relation to the coprecipitation method because it is difficult to maintain a uniform temperature profile in a large reactor.

6.3.3 SPIO Particle Functionalization and the Preparation of Magnetic Colloids

The control over the surface chemistry of iron oxide NPs for colloidal systems has two fundamental purposes. The first is to prevent the agglomeration of particles, thereby

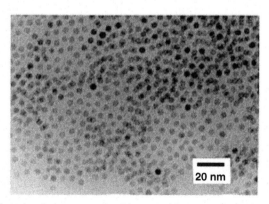

FIGURE 6.10 A TEM image of maghemite nanocrystals with a mean diameter of 4 nm. *Reproduced with the permission from T. Hyeon, et al., Synthesis of highly crystalline and monodisperse maghemitenanocrystallites without a size-selection process, J. Am. Chem. Soc. 123 (2001) 12798–12801 [9].*

guaranteeing the stability of the colloidal state. The second is to manipulate the surface of the NPs so that it acts as a platform for the anchorage of the species of interest, especially pharmaceuticals, markers, therapeutic agents, biomolecules, and others. In addition, the surface modifications are fundamental for the compatibilization of the NPs when they are in contact with or injected into biologic systems [37,43,44].

The colloidal systems based on SPIO NPs can be classified as irreversible or lyophobic. The redispersion of the particles after they are dried is only reached when the surface has been previously treated so that a strong enough repulsion exists to prevent their aggregation. In the colloid dispersion, the Gibbs free energy increases when the disperse phase is distributed throughout the dispersing phase [45]. In fact, the minimum free energy is reached when the particles form one agglomerate. Therefore, the colloidal system is thermodynamically unstable, and the barrier for coagulation is only kinetic.

The constituent particles of a colloidal dispersion are in Brownian movement and therefore susceptible to frequent collisions [45]. Thus, the irreversible colloidal systems can remain as individual particles for a long period of time (even years) as long as a mechanism prevents their irreversible aggregation during particle collision. Two stabilization mechanisms can be applied to prevent colloidal aggregation [29,45]: (1) electrostatic stabilization, which describes when the surface of the dispersion particles is electrically charged (positively or negatively) to establish a mutual electrostatic repulsion in the aggregation of particles and (2) steric stabilization, which describes when the particle surface is coated with a polymer or surfactant layer or even small molecules that prevent aggregation by steric impediment. A combination of both mechanisms is known as electrosteric stabilization, which is generally obtained with polyelectrolytes.

The iron oxide NPs obtained via coprecipitation can be directly dispersed in an aqueous acid or alkaline medium, forming stable sols for an undetermined period of time. This condition is only possible because their diameter is smaller than 10 nm, and the presence of superficial oxyhydroxide groups are susceptible to acid–base reactions. The charges introduced by the displacement of the acid–base equilibrium at the surface of the iron oxide NPs cause a strong electrostatic interparticle repulsion that maintains their separation and disperses them in the carrier liquid [11]. Fig. 6.11 illustrates how the surface charge density (ς) of the iron oxide NP varies as a function of the pH of the magnetic colloid. Observe that when $0 < pH < 6$, the surface charge density of the NPs is at its maximum because the hydroxide groups are protonated. The respective fluid is stable (sol) with positively charged NPs. When $6 < pH < 10$, the protons are removed from the surface of the NPs, making them electrically neutral. The point of zero charge (PZC) for iron oxides is $pH \approx 7.5$. Under this condition, no repulsion forces exist to inhibit the aggregation and agglomeration of the particles, which end up flocculating. At $pH > 10$, the particles act as acids because they lose the hydroxyl proton and become negatively charged. The surface charge density reaches another maximum (of an inverse sign), and the respective fluid is also stable (sol).

Alternatively, the surface can be functionalized by ionic ligands such as tartarate [46], citric acid [47–49], dimercaptosuccinic acid (DMSA), and [50–53] polyphosphates [54–56].

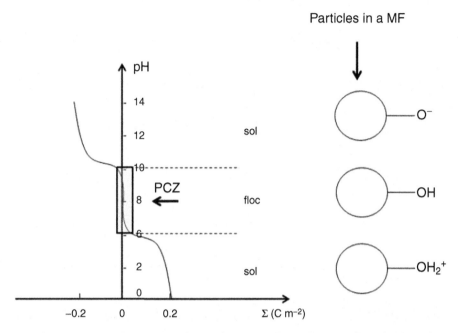

FIGURE 6.11 **A diagram showing the variation of the surface charge density (Σ, in C m^{-2}) as a function of the pH for iron oxide NPs dispersed in an aqueous medium. The evolution of the NP surface chemistry in the fluid as a function of the pH is presented next to the diagram.** *Adapted from J.C. Bacri, et al., Ionic ferrofluids: acrossing of chemistry and physics, J. Magn. Magn. Mater. 85 (1990) 27–32 [11].*

The functionalization of SPIO with citric acid or even sodium citrate enables the preparation of a stable FM biocompatible with a physiological medium (pH = 7–7.2). The DMSA molecule is particularly interesting because its two carboxylic groups are responsible for the interaction with the NP surface, whereas the two sulfhydryl groups (—SH) are used in conjugation via the disulfide bridge with biomolecules [50,52] and antibodies for tumor cell markers [50] or to intensify magnetic resonance image contrast [53]. Fig. 6.12 illustrates an iron oxide NP with DMSA molecules anchored at its surface via carboxylic groups. The free sulfhydryl groups can establish disulfide bridges, including those between adjacent DMSA molecules [52]. DMSA has also been used for the phase transfer of particles synthesized in an organic medium to an aqueous medium [53]. The functionalization with phosphate groups has proven to be promising for guaranteeing magnetic phase stability against oxidation processes. Furthermore, it offers the proper functionality for coupling with proteins and cells [54–56].

The particles obtained via coprecipitation can also be functionalized via surfactants such as oleic acid and then dispersed in organic solvents such as hexane or toluene. Maghemite dispersions functionalized with oleic acid have been added to vegetal or mineral oil for application in transformers [14]. In this application, maghemite NPs promote heat conduction and improve the performance and useful life of the transformer parts. In addition, NPs functionalized via fatty acid can undergo a second functionalization step (e.g.,

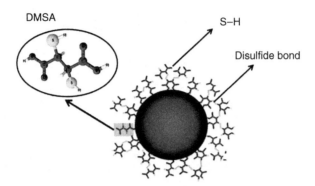

FIGURE 6.12 Schematic of an SPIO NP functionalized with DMSA.

by another fatty acid or an amine with a long carbon chain) to form a bilayer. The bilayer encapsulates the NP, allowing its dispersion in the aqueous medium, and simultaneously provides a microenvironment for the immobilization of hydrophobic species such as, pharmaceuticals and other therapeutic agents. For instance, curcumin was incorporated into dodecanoic acid bilayers and bonded to the surface of maghemite NPs to improve drug solubility and delivery efficiency [57].

The use of liposomes to encapsulate magnetic NPs and produce magnetoliposomes (MLs) has also been applied for functionalization for biomedical applications. MLs are biocompatible and physiologically stable structures that consist of a vesicle formed by phospholipids filled with magnetic NPs [58,59].

Many coatings can be used to simultaneously obtain an FM with steric or electrosteric stability, biocompatibility, and an anchorage platform for the species of interest, both for therapies and diagnosis (in vivo and in vitro). Generally, biocompatible polymers [e.g., polyethylene glycol (PEG), dextran, polyvinylpyrrolidone, polyvinyl alcohol, chitosan, gelatin, or even polypeptides] are employed [60,61].

The preparation of nanostructures with a core–shell structure is also a strategy to functionalize SPIO particles. Functionalization with a silica layer guarantees high stability for the NPs in silanol groups that act as a platform to immobilize species of biomedical interest. Using the Stöber method, silica is formed via the hydrolysis and condensation of tetraethoxysilane in an aqueous solution containing ethanol, ammonia, and the NPs [62,63].

6.4 The Structure and Physico-Chemical Properties of the SPIO Systems

The physico-chemical properties of SPIO systems are directly related to the spinel structure as well as morphological characteristics such as the size, shape, aggregation state, and chemistry of the surface of the NPs. Each characteristic is determined using different analysis techniques described in detail further.

6.4.1 Measurements of the Size and Agglomeration State

The size and shape of nanomaterials, including iron oxide NPs, are generally determined by the combined use of microscopy and X-ray diffraction techniques. The measurement of the magnetic properties can also be used for size estimates. Importantly, however, the mean diameter obtained by TEM (D_{TEM}), which is a direct measurement, differs from that calculated indirectly using the magnetic measurements (D_{M}); generally, $D_{\mathrm{TEM}} > D_{\mathrm{M}}$. The various experimental techniques provide different information about particle size, crystal size, and magnetic content [64]. Microscopy offers certain advantages, including the possibility of obtaining a large amount of digital images that can later be analyzed and provide a more realistic statistical distribution of particle size. Furthermore, with the advent of high-resolution TEMs (HRTEMs), the crystalline morphology and structure of the NPs can be determined with high accuracy (resolution ± 0.1 angstrom). In the case of colloidal systems, the preparation of the sample is relatively simple and consists of dripping the suspension into the sample holder and then waiting for the evaporation of the carrier (water or organic solvent). The parameters that describe NP size are determined using a histogram with the profile fitted to a log-normal distribution, $P(D)$, according to Eq. 6.18

$$P(D) = \frac{1}{D\sigma 4\pi^2} \exp\left[-\frac{\left(\ln D - \ln D_{\mathrm{TEM}}^2 \right)}{2\sigma^2} \right] \tag{6.18}$$

where σ is the standard deviation of the diameter (dispersion) and D_{TEM} is the mean diameter [45].

The data obtained using the X-ray diffraction technique applied to nanocrystals provide information regarding their crystalline quality, with the possibility of estimating the value of the lattice constant and the mean diameter of the crystals. The mean diameter of the crystalline domain can be determined, for instance, using the full width at half maximum (FWHM) of the most intense reflection peak following Scherrer's equation [65], considering that the crystallinity of the samples is maintained:

$$D_{\mathrm{c}} = \frac{0.9\lambda}{B\cos\theta_B} \tag{6.19}$$

where D_{c} represents the mean diameter of the crystalline domain, and λ is the wavelength of the incident X-ray. B is given by $B = (B_{\mathrm{med}}^2 - B_{\mathrm{pad}}^2)^{1/2}$, where B_{med} is the measured FWHM, B_{pad} is the FWHM of the standard sample used, and θ_B is the angle of the most intense diffraction peak [65].

Magnetization measurements can also be used to determine the size of magnetic particles [66,67]. For instance, the derivative of the magnetization curve ZFC–field cooled (FC) in relation to temperature (i.e., d[ZFC–FC]/dT) results in a distribution whose maximum can be related to the critical volume of the particle. The equation for this calculation is [68]

$$V_{\mathrm{c}} = \frac{25k_{\mathrm{B}}T}{K} \tag{6.20}$$

where V_c is the critical volume, k is the Boltzmann constant, and T is the temperature (in Kelvin) that corresponds to the maximum of the distribution. For the maghemite NP samples immobilized in self-assembled films, a significant correlation was found between the size determined using X-ray diffraction and the size obtained by calculating the critical volume with the derivative of the ZFC–FC curve [69]. The first-order Langevin function, described by Eq. 6.15, can also be used to obtain the particle size distribution. However, this approximation does not consider the interparticle interaction, and it can only be used for systems containing SPIO particles that interact weakly. The difference between the values obtained using TEM and those using magnetization measurements support the existence of a magnetically dead layer at the surface of the NP.

6.4.2 Composition and Structure

The chemical composition of an iron oxide is generally determined using traditional methods of analysis: atomic absorption spectrometry and X-ray fluorescence, both to determine the total iron content and that of other cations; colorimetric analysis (the *o*-phenanthroline method), to determine the Fe(II) content; and dichromatometry, to determine the Fe(III) content. Coordination can also be determined in parallel with microscopy techniques such as in energy dispersive spectroscopy (EDS) coupled to a scanning electron microscope. If the particle is functionalized with an organic ligand, then the coating composition can be determined using elemental analysis (e.g., CHN) [70] or by thermogravimetry [14].

The crystalline structure of NPs is determined using X-ray diffraction (the powder method). Iron oxides have a typical diffraction pattern of the spinel structure, whose maximum diffraction refers to the reflection of the (311) plane. This analysis can also be performed using electron diffraction coupled with TEM. An HRTEM can also provide images of the diffraction planes inside a single NP. However, it is difficult to distinguish magnetite from maghemite samples using only X-ray diffractograms because both materials generate similar diffractograms. To unequivocally identify each phase, it is necessary to use other techniques such as infrared spectroscopy (FT-IR), Raman scattering [71], or both. These techniques will be discussed in subsequent chapters.

The calculations predict five active Raman modes and four active bands in the infrared (IR) for the vibrational spectrum of the spinel structure. The predicted Raman modes are A_{1g}, E_g, and the three modes $T_{2g}(A_{1g} + E_g + 3T_{2g})$ [72]. The IR and Raman bands are sensitive to the oxidation states of the ions. The Raman or IR spectra depend on the preparation conditions of the iron oxides samples, especially those that are nanostructured. The IR absorption bands in the region between 200 and 900 cm^{-1} refer to the vibration of the Fe-O bond, and they are different for maghemite and magnetite. The maghemite FT-IR spectrum in this range is composed of two wide bands that, depending on the particle size and preparation method, can be subdivided into more bands. The magnetite spectrum in the same region generally presents two well-defined peaks, enabling the distinction between the two iron oxide phases [73].

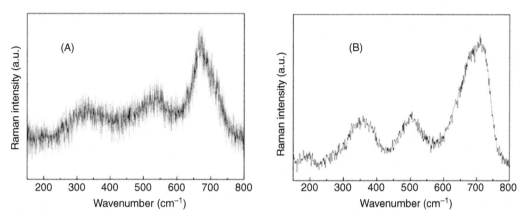

FIGURE 6.13 Raman spectra obtained from magnetite (A) and maghemite (B) NP samples prepared by coprecipitation

Raman spectroscopy complements the other techniques previously mentioned, and it is the most appropriate to unequivocally identify the magnetite and maghemite phases [72]. For instance, as previously mentioned, γ-Fe_2O_3 has an iron-deficient structure and wider Raman bands. Fig. 6.13A and B presents the Raman spectra obtained from powder samples of magnetite (A) and maghemite (B) NPs. The spectra were obtained using an argon laser (λ = 514 nm) at the lowest possible intensity to avoid degrading the sample [74]. The curve fitting (Lorentzian lines) process for the Raman spectrum of magnetite (A) shows the presence of three vibrational modes (E_g + T_{2g}, and A_{1g}) out of the five predicted for the spinel structure [72], Fig. 6.13(B) shows the characteristic Raman spectrum of maghemite, which presents wider bands and a band at approximately 720 cm^{-1} associated with the oxidation of the iron ions at the octahedral site [72,75].

The effects of the sample preparation process, such as the oxidation of magnetite to maghemite, and the passivation of the surface of cobalt ferrite particles can be studied using Raman spectroscopy via the micro-Raman configuration [8,14,36]. The sensitivity of micro-Raman spectroscopy is high enough to detect the SPIO phase in multiple layers with fine polyelectrolytes (thickness = 60 nm) [76]. Fig. 6.14 shows the Raman spectrum obtained for a multilayer film formed by positively charged maghemite NPs (p-MAG) alternated with poly(sodium styrenesulfonate).

Important information regarding the short-range structure and magnetic properties of SPIO systems can be obtained using Mössbauer spectroscopy, which is discussed in the next volume [77]. The spectra can be obtained as a function of the temperature, with or without the application of an external magnetic field. This technique is also sensitive to the relaxation phenomena. The spectra at low temperatures can be fitted with two sextets that correspond to the two sites presented by the spinel structure. In the case of the spectra obtained at the temperature of liquid helium, the surface effects can be considered in nanomaterials, introducing a third sextet in the spectral fitting [35]. For instance, the effect of the synthesis route of the maghemite NPs on the magnetic disorder

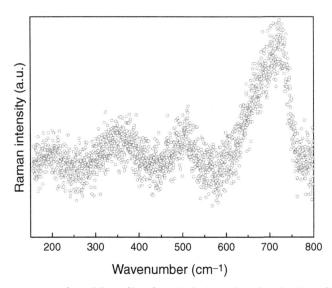

FIGURE 6.14 The Raman spectrum of a multilayer film of positively charged maghemite NPs and PSS. The presented region is limited to the transitions regarding the iron oxide phase.

can be studied using the Mössbauer spectra obtained via the application of an external magnetic field [73].

Fig. 6.15 presents the Mössbauer spectra of cobalt ferrite NPs (diameter ~8 nm) obtained at different temperatures (4, 77, 165, and 300 K) without the application of an external magnetic field [35]. The Mössbauer spectrum obtained at the temperature of liquid helium has a slightly asymmetrical, hyperfine magnetic structure, similar to that observed for the cobalt ferrite NPs available in the literature. The spectral fitting was obtained using three well-defined sextets. The first (which had the highest value for the hyperfine field) was attributed to the Fe(III) ions located at the tetrahedral site of the spinel structure. The second sextet (which presented an intermediary value for the hyperfine field) was associated with the Fe(III) ions that occupy the octahedral site. The third sextet (which had the lowest hyperfine field) was attributed to the Fe(III) ions located next to the NP surface. According to Fig. 6.15, an asymmetrical widening of the sextet lines and the appearance of a doublet occur when the temperature increases; furthermore, the intensity of the doublet increases as the sextet intensity decreases. This behavior is characteristic of superparamagnetic relaxation, and the approach used for the fitting consists of the superposition of a hyperfine magnetic field with a quadrupole doublet. For the spectrum obtained at 77 K, the fitting using two sextets and one doublet indicates 512 and 458 kOe hyperfine fields attributed to Fe(III) at sites A and B, respectively. The central doublet is attributed to the superparamagnetic NPs present in the sample at that temperature, whereas the sextets refer to the blocked NPs (i.e., those at the magnetically ordered state). The relative area under the doublet related to the fraction of NPs in the superparamagnetic regimen increases from 6% at 77 K to 32% at room temperature.

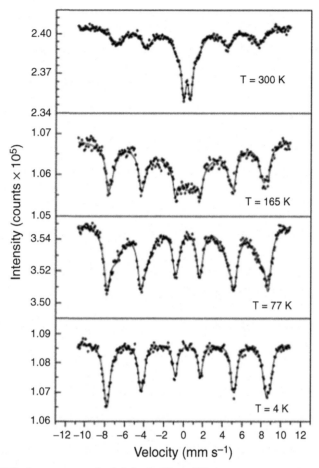

FIGURE 6.15 The typical Mössbauer spectra of cobalt ferrite NPs at different temperatures. *Reproduced with the permission from M.A.G. Soler, et al., Surface passivation and characterization of cobalt–ferrite nanoparticles Surf. Sci. 575 (2005) 12–16 [35].*

6.4.3 Surface Morphology and Properties

The properties that characterize the NP surface include the charge, surface area, and surface reactivity. Given that NPs interact with the cell membrane through its surface, these characteristics are considered as potential toxicity modulators [78].

The surface charge of the particles in an aqueous colloidal solution is determined by measuring the zeta potential [45,79]. The zeta potential (ζ) is expressed by Eq. 6.21 [45]

$$\zeta = \frac{3\mu\eta}{2\varepsilon_0 D},$$

(6.21)

where μ is the electrophoretic mobility of the particle, η is the viscosity, ε_0 is the permittivity of vacuum, and D is the dielectric (static) constant of the medium. Generally, stable solutions of charged colloids are prepared using NPs with zeta potential equal to or greater than $(+/-)$ 30 mV. The measurement of the zeta potential is based on the knowledge that every charged particle in a liquid medium moves at a constant velocity (v) under the action of an external electric field. Hence, the zeta potential value is determined by the electrophoretic mobility (μ) that, in turn, is calculated using the mean particle velocity (v) under the action of the electric field E according to Eq. 6.22 [45]

$$\mu = \frac{v}{E} \qquad (6.22)$$

The monitoring of the chemisorbed molecular layer at the surface of the SPIO particles is important for the design of magnetic colloids for different applications such as, specific cell tracking.

Raman spectroscopy has been used to characterize iron oxides for many years [80]. Only recently was this technique applied to study the microscopic details of the first NP surface coating layer (i.e., the NP surface-carrier liquid interface) [34,81,82]. Raman and photoacoustic spectroscopy can be applied together to study the formation of the disulfide bridges in DMSA-coated iron oxide NP systems [52]. Photoacoustic spectroscopy provides information about the presence of thiol groups (S—H), and the disulfide bridges (S—S) were monitored using Raman spectroscopy. This strategy enables the quantification of the [S—H]/[S—S] ratio over a wide range of the density values of DMSA molecules per NP area.

As previously discussed, the surface characteristics of NPs influence the long-term stability of the magnetic colloid, a mandatory characteristic for certain applications. For instance, if these particles are covered by organic acid, then the way they are bonded/anchored to the NP surface will influence colloidal stability. One study [14] was performed on the effect of the different oxidation conditions of NPs from magnetite to maghemite and the effect on the adsorption of oleic acid. Oleic acid–coated maghemite NPs were suspended in an insulating mineral oil and stored to study their stability as a function of time. The samples were characterized using thermogravimetry, Raman spectroscopy, and FT-IR; the results showed the significant influence of the oxidation protocol on the way that the oleic acid was anchored to the maghemite NP surface. The combined use of these techniques enabled the estimation of the relative content of the physisorbed and chemisorbed species. For instance, the suspensions prepared with NPs of smaller hydrodynamic radii and greater colloidal stability had only chemisorbed oleic acid at their surface.

The morphology of SPIOs can be investigated in detail by atomic force microscopy when they are immobilized on a solid substrate. The layer-by-layer self-assembly technique (electrostatic mode) described in detail in Chapter 4 (p. 134) is a simple and low-cost method that provides samples with different NP volumetric fractions with sufficient control [83]. The method consists of depositing a film on a solid substrate (glass, silica, mica sheets, etc.) via the successive immersion of the substrate, alternately, in cationic and anionic solutions/suspensions. The layers are adsorbed via electrostatic attraction in a spontaneous process, reaching equilibrium after a few minutes of deposition. The layer thickness depends on different physico-chemical deposition parameters, especially those associated with the solution conditions (concentration, pH, ionic force, etc.), and it can vary from 1 nm to 20–30 nm (approximately). When dispersed in an aqueous acid medium, the SPIO particles act as cations and can therefore be adsorbed with different polyanions. The film thickness varies linearly with the number of deposition cycles (bilayers); therefore, the deposition of films with different thicknesses can be applied. An example of the AFM topographical and phase images of a self-assembled film with cobalt ferrite ($CoFe_2O_4$) and poly(styrene sulfonic acid)-doped poly(3,4-ethylenedioxythiophene) NPs, PEDOT:PSS, is shown in Fig. 6.16 [84]. The images at the top refer to a sample with a ($CoFe_2O_4$/PEDOT:PSS/$CoFe_2O_4$) NP final layer, whereas the images at the bottom refer to a sample with a PEDOT:PSS final layer. The topographical images for both samples, Fig. 6.16A and C, are similar because they reflect the NP boundary, regardless of the polymer chain coating. In fact, the typical globular morphology of the NPs is predominant in both cases, with a similar mean quadratic roughness in both samples (3.3 nm for NPs at the top and 3.7 nm for the sample with the polyelectrolyte finish). However, the phase

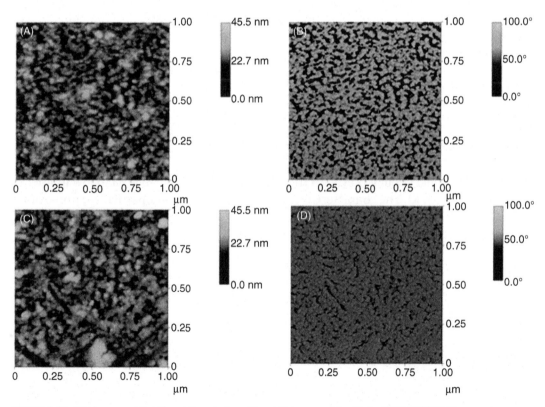

FIGURE 6.16 AFM topographical (A and C) and phase (B and D) images of the ($CoFe_2O_4$/PEDOT:PSS)$_2$ films, varying the composition of the layer at the top of the structure: (A and B) $CoFe_2O_4$ NP final layer and (C and D) PEDOT:PSS NP final layer. *Reproduced with the permission from G.B. Alcantara, et al., Morphology of cobalt ferrite nanoparticle–polyelectrolyte multilayered nanocomposites, J. Magn. Magn. Mater. 323 (2011) 1372–1377 [84].*

images have different appearances, Fig. 6.16B and D. When the last layer is formed using CoFe$_2$O$_4$ NPs, the contrast between the NPs and the polymer is highlighted, and the boundaries and empty voids between the particles are more evident. When the last layer is formed by PEDOT:PSS, the phase image seems more planar than the topographical image. This result is because the NP surface and the voids between them are filled with the polyelectrolyte with different viscoelastic properties in relation to the NPs, and they are translated by the microscope as a single-phase image.

The NP distribution along the self-assembled film thickness can be investigated using high-resolution electron microscopy of the cross-section [85]. Fig. 6.17A shows the morphology of the cross-section of a (PAni/n-MAG)$_{10}$ nanofilm, where ten refers to the number of bilayers. The NPs are homogeneously distributed throughout the film, and no stratification exists in layers. The particles are individually encapsulated by a fine polyelectrolyte layer (clear boundary around the particles). This observation, as well as those performed using AFM, corroborate the NP packing model with polyelectrolyte. Fig. 6.17B shows the amplification of the NPs (n-MAG) in the film and indicates the crystalline planes associated with the γ-Fe$_2$O$_3$ phase. The film thickness with 10 bilayers was estimated as 60 nm, confirming the value previously obtained using AFM [86]. TEM cross-sectional images obtained based on the number of bilayers indicated the systematic reduction of the particle–particle distance with the increase of bilayers [85,86]. This information is important, given that the dipolar interaction is influenced by the interparticle separation.

6.4.4 Magnetic Properties

The magnetic characteristics of the nanoparticulate iron oxides can be evaluated using magnetization measurements as a function of the field ($M \times H$ curve) and temperature. The measurements as a function of temperature are performed by cooling the sample, with or without field cooling (ZFC–FC) from room temperature (~300 K) to low temperatures (generally 2–4 K). The characteristic $M \times H$ curves of the nanoparticulate iron oxides

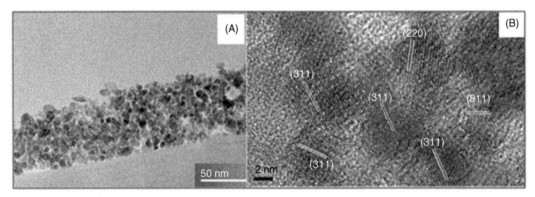

FIGURE 6.17 High-resolution electron microscopy images of the cross-section of the (PAni/n-MAG)$_{10}$ nanofilm (A) and NPs inside the film, emphasizing the crystallographic planes (B). *Adapted from M.A.G Soler et al., Assembly of γ-Fe$_2$O$_3$/ polyaniline nanofilms with tuned dipolar interaction, J. Nanopart. Res. 14 (653) (2012) 1–10 [85]*

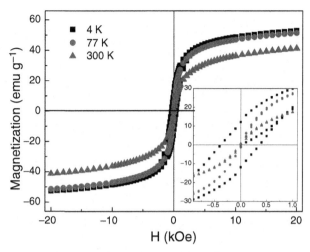

FIGURE 6.18 Magnetization curves of the POX3OA sample obtained at different temperatures as indicated. The amplified curves for the low-field region are also shown. *Reproduced with permission from W.R. Viali, et al., Investigation of the molecular surface-coating on the stability of insulating magnetic oils, J. Phys. Chem. C. 144 (2010) 179–188 [14].*

exhibit superparamagnetic behavior at temperatures higher than the blocking temperature and hysteresis for temperatures below it [14]. Fig. 6.18 shows the typical $M \times H$ curves obtained at 4, 77, and 300 K in the range of ±20 kOe for powder maghemite samples (diameter ~7 nm). The curve included in Fig. 6.18 shows the details of the hysteresis curve in the range of ±1 kOe for different temperatures. Remanence (M_R) and coercivity (H_C) were not observed at 300 K, as expected for a sample containing small particles (mean diameter = 7 nm) that do not interact. The saturation magnetization (M_S) can be evaluated by plotting a graph of magnetization as a function of the reciprocal of the applied field [87]. The saturation magnetization values obtained using the $M \times H$ curves are usually smaller than those determined using the corresponding bulk materials. In addition to depending on the structural quality of the particle, the saturation magnetization value also reflects surface effects. The reduction in the saturation magnetization value is likely because of the inclination of the spins close to the surface, disorders associated with the rupture of bonds, and the frustration of exchange interactions [88,89]. In nanoparticulate magnetic systems, the ZFC–FC curves are sensitive to the size, size distribution, morphology, and particle–particle interactions [90]. The ZFC–FC magnetization curves of the same maghemite sample with a DC field of 100 Oe, obtained in the range of 4–300 K, are presented in Fig. 6.19.

One study concerning the oxidation process of magnetite to maghemite showed that the values of M_S and T_B for samples oxidized at different degrees depend on particle size and the Fe^{3+}/Fe^{2+} ratio, regardless of the route used to oxidize the samples. After synthesis, the samples were coated with oleic acid, and the systematic reduction of T_B was observed because of the presence of the coating [14].

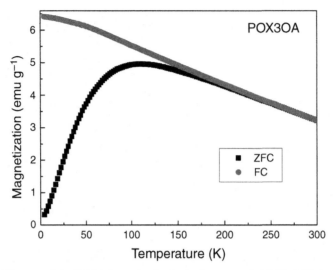

FIGURE 6.19 The ZFC–FC curves of the POX3OA sample obtained using an applied field of 100 Oe. *Reproduced with permission from W.R. Viali, et al., Investigation of the molecular surface-coating on the stability of insulating magnetic oils, J. Phys. Chem. C. 144 (2010) 179–188 [14].*

The magnetic properties of the NP arrangements depend on the intrinsic characteristics of the individual particles as well as the chemical composition, stoichiometry, crystallinity, shape, magnetic anisotropy, and their interaction [91,92]. These interparticle interactions have an important effect; thus, the control of the distances between particles should be considered in the design of magnetic nanocomposites with specific properties. One way to control the interparticle distance is by varying the volumetric fraction (or mass fraction) of NPs in the nanocomposite via self-assembly (see item 6.4.3 for a discussion). This method has been explored in studies of particle–particle interactions, enabling the production of samples with a variable fraction of NPs [76,85,86,93].

The nanocomposites obtained using the self-assembly of magnetite and maghemite NPs exhibit superparamagnetic behavior at room temperature [76,85,86,93,94]. The polyaniline/maghemite bilayer system shows a superparamagnetic behavior from room temperature to 40 K and a similar saturation magnetization in relation to free particles in powder [86,94]. Moreover, the films start to present a coercive field of 149 Oe and a saturation magnetization of 50 emu g^{-1} at 4 K; these values are comparable with those obtained using NPs in powder under the same measurement conditions (180 Oe and 55 emu g^{-1}) [94]. However, the blocking temperature of the system increases as more bilayers are deposited, indicating an increase in interparticle interactions.

One way to evaluate the level of interparticle interaction is by performing ac susceptibility measurements. The ac magnetic susceptibility of a system composed of particles that do not interact can be described using the models proposed by Dormann and Gitlemann, which consider that the time (τ) required for the inversion of the magnetization orientation of each NP follows Arrhenius behavior [Eq. 6.13] [95,96]. The dependency of the imaginary component of ac susceptibility on frequency can be used to evaluate τ_0 and ΔE

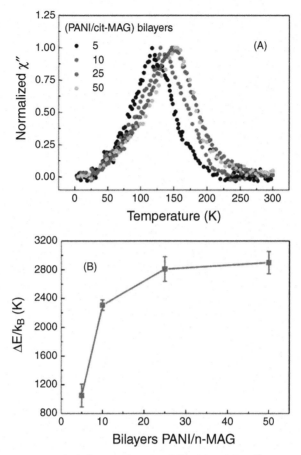

FIGURE 6.20 The imaginary component (χ'') of magnetic susceptibility (A) and the effective energy barrier (ΔE) of the (PAni/n-MAG)$_N$ sample with a variable number of bilayers ($N = 5, 10, 25,$ and 50 bilayers) (B). *Reproduced with the permission from Assembly of γ-Fe$_2$O$_3$/polyaniline nanofilms with tuned dipolar interaction, J. Nanopart. Res. 14 (653) (2012) 1–10 [85].*

using the Arrhenius curve [97]. Fig. 6.20 shows the results obtained using measurements of susceptibility for polyaniline/maghemite nanofilms. An asymptotic increase was observed in the effective energy wall ΔE with the number of bilayers deposited, where saturation is reached with approximately 25 bilayers. The increase in ΔE is attributed to the decrease of the interparticle distance that causes the densification of the film and increases the dipolar interaction between particles. In fact, the densification of the film can also be induced by increasing the NP concentration in the suspension, used for the film deposition.

6.5 Biomedical Applications

Biomedical applications of magnetic colloids can be carried out in vivo and in vitro. In the in vivo applications, the colloidal NPs are injected into the blood stream and are guided to the location of delivery with external manipulation by applying a magnetic field gradient,

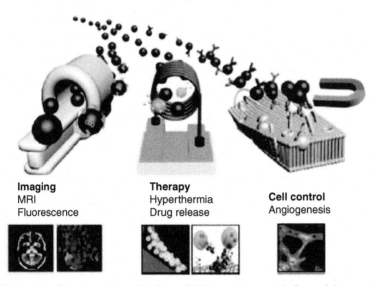

Imaging
MRI
Fluorescence

Therapy
Hyperthermia
Drug release

Cell control
Angiogenesis

FIGURE 6.21 An illustration of the potential applications of SPIO particles as a platform of theranostic agents. *Reproduced with the permission from D. Yoo, Theranostic magnetic nanoparticles, Acc. Chem. Res. 44 (2011) 863–874 [13].*

using molecular recognition to specifically bond to the target cells, or both. In some cases, the colloidal particles are alternatively injected directly into the area of interest. Once they are present in the target tissue, the complexes formed by the SPIO particles act as theranostic agents. A theranostic agent is defined as one that can perform diagnosis and therapy with a single platform [13,98,99]. The potential applications of theranostic agents using SPIO system-based platforms are shown in Fig. 6.21, including hyperthermia, controlled drug delivery, cell markers, and so on.

In particular, the iron oxide nanoparticulate systems exhibit, in addition to their biocompatibility, properties such as, superparamagnetism and a simultaneous capability to reach the target cell, obtain an image, and perform therapy; these characteristics make the platforms ideal for theranostic complexes. Although SPIO systems were initially used as $T2$ contrast agents to improve the sensitivity of magnetic resonance signals [100], they are currently recognized as efficient for other image modalities, as platform for therapeutic drugs, and/or ligands to reach the target [44,101–105].

The control of biodelivery is one of the major challenges in the design of SPIO-based complexes for control drug delivery (DDS) systems. Once they are injected, NPs can be quickly eliminated from the blood stream via the reticuloendothelial system (RES), which limits biomedical applications. The internalization and the subcell processing of the drugs are also critical issues in the design of drug delivery systems. To extend the permanence time in the bloodstream and reach target cells, it is necessary to break biologic barriers. These biologic barriers include the blood, liver, spleen, kidneys, blood-brain barrier, and tumor vasculature. This problem is addressed by designing the proper surface coating for the SPIO particle to block the adsorption of blood components and improve the colloidal stability. PEG is one of the most commonly used materials to protect the SPIO complex. In addition, to avoid elimination via the filtering organs, the hydrodynamic size of the NP should be small because the RES absorbs materials larger than 100 nm [44].

Suspended SPIO particles that take the form of stable biocolloids were approved by the US Food and Drug Administration (FDA) in 1996 [106]. Since then, these particles have been used in clinical applications, for example, in diagnosis to improve the contrast of magnetic resonance imaging. SPIO-based colloids can be purchased from many companies such as Endorem (Guerbet, Sulzbach, Germany) and Resovist (Bayer HealthCare Pharmaceuticals, Berlin, Germany). Outside of Europe, Endorem is commercialized under the name Feridex [49]. In an attempt to detect diseases during the initial stage, NPs can be directed toward the lymph nodes and internalized by the macrophages, resulting in intra-cellular trapping and consequently improving the contrast of magnetic resonance imaging. This procedure was proposed to detect metastasis in patients with testicular [107] or gastric cancer [108].

Cancer treatment nanotherapy using SPIO complexes can be performed using magneto hyperthermia (MHT), chemotherapy, or gene therapy [44]. In addition to super-paramagnetism, which allows the performance of MHT, SPIO particles can be used as an anchorage for target species, such as gene therapy agents, therapeutic proteins, and chemotherapeuticals, to form the multifunction system illustrated in Fig. 6.22.

Hyperthermia in cancer therapy refers to the generation of heat at the tumor location [98]. The heating induced by the magnetic NPs present in the tumor is known as local

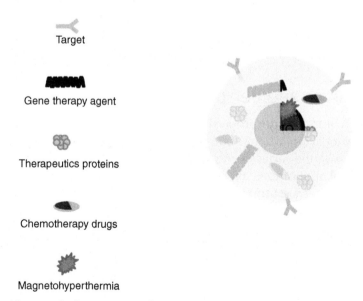

Target

Gene therapy agent

Therapeutics proteins

Chemotherapy drugs

Magnetohyperthermia

FIGURE 6.22 **The architecture of a theranostic nanotherapeutic system using an SPIO particle as a platform.** *Modified from F.M. Kievit, M. Zhang, Surface engineering of iron oxide nanoparticles for targeted cancer cells, Acc. Chem. Res. 44 (10) (2011) 853–862 [44]*

MHT. The principle of MHT consists of the impregnation of magnetic NPs in the target tissue, followed by the generation of heat via the application of an alternating magnetic field. This thermal energy dissipates from the magnetic NPs by aligning the magnetic moments of the monodomains via the Néel and Brownian relaxation processes in response to the application of an alternating magnetic field. The local temperature can reach 40–42°C, which is enough to destroy the target cells [72,109,110]. Research has also verified that MHT therapy induces an antitumor immunity effect [111]. SPIO complex-assisted chemotherapy seeks to inhibit the growth of cancer cells, delivering agents that inhibit their cell function. The multifunctionality of magnetic NPs improves drug delivery, enables the local administration of higher doses, and minimizes induced side effects when the drug is administered in the conventional manner [112]. The transfection of cells using magnetic markers is assisted by noninvasive MRI and offers a way to improve the experimental control of cell therapies [113].

Another therapy that has been coupled to SPIO particles is photodynamic therapy (PDT), which has been used to treat a variety of cancerous and noncancerous diseases with minimal side effects. The use of the photosensitizers adsorbed in SPIO particles created a new class of complex materials that enable the combination of PDT and MHT [114,115]. This multifunctional magnetic complex can simultaneously increase the MRI contrast and allow the real-time visualization of tumor destruction, combining MRI contrast improvement with localized PDT [116].

Magnetic separation is one example of the in vitro application of SPIO particles. This technique consists of coupling the target material with an SPIO particle, and then, the medium is separated using magnetic field gradients [72]. This coupling depends on the functionalization of the NP. SPIO particles coated with antibodies can be used to isolate and separate immunogenic (or even functionalized) cells to recognize and isolate DNA.

6.6 Conclusions and Perspectives

The possibility of controlling the properties of NPs under the action of an external magnetic field is clearly the most important property of magnetic NPs. In particular, the biocompatibility associated with the surface control of SPIO particles makes them promising materials for biomedical applications. Biocompatible magnetic colloids used to improve MRI image contrast have been commercialized for some time. In addition, different strategies have been adopted to functionalize the SPIO surface and obtain platforms to couple a variety of biomedical devices to promote their localized and controlled delivery. Currently, efforts have focused on the development of SPIO platforms to couple bioagents, resulting in multifunction complexes adapted to simultaneously perform diagnosis and therapy (i.e., "theranostics"). Theranostics are designed to perform diagnosis and therapy more efficiently than conventional methods, improving the cost, side effects, and benefit relationships. Different agents can be included in the system composition, generally at the surface or on the shell (in the case of nanoparticulate core–shell systems) of SPIOs.

The magnetic behavior of SPIO systems can also be used in dozens of applications other than biomedicine, especially in microelectronics, energy, catalysis, and sensors. How to control the interparticle interaction effects in a system of magnetic NPs and use them to execute desired tasks for a device remain major questions. The breakthroughs in experiments, theory, and simulation have partially answered these questions; hence, this field of study is leading the way to other applications of magnetic nanomaterials. SPIO systems require relatively simple preparations and are of low cost via synthesis routes that enable the total control of the intrinsic properties of NPs. The literature provides much information about the chemistry of iron, the major actor in this scenario. Moreover, the support of the chemistry of colloids has been fundamental for the development of these systems.

SPIO systems will certainly be part of the composition of most intelligent biomedical devices, which should become increasingly present in our daily lives. In addition, new electronic devices and other applications will also benefit from the properties of SPIO systems that, by themselves, open new opportunities for basic research and the development of new products.

List of Symbols

L	Orbital angular momentum
J	Exchange integral
j	Total angular momentum
τ_N	Néel relaxation time
τ_B	Brownian relaxation time
τ_{eff}	Effective relaxation time
ζ	Zeta potential
n	Principal quantum number
T_c	Curie temperature
M_S	Saturation magnetization
M_r	Remanent magnetization
T_d	Tetrahedral coordination
O_h	Octahedral coordination
D_{TEM}	Mean diameter obtained by transmission electron microscopy
D_M	Mean diameter calculated indirectly with the magnetic measurements
T_B	Blocking temperature

References

[1] M.M. Walker, et al. Structure and function of the vertebrate magnetic sense, Nature 390 (1997) 371–376.

[2] D. Sellmeyer, R. Skomski, Advanced Magnetic Nanostructures, Springer, New York, NY, (2006).

[3] D. Ling, et al. Chemical synthesis and assembly of uniformly sized iron oxide nanoparticles for medical applications, Acc. Chem. Res. 48 (5) (2015) 1276–1285.

[4] T.T. Thuy, et al. Next generation magnetic nanoparticles for biomedical applications, in: N.T.K. Thanh (Ed.), Magnetic Nanoparticles From Fabrication to Clinical Applications, Taylor & Francis Group, Boca Raton, FL, 2012.

[5] M.A.G. Soler, et al. Superparamagnetic iron oxides, Encyclopedia of Nanoscience and Nanotechnology, 23, American Scientific Publishers, Valencia, California, 2011, pp. 513–532.

[6] M.A. Willard, et al. Chemically prepared magnetic nanoparticles, Inter. Mater. Rev. 49 (2004) 125–170.

[7] Y.S. Kang, et al. Synthesis and characterization of nanometer-size Fe_3O_4 and γ-Fe_2O_3 particles, Chem. Mater. 8 (1996) 2209–2211.

[8] M.A.G. Soler, et al. Aging investigation of cobalt ferrite nanoparticles in low pH magnetic fluid, Langmuir 23 (2007) 9611–9617.

[9] T. Hyeon, et al. Synthesis of highly crystalline and monodisperse maghemitenanocrystallites without a size-selection process, J. Am. Chem. Soc. 123 (2001) 12798–12801.

[10] S. Bedanta, W.J. Kleemann, Superparamagnetism, J. Phys. D 42 (2009) 013001.

[11] J.C. Bacri, et al. Ionic ferrofluids: acrossing of chemistry and physics, J. Magn. Magn. Mater. 85 (1990) 27–32.

[12] J.P. Jolivet, et al. Iron oxide chemistry. From molecular clusters to extended solidnetworks, Chem. Commun. (2004) 481–487.

[13] D. Yoo, et al. Theranostic magnetic nanoparticles, Acc. Chem. Res. 44 (2011) 863–874.

[14] W.R. Viali, et al. Investigation of the molecular surface-coating on the stability of insulating magnetic oils, J. Phys. Chem. C 144 (2010) 179–188.

[15] G. Li, et al. Spin valve biosensors: signal dependence on nanoparticle position, J. Appl. Phys. 99 (2006) 08P107.

[16] X. Teng, et al. Platinum-maghemite core–shell nanoparticles using a sequential synthesis, Nano Lett. 3 (2003) 261–264.

[17] C.J. Belle, et al. Size dependent gas sensing properties of spinel iron oxide nanoparticles, Sens. Actuat. B 160 (2011) 942–950.

[18] B.D. Cullity, C.D. Graham, Introduction to Magnetic Materials, second ed, John Wiley & Sons, New Jersey, (2009).

[19] N. Spaldin, Magnetic Materials, Cambridge University Press, Cambridge, UK, (2006).

[20] K.H. Buschow, F.R. Boer, Physics of Magnetismand Magnetic Materials, Kluwer Academic/Plenum Publishers, New York, (2003).

[21] A.P. Guimarães, Magnetism and Magnetic Resonance in Solids, John Wiley & Sons, New York, (1998).

[22] D. Ortega, Structure and magnetism in magnetic nanoparticles, in: N.T.K. Thanh (Ed.), Magnetic Nanoparticles From Fabrication to Clinical Applications, Taylor & Francis Group, Boca Raton, FL, 2012.

[23] S. Krupicka, P. Novák, , in: E.P. Wohlfarth (Ed.), Oxide Spinels, Ferromagnetic Materials, vol. 3, North-Holland Publishing Company, Amsterdan, 1982.

[24] W.D. Callister, Materials Science and Engineering: an Introduction, seventh ed., John Wiley & Sons, New York, (2007).

[25] A.P. Guimarães, Principles of Nanomagnetism, Springer, Berlin, (2009).

[26] Y.-W. Jun, et al. Nanoscaling laws of magnetic nanoparticles and their applicabilities in biomedical sciences, Acc. Chem. Res. 41 (2008) 179–189.

[27] M.L. Néel, Théorie du traînage magnétique des ferromagnétiques en grains fins avec applications aux terres cuites, Ann. Geophys. 5 (1949) 99–136.

[28] W.C. Nunes, et al. Temperature dependence of the coercive field in single-domain particle systems, Phys. Rev. B 70 (2004) 014419.

[29] R.E. Rosensweig, Ferrohydrodynamics, Cambridge University Press, Cambridge, (1985).

[30] W.F. Brown, Thermal fluctuations of a single-domain particle, J. Appl. Phys. 34 (1963) 1319–1320.

[31] M.I. Shliomis, Y.L. Raikher, Experimental investigations of magnetic fluids, IEEE Trans. Magn. 16 (1980) 237–250.

[32] D.A.J. Herman, et al. How to choose a precursor for decomposition solution-phase synthesis: the case of iron nanoparticles, Nanoscale 7 (2015) 5951–5954.

[33] J.A. Gomes, et al. Synthesis of core–shell ferrite nanoparticles for ferrofluids: chemical and magnetic analysis, J. Phys. Chem. C 112 (2008) 6220–6227.

[34] P.C. Morais, et al. Raman study of ionic water-based copper and zinc ferrite magnetic fluids, J. Magn. Magn. Mater. 201 (1999) 105–109.

[35] M.A.G. Soler, et al. Surface passivation and characterization of cobalt–ferrite nanoparticles, Surf. Sci. 575 (2005) 12–16.

[36] T.F.O. Melo, et al. Investigation of surface passivation process on magnetic nanoparticles by Raman spectroscopy, Surf. Sci. 600 (2006) 3642–3645.

[37] A.H. Lu, et al. Magnetic nanoparticles: synthesis, protection, functionalization, and application, Angew. Chem. Int. Ed. Engl. 46 (2007) 1222–1244.

[38] J.P. Jolivet, et al. Design of metal oxide nanoparticles: control of size, shape, crystalline structure and functionalization by aqueous chemistry, C.R. Chimie 13 (2010) 40–51.

[39] S.G. Kwon, T. Hyeon, Colloidal chemical synthesis and formation kinetics of uniformly sized nanocrystals of metals, oxides, and chalcogenides, Acc. Chem. Res. 41 (2008) 1696–1709.

[40] S. Sun, H. Zeng, Size-controlled synthesis of magnetite nanoparticles, J. Am. Chem. Soc. 124 (2002) 8204–8205.

[41] R.H. Gonçalves, et al. Magnetite colloidal nanocrystals: a facile pathway to prepare mesoporous hematite thin films for photoelectrochemical water splitting, J. Am. Chem. Soc. 133 (2011) 6012–6019.

[42] J. Park, et al. Ultra-large-scale syntheses of monodisperse nanocrystals, Nat. Mater. 3 (2004) 891–895.

[43] M.K. Yu, et al. Targeting strategies for multifunctional nanoparticles in cancer imaging and therapy, Theranostics 2 (2012) 3–44.

[44] F.M. Kievit, M. Zhang, Surface engineering of iron oxide nanoparticles for targeted cancer cells, Acc. Chem. Res. 44 (10) (2011) 853–862.

[45] R.J. Hunter, Foundations of Colloid Science, Claredon Press, Oxford, (1986).

[46] S. Neveu, Size-selective chemical synthesis of tartrate stabilized cobalt ferrite ionic magnetic fluid, J. Coll. Int. Sci. 255 (2002) 293–298.

[47] P.C. Morais, et al. Preparation and characterization of ultra-stable biocompatible magnetic fluids using citrate-coated cobalt ferrite nanoparticles, Thin Solid Films 515 (2006) 266–270.

[48] S. Kückelhaus, et al. Optical emission spectroscopy as a tool for the biodistribution investigation of cobalt-ferrite nanoparticles in mice, J. App. Phys. 97 (2005) 10Q910.

[49] K. Andreas, et al. Highly efficient magnetic stem cell labeling with citrate-coated superparamagnetic iron oxide nanoparticles for MRI tracking, Biomaterials 33 (2012) 4515–4525.

[50] N. Fauconnier, et al. Thiolation of maghemite nanoparticles by dimercaptosuccinic acid, J. Coll. Int. Sci. 194 (1997) 427–433.

[51] C.R.A. Valois, et al. The effect of DMSA-functionalized magnetic nanoparticles on transendothelial migration of monocytes in the murine lung via a β2 integrin-dependent pathway, Biomaterials 31 (2010) 366–374.

[52] M.A.G. Soler, et al. Spectroscopic study of maghemite nanoparticles surface-grafted with DMSA, J. Phys. Chem. A 115 (2011) 1003–1008.

[53] Y.M. Huh, et al. In vivo magnetic resonance detection of cancer by using multifunctional magnetic nanocrystals, J. Am. Chem. Soc. 127 (2005) 12387–12391.

[54] T.J. Daou, et al. Phosphate adsorption properties of magnetite-based nanoparticles, Chem. Mater. 19 (2007) 4494–4505.

[55] C. Yee, et al. Self-assembled monolayers of alkanesulfonic and-phosphonic acids on amorphous iron oxide nanoparticles, Langmuir 15 (1999) 7111–7115.

[56] R. Frantz, et al. Phosphonate derivatives of pyridine grafted onto oxide nanoparticles, Tetrahedron Let. 43 (2002) 9115–9117.

[57] F.F. Souza, et al. Curcumin associated magnetite nanoparticles inhibit in vitro melanoma cell growth, J. Nanosci. Nanotechnol. 11 (2011) 7603–7610.

[58] M. De Cuyper, M. Joniau, Magnetoliposomes, Eur. Biophys. J. 15 (1988) 311–319.

[59] M.A.G. Soler, et al. Raman spectroscopy of magnetoliposomes, J. Magn. Magn. Mater. 252C (2002) 415–417.

[60] A.K. Gupta, M. Gupta, Synthesis and surface engineering of iron oxide nanoparticles for biomedical applications, Biomaterials 26 (2005) 3995–4021.

[61] A.F. Thünemann, et al. Maghemite nanoparticles protectively coated with poly(ethyleneimine) and poly(ethylene oxide)-block-poly(glutamic acid), Langmuir 22 (2006) 2351–2357.

[62] M.P.S. Almeida, et al. Preparation and size-modulation of silica-coated maghemite nanoparticles, J. Alloy. Comp. 500 (2010) 149–152.

[63] R. Pedroza, et al. Raman study of nanoparticle-template interaction in a $CoFe_2O_4/SiO_2$-based nano-composite prepared by Sol–gel method, J. Magn. Magn. Mater. 289 (2005) 139–141.

[64] J. Popplewell, L. Sakhnini, The dependence of the physical and magnetic properties of magnetic fluids on particle size, J. Magn. Magn. Mater. 149 (1995) 72–78.

[65] B.D. Cullity, Elements of X-Ray Diffraction, Addison Wesley Pub. Co, New York, (1978).

[66] R.W. Chantrell, et al. Measurements of particle size distribution parameters in ferrofluids, IEEE Trans. Magn. 14 (1978) 975–977.

[67] K. ÓGrady, A. Bradbury, Particle size analysis in ferrofluids, J. Magn. Magn. Mater. 39 (1983) 91–94.

[68] H. Mamiya, et al. Extraction of blocking temperature distribution from zero-field-cooled and field-cooled magnetization curves, IEEE Trans. Magn. 41 (2005) 3394–3396.

[69] L.G. Paterno, et al. Fabrication and characterization of nanostructured conducting polymer films containing magnetic nanoparticles, Thin Solid Films 517 (2009) 1753–1758.

[70] M.A.G. Soler, et al. Study of molecular surface coating on the stability of maghemite nanoparticles, Surf. Sci. 601 (2007) 3921–3925.

[71] N.B. Colthup, L.H. Daly, S.E. Wiberley, Introduction to Infrared and Raman Spectroscopy, third ed., Academic Press, San Diego, CA, (1990).

[72] M.A.G. Soler, Q. Fanyao, Raman spectroscopy of iron oxide nanopartices, Raman Spectroscopy for Nanomaterials Characterization, Springer, Berlin, 2012, pp. 379–416.

[73] P. Tartaj, The preparation of magnetic nanoparticles for applications in biomedicine, J. Phys. D 36 (2003) R182–R197.

[74] S.W. Da Silva, et al. Stability of citrate-coated magnetite and cobalt–ferrite nanoparticles under laser irradiation: a Raman spectroscopy investigation, IEEE Trans. Magn. 39 (2003) 2645–2647.

[75] G.V.M. Jacintho, et al. Structural investigation of MFe_2O_4 (M = Fe, Co) magnetic fluids, J. Phys. Chem. C 113 (2009) 7684–7691.

[76] L.G. Paterno, et al. Tuning of magnetic dipolar interactions of maghemite nanoparticles embedded in polyelectrolyte layer-by-layer films, J. Nanosci. Nanotechnol. 8 (2012) 6672–6678.

[77] D.P.E. Dickson, F.J. Berry, Principles of Mössbauer Spectroscopy, in: D.P.E. Dickson, F.J. Berry (Eds.), Mössbauer Spectroscopy, Cambridge University Press, Cambridge, 1986.

[78] A. Sharma, et al. Toxicological considerations when creating nanoparticle-based drugs and drug delivery systems, Expert Opin. Drug Metab. Toxicol. 8 (2012) 47–69.

[79] R.M. Pashley, M.E. Karaman, Applied Colloid and Surface Chemistry, John Wiley & Sons Ltd, Chichester, (2004).

[80] W.B. White, B.A. De Angelis, Interpretation of the vibrational spectra of spinels, Spectrochim. Acta Part A 23A (1967) 985–995.

[81] P.C. Morais, et al. Raman investigation of uncoated and coated magnetic fluids, J. Phys. Chem. A 104 (2000) 2894–2896.

[82] P.C. Morais, et al. Raman spectroscopy in magnetic fluids, Biomol. Eng. 17 (2001) 41–49.

[83] M.A.G. Soler, L.G. Paterno, Layer-by-layer assembly of magnetic nanostructures, J. Nanofluids 1 (2012) 101–119.

[84] G.B. Alcantara, et al. Morphology of cobalt ferrite nanoparticle–polyelectrolyte multilayered nanocomposites, J. Magn. Magn. Mater. 323 (2011) 1372–1377.

[85] M.A.G. Soler, et al. Assembly of γ-Fe_2O_3/polyaniline nanofilms with tuned dipolar interaction, J. Nanopart. Res. 14 (653) (2012) 1–10.

[86] L.G. Paterno, et al. Layer-by-layer assembly of bifunctional nanofilms: surface-functionalized maghemite hosted in polyaniline, J. Phys. Chem. C 113 (2009) 5087–5095.

[87] A.D. Franklin, A.E. Berkowitz, The approach to saturation on dilute ferromagnetics, Phys. Rev. 89 (1953) 1171.

[88] K.V. Shafi, et al. Sonochemical preparation and size-dependent properties of nanostructured Co-Fe_2O_4 particles, Chem. Mater. 10 (1998) 3445–3450.

[89] L.D. Tung, et al. Magnetic properties of ultrafine cobalt ferrite particles, J. Appl. Phys. 93 (2003) 7486–7488.

[90] L. Machala, et al. Amorphous iron(III) oxides: a review, J. Phys. Chem. B 111 (2007) 4003–4018.

[91] G.C. Papaefthymiou, Nanoparticle magnetism, Nano Today 4 (2009) 438–447.

[92] A.F. Rebolledo, et al. Signatures of clustering in superparamagnetic colloidal nanocomposites of an inorganic and hybrid nature, Small 4 (2008) 254–261.

[93] B.P. Pichon, et al. Magnetotunable hybrid films of stratified iron oxide nanoparticles assembled by the layer-by-layer technique, Chem. Mater. 23 (2011) 3668–3675.

[94] L.G. Paterno, et al. Magnetic nanocomposites fabricated via the layer-by-layer approach, J. Nanosci. Nanotech. 10 (2010) 2679–2685.

[95] J.L. Dormann, et al. Magnetic relaxation in fine-particle systems, Adv. Chem. Phys. 98 (1996) 283.

[96] J.I. Gittleman, et al. Superparamagnetism and relaxation effects in granular Ni-SiO_2 and Ni-Al_2O_3 films, Phys. Rev. B 9 (1974) 3891–3897.

[97] M.A. Novak, et al. Relaxation in magnetic nanostructures, J. Magn. Magn. Mat. 294 (2005) 133–140.

[98] C.S.S.R. Kumar, F. Mohammad, Magnetic nanomaterials for hyperthermia-based therapy and controlled drug Delivery, Adv. Drug Deliv. Rev. 63 (2011) 789–808.

[99] S. PARVEEN, et al. Nanoparticles: a boon to drug delivery, therapeutics, diagnostics and imaging, Nanomedicine: Nanotechnol. Biol. Med. 8 (2012) 147–166.

[100] K. Hola, et al. Tailored functionalization of iron oxide nanoparticles for MRI, drug delivery, magnetic separation and immobilization of biosubstances, Biotechnol. Adv. 33 (2015) 1162–1176.

[101] M.W. Freeman, et al. Magnetism in medicine, J. Appl. Phys. 31 (1960) 404S–405S.

[102] J. Dobson, Gene therapy progress and prospects: magnetic nanoparticle-based gene delivery, Gene Therapy 13 (2006) 283–287.

[103] S. Moritake, et al. Functionalized nano-magnetic particles for an in vivo delivery system, J. Nanosci. Nanotechnol. 7 (2007) 937–944.

[104] C. Alexiou, et al. In vitro and in vivo investigations of targeted chemotherapy with magnetic nanoparticles, J. Magn. Magn. Mater. 293 (2005) 389–393.

[105] O.A. Garden, et al. A rapid method for labelling CD4+ T cells with ultrasmall paramagnetic iron oxide nanoparticles for magnetic resonance imaging that preserves proliferative, regulatory and migratory behaviour in vitro, J. Immunol. Methods 314 (2006) 123–133.

[106] J. Oh, et al. Detection of magnetic nanoparticles in tissue using magneto-motive ultrasound, Nanotechnology 17 (2006) 4183–4190.

[107] M.G. HarisinghanI, et al. Pilot study of lymphotrophic nanoparticle-enhanced magnetic resonance imaging technique in early stage testicular cancer: a new method for noninvasive lymph node evaluation, Urology 66 (2005) 1066–1071.

[108] Y. Tatsumi, et al. Preoperative diagnosis of lymph node metastases in gastric cancer by magnetic resonance imaging with ferumoxtran-10, Gastric Cancer 9 (2006) 120–128.

[109] I. Hilger, et al. Towards breast cancer treatment by magnetic heating, J. Magn. Magn. Mater. 293 (2005) 314–319.

[110] E. Pollert, K. Záveta, Manocrystalline oxides in magnetic fluid hyperthermia, in: N.T.K. Than (Ed.), Magnetic Nanoparticles From Fabrication to Clinical Applications, Taylor & Francis Group, Boca Raton, FL, 2012.

[111] M. Yanase, et al. Antitumor immunity induction by intracellular hyperthermia using magnetite cationic liposomes, Jpn. J. Cancer Res. 89 (1998) 775–782.

[112] M.L.B. Carneiro, et al. Free rhodium (II) citrate and rhodium (II) citrate magnetic carriers as potential strategies for breast cancer therapy, J. Nanobiotechnol. 9 (11) (2011) 3–17.

[113] A. Del Campo, et al. Multifunctional magnetite and silica–magnetite nanoparticles: synthesis, surface activation and applications in life sciences, J. Magn. Magn. Mater. 293 (2005) 33–40.

[114] P.P. Macaroff, et al. Evaluation of new complexes of biocompatible magnetic fluid and third generation of photosensitizer useful to cancer treatment, IEEE Trans. Magn. 41 (2005) 4105–4107.

[115] R. Di Corato, Combining magnetic hyperthermia and photodynamic therapy for tumor ablation with photoresponsive magnetic liposomes, ACS Nano 9 (2015) 2904–2916.

[116] R. Kolpeman, et al. Multifunctional nanoparticle platforms for in vivo MRI enhancement and photodynamic therapy of a rat brain cancer, J. Magn. Magn. Mater. 293 (2005) 404–410.

6

Electrochemical Sensors

F.R. Simões*, M.G. Xavier**

*INSTITUTE OF MARINE SCIENCES, FEDERAL UNIVERSITY OF SÃO PAULO,
SANTOS, SÃO PAULO, BRAZIL; **CENTER OF BIOLOGICAL AND NATURE
SCIENCES, FEDERAL UNIVERSITY OF ACRE, RIO BRANCO, ACRE, BRAZIL

6.1 Introduction

A sensor is a device that responds to a physical stimulus, such as heat, light, sound, pressure, magnetism, or movement, and transmits a resulting electrical impulse as a means of measuring the change of any intrinsic property of the constituent material [1]. The origin of the word sensor comes from the Latin *sentire*, to feel. Semantically, sensors have the attribute of feeling into their surrounding environment to define a coupling relationship. Electrochemical sensors, in particular, are a class of chemical sensors in which an electrode is used as a transducer element in the presence of an analyte.

Modern electrochemical sensors use several properties to detect various parameters in our everyday lives, whether physical, chemical, or biological parameters. Some examples are environmental monitoring, health and instrumentation sensors, and sensors related to machines, such as automobiles, airplanes, mobile phones, and technology media. In recent decades, modern sensing systems have benefited from advances in microelectronic and microengineering [2], mainly through the manufacturing of even smaller sensors with more sensitivity and more selectivity, and with lower production and maintenance costs.

In this context, the incorporation of techniques and concepts related to nanoscience and nanotechnology (N&N) has a great impact on the fast development of nanosensors, mainly because of the surface versus volume ratio, which is one of the most significant characteristics of the nanometric order. However, this context is far from meeting the growing consumer expectations of societies with social inequalities, rendering the issue of knowledge democratization, and new technologies, including N&N, a great challenge for science in the 21st century, causing the social and political focus to be increasingly present in the scientific context [3].

Currently, the paradigms that involve research and development of electrochemical sensors allow the study of new materials, applications of samples of different natures, new manufacturing methods, and strategies to enhance selectivity and detection limits. Associated with N&N, these paradigms involve interdisciplinary relationships, highlighting scientists' roles in the scientific revolution of the last few decades.

Historically, the use of electrochemical sensors had its beginnings in the 1950s, through the monitoring of industrial oxygen [4]. Labor laws related to the health and safety of the workers required the monitoring of toxic gases and fuels in confined spaces, prompting a burst of research into electrochemical sensors that could exhibit good selectivity for the detection of different gases. Leland C. Clark proposed the oxygen sensor concept with the use of two electrodes of a cell with an oxygen permeable membrane separating the electrodes and the electrolyte solution. The oxygen diffused through the membrane and was reduced in the indicator electrode. This resulted in a current proportional to the concentration of oxygen in the sample. The Clark oxygen sensor found wide application in medicine and environmental and industrial monitoring. In 1963, during the Cold War, a different oxygen sensor was developed and used for monitoring water quality in the former Union of Soviet Socialist Republics. Despite its commercial popularity, the electric current signal obtained in the Clark oxygen sensors was unstable, and the oxygen analyzers required frequent precalibration, which hindered their use [5].

The determining factor for the contemporary development of electrochemical sensors was the development of the scanning tunneling microscope (STM) and the atomic force microscope (AFM) in 1981 and 1986, respectively, by the Binnig and Rohrer teams [6,7]. These advances in scientific instrumentation enabled the development of new products and processes based on the ability of these instruments to visualize material structures with dimensions on the order of 1.0×10^{-9} m (1 nm), resulting in the so-called nanoscience and nanotechnology. Nanoscience shortly became a very broad and interdisciplinary research field, as it is based on various types of materials (polymers, ceramics, metals, semiconductors, composites, and biomaterials) structured on the nanometer scale to form nanoscale structures, such as nanoparticles, nanotubes, nanofibers, and nanospheres. Based on material behavior at the nanometric scale, promising new possibilities arose in the areas of research and applications for electrochemical sensors and electroanalytical methods.

Table 6.1 Main Electrochemical Methods, Monitored Electrical Properties, and Respective Units

Electrochemical Methods	Monitored Electrical Properties	Units
Potentiometry	Potential difference (volts)	V
Conductometry	Resistance (ohms)	Ω
Amperometry and voltammetry	Current (amps) as a function of applied potential	I
Coulometry (Q)	Current as a function of time (coulombs)	$C = I \cdot s$
Capacitance (C)	Potential load (farads)	$F = C \cdot V^{-1}$

6.2 Electroanalytical Methods

6.2.1 Principles

The signal from an electrochemical sensor is usually derived from an electrical response given the presence of an analyte. Table 6.1 lists some of the main electrochemical methods and their respective monitored electrical signals [8].

Table 6.1 indicates that in general, electrochemical responses monitored by different methods are based primarily on potential, resistance, and electrical current. Coulometry and capacitance methods, for example, have their responses derived from potential, resistance, and electrical current.

When we treat electrochemical techniques as methods of analysis, we can divide them into two main groups: *interfacial methods* and *noninterfacial methods* [8] (measuring the solution as a whole). From the examples mentioned in Table 6.1, *conductometry* is a *noninterfacial* method, as it is based essentially on a cell of known size, with two equidistant electrodes that measure the electrical conductance (essentially, the resistance) of a solution as a whole. When applying an alternating current signal, there is no electrode polarization and the response is given in terms of the electric resistance of the solution as a function of the cell constant, which is based on the surface area of the electrodes, their spacing, and the solution volume in the electrochemical cell [8]. In turn, the *interfacial methods* are methods that respond directly or indirectly to the presence of the analyte on the electrode surface (sensory unit), resulting in a disturbance of an electric signal that can be measured [8]. We can divide the interfacial methods into two main groups: the *static methods* and the *dynamic methods*. The static methods are defined as those in which there is no disturbance and the electric current is zero ($i = 0$), and the dynamic methods are those that exploit a redox reaction; that is, electron transfer occurs between the electrode and the analyte [8]. Importantly, dynamic methods employ current flow (and therefore charge transfer reactions involving electron transport) and can thus be used both for analytical determination (sensors) and in the manufacturing of nanostructured materials (Chapter 3, Volume 1, of this collection). This chapter primarily addresses the use of interfacial methods used in electroanalytical determination (sensors).

FIGURE 6.1 Bench pH meter and glass electrode (Metrohm). *Adapted from http://www.metrohm.com/en-us/products-overview/ph_ion-measurement/912_913_914-ph_cond-meters/%7B100FAF11-60F3-4D90-B244-0E438515D92D%7D*

6.2.2 Potentiometry

Potentiometry is a static interfacial method and the most widely used electrochemical method in analytical applications [8]. A notable example of potentiometry is the pH meter (Fig. 6.1), which is usually based on glass membrane electrodes as indicators of the concentration (more accurately, the activity) of the H^+ ions in solution.

In general, a pH meter compares the activity of the H^+ ions on both sides of the glass membrane for which the internal activity (concentration) is fixed and known, and the external activity (analysis solution) is measured. The responses are obtained in the potential difference (V) of the solution on both sides of the membrane and are converted to a pH scale based on the Nernst equation, which relates potential and concentration [8]:

$$E = E^0 + \frac{RT}{nF} \ln \frac{[RED]}{OXI}$$
(6.1)

where E, cell potential; E^0, standard potential of a half-reaction; R, universal gas constant; T, temperature; n, number of electrons (eq. mol^{-1}) involved in the half-reaction; F, Faraday constant; [RED] = activity of the reduced species; and [OXI] = activity of the oxidized species.

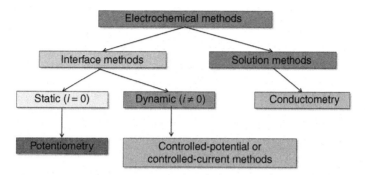

FIGURE 6.2 Main electrochemical methods and their subdivisions. *Adapted from F.J. Holler, D.A. Skoog, S.R. Crouch, Principles of Instrumental Analysis, sixth edition, Thomson Brooks/Cole, 2007 [8].*

Dynamic interfacial methods are those in which the surface of the working electrode, indicator electrode, or sensor receives an electric stimulus with a nonzero current ($i{\neq}0$). Among these, we can highlight amperometry, voltammetry, and coulometry as controlled potential techniques, and the coulometric titrations with controlled current. Fig. 6.2 shows the main electrochemical methods and their subdivisions.

The fundamental principle of any electrochemical sensor is the recognition of the analyte through the active layer of the material that composes it (Fig. 6.3) and the subsequent signal transduction to the recording equipment. Thus, by observing the diagram of electrochemical methods (Fig. 6.3), we can conclude that an electrochemical sensor is based on interface methods.

The electrochemical sensors of analytes in solution consist, in general, of electrochemical cells specifically set between an active material of the working electrode (sensory unit), the analyte, the reaction medium, and the set of electrodes used. The interface methods are electrolyte processes in which the polarization of the electrodes (cathode and anode)

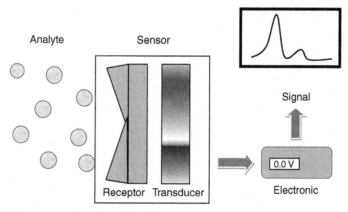

FIGURE 6.3 General diagram of an electrochemical sensor.

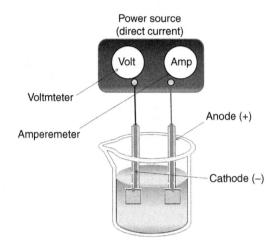

FIGURE 6.4 Electrochemical cell in the system of two electrodes.

induces an electrochemical reaction. In a conventional electrolyte cell with two electrodes, there is a source, a voltmeter, an anode, and a cathode (Fig. 6.4).

Despite the development of electrochemical analysis methods, there was a need for potential control. As previously mentioned, according to Eq. (6.1) (Nernst equation), the potential of an electrochemical cell is governed by the activity (concentration) of the analyte. As a reaction takes place, with the decrease in the analyte activity, the cell potential changes. The potential control also circumvents the problems related to the application of *overpotential* (η), which is the additional potential in relation to the thermodynamic potential to overcome the effects of electric and kinetic resistance (activation energy) [8]. The equipment that was derived from this need is called a potentiostat, which maintains the cell potential constant based on the constant monitoring of the reversible reaction potential of a *reference electrode*. All electrochemical potentiometric, voltammetric, and amperometric sensors use reference electrodes.

A reference electrode is an electrode with a known, constant potential and is insensitive to the composition of the solution under study. The ideal reference electrode must be based on a reversible reaction that obeys the Nernst equation and should preferably have low dependence on temperature [8]. Some reference electrodes meet these requirements well. The most common electrodes are the *normal (or standard) hydrogen electrode (NHE or SHE)*, *saturated calomel electrode (SCE)*, and *saturated silver/silver chloride electrode* ($E_{Ag/AgCl}$). The NHE is the historical and universal reference and can be easily constructed in the laboratory; however, its application is limited to certain applications, as it must be prepared each time it is used. The metal electrodes (SCE and $E_{Ag/AgCl}$) play a key role in the performance and applications of the current electrochemical sensors, as they are more robust and can be stored. They consist of a metal and a soluble salt of this metal in the electrode body. Table 6.2 shows the reference electrodes, their compositions, and potential constants at 25°C.

Table 6.2   Types of Reference Electrodes, Their Composition, and Potential at 25°C

Reference Electrode	Reversible Reaction	Standard Potential at 25°C
NHE	$2H^+ + 2e^- \rightleftarrows H_2$	$E^0 = 0.0$ V (by definition)
SCE	$Hg_2Cl_{2(S)} + 2e^- \rightleftarrows Hg_{(l)} + 2Cl^-$	$E^0 = +0.244$ V (vs. NHE)
$E_{Ag/AgCl}$	$AGCl_{(S)} + e^- \rightleftarrows Ag_{(S)} + Cl^-$	$E^0_{AgCl} = +0.199$ V(vs.NHE)

NHE, Normal hydrogen electrode; SCE, saturated calomel electrode.

In addition to using the reference electrode in its settings so that there is polarization of the working electrode (indicator electrode or sensory unit), the voltammetric and amperometric techniques (dynamic analysis methods) are based on a system of three electrodes (Fig. 6.5). The electric current of the electrochemical system is monitored at a constant potential (amperometric techniques) or as a function of the variation of the analysis potential (voltammetric techniques).

A conventional electrochemical cell of three electrodes consists of a *working electrode*, on which there is oxidation and reduction reactions of the species involved, the *counter electrode (or auxiliary electrode)*, which serves to complete the electric circuit of the electrochemical system, and the *reference electrode*, which provides a reference for the assessment of other measured parameters [8].

6.2.3 Chronoamperometry

The potentiostatic technique of *chronoamperometry* measures the current flowing through the working electrode as a function of time as a constant potential is applied to the working electrode [9]. The current flow is correlated with the concentration of the oxidized or reduced species on the surface of the working electrode through the Cottrell equation:

$$It = \frac{nFAc_O^0 D_O^{1/2}}{\pi^{1/2}t^{1/2}} = bt^{1/2}i_i \tag{6.2}$$

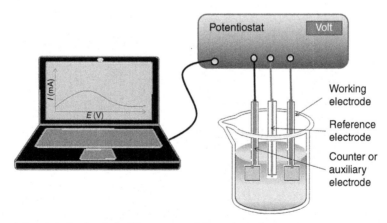

FIGURE 6.5 Conventional electrochemical cell in a system of three electrodes.

where i_t is the current at time t (s), n is the number of electrons (eq. mol^{-1}), F is Faraday's constant (96,485 C eq.$^{-1}$), A is the geometric area of the electrode (cm^2), c_0 is the concentration of the oxidized species (mol cm^{-3}), and D_0 is the diffusion coefficient of the oxidized species ($cm^2\,s^{-1}$).

Chronoamperometry allows processes to be studied in which the current is directly measured as a function of a constant potential applied to the electrochemical system. Given that the current detected in electrochemical experiments is governed by two processes, the mass transfer (corresponding to the transfer of species from the solution to the electrode–solution interface) and the charge transfer (corresponding to the electron transfer in the working electrode surface), the occurrence of chemical reactions that can occur simultaneously with the main processes, such as protonation, polymerization, adsorption, desorption, and crystallization, is studied to gain an understanding of the synthesis processes and characterization of materials through charge and mass transfer mechanisms (oxidation and reduction reactions that occur at the electrode-solution interface) [9]. During the 1960s, the potentiostatic chronoamperometry technique was used in several studies to explain the process of the nucleation and growth of polymers considering them as crystalline materials and therefore suitable for the nucleation of crystals with controlled nucleation [10].

Fig. 6.6 shows the chronoamperometry performed in a simultaneous determination in an aqueous solution of a mixture of nitrite and nitrate using an electrochemical sensor based on a composite electrode of silver (Ag), zeolite (ZE), and graphite-epoxy (GE).

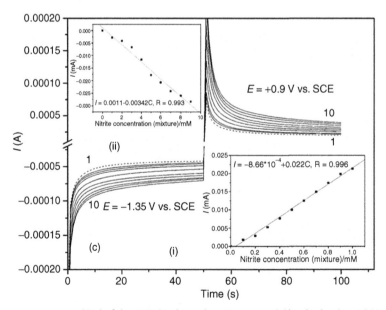

FIGURE 6.6 Chronoamperograms (CAs) of the AgZEGE electrode at two potential levels, that is, −1.35 V and +0.9 V versus SCE, recorded in 0.1 M Na_2SO_4 supporting electrolyte with the addition of a mixture of 0.1–1 mM nitrite and 1–10 mM nitrate. Insets: Calibration plots corresponding to current readings from CAs at 50 s [11].

The authors observed a linear dependence of the nitrate and nitrite concentrations in the concentration ranges of 1–10 mM for nitrate and 0.1–1 mM for nitrite [11]. Historically, the maximum current peaks followed by a sharp drop in current shown in Fig. 6.6 are associated with the charging process of the electrical double layer, followed by its evolution featuring a maximum peak until its gradual decrease over time [9].

Despite the difficulty in explaining such phenomena on the surface of the working electrode, in 1974, the pioneering work of Hills, Schiffrin, and Thompson [12] associated the maximum current value found in the chronoamperogram with equations related to the nucleation and growth model of metals. In later work, Scharifker and Hills demonstrated that the growth of the metal core was not described by linear ion diffusion, as previously thought, but by diffusion of spherical nuclei. This description of the growth of individual nuclei was later confirmed by computer simulation and monitoring of metal core growth in microelectrodes.

6.2.4 Voltammetric Methods

Historically, voltammetric methods were developed from the discovery of polarography in 1922 by the Czech chemist Jaroslav Heyrovsky, who received the chemistry Nobel Prize in 1959 [13]. Polarography studies electrolysis solutions and substances that are reduced or oxidized in a dropping mercury electrode and a reference electrode. The potential between these electrodes is varied, and the resulting changes in the current flow are measured. By plotting the changes in current flow against the potential variation, a current versus potential polarographic graph is obtained ($I \times E$) [8].

During the 1960s and 1970s, theories, methods, and instrumentation were developed for the voltammetry field, thus increasing the sensitivity and repertoire of electroanalytical methods [14]. The common characteristic of all voltammetric techniques is that they involve the application of a potential (E) on an electrode and the monitoring of the resulting current (I) flowing through the electrochemical cell. In many cases, the applied potential is varied or the current is controlled for a period of time (t). Thus, all voltammetric techniques can be described as a function of potential, current, and time (E, I, and t). The analytical advantages of the many voltammetric techniques include the following: excellent sensitivity with the detectable concentration range of organic and inorganic species; a large number of useful solvents and electrolytes; a wide range of temperatures; fast analysis times (seconds); simultaneous determination of several analytes; the ability to determine kinetic parameters and estimate unknown parameters; the ease with which different potential wavelength shapes can be generated; and the measurement of small currents [8].

The potential sweep methods, also known as voltammetric methods, consist of the application of a potential varying continuously with time on a working electrode, which leads to the occurrence of oxidation or reduction reactions of electroactive species in the solution (faradaic reactions), in accordance with the adsorption of species with the potential and a capacitive current due to the electrical double layer. The observed current is therefore different from the current in the steady state. These methods are commonly used

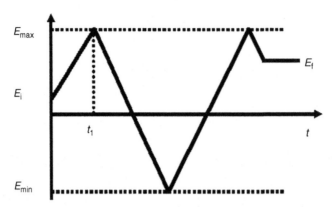

FIGURE 6.7 Potential applied as a function of time in cyclic voltammetry (CV) (E_{inic}, initial potential; E_f, final potential; $E_{máx}$, maximum potential; E_{min}, minimum potential). Sweep speed $v = dE/dt$ [12].

for the study of processes occurring in the working electrode and can be used with linear, pulse, and cyclic sweep, in addition to cyclic voltammetry (CV). Their main use has been for the diagnosis of electrochemical reaction mechanisms, for the identification of species present in solution, and for semiquantitative analysis of reaction speeds [15], and in addition to these applications, these methods are also widely used for the measurement of kinetic and constant rates, the determination of adsorption processes in surfaces, the study of electron transfers and reaction mechanisms, the determination of thermodynamic properties of solvated species, essential studies of oxidation and reduction processes in several ways, and the determination of complexing values and coordination.

6.2.4.1 Cyclic Voltammetry

In linear sweep voltammetry, the potential sweep is performed in only one direction, stopping at a chosen value E_f, for example, for $t = t_1$. The sweep direction can be positive or negative, and the sweep speed, at first, can take any arbitrary value [9]. In CV, when reaching $t = t_1$, the sweep direction is reversed and changed until E_{min} is reached, then reversed and changed to $E_{máx}$, and so on, generating cycles with several sweeps. The basic diagram involving the application of a potential sweep is shown in Fig. 6.7.

In a cyclic voltammogram, the most analyzed parameters are the following:

- initial potential, E_i;
- initial sweep direction;
- sweeping speed, v;
- maximum potential, $E_{máx}$;
- minimum potential, E_{min}; and
- final potential, E_f.

An example of a cyclic voltammogram, that is, the measured current response as a function of the potential for a reversible system, is shown in Fig. 6.8. A faradaic current (I_f), resulting from the electrode reaction is recorded in the potential window during the

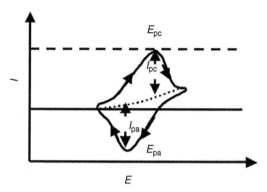

FIGURE 6.8 Cyclic voltammogram for a reversible system, where *E* is the potential and *I* is current. *Adapted from A.M.O. Brett, C.M. Brett, A Electrochemistry Principles, Methods, and Applications, Oxford University Press, Coimbra, 1994 [9].*

reaction of the analyte on the electrode surface. There is also a capacitive contribution (I_c) because when sweeping the potential, the charge of the electrical double layer (C_d) changes. This contribution increases with increasing sweeping speed. The total current is a sum of the capacitive and faradaic currents [9]. Both currents tend to increase with increasing sweeping speed. This relationship limits the technique sensitivity because the high capacitive current can interfere in the sensitivity of the faradaic current, which, in the linearity region, is proportional to the analyte concentration. The potential obeys the Nernst equation (Eq. 6.1) and is therefore characteristic of a given redox process or an analyte. Thus, CV can be used for quantitative determinations; because of its limitations, however, it is more generally used for exploratory purposes, that is, to determine the redox process of different analytes. To minimize the contribution of the capacitive current and therefore increase the sensitivity of voltammetric methods, potential impulse (or pulse) methods were developed, including pulse voltammetry and square wave voltammetry (SWV) [9].

Fig. 6.9 shows CV results in the determination of nitrite and nitrate using the composite electrode of silver, zeolite, and graphite-epoxy (AgGE). As expected, the AgZEGE electrode enables characterization of the redox peaks of the Ag^0/Ag^+ pair. Nitrite exhibits a well-defined oxidation peak at approximately 0.9 V compared with the SCE reference electrode. This peak can be attributed to the oxidation process from nitrite to nitrate. Furthermore, during the cathodic direction sweep, no cathodic peak appears, and an increase of cathodic current is observed starting at -1.0 V (vs. SCE) that corresponds to the reduction of nitrate to nitrite. Finally, successive potential sweeps using the AgZEGE electrode in 0.1 M Na_2SO_4 supporting electrolyte and the mixture of nitrate (0.1 – 1 mM) and nitrate (1 – 10 mM) reveal the linear dependence of the anodic and cathodic currents on the concentrations. These dependences form the basis for the use of the amperometric method (Fig. 6.6) [11].

6.2.4.2 Differential Pulse Voltammetry

Differential pulse voltammetry (DPV) is a technique that involves applying amplitude potential pulses on a linear ramp potential. In DPV, a base potential value is chosen at which

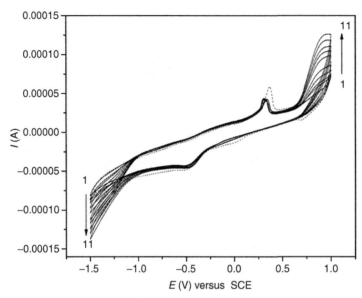

FIGURE 6.9 Cyclic voltammograms of the AgZEGE electrode recorded in 0.1 M Na$_2$SO$_4$ supporting electrolyte with the addition of a mixture of 0.1–1 mM nitrite and 1–10 mM nitrate. Scan rate: 50 mVs^{-1} [11].

there is no faradaic reaction and is applied to the electrode. The base potential is increased between pulses with equal increments. The current is immediately measured before the pulse application and at the end of the pulse, and the difference between them is recorded. Fig. 6.10 shows the pulse shape in DPV [9].

DPV is a differential technique similar to the first derivative of a linear voltammogram in which the formation of a peak is observed for a given redox process. In the linear sweep technique, the voltammogram has a shape similar to a wave, and the first derivative originates a peak. In linear sweep voltammetry, as in *polarography* (dropping mercury

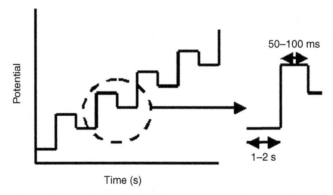

FIGURE 6.10 Diagram of the application of pulses in the differential pulse voltammetry (DPV) technique. *Adapted from A.M.O. Brett, C.M. Brett, A Electrochemistry Principles, Methods, and Applications, Oxford University Press, Coimbra, 1994 [9].*

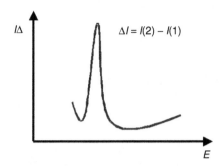

FIGURE 6.11 Typical response of a differential pulse voltammogram. *Adapted from A.M.O. Brett, C.M. Brett, A Electrochemistry Principles, Methods, and Applications, Oxford University Press, Coimbra, 1994 [9].*

electrode), the qualitative information of an analyte is given by the half-wave potential ($E_{1/2}$), which corresponds to the potential at half the wave height. Similarly, in DPV, the peak potential, E_p, can be approximately identified with $E_{1/2}$. Increasing the irreversibility, E_p deviates from $E_{1/2}$ as the base of the peak widens and its height decreases. The DPV is therefore a graph of differences between measured currents and applied potentials (Fig. 6.11) [9].

The best responses obtained with the use of DPV, compared to normal pulse voltammetry, are on solid electrodes [9], especially those involving organic compounds. Since they are usually adsorbed by the electrode, it is possible that a differential technique discriminates the effects that are more or less constant before and after the application of pulses.

In general, pulse techniques, such as DPV, are more sensitive than the linear sweep methods because there is minimization of the capacitive current. In turn, CV is most commonly used for exploratory purposes. Thus, in general, it is not unusual in sensor development to use both techniques because CV provides essential information, such as the process reversibility and types of redox processes present in the analysis (matrix, analyte, and electrode), whereas the pulse techniques are used for quantitative determinations. For instance, DPV was developed for detecting methyl parathion (MP) using a glassy carbon electrode modified with a nanocomposite film based on multiwalled carbon nanotubes (MWCNTs) and polyacrylamide (PAAM) [16]. The experimental results demonstrated that the MWCNT-PAAM/GCE nanocomposite film exhibited strong adsorption and high affinity with MP in environmental samples. Fig. 6.12 shows the results obtained from CV, used for verification of the redox behavior of the pesticide, and DPV, used for the analytical determination. The authors obtained a linear calibration curve in the concentration range of 5.0×10^{-9} to 1.0×10^{-5} mol L^{-1}, with a detection limit of 2.0×10^{-9} mol L^{-1}.

6.2.4.3 Square Wave Voltammetry

SWV is one of the fastest and most sensitive pulse voltammetry techniques. The detection limits can be compared with those of chromatographic and spectroscopic techniques. In addition, the analysis of the characteristic parameters of this technique also enables the

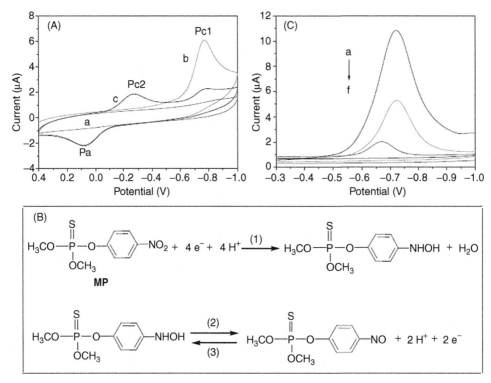

FIGURE 6.12 CV of MWCNTs–PAAM/GCE in the absence (a) and presence (b and c) of 5.0×10^{-6} mol L^{-1} MP; (B) mechanism of the electrochemical reaction of MP with MWCNTs–PAAM/GCE; (C) DPV of MWCNTs–PAAM/GCE (a and d), PAAM/GCE (b and e), and GCE (c and f) in the presence (a, b, and c) and absence (d, e, and f) of 1.0×10^{-5} mol L^{-1} MP; supporting electrolyte, 0.2 M PBS (pH 7.0); scan rate, 100 mV s^{-1}; adsorption time, 5 min [17].

evaluation of the kinetics and mechanism of the electrode process under study [9,17]. In SWV, the shape of the potential current curve is derived from the application of potentials of height ΔE (pulse amplitude), which vary according to a potential step E_{step} (in mV) and τ duration (period). On the potential–time curve, the pulse width ($\tau/2$) is denoted by t, and the frequency of pulse application is denoted by f and is given by ($1/t$). The electric currents are measured at the end of the direct (I_1) and reverse (I_2) pulses, and the signal is obtained as an intensity of the resulting differential current (ΔI); this technique offers excellent sensitivity and high rejection to capacitive currents. This measurement precedes an initial time (t_i) at which the working electrode is polarized at a potential for which the redox reaction does not occur [17].

Fig. 6.13 shows details of the potential application of SWV, with the definition of the used parameters, whereas Fig. 6.14 shows the theoretical voltammograms associated with (A) a reversible system and (B) an irreversible system, with observed separation of direct, reverse, and resulting currents, and both voltammetric profiles have similarities to those obtained in square wave polarography.

Current–potential curves display well-defined profiles and are usually symmetrical. This is due to all currents being measured only at the end of each semiperiod and the

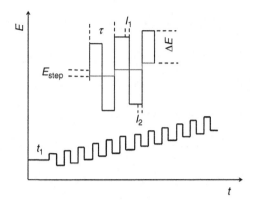

FIGURE 6.13 Application of potentials in square wave voltammetry (SWV). *Adapted from A.M.O. Brett, C.M. Brett, A Electrochemistry Principles, Methods, and Applications, Oxford University Press, Coimbra, 1994 [9].*

variations in the height and width of the potential pulse being always constant for a determined potential range [17]. Thus, electrochemical techniques can be used in the synthesis and characterization of materials through voltammetric methods that relate the current and the electric potential in the electrochemical cell. In amperometric sensors, for instance, a constant electric potential is applied to the electrochemical cell and a corresponding current appears because of the redox reactions that occur on the surface of the working electrode. This current can be used to quantify the reactions involved. Amperometric sensors can be operated through CV, another powerful technique for the synthesis and characterization of different electroactive species, and the relationship between the current–potential characteristic of each oxidation or reduction reaction involved. In general, voltammetric sensors are used in the detection of species in redox reactions that occur in the electrochemical cell. The SWV technique has also been used in the development of sensors and biosensors because of its high sensitivity and selectivity [17]. It is currently of great interest to the pharmaceutical industry for the use of biomarkers in the

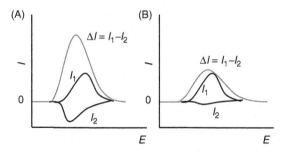

FIGURE 6.14 Schematic square wave voltammogram, where (A) represents a redox process of a reversible system and (B) represents that of an irreversible system. *Adapted from A.M.O. Brett, C.M. Brett, A Electrochemistry Principles, Methods, and Applications, Oxford University Press, Coimbra, 1994 [9].*

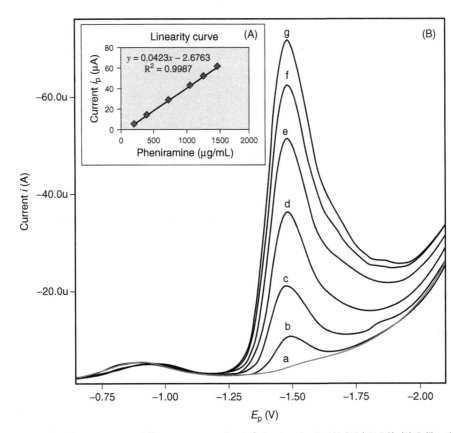

FIGURE 6.15 (A) Plot of current versus different concentrations of pheniramine in 1% SLS (pH 5.3). (B) Calibration of square-wave voltammetric peak current response of pheniramine at different concentrations in 1% SLS (pH 5.3): (a) blank, (b) 200 µg mL^{-1}, (c) 400 µg mL^{-1}, (d) 750 µg mL^{-1}, (e) 1100 µg mL^{-1}, (f) 1300 µg mL^{-1}, and (g) 1500 µg mL^{-1}.

detection of disease, environmental pollutants, such as heavy metals, and other chemical contaminants that are part of the environmental liability in contemporary societies. Fig. 6.15 shows the SWV results for the quantification of pheniramine in a pharmaceutical formulation using a glassy carbon electrode modified with MWCNTs in the presence of sodium lauryl sulfate [18]. Experimental results suggest that pheniramine in anionic surfactant solution has an electrocatalytic effect, resulting in a significant increase of the peak current. The authors observed a linear dependence of the peak current with the analyte concentration (pheniramine) in the range of 200–1500 µg mL^{-1}, with a correlation coefficient of 0.9987. The detection limit was 58.31 µg mL^{-1}.

The efficiency and sensitivity of SWV are used in the detection of food contaminants (bacteria, viruses, and parasites) and to verify the therapeutic ingredients of dietary supplements. Other SWV applications are to research the mechanism of enzymatic kinetics and the development of new methods to improve the surfaces of nanomaterial used in high sensitivity sensors and biosensors. Thus, the SWV technique is a powerful tool in the creation of diagnostic devices and environmental/food monitoring with high sensitivity and selectivity for enzymatic studies [19].

6.2.5 Electrochemical Impedance Spectroscopy

Electrochemical impedance spectroscopy (EIS) is a technique used in the analysis of electrochemical processes that occur in the electrode/electrolyte solution interface. This is a method of identification and determination of parameters from a model developed based on the frequency response of the electrochemical system under study. A frequency response analyzer coupled with an electrochemical interface is used in such experiments, which measures the current response of the system as it changes the frequency of an input sinusoidal signal [9,20,21], $E = E_o \sin(\omega = 2\pi f)$, that is applied to an unknown sample; the response is analyzed through a correlation with the resulting current $I = I_o \sin(\omega t + \varphi)$, where φ is the phase angle displacement.

The impedance (Z) is the vector sum given by [9]

$$Z = R - jX_c \tag{6.3}$$

where j is equal to $\sqrt{1}$, (Fourier) R is resistance, X_c is the capacitive reactance and equal to a$1/\omega C$ (measured in ohms) with $\omega = 2\pi f$, and C is the capacitance of the electric double layer. Therefore, it is noted that R is the real part (Z') and $-jX_c$ is the imaginary part (Z'') of the impedance.

The complex impedance, represented by the displacement vector $Z(\omega)$, is obtained by changing the alternating signal ω, with these parameters, the modulus $Z_0 = E_0/I_0$ and the phase angle φ can be calculated to determine the real and imaginary impedance values, as shown in Fig. 6.16.

An example of a simple equivalent circuit of an electrochemical cell, a resistor in parallel with a capacitor (RC), is depicted in Fig. 6.17. The typical resulting spectrum is shown, consisting of a semicircle in the complex impedance plane, with the corresponding imaginary opposite $Z'' = -Im = Z_0 \sin(\omega)$ conventionally represented on the y axis and the real component $Z' = R$ and $Z = Z_0 \cos(\omega)$ on the x axis.

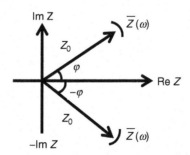

FIGURE 6.16 Vector representation of the complex impedance *Z*. *Adapted from A.M.O. Brett, C.M. Brett, A Electrochemistry Principles, Methods, and Applications, Oxford University Press, Coimbra, 1994 [9].*

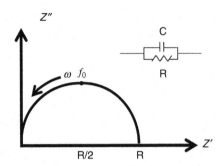

FIGURE 6.17 Schematic representation of the impedance analysis response of a parallel RC circuit. *Adapted from A.M.O. Brett, C.M. Brett, A Electrochemistry Principles, Methods, and Applications, Oxford University Press, Coimbra, 1994 [9].*

The frequency at the top of the semicircle indicates the resonance condition of the circuit [9,20,21], given by the expression $2\pi f_o RC = 1$. Impedance spectroscopy is often used to study films deposited on electrodes because it can distinguish the different conductivity processes that occur in the material. The results are often associated with an electrical circuit, in which it is possible to distinguish and calculate parameters, such as the electrolyte resistance, ionic conductivity, capacitance of the electrical double layer, and electron transfer resistance [9].

Although minimally used for analytical purposes, capacitance can be used for this purpose, enabling capacitive sensors. When an analyte interacts with the receptor (Fig. 6.2), there may be a change in its specific capacitance, and this response change can be interpreted as an analytical signal. There are currently several types of commercial sensors that use capacitance as a measure. Among them are analyte sensors, such as gas sensors, and especially physical sensors, such as of presence or of touch. Further on, we will discuss the sensory network termed the electronic tongue (ET), which relies on electrochemical impedance spectra, whose signals are results obtained from the capacitance values obtained for each sensor (sensory unit).

6.2.6 Electronic Tongue: Concepts, Principles, and Applications

As previously described, chemical sensors are widely used analytical tools. The first ion selective electrode (ISE) based on potentiometric measures, which appeared in 1909, was the pH meter [22]. This sensor, with minor modifications, is still the most commonly used one for liquids. From the beginning, sensor development has always been based on the selectivity of the analytes. Other ion-selective sensors have been developed, and dozens are commercially available. However, the number of existing analytes that could be sensed, combined with the numerous variables in complex matrices, is incalculable, which makes it necessary to develop less specific sensors. One way to increase the analytical application of sensors is to use a combined matrix of sensors and more sophisticated techniques of data manipulation so that the combination of responses constitutes a "fingerprint," not only of an analyte but also of the combined form between analytes and their matrices, for instance, a pesticide in a particular soil water matrix [23].

Environmental samples are highly complex matrices because of the many substances that may be present; they are different from one environment to another in addition to having seasonal characteristics. If we analyze samples collected at different locations or days from a given lake or dam, we can easily observe different characteristics among them.

The idea of combining responses from different sensors to obtain this "fingerprint" was initially implemented for gas sensors, which are now known as electronic noses [24]. Subsequently, several researchers developed the principle of the electronic nose method for liquid samples, which became known as the ET.

The electronic nose principle was inspired by sensorial biological systems, for example, the mammalian olfactory power. The olfactory system consists of a large number of nonspecific receptors that react to volatile compounds and transfer nervous stimuli to the brain, where a neural network processes the signal pattern. Any animal with olfactory power is capable of recognizing thousands of odors with very low detection limits, despite the high sensitivity and low selectivity of each individual receiver. However, the high performance of olfactory systems is given by the combined response of several sensor units and subsequent conjoint treatment of the responses in the nervous system. Similarly, taste employs an operational principle and interpretation of responses similar to the olfactory system, except that taste involves, in general, significantly fewer receptors.

As with the biological tongue, the ET operates in a liquid medium, but with a much greater power of sensitivity and selectivity. In terms of performance, these advantages allow the ET to be closer to the olfactory system than the gustatory system. The ET can also be used for the quantitative detection of a variety of dissolved compounds, including volatile substances that are responsible for odors but that originate from both liquid and solid phases. A common feature of all electronic noses and tongues is the use of a matrix of nonspecific sensors combined with data processing by pattern recognition methods.

The most often used data processing system is the artificial neural network (ANN) [25], in which the algorithms are based on the learning and recognition processes of the human brain [26]. The most traditionally used methods are statistical methods, such as the principal component analysis (PCA) method, multivariate regression techniques, and factorial analysis [27,28].

Since the mid-1980s, there have been several studies describing the combination of signals from a sensor matrix in liquid media with complex matrices, using several mathematical models for pattern recognition [29–34]. The first studies were conducted with potentiometric sensors to improve the performance of ion-selective conventional electrodes. The first study that can be considered as a multisensory approach for liquid detection was the one by Otto and Thomas in 1985 [29].

In a simplified manner, the ETs respond to the substances responsible for originating the basic flavors analogously to the biological system and can differentiate beverages with similar flavors. For instance, some of them can distinguish between different brands of bottled water or wine brands and harvests, in addition to detecting metal ions in water. There are several detection methods used in ETs. Potentiometry, CV, and alternating current measurements are examples of applications. However, the method choice can affect the sensor performance. For example, measurements of admittance (Y) or its reciprocal, impedance (Z), using sensors based on polymer films are particularly sensitive [35], with very low detection methods for sucrose and other substances related to flavor. Since then, the ET has been used for measurements of several flavors [36–38]. Other advantages of the use of impedance, admittance or even capacitance (which uses a specific frequency) combined with polymer films, compared with potentiometric sensors, are the experimental simplicity and better performance for nonpolar substances, such as caffeine and sucrose. Unlike the sensors based on voltammetry or potentiometry techniques, the solutions to be analyzed do not require any pretreatment and do not use a reference electrode. In general, the basic requirement to assess the performance of the ET is the ability to differentiate substances that create the basic flavors. A single material likely be sensitive to all basic tastes; therefore, a combination of sensors is usually used. For example, Embrapa Agricultural Instrumentation, in São Carlos, in the state of São Paulo, developed the ET for different applications [39]. The developed ET consists of a network of sensory units based on different conductive polymers deposited on substrates of interdigitated gold electrodes, as shown in Fig. 6.18.

The sensory units consist of nanostructured films of conductive polymers, such as polyaniline and their derivatives or polypyrrole, deposited on interdigitated microelectrodes. These polymers have the ability to reversibly change their optical and electrical properties when the characteristics of the surrounding environment are modified [40,41], such as the chemical composition, pH, and ionic strength. Different beverage types have been analyzed with the ET based on conductive polymers, including mineral water, coconut water [42], coffee and tea [43], wine [44,45], and sugarcane juice [46]. In mineral waters, taste characteristics measured with the ET might be associated with their physicochemical composition. With this system, it was possible to detect the presence of heavy metals [47] and pesticides [48] in these beverages. In wines, the ET proved able to identify them

FIGURE 6.18 (A) Sensor network, (B) diagram of the interdigitated sensor, and (C) sensory unit with a PANI and carbon nanotube film used in the electronic tongue (ET) of Embrapa Agricultural Instrumentation.

according to the harvest, producer, and grape type used for their production. The coffee quality was also systematically investigated with the ET, which led to the technological entrepreneurship project *"Development of an electronic tongue prototype for the evaluation of coffee quality"* with the support of the Brazilian Association of the Coffee Industry (Associação Brasileira de Indústrias de Café—ABIC). There was a good correlation between the sensor results and scores awarded by tasters.

ETs built with electrochemical, optical, or mass sensors often have the disadvantage of low specificity. Thus, in the search for specificity, the ET has also been developed in the manufacturing of biosensors [49–54]. For example, Hu et al. [53] recently evaluated the applicability of germ cells from male mice testes in the manufacture of a cell-based impedance sensor (CIS) to increase the specificity for the bitter taste, which can have many variations as a result of the wide variety of bitter compounds with diverse chemical structures and functions. The cell impedance response profiles of the germ cells for four bitter compounds were analyzed, assessing the response intensity under different concentrations. Finally, the impedance responses for the five basic tastes were examined to assess the biosensor-based detection performance for bitter compounds. The results revealed that this hybrid bitter biosensor could respond to these bitter compounds in a dose-dependent manner. The quinine detection limit was 0.125 μM. The authors stated that this sensor based on germ cells could specifically detect bitter compounds between the five flavor stimuli, which can provide a promising and valuable approach in the detection of several bitter compounds.

Currently, the ET is also a commercial product. The company Alpha MOS manufactures and sells the "ASTREE electronic tongue" device using chemically modified field effect transistor technology and potentiometric measurements with seven selective liquid sensors for ionic, neutral, and chemical compounds responsible for flavor [55].

List of Symbols

ΔE	Potential pulse amplitude
A	Geometric electrode area
Au	Gold
C	Capacitance
c	Concentration
C	Coulomb
C_d	Capacitance of electric double layer
cm^2	Square centimeter
cm^3	Cubic centimeter
D_0	Diffusion coefficient
dE	Differential compared to potential
dt	Differential compared to time
e$^-$	Electron
E	Potential
E^0	Standard potential
$E_{1/2}$	Half-wave potential
$E_{Ag/AgCl}$	Standard potential of silver/silver chloride saturated electrode

E_f	Final potential
E_i	Initial potential
E_{inic}	Initial potential
$E_{máx}$	Maximum potential
$E_{mín}$	Minimum potential
E_p	Peak potential
E_{pa}	Anodic peak potential
E_{pc}	Cathodic peak potential
eq.	Equivalent
f	Frequency
F	Farad
HCl	Hydrochloric acid
I or i	Current
I_f	Faradaic current
I_c	Capacitive current
Im Z	Imaginary impedance
I_{pa}	Anodic peak current
I_{pc}	Cathodic peak current
jX_c	Imaginary part (Z'') of impedance
m	Milli (1×10^{-3})
mol	Mol
mol L^{-1}	Mol per liter
n	Number of electrons (eq. mol^{-1})
°C	Degrees Celsius
PANI	Polyaniline
Q	Charge
R	Universal constant of gases
R	Electrical resistance/real part (Z') of the electrochemical impedance
RC	Resistance in parallel with a capacitor R
Z	Real impedance
s	Seconds
T	Temperature
t_1	Initial time
V	Volt

References

[1] Merriam-Webster Dictionary, www.merriam-webster.com/dictionary/sensor

[2] J. Brugger, Nanotechnology impacts on sensors, Nanotechnology 20 (43) (2009) 2.

[3] B. Latour, Pandora's Hope: Essays on the Reality of Science Studies, Harvard University Press, Cambridge, MA, 1999. 324 p.

[4] S.K. Islam, S. Haider, M. Hafiqul, Sensors and Low Power Signal Processing, Springer Science+ Business Media, Springer, US., 2010.

[5] L. Neia, Milestones in the 50-year history of electrochemical oxygen sensor development, ECS Trans. 2 (25) (2007) 33–38.

[6] G. Binning, et al., Atomic force microscope, Phys. Rev. Lett. 56 (1986) 930.

[7] G. Binning, et al., Scanning tunneling microscope, Phys. Rev. Lett. 49 (57) (1982) 1082.

[8] F.J. Holler, D.A. Skoog, S.R. Crouch, Principles of Instrumental Analysis, sixth edition, Thomson Brooks/ Cole, 2007.

[9] A.M.O. Brett, C.M. Brett, A Electrochemistry Principles, Methods, and Applications, Oxford University Press, Coimbra, 1994.

[10] A.S. Michaels, Nucleation Phenomena: A Symposium, American Chemical Society, Washington D.C., 1966. 89 p.

[11] F. Manea, A. Remes, C. Radovan, R. Pode, S. Picken, J. Schoonman, Simultaneous electrochemical determination of nitrate and nitrite in aqueous solution using Ag-doped zeolite-expanded graphite-epoxy electrode, Talanta 83 (2010) 66–71.

[12] G.J. Hills, et al., Electrochemical nucleation from molten salts. I-Diffusion controlled electrodeposition of silver from alkali molten nitrates, Electrochim. Acta. 19 (1974) 657–670.

[13] K. Kalcher, et al., Sensing in Electroanalysisvol. 6 University Press Centre, Pardubice, Czech Republic, 2011.

[14] J.O'm. Bockris, A.K.N. Reddy, third ed., Modern Electrochemistry—An Introduction to an Interdisciplinary Area, vols. 1 and 2 Plenum Press, New York, 1977.

[15] O.P. Amarante Jr., et al., Breve revisão de Métodos de Determinação de Resíduos do Herbicida ácido 2,4-Dicloro-fenoxiacético (2,4-D), Quím. Nov. 26 (2) (2003) 223–229.

[16] Y. Zeng, D. Yu, Y. Yu, T. Zhou, G. Shi, Differential pulse voltammetric determination of methyl parathion based on multiwalled carbon nanotubes-poly(acrylamide) nanocomposite film modified electrode, J. Hazard. Mater. 217–218 (2012) 315–322.

[17] D. Sousa, et al., Square wavevoltammetry. Part I: theoretical aspects, Quím. Nov. 26 (2) (2003) 81–89.

[18] R. Jainn, S. Sharma, Glassy carbon electrode modified with multi-walled carbon nanotubes sensor for the quantification of antihistamine drug pheniramine in solubilized systems, J. Pharm. Anal. 2 (1) (2012) 56–61.

[19] A. Chen, B. Shaha, Critical review electrochemical sensing and biosensing based on square wave voltammetry, Anal. Methods 5 (2013) 2158–2173.

[20] P. Agarwal, et al., Application of measurement models to impedance spectroscopy: II determination of the stochastic contribution to the error structure, J. Electrochem. Soc. 142 (12) (1995) 4149–4158.

[21] D.D. Macdonald, et al., Characterizing electrochemical systems in the frequency domain, Electrochim. Acta. 43 (1998) 87–107.

[22] F. Haber, Z. Klemensiewics, Ober elektrische. Phasengrenzkrafte (On the electrical forces at phase borders), Phys. Chem. 64 (1909) 385–392.

[23] F.R. Simões, et al., Conducting polymers as sensor materials for the electrochemical detection of pesticides, Sens. Lett. 4 (2006) 319.

[24] J.W. Gardner, P.N. Bartlett (Eds.), Sensors and Sensory Systems for an Electronic Nose, Nato ASI Series E: Applied Sciences, vol. 212, Kluwer, Dordrecht, 1992.

[25] Neural Computing, Neural Ware, Pittsburgh, USA, 1997.

[26] D.E. Rumelhart, J.L. Mcclelland, Parallel Distributed Processing, vols. 1–2, The MIT Press, Cambridge MA, 1986.

[27] K. Esbensen, Multivariate Analysis in Practice, fifth ed., Camo ASA/CAMO Process AS, Norway, 2001.

[28] H. Martens, T. Naes, Multivariate Calibration, Wiley, New York, 1989.

[29] M. Otto, J.D.R. Thomas, Model studies on multiple channel analysis of free magnesium, calcium, sodium, and potassium at physiological concentration levels with ion-selective electrodes, Anal. Chem. 57 (1985) 2647–2651.

[30] K. Beebe, B. Kowalski, Nonlinear calibration using projection pursuit regression—application to an array of ion-selective electrodes, Anal. Chem. 60 (1988) 2273–2276.

[31] K. Beebe, D. Uerz, J. Sandifer, B. Kowalski, Sparingly selective ion-selective electrode arrays for multicomponent analysis, Anal. Chem. 60 (1988) 66–71.

[32] W.E. Van Der Linden, M. Bos, A. Bos, Array of electrodes for multicomponent analysis, Anal. Proc. 26 (1989) 329–331.

[33] M. Bos, et al., Processing of signals from an ion-selective electrode array by a neural network, Anal. Chim. Acta. 233 (1990) 31–39.

[34] M. Hartnett, D. Diamond, P.G. Barker, Neural network based recognition of flow-injection patterns, Analyst 118 (1993) 347–354.

[35] A. Riul, et al., Artificial taste sensor: efficient combination of sensors made from Langmuir-Blodgett films of conducting polymers and a ruthenium complex and self-assembled films of an azobenzene-containing polymer, Langmuir 18 (2002) 239–245.

[36] K. Toko, A taste sensor, Meas. Sci. Technol. (1998) 1919–1936.

[37] K. Toko, Electronic sensing of Tastes, Electroanalysis 10 (1998) 657–669.

[38] S. Ezaki, H. Kunihiro, Detection of mutual interaction between taste substances by impedance measurements of lipid/polymer membranes, Sens. Mater. 11 (1999) 447–456.

[39] P. Prizolatto et al., Uso de eletrodos interdigitados de ouro revestidos com filmes poliméricos para a detecção de pesticidas em água por impedância eletroquímica, Série Documentos, EMBRAPA, 2006.

[40] K.P. Persaud, Polymers for chemical sensing, Mater. Today (2005) 38.

[41] E.S. Medeiros, et al., Potential application of conjugated polymers in sensor systems, in: C. Graimes (Ed.), Encyclopedia of Sensors, ASP, USA, 2005, pp. 43.

[42] A. Riul Jr., et al., Nano-assembled films for taste sensor application, Artif. Organs 27 (5) (2003) 469–472.

[43] A. Riul Jr., et al., An electronic tongue using polypyrrole and polyaniline, Synth. Met. 132 (2003) 109–116.

[44] A. Riul Jr., et al., An artificial taste sensor based on conducting polymers, Biosens. Bioelectron. 18 (2003) 1365–1369.

[45] A. Riul Jr., et al., Wine classification by taste sensors made from ultra-thin films and using neural networks, Sens. Actuat. B Chem. 98 (1) (2004) 77–82.

[46] M. Ferreira, et al., High-performance taste sensor made from Langmuir–Blodgett films of conducting polymers and a ruthenium complex, Anal. Chem. 75 (2003) 953–955.

[47] P.A. Antunes, et al., The use of Langmuir–Blodgett films of a perylene derivative and polypyrrole in the detection of trace levels of Cu 2+ ions, Synth. Met. 148 (1) (2005) 21–24.

[48] C.J.L. Constantino, et al., Nanostructured films of perylene derivatives: high performance materials for taste sensor applications, Sens. Lett. 2 (2) (2004) 95–101.

[49] H. Guohua, M. Shanshan, C. Qingqing, C. Xing, Sweet and bitter tastant discrimination from complex chemical mixtures using taste cell-based sensor, Sens. Actuat. B Chem. 192 (2014) 361–368.

[50] J.Y. Heras, D. Pallarola, F. Battaglini, Electronic tongue for simultaneous detection of endotoxins and other contaminants of microbiological origin, Biosens. Bioelectron. 25 (2010) 2470–2476.

[51] G. Hui, S. Mi, S. Ye, J. Jin, Q. Chen, Z. Yu, Tastant quantitative analysis from complex mixtures using taste cell-based sensor and double-layered cascaded series stochastic resonance, Electrochim. Acta. 136 (2014) 75–88.

[52] G.-H. Hui, S.-S. Mi, S.-P. Deng, Sweet and bitter tastants specific detection by the taste cell-based sensor, Biosens. Bioelectron. 35 (2012) 429–438.

[53] L. Hu, J. Xu, Z. Qin, N. Hu, M. Zhou, L. Huang, P. Wang, Detection of bitterness in vitro by a novel male mouse germ cell-based biosensor, Sens. Actuat. B Chem. 223 (2016) 461–469.

[54] T.-H. Wang, G.-H. Hui, S.-P. Deng, A novel sweet taste cell-based sensor, Biosens. Bioelectron. 26 (2010) 929–934.

[55] ASTREE electronic tongue, http://www.alpha-mos.com/analytical-instruments/astree-electronic-tongue.php

7

Molecular Modeling Applied to Nanobiosystems

A.M. Amarante*, G.S. Oliveira*, J.C.M. Ierich*, R.A. Cunha**,
L.C.G. Freitas†, E.F. Franca**, F. de Lima Leite*

*NANONEUROBIOPHYSICS RESEARCH GROUP, DEPARTMENT OF PHYSICS, CHEMISTRY
& MATHEMATICS, FEDERAL UNIVERSITY OF SÃO CARLOS, SOROCABA, SÃO PAULO,
BRAZIL; **INSTITUTE OF CHEMISTRY, FEDERAL UNIVERSITY OF UBERLANDIA—UFU,
UBERLANDIA, MINAS GERAIS, BRAZIL; †CHEMISTRY DEPARTMENT, UFSCAR, SÃO CARLOS,
SÃO PAULO, BRAZIL

CHAPTER OUTLINE

7.1 Introduction

7.1.1 Molecular Modeling

Given the physicochemical properties of matter and the laws of physics, it is possible to create models that facilitate an understanding and can predict the intrinsic characteristics of chemical and biological systems [1]. Thus, molecular modeling is a method that involves the application of specific theoretical models for the specific properties of the system of interest. Through the use of the concepts of atoms and molecules, it is possible to predict the behavior of real systems in the areas of biology, chemistry, pharmaceuticals, and materials science [2,3]. It is also possible to describe the molecular structures and their physicochemical properties using molecular modeling [2,4].

Using theoretical models to describe the properties of matter, it is possible to idealize the real world. However, there are limitations to the use of these models. Molecular modeling is a tool by which it is possible to represent and manipulate the structural forms of molecules. This tool fosters the understanding of chemical reactions and physical phenomena and helps to establish relationships between the structure of matter and its properties [1]. For each system of interest, it is possible to model a representation that contains its characteristics of interest and to extract relevant information about the real system. For each situation studied, there is a specific model that is tailor-made. Thus, for a given property of the molecule, a specific model that displays this property is adopted. There can exist various models for a single molecule and various representations for a single model. The computer simulation tool also seeks to construct models from the physical properties of the study object, which can define the level of knowledge about the structure of matter [3,5–7].

7.1.2 Computer-Assisted Molecular Modeling

To understand the behavior of real systems, molecular modeling encompasses a number of computational tools and methods that are used to describe and predict molecular structures, reaction properties, thermodynamic properties, etc. [1,8]. Therefore, computer-assisted molecular modeling studies atomistic properties through the use of computational chemistry and graphic visualization techniques. This is a branch of science that studies complex chemical and biological systems with a large number of particles on the order of 10^{23}. This technique can perform specific calculations and create models that would otherwise be impossible to obtain [9,10].

With the advances in computational resources, both software and hardware, molecular computer modeling has undergone surprising development in the last few decades. In the past, this technique was exclusive to the groups of scientists and researchers responsible for the development of the software components utilized. Today, various modeling programs and resources are available online at no cost, allowing the scientific community to explore the physicochemical data of molecular structures corresponding to their systems of interest [7]. Some of the main software components available for free will be discussed later in this chapter.

It is worthwhile to emphasize that the field of computational chemistry and computer-assisted molecular modeling were recognized in 1998. This resulted in the awarding of the Nobel Prize in chemistry to John Pople and Walter Kohn [11] for their contributions to quantum chemistry, and more recently, in 2013, Martin Karplus, Michael Levitt, and Arieh Warshell were awarded for their contributions to the development of new methods for modeling systems with many atoms [12].

7.2 Basic Representation Types

The discovery in 1874 of carbon's tetrahedral arrangement in organic compounds allowed the initial framework for molecular modeling, abandoning other imaginable molecular and matter structures that differ from each other primarily by their sizes [13]. The Lewis structure, or structural dot diagram, emerged in 1916, when it was proposed by Gilbert Lewis. This structure has been used until now as the most basic form of modeling, emphasizing the sequence of atoms and the types of bonds present in a chemical or biological structure [14,15].

With the proposal of the Lewis structure and the discovery of the structural geometric arrangements of certain molecules, molecular modeling began to take shape, and different types of representations began to emerge. Approximately 60 years after Gilbert Lewis, Barton formulated the concept of conformational analysis, which refers to molecules being represented by atomic arrangements with specific energies depending on their spatial locations [3,4]. With modern crystallography based on X-ray diffractometry (XRD) (second half of the 20th century) and other structural analysis techniques, a better understanding of the three-dimensional structures of molecules became possible, thus obtaining bonding patterns, angles, and other intrinsic properties of each molecule [16].

All this information about molecular structures led to the development of basic representation types used for molecular modeling. Three of the main representation types are presented in Fig. 7.1, where the structure of benzene is represented by sticks, balls-and-sticks, and van der Waals volumes.

0.1 nm

FIGURE 7.1 Different representations of the benzene molecule: sticks, balls-and-sticks, and van der Waals volume.

The three models presented possess advantages in simplicity, design, and presentation. They highlight important information about the molecular geometry and the atoms involved. The bond types between the atoms are data that allow theoretical predictions about the physical and chemical properties of matter [14]. In some situations, the models may be limited by their qualitative character [9,17]. Other representations, which will be discussed later, will provide more detailed information, such as the type of protein folding, electronic properties, and surface properties.

7.3 Biomolecules and Protein Modeling

7.3.1 Biomolecules

Biomolecules are chemical compounds synthesized by living organisms. These include proteins, the structures of which are widely studied using molecular modeling [18,19]. Due to their high degree of complexity, more dynamic models that are able to portray the chemical and physical properties of proteins are required. In this context, molecular modeling features powerful computational tools specific to these types of structures [20].

7.3.2 Proteins

Proteins are the most abundant biological macromolecules in the cell [19]. They are found with an enormous variety of types and molecular masses, resulting in a great diversity of biological functions. A protein is defined as an amino acid sequence or a polymer, with each amino acid joined by a specific type of covalent bond known as a peptide bond. The great majority of proteins are composed of only four chemical elements: C, H, O, and N [19,21,22].

Amino acids are relatively simple monomeric subunits that provide the basic structure of proteins. All proteins—those originating from the simplest and most ancient bacterial strains and those of the most complex life forms—are constructed primarily from the same set of 20 amino acids [17,22]. Cells can produce proteins with diverse properties and activities simply by the joining the same 20 amino acids in different combinations and sequences [22]. As a consequence, organisms can produce various types of products, such as enzymes, hormones, antibodies, transporters, and fibers, creating elements such as feathers, webs, and horns. Among these protein products, enzymes are the most varied and specialized [19,22].

All 20 common amino acids are α-amino acids. They have a carboxyl group and an amino group bound to the same α carbon atom. These amino acids differ from each other in their side chains or R groups, which vary in structure, charge and affect the amino acids' solubility in water [19,22]. Simple functional models can be used to represent these structures, and in this case the ball-and-stick model is the most useful (Fig. 7.2).

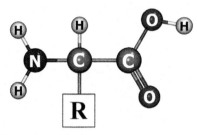

FIGURE 7.2 General representation of an amino acid.

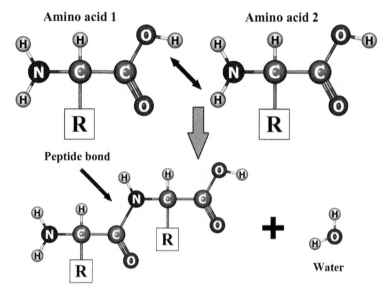

FIGURE 7.3 Schematic representation of the union between two amino acids by a peptide bond.

The bond between the amino and the carboxyl groups between two amino acids is called a peptide bond (Fig. 7.3). The formation of this bond results in the loss of a water molecule, generating what is called an initial amino acid molecule [17].

7.3.3 Protein Structure

Given the complexity of the protein structure [4], it is commonly described by four basic structure levels (Fig. 7.4) [18]:

- *Primary structure*: the sequence of amino acids joined by peptide bonds, including any disulfide bonds [23].
- *Secondary structure*: the polypeptide resulting from the primary structure; can be a spiral or a helix [23].

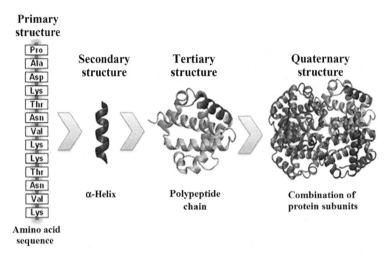

FIGURE 7.4 **Types of modeling of basic structures formed by proteins.** Modeled based on PDB ID 3HHB [24]. *Adapted from D.L. Nelson, M.M. Cox, Lehninger Principles of Biochemistry, W.H. Freeman, New York, 2012, 1340 p. [22].*

- *Tertiary structure*: various helices or spirals that describe all aspects of the three-dimensional folding of the polypeptide chain [19].
- *Quaternary structure*: occurs when two polypeptide chains are joined together [19].

The protein represented in Fig. 7.4 is hemoglobin, an oxygen-carrier protein of the circulatory system [25]. The majority of polypeptide chain links in proteins occur via disulfide bonds and hydrogen bonds. The three-dimensional structure, together with the sequence of amino acids in the molecule, gives the protein its specific function [18]. These specific functions often consist of catalyzing chemical reactions. Hemoglobin belongs to a class of proteins called enzymes [19,22].

7.3.4 Types of Protein and Biomolecule Representations

As mentioned previously, molecular modeling consists of creating theoretical models that represent the structures and certain properties of molecules. Thus, different properties, such as the atomic arrangement, spatial disposition of the molecule, and functional groups, can be highlighted in different representation models. The goal of the models is to emphasize certain important properties of molecules. Often, these properties can be applied in a manner distinct from reality but can represent the study object. To better elucidate these concepts, different representation forms of enzymatic structures obtained from the repository of crystallographic structures will be exemplified in this chapter. First, we will consider representations for acetyl-coenzyme A carboxylase (ACCase), which participates in the synthesis of lipids in plants and animals and is widely studied by the pharmaceutical industry and in technology research [26].

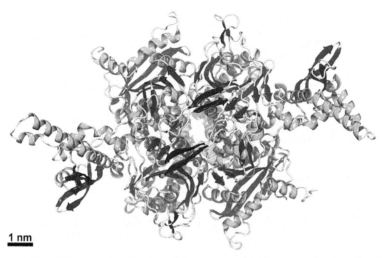

1 nm

FIGURE 7.5 Representation of the secondary structure of the enzyme acetyl-coenzyme A carboxylase (ACCase). Modeled based on PDB ID 3FF6 [27].

The model represented in Fig. 7.5 shows the secondary structure of ACCase. This type of representation emphasizes the structural properties of a protein, the secondary arrangements and folds, which are represented mainly as spiral and straight arrow shapes corresponding to α-helices and β-sheet folds, respectively [23]. It is important to note that α-helices are also commonly represented as cylinders. The type of representation of the secondary structure is very common for proteins because of their three-dimensional arrangement provides information about its structural behavior.

For a given molecule, different types of representation can generate vast models to better understand its 3D representation. In Fig. 7.6, the same protein is represented in the form of sticks. This model emphasizes certain residues of specific amino acids that compose the protein. Often, names, numbers, or different colors can be used for each amino acid. This representation is widely used, especially when one wishes to highlight the amino acids that participate in a given bond or chemical reaction.

The stick model presented previously is widely used in studies of chemical elements and some bond types in a molecule. Fig. 7.7 shows in detail the atoms that make up the enzyme ACCase.

When studying surface effects, the previous representations do not emphasize properties, such as charge and mass distribution, spatial geometry, and possible reaction sites. Fig. 7.8 shows one of these possible surface representations. In this example, the atomic mass distribution representation was used (in atomic mass units, amu) for the surface of the ACCase enzyme.

Many enzymes possess nonprotein parts, that is, parts that are not amino acids. These parts, called cofactors, may be of organic character, in the case of coenzymes, or inorganic

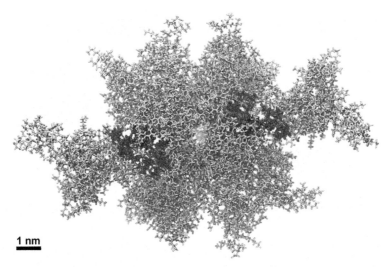

FIGURE 7.6 Stick representation of the enzyme ACCase. The darker residues indicate the entrance of the enzymatic active site. Modeled based on PDB ID 3FF6 [27].

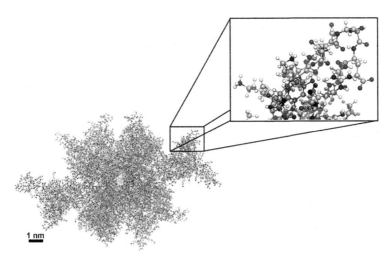

FIGURE 7.7 Ball-and-stick representation of the atoms constituting the enzyme ACCase. Different colors indicate the elements (from lighter to darker: hydrogen, carbon, oxygen, and nitrogen). Modeled based on PDB ID 3FF6 [27].

character. Given the need to represent structures with different properties within the same model, a mixed representation, which is a combination of the representations shown previosuly, is common. Fig. 7.9 shows an example of different types of representations within the same model for the enzyme acetolactate synthase (ALS). ALS is a protein that participates in the synthesis of amino acids in plants, fungi, and bacteria [28]. In this model, the secondary structure for each of the two tertiary protein structures is represented, with the stick representation used for coenzymes and the ball representation (without scale) used for the ions Mg^{2+} and K^+.

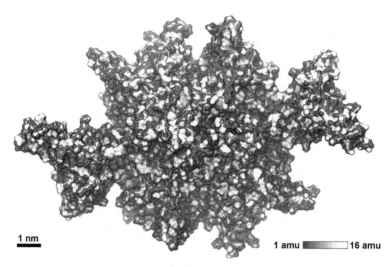

FIGURE 7.8 Representation of the enzyme ACCase based on the surface charge distribution. Gradient: 1–16 amu. Modeled based on PDB ID 3FF6 [27].

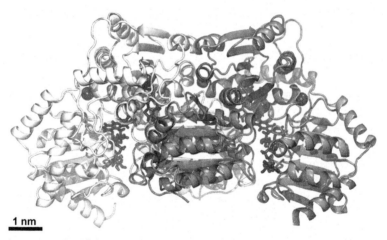

FIGURE 7.9 Mixed representation of the enzyme enzyme acetolactate synthase (ALS). Modeled based on PDB ID 1T9D [29].

7.3.5 Biological Structure Databases

The description of the structure of biomolecules may provide a foundation for the understanding of their properties, behavior in biological systems, and, consequently, the specific function(s) they perform [22]. Furthermore, the molecular configuration and function form the basis for the development of new technologies, such as inhibitors, medication, and vaccines [30]. It is important that the molecular structure of biomolecules be accessible to the scientific community, thus collaborating with and stimulating associated studies.

Currently, numerous data on biomolecular structures have been produced, gathered, organized, and made available in specific databases.

In particular, protein structures are available in a wide range of databases that can store the primary structure (sequence of amino acids that make up the protein), information about its function, or even experimentally resolved three-dimensional structures. These databases include the following: (1) Universal Protein Resource (UniProt) [31,32], which contains the amino acid sequences that form the protein, a description of its function, the characterization of the protein domains, and posttranslational modifications; (2) Protein, of the National Center for Biotechnology Information [33], which stores protein sequencing data; (3) The system of enzymatic data BRENDA (Braunschweig Enzyme Database) [3,34,35], which stores sets of data that have been obtained primarily from the literature reflecting the function, structure, production, preparation, and applications of enzymes; (4) PROSITE [36,37], which stores protein data that describe the protein families, domains, and functional sites (integrated into UniProt); and (5) the Protein Data Bank (PDB) [38], the largest repository of three-dimensional biomolecule structures obtained experimentally [8,39,40].

In particular, PDB is of great importance for protein structure prediction methods that use computer modeling, especially for comparative modeling or homology modeling, which uses protein structural data stored in this repository as a reference for predicting the structure of a target protein. After determining the crystallographic structure by XRD or nuclear magnetic resonance (NMR), researchers make their data available to PDB, providing the three-dimensional data obtained and information about the structure so that other researchers can use the coordinates generated by the PDB file to obtain static and dynamic information about the molecule [41]. Queries for models are performed by field and may use the macromolecule name, research author, PDB file identification (ID) number, or even the amino acid sequence of the molecule of interest. The molecular structure can be visualized on the database web page, and the PDB structure files are opened by specific text editing and processing programs. PDB files can be read by biomolecular programs to characterize the structure of the macromolecule using the set of three-dimensional coordinates for each constituent atom [30].

7.3.6 Protein Structure Visualization, Manipulation, and Analysis Tools

Given the great technological advances and the improvement and sophistication of computers, software components have been improved and have become important tools for the study and solving of problems in scientific investigations [42,43]. Just as the quantity of available biomolecule structural data is large, the number of programs for reading, displaying, or editing the structural data is great. Beyond editing and displaying the macromolecular structures, these programs may also aid in the study of their behavior within a biological system and their interactions with the medium, such as other molecules, enzymatic substrates, and inhibitors.

In addition to the ability of a program to solve problems, it is fundamental that it also has a simple user interface that is easy to understand and clearly displays the functions and dynamics of the program to ensure efficient computer–user communication. It is of utmost importance that programs make this interaction as easy as possible by using simple command sequences the user can successfully execute. The role of computers in editing, producing, and visualizing protein structures is ever growing, especially for protein structure prediction methods, which are performed by the application of a set of stages based on the use of specific programs [43]. Thus, given their importance for the visualization, manipulation, and interpretation of protein structures, some of the main programs will be studied in further detail.

The visual molecular dynamics (VMD) package [44], developed and distributed by the Theoretical and Computational Biophysics Group of the University of Illinois at Urbana-Champaign and the Beckman Institute, is a free visualization, modeling, and analysis program for three-dimensional structures of macromolecules, such as proteins and nucleic acids. VMD was written in the C and C++ languages, with versions available for most operating systems (Mac OS X, Unix, and MS-DOS). It features an editing package that includes several molecule rendering and coloring methods. With its capacity to generate high-definition images and movies, this software allows one to analyze the behavior of the macromolecular structure over time through molecular dynamics (MD) simulations, which highlights its vast applicability in scientific research. In addition, it has a fairly intuitive graphical user interface, which allows it to be used as a teaching tool [45–47].

Another interesting program for visualizing and manipulating the three-dimensional structures of macromolecules is PyMol [48]. Developed by Warren Lyford De Lano, PyMol is a multiplatform system written in the Python and C languages. In 2006, some versions ceased to be freely distributed except for academic use. PyMol has good performance, especially during complex tasks, such as surface representation, but it may require improved hardware, including graphics acceleration. This software offers the necessary resources for high-resolution image generation with excellent quality and richness of detail, and it is widely utilized in scientific publications. Furthermore, the program also provides applicability in MD analyses, allowing the user to produce videos of the molecular behavior over time. However, this program has a complex user interface, which may complicate the use of its available resources [49,50].

The JMol software package [51,52] is also widely used in scientific research for the visualization, analysis, and editing of biomolecule models. Coded in Java, it is compatible with most operating systems. The tools provided for the user do not demand advanced knowledge of informatics; its graphical interface is simple, allowing for easy use of the resources and making this software a powerful teaching tool [53]. Rendering three-dimensional models does not require advanced hardware resources. JMol possesses integration components with other Java applications, which ensures greater software functionality. For example, the JMol Applet integrates JMol into webpages, allowing for the direct visualization of structures in databases. In addition, it is possible to execute animations and molecular vibrations, both of which are useful in MD simulations [54].

Another option for the analysis of molecular structures widely used by the scientific community is UCSF Chimera [55,56], or simply Chimera. It was developed by the Resource for Biocomputing, Visualization, and Informatics of the University of California—San Francisco. This program is free for noncommercial purposes. It was coded in Python, with some libraries written in C and C++, and it is compatible with MS-DOS, Mac OS X, and Linux. Chimera executes high-quality rendering, despite sometimes demanding more intense graphics processing, depending on graphics acceleration. In part, this may be a consequence of the use of the Python language, which may negatively impact speed and memory usage. Nonetheless, these problems can be circumvented with hardware improvements, as well as algorithmic changes in the software programming language [55]. In addition, the program has a simple graphical interface with evident basic functions. Thus, it is easy for beginners to use the program and also practical for advanced users [33].

The Swiss-PDB Viewer, or Deep Viewer [57], is also widely used in biomolecule visualization and manipulation tasks. Developed and made freely available by the Swiss Institute of Bioinformatics, this program is extremely versatile, combining several study, comparison, and error-searching tools in three-dimensional models, and it has a great impact in molecular modeling by homology. This program is widely integrated with internet servers, working in association with ExPASy (Expert Protein Analysis System) and performing homology searches on international databases [57]. In addition, all functions are made available in an intuitive and simple graphical interface, which dispenses with the need for previous programming knowledge [30,58].

7.4 Molecular Computer Modeling Methods Applied To Biomolecules

7.4.1 Classical Methods

7.4.1.1 Molecular Dynamics

MD is a powerful and versatile technique for the study of biological macromolecules that is based on classical mechanics. Experimentally, it is possible to measure properties that are the sum of all possible entropic and energetic microstates that a molecule can occupy. However, even with the constant evolution of experimental methods, it is necessary to use theoretical models to predict and understand the molecule-level properties of the experiments [59–61]. Thus, experimental methods can be used for the validation and improvement of the properties, which may be measured through theoretical methods developed to make predictions regarding that system.

For molecular systems composed of hundreds to millions of atoms, such as proteins in aqueous media, it is impossible to provide a precise characterization at the quantum level. Classical physics, however, offers a good approximation for modeling these systems because the molecular orbitals are well defined, and consequently, the energy difference between the highest occupied molecular orbital (HOMO) and the lowest unoccupied molecular state (LUMO) is often many times larger than $k_B T$ (where k_B is the Boltzmann constant and T the temperature in Kelvin) [60].

7.4.1.1.1 MOLECULAR DYNAMICS AND RELATED POTENTIALS

The first approximation of molecular mechanics is to consider atoms as spheres endowed with charge with positions defined in space by a set of harmonic potentials. Thus, the temporal evolution of the molecular structure depends on the linear (V_d), angular (V_θ), and torsional (V_φ) terms that describe the interactions associated with the molecule being simulated [62,63]. Similarly, the nonbonding interactions, or intermolecular interactions, are described by Lennard-Jones (V_{LJ}), or van der Waals, and Coulomb (V_C) potentials. The sum of the potentials associated with the bonding and nonbonding interactions results in a full expression of the total energy potential of the system (V_{total}), which depends only on the atomic positions (Eq. 7.1) [7].

$$V_{total} = V_d + V_\theta + V_\varphi + V_{LJ} + V_C \tag{7.1}$$

In other words, the molecules are described based on nondeformable spheres possessing charges bound by "springs" (harmonic potentials). These springs are appropriately calibrated to describe the vibrations, chemical bonds, and energy barriers associated with the conformational, torsional, and angular transitions that reconstruct the behavior of the molecule to be mimicked. The sum of the bonding and nonbonding potentials comprises what is known as the system force field [63]. Therefore, each function of potential energy seeks to describe the structural behavior characteristic of the molecule.

Harmonic (or linear) bonding potential: In molecular mechanics methods, covalent bonds are described as a classical harmonic oscillator. The stretching between two atoms is defined by the linear harmonic potential V_d, which describes the energy associated with deviations from the equilibrium distance d_0 (Eq. 7.2) [64].

$$V_d = \frac{1}{2} k_d (d - d_0)^2 \tag{7.2}$$

where d is the bond length between two atoms i and j (Fig. 7.10), and k_d is the elastic force constant.

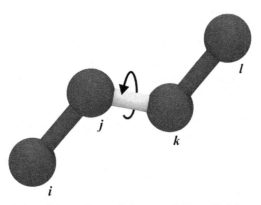

FIGURE 7.10 Geometric representation of the torsion angle between *i, j, k,* and *l*. *Adapted from W. Cornell, et al., A second generation force field for the simulation of proteins, nucleic acids, and organic molecules, J. Am. Chem. Soc. 117 (3) (1995) 5179–5197 [59].*

The distance d_0 and the force constant are the free parameters (to be determined) of the force field. These parameters can be obtained experimentally by infrared spectroscopy or through quantum mechanics. The constant k_d varies according to the stiffness of the chemical bond. As an example, the constant k_d for a simple carbon–carbon bond C—C is 317 kcal mol^{-1} Å^2, whereas for a double bond C=C, k_d is 570 kcal mol^{-1} Å^2 [59].

Angular harmonic potential: This potential describes the potential energy related to the relative deviations from the reference angle. Analogously to the linear harmonic potential, the angular harmonic potential can be defined by Eq. 7.3 [59].

$$V_\theta = \frac{1}{2}k_\theta(\theta - \theta_0)^2 \tag{7.3}$$

Torsional (or dihedral) potential: Considering a set of four bound atoms i, j, k, and l, as shown in Fig. 7.10, the dihedral angle φ is defined as the angle between the planes formed by the atoms ijk and jkl.

The torsional potential V_φ can be understood as an explicit representation of the energy barrier attributed to the rotation of atoms i and l around the j–k bond, as defined by Eq. 7.4 [59].

$$V_\varphi = k_\varphi[1 + \cos(n\varphi + \delta)] \tag{7.4}$$

Where k_φ is the constant defining the height of the energy barrier for a possible torsion, n is the number of minima in the potential energy curve by the torsion angle, φ is the dihedral angle of the central bond in relation to four atoms, and δ is the phase difference of the dihedral angle φ, which may be 0° or 180°, depending on whether this value is a maximum or a minimum point.

The experimental determination of torsional potentials is not trivial. In defining such parameters for new molecules it is common to use dihedral potentials similar to those of a known and previously described molecule. Another strategy for determining the parameters k_φ, n, and δ is to optimize the parameters for simple molecules and extrapolate to complex molecules with similar atomic groupings. The potential energy curve can be obtained through quantum mechanics calculations, which is a great advantage given the independence from experimental values [7,64].

Lennard-Jones (or van der Waals) potential: The Lennard-Jones potential is used to describe intramolecular van der Waals interactions between bonded and nonbonded atoms. This potential is generally used for describing interactions between atoms within the same molecule if they are separated by three or more covalent bonds [63]. The expression for the potential energy corresponding to this type of interaction between two atoms (i and j) is given by Eq. 7.5 [65].

$$V_{LJ} = 4\varepsilon\left[\frac{\sigma^{12}}{r_{ij}^{12}} - \frac{\sigma^6}{r_{ij}^6}\right] \tag{7.5}$$

where r_{ij} is the distance between atoms i and j (Fig. 7.10), ε is the depth of the potential energy well, and σ is the Lennard-Jones diameter, which depends on the chemical species.

There are other approaches to the description of this type of interaction; however, the form in Eq. 7.5 is the simplest and most general [59,65].

Coulomb potential: Purely electrostatic interactions between the charges of two atoms i and j (Fig. 7.10) are described by the Coulomb potential energy (Eq. 7.6) [59].

$$V_C = \frac{q_i q_j}{4\pi\varepsilon_0 \varepsilon r_{ij}}$$

(7.6)

where r_{ij} is the distance between the atoms i and j, q_i and q_j are the atomic charges, ε_0 is the vacuum permittivity, and ε is the dielectric constant of the medium.

To apply the equations describing the potential energies corresponding to the interactions present in molecular systems, it is necessary to determine several parameters. As previously mentioned, these parameters can be obtained through calculations or with experimental techniques, such as XRD, infrared spectroscopy, and NMR [59].

7.4.1.1.2 MOLECULAR DYNAMICS SIMULATION METHODS

MD methods are based on the laws of classical mechanics [22]. The forces acting on each atom are based on the set of parameters determining the force field and the atomic coordinates and velocities, which are mapped throughout the entire process. In an MD simulation, the force F acting on a given atom is expressed as a function of time and is equal to minus the gradient of the potential energy V with respect to the position r_i of each atom, as expressed by Newton's law (Eq. 7.7) [66].

$$-\frac{dV}{dr_i} = m\frac{d^2 r_i}{dt^2}$$

(7.7)

In general terms, in MD methods, the classical equations of motion are numerically and iteratively integrated for each atom in the system at a given time. At each step, the integration process is repeated considering the atomic positions and obeying a given time increment (Δt) determined by the user, which in general is on the order of femtoseconds (10^{-15} s). Such small timescales are necessary to show the fastest normal modes, that is, the oscillation frequency of the chemical bonds. In simple terms, if the value of Δt is large, some important molecular events taking place at small timescales would not be shown, creating computational artifacts that would result in errors that may compromise any result of the simulated model. Such artifacts can generate significant errors in the force calculations and consequently in the sampling of the canonical distribution of the system microstates [67].

Another basic principle of MD is the relationship between the theoretical and experimental data, which can be established based on Boltzmann's ergodic hypothesis [68]. With this hypothesis, it is possible to show that the mean value of a property measured for a small number of particles over a long period of time is equivalent to the means obtained for a system of many particles over a short-time interval. Based on this observation, MD simulations with few particles (compared to any real system) must have sufficiently long trajectories for the system to visit a sufficient number of representative configurations to satisfy this principle. This theoretical argument validates the means obtained for the thermodynamic properties of systems simulated over sufficiently long-time intervals, in an ergodic sense [67].

Currently, there are several available algorithms to numerically integrate the equations of motion, with the most common being the Verlet-type algorithms [65]. The Verlet, Velocity-Verlet [69], and leap-frog [70,71] algorithms are the most widely used because they have a lower computational cost in terms of RAM consumption and processing time [65,72]. In general, the choice of algorithm is based on two important variables: computational cost and numerical integration precision [73]. The Velocity-Verlet algorithm uses the velocities of the particles to calculate the force, and with this, it updates the positions of the atoms. In this manner, the inter- and intramolecular forces are calculated at each MD step, making the calculation of the kinetic and potential energies more computationally convenient compared to the Verlet algorithm, for example, which depends only on the previous and current atomic positions to determine the next ones. Other algorithms achieve much better precision in terms of the conservation of the total system free energy, although they have a much higher computational cost, such as the Beeman algorithm [74]. The condition for choosing a lighter algorithm (in terms of processing) over one that is more precise is that with lower computational cost, the simulations may achieve longer timescales. Thus, they represent more entropic and energetic microstates of the system, and allow the calculation of the ergodic means of the properties that can safely be compared with experimental values [73].

It is currently possible to perform MD simulations that vary between hundreds of nanoseconds and a few microseconds, depending on the number of atoms in the system [75]. However, even with the temporal evolution obtained through the computation, the timescales that can be simulated are not sufficient to obtain the full information on many important phenomena, such as protein and DNA folding [75]. A number of strategies have been used for the purpose of increasing the time step (Δt), thus obtaining longer simulations. Among the fastest processes occurring in molecular systems are molecular bond vibrations, especially from hydrogen bonds [76]. Since these vibrations have some influence on several molecular properties, some algorithms keep them fixed, such as SHAKE [77], RATTLE [78], and LINCS [79].

In parallel with these strategies, there are alternative MD methodologies that seek to accelerate the molecular processes, modifying the probability distribution of the system states by introducing an additional term to the potential energy. In cases where the atomic behavior can be ignored, coarse-grained models can be utilized, which consider groups of atoms as particles [80]. These groups are appointed based on judicious analyses of the molecule's chemical behavior, respecting the most important molecular properties for a given property. In addition, it is possible to group an amino acid or monomer units as a single particle [81]. This grouping can be performed based on a better sampling of the conformational states of the system at hand, driving it to other potential energy minima, and altering the relative probabilities of some states that are defined by variables that describe a reaction coordinate. Some of these methodologies include local elevation search, replica exchange, parallel tempering, simulated annealing, and metadynamics [82–84].

Other problems that appear in MD simulations are the use of solvents and the choice of model to use. The use of solvent(s) in a given simulation increases its computational cost; however, it allows the precise description of several properties that are directly dependent on the interaction with solvents [76]. The most widely used water models are SPC, TIP3P, TIP4P, and TIP5P [85–87]. An increase in the number of particles in a simulation may lead to several problems related to the estimation of the forces between them due to possible surface effects [67], that is, boundary effects in the system due to the tendency of the molecules to collide with the walls of the simulation box. An approximation that sidesteps this problem is the use of periodic boundary conditions [87], where the system is periodic in a two-dimensional representation, and the particles can enter or exit the simulation box through any of the six faces of the cube. With this method, the surface effects are canceled, and the simulation mimics the system in solution.

In summary, computer simulations of systems with thousands of atoms can be created through the use of MD, considering phenomena that take place in a few nanoseconds, without breaking or creating chemical bonds. Any thermodynamic property of the system at hand can be obtained if the simulation considers a sufficiently large number of energy states to satisfy the ergodic hypothesis. Using this powerful technique, it is possible to analyze adsorption processes, interactions, and conformational molecular behaviors in biological systems [87–92].

7.4.1.2 Molecular Docking

Molecular docking is a method that predicts the preferential orientation of a molecule with respect to another when both form stable complexes [93,94]. In general, a larger molecule is chosen as a receptor, for example, an enzyme, and a smaller molecule is chosen as a ligand. The ligand molecule can be a pharmaceutical, herbicide, or another specific molecule whose interaction with the protein one wishes to study [95,96]. The characteristics of this type of calculation are presented in the following sections.

7.4.1.2.1 RECEPTOR–LIGAND MOLECULAR DOCKING

The receptor–ligand molecular docking technique is based on a prediction of the bond type of a smaller molecule, the ligand, in the bonding region of a macromolecule, the receptor. In this way, it is possible to quantify the receptor–ligand affinity [97]. The protein–ligand molecular interaction involves the combination of entropic and enthalpic effects, which can be predicted in Eq. 7.8 through the Gibbs free binding energy (ΔG_{lig}), which itself is related to an experimental inhibition constant k_i [96,98].

$$\Delta G_{lig} = \Delta H - T\Delta S = RT \ln k_i \qquad (7.8)$$

where ΔH is the enthalpy variation, T the absolute temperature, ΔS the entropy variation, and R the universal gas constant.

The molecular docking method has many applications in the recognition of pharmaceuticals, the study of diseases, the development of new pesticides, and the study of sensors [10,99]. Fig. 7.11 shows an example of a receptor–ligand interaction; the reverse

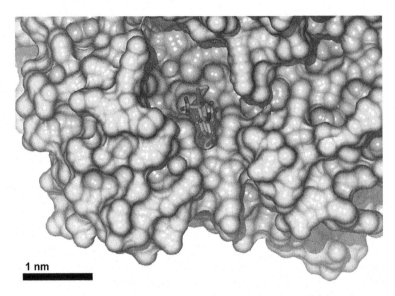

FIGURE 7.11 Reverse transcriptase of HIV-1 virus complexed with the pharmaceutical Efavirenz. Modeled based on PDB ID 1FK9 [99].

transcriptase of human immunodeficiency virus (HIV)-1 is in complex with the pharmaceutical DMP-266 (efavirenz) [99].

The description of the receptor–ligand system using computer-assisted molecular modeling techniques, such as molecular docking, has as a main motivation the reduction of the time and high costs involved with developing new medications, sensors, herbicides, and other pharmaceuticals and technologies [95,96,100].

7.4.1.2.2 GENETIC ALGORITHMS

The molecular docking method requires computational techniques with the potential to investigate a large number of possible solutions to identify the optimum solution, that is, the solution that best meets the system needs [96,101]. The first docking algorithms were developed at the beginning of the 1980s when this was a highly active area of research [101]. From the algorithms employed for the construction of models using docking, genetic algorithms stood out. Genetic algorithms are evolutionary algorithms belonging to a class of stochastic global optimization methods inspired by the evolution of species and Darwin's theory of survival of the fittest [102,103]. These algorithms are mainly used for docking flexible ligands, that is, ligands to which rotational degrees of freedom are attributed [104]. New configurations will be generated via a process called "gene exchange," in which two generating individuals provide random changes to the values of the transmitted genes. The process is iterative; the population evolves into favorable energy states until a stopping criterion is satisfied [102,103]. Currently, these algorithms are the mathematical tools most used by molecular docking programs to execute receptor–ligand system optimization calculations to evolve multiple ligand subpopulations [105]. In the case of the program AutoDock, an additional hybrid genetic algorithm called Lamarckian performs local searches [100]. At each generation, a portion of the population is chosen randomly for a local search. Thus, the chosen individual substitutes the previous one, making an allusion to Lamarck's theory of inheritance [96,104,106,107].

7.4.2 Hybrid Methods (Quantum Mechanics/Molecular Mechanics)

7.4.2.1 Introduction to Hybrid Methods

The quantum description of biomolecules can be performed through a combination of molecular and quantum mechanics modeling methods. These hybrid methods are applied to the description of specific protein–ligand interactions and are commonly used in very large systems, such as proteins (more than 10,000 atoms), to provide more consistent results [63]. The term that refers to a combination of quantum mechanics and molecular mechanics methods is QM/MM (quantum mechanics/molecular mechanics). It was introduced in 1976 by Washel and Levitt, who presented the first semiempirical model applied to an enzymatic reaction [108]. This model was used as a starting point for the most refined methods that exist today. QM methods are used for describing the active site at which chemical reactions or electronic excitations take place, whereas MM methods are employed for capturing the effects of the environment on the active site [63].

The QM/MM approximation is a valuable tool for investigating large systems, such as inorganic, organic, and organometallic compounds and solids [109–111]. Some methodological practices and applications will be discussed in Sections 7.4.2.2 and 7.4.2.6, respectively.

7.4.2.2 Simulation of Biomolecular Systems

Studying large systems, such as enzymes, requires realistic structures as a starting point. This is possible when molecular structures obtained with XRD and NMR are used [112]. A few small modifications in the experimental structures are frequent, for example, the exchange of an inhibitor for a substrate or the substitution of specific residues to generate a starting structure of interest [113]. Often the structures generated by experiments are stored in websites, which are three-dimensional databases of proteins and nucleic acids (Section 7.3.5), and most of the time, these structures are incomplete, which can cause errors when they are used as a starting point in computer simulation programs. In this context, the chemical adjustment of the molecular structure by programs that computationally infer the possible incomplete amino acid sequence becomes necessary.

Specific criteria are used for choosing methods to describe the QM and MM regions. The QM region is treated at the quantum level, whereas the MM region is described by force fields [62]. The interface between these regions can be fixed at a defined region (Fig. 7.12), or the boundary can be allowed to move during the course of the simulation to describe processes, such as changes in the active site and/or the movement of solvated ions [114]. The choice of the QM region takes into account precision and reliability versus the computational effort. [115] In spite of their computational cost, ab initio methods are the most

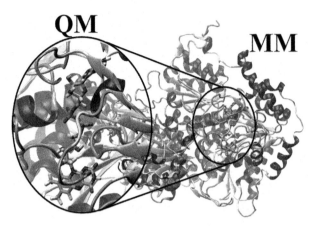

FIGURE 7.12 Description of the molecular system with the interfaces of the quantum mechanics *(QM)* and molecular mechanics *(MM)* regions.

widely used and continue to be the most important in quantum characterization [116,117]. However, the method using density functional theory (DFT) [67,118] is considered an alternative and sophisticated method in many contemporary studies for providing high-precision results with a shorter computing time compared to ab initio methods. In the MM region, there are force fields available for biomolecular applications, such as CHARMM [119,120], AMBER [121,122], GROMOS [123,124], and OPLS [125,126]. For the treatment of explicit solvents, such as water, the TIP3P or SPC models are available, as described in Section 7.4.1.1. In Fig. 7.12, it is possible to differentiate the QM and MM regions.

One of the difficulties in the treatment of QM/MM is that the MM methods do not possess molecular parameters for the description of organometallic compounds and solids containing heavy metals. This limitation occurs primarily because the MM region is not polarizable, which can result in a loss of precision in the QM/MM results [127]. Some polarizable force fields have been developed using different classical models, such as dipoles, charge fluctuations, or charges in mass-spring models [110].

7.4.2.3 Programs for Using the QM/MM Methodology

Contemporary computer applications employing the QM/MM methodology are available in different computational packages and include the following programs: MOPAC7 [128], GROMACS [129] compiled with Gaussian03 source code [130], NWChem [7], TUR-BOMOLE [131], ORCA [132], and GAMESS-UK [133], which uses force fields, such as CHARMM [119,120], GROMOS [64,123,124], and AMBER [59,121,122], as well as GULP [49], DIADORIM [5], and DICE [134,135].

7.4.2.4 Theoretical Foundations and Mathematical Description

Classical and quantum Hamiltonians take into account different system properties, and the classical Hamiltonian of a classical molecular system can be written as a function of position and moment [62]. With H being the total energy of the system, the Hamilton equations of motion lead to Newton's equations of motion. Quantum Hamiltonians take into account the nuclear and electronic density contributions of the system. In comparison, the contributions from each are quite different in Hamiltonians of both methodology types, according to Eqs. 7.9 and 7.10 [62,63].

$$\hat{H} = \sum_{\text{bonds}} k_r(r-r_{\text{eq}})^2 + \sum_{\text{angles}} k_\theta(\theta-\theta_{\text{eq}})^2 + \sum_{\text{dihedral}} \frac{V_\varphi}{2}[1+\cos(n_\varphi-\varphi)] + \sum_{i<j}\left[\frac{A_{ij}}{R_{ij}^{12}}-\frac{B_{ij}}{R_{ij}^6}\right] + \sum_{i<j}\frac{q_i q_j}{\varepsilon R_{ij}} \quad (7.9)$$

The classical Hamiltonian describes $H = E_{\text{covalent}} + E_{\text{noncovalent}}$

From left to right, the terms in Eq. 7.9 refer to bonds, angles, dihedral angles, the Lennard-Jones potential (van der Waals forces), and electrostatic forces (Section 7.4.1). In contrast, the quantum Hamiltonian is described by Eq. 7.10 [62,63].

$$\hat{H} = -\sum_N \frac{\hbar^2}{2m_N}\nabla_N^2 - \sum_i \frac{\hbar^2}{2m_e}\nabla_i^2 - \sum_i\sum_j \frac{Z_i e^2}{4\pi\varepsilon_0|R_i-r_j|} + \frac{1}{4\pi\varepsilon_0}\sum_{i<j}\frac{e^2}{|r_i-r_j|} + \frac{1}{4\pi\varepsilon_0}\sum_{A<B}\frac{Z_A Z_B e^2}{R_A R_B} \quad (7.10)$$

The quantum Hamiltonian in this equation contains, from left to right, the kinetic (nucleus), kinetic (electrons), attraction (nucleus–electrons), repulsion (electron–electron), and repulsion (nucleus–nucleus) contributions.

7.4.2.5 Electrostatic Interactions

The electrostatic interactions in biomolecular systems can be described through quantum mechanics (solving Schrödinger's equation—Eq. 7.10) or by solving the Poisson–Boltzmann equation (Eq. 7.11). In this equation, we have information about the charge distribution in different regions of the proteins and amino acids resulting from long-range force interactions, which are essentially of electrostatic nature [62,63,110]. It is important to note that for systems with more than 10,000 atoms, the computational cost of using the quantum method for the calculations becomes very high. Due to this, a good strategy is to use this approach only for evaluating the electrostatic interactions at the active site and, in the rest of the system, to use the electrostatic potential through the numerical solution of the Poisson–Boltzmann equation [136].

$$\vec{\nabla}\cdot\left[\varepsilon(\vec{r})\vec{\nabla}\psi(\vec{r})\right] = -4\pi\rho^{\text{f}}(\vec{r}) = 4\pi\sum_i C_i^\infty Z_i q\lambda(\vec{r})\exp\left[\frac{-z_i q\psi(\vec{r})}{k_{\text{B}}T}\right] \quad (7.11)$$

where $\vec{\nabla}$ is the divergence operator, $\varepsilon(\vec{r})$ is the position-dependent dielectric function, $\vec{\nabla}\psi(\vec{r})$ is the gradient of the electrostatic potential, $\rho^{\text{f}}(\vec{r})$ is the charge density of the solute, C_i^∞ is the concentration of ion i at an infinite distance from the solute, z_i is the ionic charge, q is the proton charge, k_{B} is Boltzmann's constant, T is temperature, and $\lambda(\vec{r})$ is an accessibility factor for the position dependence of position r for the ions in solution. If the potential is not too large compared to kT, the equation can be linearized to solve it more efficiently, leading to the Debye–Huckel equation, which is simpler and can be described by Eq. 7.12 [62,63].

$$\vec{\nabla}\cdot\left[\varepsilon(\vec{r})\vec{\nabla}\psi(\vec{r})\right]=\varepsilon k^{2}\varphi(\vec{r})-4\pi\rho(\vec{r}) \qquad (7.12)$$

where k is the Debye–Huckel or ionic constant of the medium. The shape of the dielectric interface and the exclusion of the ions contained inside the protein are introduced into the calculation through the dependence of the constants ε and k on the vector $\vec{r}$. In conditions where $k = 0$, the ions are absent, and the Debye–Huckel equation can be reduced to the Poisson equation [62,63].

7.4.2.6 Applications of QM/MM in Molecular Systems

Hybrid QM/MM methods are not only valid for research on chemical reactions at the active site in large systems but can also be employed for the investigation of electronic processes, such as electron excitations [110]. Typical biomolecular studies can employ DFT/MM calculations with a standard force field, electrostatic forces, and an approach where boundary atoms are included with the possibility of changing the atomic charge [110,137]. The QM/MM scheme will depend on the treatment level given to the system and also on the type of system. With the combined QM/MM methodologies, and with the data obtained from the interactions observed through experimental XRD techniques, it is possible to define the residue contributions of the intermolecular interactions and system interaction energy, and to obtain the system trajectory in three-dimensional over time. In this context, the rational planning of drugs by computational techniques appears to be an alternative for the synthesis of new pharmaceuticals, for which traditional approaches have not produced satisfactory results.

The combination of QM/MM methods was developed for calculations with proteins, and it has been used for the study of enzymatic reactions, especially enzymatic catalysis, which is concerned with the change of the reagents' bond free energy, transition state, and products [26]. In this method, the ligand is explicitly treated by a QM model [26], whereas the protein environment and the solvent are represented by an MM force field.

The classical dynamics equations are adequate for many systems, including biomolecules. This is because, through an MD study, the different conformations of a system can be determined in addition to evaluating their time evolution [63]. MD is a technique in which the motion of a particle system is studied through simulation, and thus, it can be employed in systems of electrons, atoms, or molecules [138]. The first MD simulation involving proteins was performed in 1987 [139]. Currently, many studies can be found in the literature involving this type of simulation with proteins, mainly for validating experimental data, such as XRD and NMR. The interaction potential between particles and the equations of motion governing their dynamics are prerequisites for MD performance. The potential may vary from simple, as in the gravitational potential for interactions between stars, to complex, composed of various terms to describe the interactions between atoms and molecules [115]. For some problems, such as chemical reactions involving tunneling, electron transfer, and bond breaking and formation, quantum corrections are necessary [138]. Thus, the use of QM/MM methods combined with MD is appropriate for the study of reactions involving enzymatic systems [115].

A typical procedure that employs QM/MM is the structural refining of large biomolecular systems. The basic idea is to use QM/MM instead of only MM, which is the model that is refined against experimental data [114]. This method is advantageous because, around the active site, the standard force fields are less reliable for inhibitors and substrates [62]. This approximation has been applied for the refining of XRD and NMR data [140].

7.4.3 Structural Characterization

The solving and publication of the three-dimensional structures of biomolecules are essential to propel the development of new studies and technologies. However, the process of structural determination is extremely complicated and may require computational methods, which is the scope of this chapter, in addition to advanced experimental methods including sophisticated technologies, such as XRD and NMR. The structural assignment of biomolecules is performed to analyze the nanostructure of the material, which involves all the atoms and atomic arrangements describing the conformation of the molecular system [141]. The application of structural determination methods, both experimental and computational, culminates in the generation of computer graphical approximations of the structures of macromolecules, making it possible to construct computer models in three dimensions through the treatment of data obtained and/or collected [30].

7.4.3.1 X-Ray Diffraction

In Chapter 5: *Low-Dimensional Nanoparticle Systems* of Volume 1: *Nanostructures—Principles and Applications, in the Nanoscience and Nanotechnology Collection*, the reader was introduced to some XRD concepts for nanomaterial applications. The XRD technique also has vast applicability for the description of biomolecules. XRD is one of the main techniques used for the description of three-dimensional structures, especially for proteins [30]. The efficacious application of XRD in the structural study of biomolecules depends on the precise atomic organization characteristic of a crystalline solid. In nature, it is possible for atoms to organize as amorphous solids, described by the organization of atoms in short distances without exhibiting periodicity. It is also possible for atoms to be highly organized as crystalline solids, characterized by the formation of three-dimensional patterns of atomic ordering that repeat periodically along the material structure [141]. The latter situation is of great scientific relevance for the study of macromolecules by XRD because X-rays can specifically describe the atomic arrangement that forms a biological structure [111].

Since the German scientist Max Von Laue began using XRD in 1912, when he observed that crystals could diffract X-ray beams according to the specific atomic organization pattern of the material [143], this technique has been employed in the structural description of biomolecules. Today, the improvement and sophistication of diffractometers—specific apparatuses for the application of XRD (Fig. 7.13A)—as well as the development of new computational methods for crystallographic data analysis and interpretation, have impacted crystallography by increasing the number of obtained macromolecular models. Currently, the PDB repository has approximately 103,265 structures resolved by XRD, with

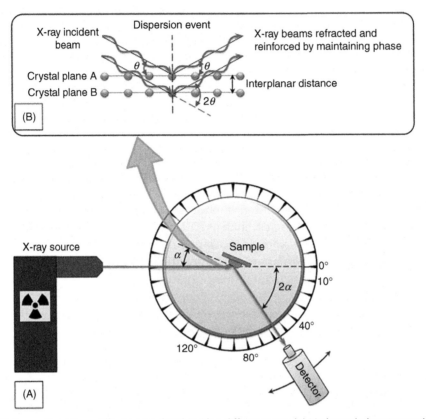

FIGURE 7.13 Schematic of X-ray diffractometry (XRD) inside a diffractometer. (A) Mathematical representation of the incident angles in a dispersion even. (B) Apparatus for determining dispersion angles, X-ray source, and detector.

close to 8680 structures deposited in 2015 [144]. One example of a recently submitted structure is a bacterial transferase complexed with antibacterials of the aminoglycoside class (PDB ID 5C4K), determined by Kaplan et al. [145]. Another example is the structure of the methyltransferase of the dengue virus complexed with a potential inhibitor (PDB ID 5CUQ), resolved by Brecher et al. [146].

Briefly, because the wavelength of X-rays is comparable to the atomic spacing in a crystalline structure, the X-rays can be diffracted in such a way as to describe the atomic positioning [16]. Inside the diffractometer, the incident X-rays are scattered by the sample structure, and the relationship between the phases of the scattered beams describes the diffraction events [147]. If after being scattered, the waves are out of phase, they interfere destructively and cancel each other out. In contrast, if after the scattering event, the waves remain in phase, they will interfere constructively; that is, they reinforce each other, characterizing the phenomenon of diffraction (Fig. 7.13B) [16,148].

The diagram in Fig. 7.13 describes the positions of the components of a diffractogram (A). When, after the scattering event, the X-ray beams remain in phase, their amplitudes are summed and reinforce each other, and diffraction occurs (B). The X-ray beam generated at the source is directed toward the sample, and the angles at which diffraction occurs are recorded. The sample and the detector are coupled such that a rotation α of the sample is equivalent to a rotation of 2α for the detector [149]. A large portion of the radiation scattered by the sample is canceled out, and a minority of scattered X-ray beams reinforce each other, undergoing diffraction. The mutual reinforcement of the beams occurs when conditions satisfy Bragg's law, as described in Volume 1: *Nanostructures—Principles and Applications*, Chapter 5: *Low-Dimensional Nanoparticle Systems* in this collection.

When diffracted, the beams reveal the structural characteristics of the sample (types of atoms that constitute it, their arrangement, and the crystal geometry). When recorded by the detector, these generate a diffraction pattern, or diffractogram, that indicates the angles at which diffraction occurs as well as the intensity of the diffracted beams [150].

The diffractogram data are the basis for determining the material structure. Through the treatment of these data, the structural model of the molecule is obtained. This final part of the crystallographic study is described by three main stages: the resolution of the structure using mathematical methods for determining the real space in the structure and the computer-assisted reconstruction of an image of the atomic ordering in the macromolecule; refining of the model based on the electron density map of the crystalline structure; and the extraction of the final model, which contains the atomic coordinates of the studied structure [30]. For a deeper review of XRD concepts, we refer the reader to Volume 3: *Nanocharacterization Techniques—Principles and Applications*, of the collection *Nanoscience and Nanotechnology*, Chapter 5: *Diffraction and Scattering of X-Rays by Nanomaterials*.

7.4.3.2 Nuclear Magnetic Resonance

Another technique for determining the structures of biomolecules involves the application of NMR. In this technique, the samples are generally analyzed in solution and are thus as close as possible to biological conditions. In addition, this technique does not require the crystallization of the sample, which is a critical step in XRD. Currently, the PDB contains close to 11,260 three-dimensional structures of proteins and other biomolecules resolved by NMR, and this number is growing [151]. In 2015, close to 440 structures resolved by NMR were deposited into the PDB; one of the most recent is the structure of a zinc-dependent metalloprotease of *Clostridium difficile* (PDB ID 2N6J), determined by Rubino et al. [152]. Another example is the structure of an acid-resistant periplasmic bacterial chaperone (PDB ID 2MYJ), determined by Ding et al. [153].

NMR is based on the interaction of electromagnetic radiation with the sample. More precisely, this technique uses the magnetic property displayed by some atomic nuclei of the molecule for the structural study of the material [154,155]. Certain nuclei, such as ^{1}H, ^{13}C, and ^{15}N, exhibit a nuclear spin, a property described by quantum mechanics, whose orientation is given by a characteristic angular momentum [22,156]. In the absence of a magnetic field, the nucleus will be oriented randomly. However, in the presence of an external magnetic field, the magnetic field generated around the nucleus forces it to assume a specific orientation [157]. In a state of low energy, the α state, the nuclear orientation is parallel to the applied magnetic field. However, in a higher-energy state, the β state, the nuclear orientation is antiparallel (opposite) to that of the external magnetic field [157]. The energy difference between states α and β is proportional to the necessary energy for a nucleus to change its angular momentum, that is, to go from a low-energy to a high-energy state [156]. This difference between states α and β is proportional to the intensity of the external magnetic field [158]. Under an intense magnetic field, the difference in energy between states would be greater. As a consequence, the difference between the number of nuclei in state α and the number in state β can also be larger [156]. This difference results in increased spectrometer sensitivity and spectral resolution, allowing devices with intense magnetic fields to describe more complex molecular systems [158].

In an NMR spectrum, the sample in solution is subjected to a high-intensity magnetic field (B_h), which will generate different energy levels, depending on the different orientations of the nuclei aligned with the field. Then, a series of electromagnetic energy pulses in the radiofrequency range is applied. These pulses drive the system out of equilibrium. The radiofrequency electromagnetic radiation will supply energy to the nuclei, leading to a change in their orientation. In the absence of radiofrequency radiation, equilibrium will be restored in the system, the nuclei will return to their original positions, and they will emit characteristic radiation (Fig. 7.14) [159–161]. The emitted radiation is detected by the apparatus, producing a free-induction decay (FID) signal [162]. Due to the complexity of the FID signal, it undergoes a mathematical treatment using the Fourier transform, thus producing data that will generate a plot of the resonant-radiation intensity versus frequency, which is characteristic of the material [161,163].

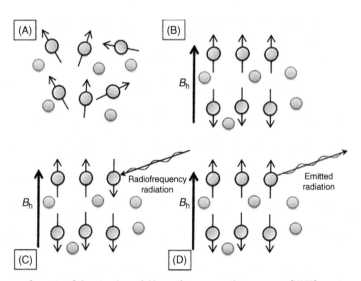

FIGURE 7.14 Sequence of events of the atomic nuclei in nuclear magnetic resonance (NMR) spectroscopy.

The atomic nucleus emission frequency is given by the ratio between the nuclear angular momentum and magnetic moment: the so-called gyromagnetic ratio. However, this frequency may suffer interference from the biomolecule's chemical environment. Therefore, to interpret the NMR spectrum, a reference substance is added to the sample. The positions of the sample spectra are given in relation to their distance from the reference signal, thus characterizing the chemical displacement (in parts per million) [156]. Through the interaction between the nuclear magnetic moment and the external magnetic field, it is possible to characterize the environment around the atomic nucleus of interest and thus the structure and dynamics of the studied biomolecule. Therefore, it is possible to obtain from the NMR spectrum the following: (1) the chemical environment of the atom studied via the chemical displacement, (2) a quantification of the atom in each type of environment through the area of each signal, and (3) the number of atoms in different chemical environments according to the number of different signals compared to the control (reference) [160,161].

The sequence of Fig. 7.14 summarizes the NMR technique: (A) some atomic nuclei have spin, which, in the absence of an external magnetic field, is randomly oriented; (B) when an external magnetic field is applied, the nuclei with *spin* ≠ 0 assume specific orientations with different energy levels; (C) the radiofrequency radiation is used to disturb the system, changing the orientation of the nuclei; and (D) as they return to their original positions, the nuclei release the excess energy with a characteristic signal that will produce the sample NMR spectrum [159,162].

7.4.3.3 Protein Modeling by Homology

In spite of their heavy participation in the determination of biomolecule tridimensional structures, experimental methods such as XRD and NMR may present some problems that complicate and, in some cases, altogether prevent their use, such as difficulties in the isolation and purification of the biomolecule of interest or the characterization and production of crystals of appropriate quality for the analysis [142,164]. Thus, other methods for structural prediction have gained importance as alternative tools for describing biomolecular structures, including homology modeling, also called comparative modeling, which is considered one of the most reliable prediction techniques [164–166].

The goal of protein modeling techniques is the prediction of the three-dimensional protein structures based on their primary structure (sequence of amino acids that forms the protein) to obtain protein models of comparable quality to those generated by experimental techniques [167]. Homology modeling is based on the premise that homologous protein structures maintain a high degree of similarity, which can justify the construction of three-dimensional models [168]. This is possible because homologous proteins evolved from a common ancestor and may have undergone mutations at any given time, culminating in differences between them [169]. Such proteins preserve numerous similar or identical sections, which can be classified as patterns. The internal regions of the structure are better preserved, whereas regions closer to the surface—the folds or loops that connect elements of the secondary structure—are more susceptible to changes [170]. Thus, these similarity patterns form the basis for constructing the structure of an unknown protein by starting from the structure of a homologous protein whose structure has already been determined [168].

Compared to experimental methods, the construction of a three-dimensional model by homology modeling does not require the use of laboratory routines for sample preparation and analysis. However, it demands specialized computational programs and tools, and queries in the various available databases because it utilizes preexisting structural data for the modeling [171]. In general terms, the stages involved in the homology modeling process are as follows: (1) search for and selection of proteins that will serve as a template, (2) alignment of the template protein and the target protein, (3) construction of the target-molecule model, and (4) model validation [172]. Each stage of the homology modeling is based on a specific computational method [172].

The identification and selection of the templates to be used are aided by tools that perform searches in structural databases for structures that are homologous to the molecule of interest. An example of a tool used in this process is BLAST (basic local alignment search tool), which finds homologous protein structures, identifying similar regions between the catalogued structures and calculating their degree of similarity [45,173,174]. Since the program Swiss-PDB Viewer is integrated with specific servers, it can also aid the search for templates. The most adequate structures (generally those with the greatest degree of similarity) are chosen to construct modeling templates [175].

The model is constructed based on the alignment performed between the template sequences and the target sequence. The sequences are superimposed, and by displacement and repositioning of the residues (with the introduction of gaps, representing residue deletions), one obtains the symmetry between them. Such a process reveals the similarity between the sequences, indicating preserved and variable regions. The sequence alignment stage is crucial for homology modeling because it directly implies the quality of the constructed model [176]. With the sequences aligned, the stage of model construction begins with the more preserved regions, based on the template protein structural information. Then, the variable regions of the molecule, such as the loops and side chains, are modeled according to predictions based on amino acid sequences in these regions or by comparison with models for loops and side chains available in the PDB [177]. This stage can be performed by the programs discussed in Section 7.3.5 of the present chapter, such as Swiss-PDB Viewer.

Once the target-protein model has been obtained, it is essential that it is checked for possible errors during the modeling process. The stage of evaluating the model characteristics—the validation—involves the search for structural errors as well as the evaluation of the stereochemical quality of the model [178]. In addition, the evaluation of the interactions between the model obtained and the medium must be performed by MD simulations, which follows the behavior of the molecular system through time [179–181]. Such evaluations can be mediated by the same programs used in the model construction. In the end, the predicted three-dimensional model of the protein is obtained.

7.5 Some Recent Applications

7.5.1 Applications in the Pharmaceutical Industry

7.5.1.1 Pharmaceutical Development

Computational molecular modeling is an important tool for detailing the mechanisms of molecular recognition between proteins and inhibitors, which can be key to the understanding and development of new pharmaceuticals for the treatment of a number of diseases. Thus, computational molecular modeling has a fundamental role in pharmacology and medicine because it describes the structures of biomolecules and pathogens [182,183]. It has been one of the most important advances in the planning and discovery of new pharmaceuticals, allowing researchers to obtain compounds with specific properties that can influence the protein–ligand interaction in the protein-inhibitor case.

The molecular docking example cited in Section 7.4.1 illustrates the site of inhibition of the reverse transcriptase of HIV-1. The virus is susceptible to mutations, which is why, at present, a cure or prophylactic immunization for HIV is unavailable, making this disease a potential target for the study of new techniques and the design of new drugs [184–186]. As observed in Fig. 7.11, the HIV-1 reverse transcriptase was studied as a protein–ligand complex (Efavirenz pharmaceutical). Computational molecular modeling is essential at this level of study because it utilizes the methods mentioned in Section 7.4 for the development of new drugs for combating AIDS and other diseases [99,187].

7.5.1.2 Rational Pharmaceutical Design

The screening technique used until the 1980s consisted of randomly testing pharmaceuticals in microcells, a type of "trial and error" method that was inefficient for discovering new pharmaceuticals [188]. Therefore, knowing the protein-inhibitor mechanism of action is one of the most efficient and least costly ways of developing new pharmaceuticals [187–190]. Some of the most computationally efficient modeling methods include the combination of molecular docking, MD, and QM techniques. The results from these combinations allow the elaboration of similarity indices that can be related to pharmacological activity [191].

Through the performance of various tests involving a large number of inhibitor ligands, with time requirements and costs lower than practical blind screening methods, computer modeling aids in the screening of new pharmaceuticals, considerably accelerating the development of new drugs [99]. All the computer modeling methods combined, together with final in vitro, in vivo, and clinical tests, give rise to the development of a pharmaceutical, according to the stages of rational design shown in Fig. 7.15 [192,193].

7.5.2 Technological Applications

7.5.2.1 Nanoneurobiophysics: New Perspectives for the Application of Nanotechnology to Biomedical Studies

Recently, a new field of study has been outlined—nanoneurobiophysics—which seeks to investigate important molecular mechanisms in chronic and neurodegenerative disorders,

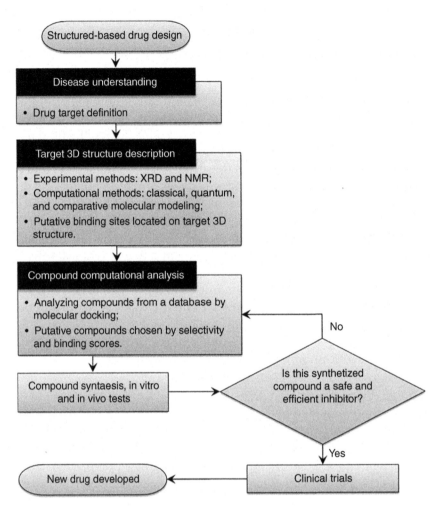

FIGURE 7.15 Flowchart of structure-based rational pharmaceutical design.

and to develop new diagnosis and treatment methods for these disorders based on nano-technology [194]. In this context, the description of the mechanisms on which neurological disorders are based is performed at a molecular scale, made possible by the combined application of experimental and theoretical-computational methods. More precisely, the application of these methodological combinations for diagnostic and therapeutic purposes is outlined by the development of nanobiosensors. These are nanoscale sensing devices, whose sensing is mediated by high-specificity biomolecules, such as enzymes and antibodies [194,195].

The chemical modification of atomic force microscopy (AFM) tips and nanoparticle surfaces with specialized biomolecules is a powerful method that enables the creation of highly specific sensing devices [195–202]. Today, nanobiosensors are being developed for the detection of environmental pollutants, as well as for immunoassays involving the functionalization of nanoparticles, opening new possibilities for the application of such methods in the molecular detailing of complex diseases, such as multiple sclerosis and neuromyelitis optica [91,195,198–202]. In this context, the application of molecular computer modeling appears promising, especially at the development stages of the aforementioned biosensors. These nanodevices require an atomistic vision for the results to be correctly interpreted. This vision is made possible by the application of computational approaches. Recent studies point toward the wide applicability of computational approximations and approaches as support for understanding questions that involve both the functionalization of the surfaces of AFM tips and nanoparticles, and the biomolecular interactions involved in the detection process [89,91,195,198,199,201,203]. Thus, given the high characteristic sensitivity of the developed nanobiosensors, it is possible to detect with high precision enzyme-inhibiting herbicides, antigens, toxins, and other pathogens. Therefore, the combined use of computer techniques and experimental methods makes it possible to understand the molecular mechanisms involved in a system of interest.

7.5.2.2 Computational Methods Applied to the Development of Nanobiosensors

As discussed in the previous section, the application of sensing nanotechnologies is one of the purposes of the emerging field of nanoneurobiophysics [194]. In this context, the development of nanobiosensors based on the functionalization of both AFM and nanoparticle surfaces for the study and detection of complex diseases can be based on computer MD and modeling methods [89,91,203].

In further detail, within the context of the use of AFM tips and functionalized nanoparticles and, consequently, the construction of a nanobiosensor, knowledge and a specific description of the molecular events involved become crucial [89,91,200]. This information can be obtained by the application of computer modeling methods, molecular docking, MD, and hybrid QM/MM calculations developed to provide a background for the interaction types involved in this type of experiment [91,195,198,199,203]. Thus, specific computational techniques are being refined with advances in research and experimental techniques and the development of computers. This has made it possible to rationally develop specific sensors based on functionalized surfaces [198,204].

Computational applications in the development of nanobiosensors have proceeded along a series of paths: (1) the application of homology modeling techniques in the refinement of experimentally described structures, for the subsequent study of immobilization and functionalization of surfaces [89,181], (2) the use of molecular docking approaches for the study of receptor–ligand bonds (enzyme–substrate and/or antibody–antigen) during the sensing process [89,195,199,203], (3) the parameterization of force fields for MD simulation of molecules involved in the immobilization and functionalization of surfaces with biologic compounds [91], (4) computer simulations for the study of the structural stability of protein biomolecular systems [89,181,195] and for the analysis of biomolecule orientations during surface functionalization and the consequent production of nanobiosensors [91,203], and (5) the establishment of protocols for the study of surface biomolecule immobilization [91,199,203].

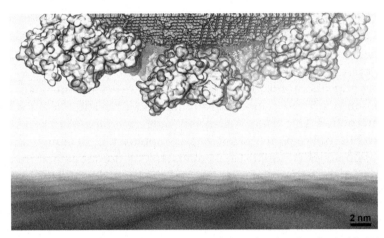

FIGURE 7.16 Representation of the atomic force microscopy (AFM) tip surface functionalized with ALS enzymes. Modeled based on PDB ID 1T9D [29].

Thus, computational methods are powerful alternatives for the high-resolution description of molecular events involved both in the development of nanobiosensors and during the sensing process mediated by such devices. As an example, Fig. 7.16 shows the computerized molecular model of the surface of an AFM tip with an approximate radius of 20 nm functionalized with ALS enzymes (enzyme presented in Fig. 7.9). The enzymes are modeled, highlighting their corresponding surfaces, and affixed to the tip surface by organic thiols, represented as sticks. The ALS enzymes remain active, turning the tip surface as a whole into a specific nanobiosensor for the detection of herbicides (e.g., metsulfuron-methyl and imazaquin, used worldwide) [205,206]. Thus, computer-assisted molecular modeling allows the description of the approximate orientation that the biomolecules assume when immobilized on a surface to form the nanobiosensor.

7.6 Final Considerations

This chapter presented the basic and historical concepts of computer-assisted molecular modeling. Concepts were introduced, including the definitions of biomolecules and proteins, discussing some of the main representation forms of protein models, and example applications of these models with the use of advanced computational techniques were considered. Biomolecule manipulation and visualization programs were also presented, and online databases were indicated, which the reader may consult to obtain previously described structures of biomolecules and biological systems.

The computer-assisted molecular modeling techniques applied to biomolecules encompass three main types of methods: classical (MD and molecular docking), hybrid (QM/MM), and structural characterization methods (homology calculation). All of these tools and knowledge allow one to develop their own study and to model biological systems through various theoretical approximations.

Finally, future perspectives for computer-assisted molecular modeling are very promising. With the increase in processing capabilities and the continuous development of software related to this area, models developed for biomolecules will become more representative of the real systems, and the range of applications will become fairly extensive.

List of Symbols

θ	Angle between two consecutive chemical bonds
$\lambda(\bar{r})$	Accessibility fator from the dependente-position of r above ions in solution
$\varepsilon(\bar{r})$	Dielectric position dependency
z_i	Charge of ion i
$\bar{\nabla}$	Divergente operator
$\rho^f(\bar{r})$	Charge density of the solute
$\bar{\nabla}\psi(\bar{r})$	Electrostatic potential gradient
θ_0	Equilibrium bond chemical distance
c_i^∞	Ion concentration (i) from a infinite distance of the solute
k_B	Boltzmann constant
k_e	Restitution equilibrium angle between two chemical bond
ΔG_{lig}	Gibbs binding free energy
ΔH	Enthalpy variation
ΔS	Entropy variation
Δt	Increment in time
Å	Angstron
B_h	High intensity magnetic field
d	Interplanar spacing between the planes to a given diffracted X-ray beam
d	Bond lenght between two atoms i and j
d_0	Equilibrium lenght between two atoms i and j
F	Active force in a determined atom
H	Hamiltonian of total energy of a molecular system
K	Debye-Huckel constant
$kcal/mol$	Kilocal per mol
k_d	Elastic force constant
k_i	Inhibition constant
k_φ	Constant to define energetic height barrier for chemical bond torsion
q_a	Proton charge
q_i	Atom charge i
q_j	Atom charge j
R	Universal gases constant
r_i	Atom position i
r_{ij}	Distance between atoms i and j
s	Seconds
T	Temperature
V	Potential energy
V_θ	Angular potential
V_C	Coulomb potential
V_d	Linear potential
V_{LJ}	Lennard-Jones or van der Waals potential
V_{total}	Total potential energy

$\mathbf{V}_{\varphi}$	Torsional potentials
δ	Phase difference of the dihedral angle φ
ε	Environment dielectric constant
ε	Deep potential energy
ε_0	Vacuum permittivity
λ	Wavelength
σ	Lennard-Jones diameter
φ	Dihedral angle of the central bond in a sequence of four atoms
$\mathbf{m}$	Mass of a determined atom i
$\mathbf{n}$	Number of minimal for chemical bond torsion

References

[1] T. Schlick, Molecular Modeling and Simulation: An Interdisciplinary Guide, second ed., Springer, New York, 2010.

[2] R. Casadesús, Molecular modeling, in: A.L.C. Runehov, L. Oviedo (Eds.), Encyclopedia of Sciences and Religions, Springer Netherlands, Dordrecht, 2013, pp. 1344–1346.

[3] J.M. Berg, J.L. Tymoczko, L. Stryer, Biochemistry, fifth ed., WH Freeman, New York, 2002, 1100 p.

[4] F. Rausch, et al. Protein modeling and molecular dynamic studies of two new surfactant proteins, J. Cheminform. 5 (1) (2013) 1–11.

[5] L.C.G. Freitas, DIADORIM: a Monte Carlo Program for liquid simulations including quantum mechanics and molecular mechanics (QM/MM) facilities: applications to liquid ethanol, J. Braz. Chem. Soc. 20 (8) (2009) 1541–1548.

[6] National Research Council (US) Committee on Challenges for the Chemical Sciences in the 21st Century. Beyond the Molecular Frontier: Challenges for Chemistry and Chemical Engineering, first ed., National Academic Press, Washington (DC), 2003, 224 p.

[7] M. Valiev, et al. NWChem: a comprehensive and scalable open-source solution for large-scale molecular simulations, Comput. Phys. Commun. 181 (9) (2010) 1477–1489.

[8] M. Scheer, et al. BRENDA, the enzyme information system in 2011, Nucleic Acids Res. 39 (Suppl. 1) (2011) D670–D676.

[9] K.I. Ramachandran, G. Deepa, K. Namboori, Computational Chemistry and Molecular Modeling: Principles and Applications, first ed., Springer, Heidelberg, 2008, 398 p.

[10] A. Bortolato, et al. Molecular docking methodologies, in: L. Monticelli, E. Salonen (Eds.), Biomolecular Simulations: Methods and Protocols, Springer, New York, 2013, pp. 339–360.

[11] Nobel Media AB, Press release: The 1998 Nobel Prize in Chemistry, 2016, http://www.nobelprize.org/nobel_prizes/chemistry/laureates/1998/press.html.

[12] Nobel Media AB. The Nobel Prize in Chemistry 2013—press release, 2014, http://www.nobelprize.org/nobel_prizes/chemistry/laureates/2013/press.html.

[13] A.C. Sudik, et al. Design, synthesis, structure, and gas (N_2, Ar, CO_2, CH_4, and H_2) sorption properties of porous metal-organic tetrahedral and heterocuboidal polyhedra, J. Am. Chem. Soc. 127 (19) (2005) 7110–7118.

[14] M.S. Lee, et al. New analytic approximation to the standard molecular volume definition and its application to generalized born calculations, J. Comput. Chem. 24 (11) (2003) 1348–1356.

[15] D. Young, Computational Chemistry: A Practical Guide for Applying Techniques to Real World Problems, first ed., John Wiley & Sons, New York, 2004, 408 p.

[16] W.D. Callister, D.G. Rethwisch, Materials Science and Engineering: An Introduction, ninth ed., Wiley, Hoboken, 2014, 984 p.

[17] T.E. Brown, et al. Chemistry: The Central Science, thirteenth ed., Prentice Hall, Upper Saddle River, 2014, 1248 p.

[18] P. Atkins, L. Jones, Chemical Principles: The Quest for Insight, fifth ed., W.H. Freeman, New York, 2010.

[19] D. Voet, J.G. Voet, Biochemistry, fourth ed., Wiley, Hoboken, 2010, 1248 p.

[20] O. Schueler-Furman, et al. Progress in modeling of protein structures and interactions, Science 310 (5748) (2005) 638–642.

[21] A.J. Doig, R.L. Baldwin, N- and C-capping preferences for all 20 amino acids in α-helical peptides, Protein Sci. 4 (7) (1995) 1325–1336.

[22] D.L. Nelson, M.M. Cox, Lehninger Principles of Biochemistry, sixth ed., W.H. Freeman, New York, 2012, 1340 p.

[23] M.J. Zvelebil, J.O. Baum, Understanding Bioinformatics, first ed., Garland Science, New York, 2008, 772 p.

[24] G. Fermi, et al. The crystal structure of human deoxyhaemoglobin at 1.74 A resolution, J. Mol. Biol. 175 (2) (1984) 159–174.

[25] T. Yokoyama, et al. Novel mechanisms of pH sensitivity in tuna hemoglobin: a structural explanation of the root effect, J. Biol. Chem. 279 (27) (2004) 28632–28640.

[26] H. Zhang, B. Tweel, L. Tong, Molecular basis for the inhibition of the carboxyltransferase domain of acetyl-coenzyme-A carboxylase by haloxyfop and diclofop, Proc. Natl. Acad. Sci. USA 101 (16) (2004) 5910–5915.

[27] K.P. Madauss, et al. The human ACC2 CT-domain C-terminus is required for full functionality and has a novel twist, Acta. Crystallog. D Biol. Crystallog. 65 (Pt. 5) (2009) 449–461.

[28] J.A. Mccourt, R.G. Duggleby, Acetohydroxyacid synthase and its role in the biosynthetic pathway for branched-chain amino acids, Amino Acids 31 (2) (2006) 173–210.

[29] J.A. Mccourt, et al. Elucidating the specificity of binding of sulfonylurea herbicides to acetohydroxy-acid synthase, Biochemistry 44 (7) (2005) 2330–2338.

[30] G. Rhodes, Crystallography Made Crystal Clear: A Guide for Users of Macromolecular Models, third ed., Academic Press, San Diego, 2010.

[31] R. Apweiler, et al. Update on activities at the Universal Protein Resource (UniProt) in 2013, Nucleic Acids Res. 41 (D 1) (2013) D43–D47.

[32] C.H. Wu, et al. The Universal Protein Resource (UniProt): an expanding universe of protein information, Nucleic Acids Res. 34 (Database issue) (2006) D187–D191.

[33] K.D. Pruitt, et al. RefSeq: an update on mammalian reference sequences, Nucleic Acids Res. 42 (Database issue) (2014) D756–D763.

[34] I. Schomburg, et al. BRENDA in 2013: integrated reactions, kinetic data, enzyme function data, improved disease classification: new options and contents in BRENDA, Nucleic Acids Res. 41 (Database issue) (2013) D764–D772.

[35] I. Schomburg, et al. BRENDA: a resource for enzyme data and metabolic information, Trends Biochem. Sci. 27 (1) (2002) 54–56.

[36] C.J. Sigrist, et al. New and continuing developments at PROSITE, Nucleic Acids Res. 41 (Database issue) (2013) D344–D347.

[37] C.J. Sigrist, et al. PROSITE: a documented database using patterns and profiles as motif descriptors, Brief Bioinform. 3 (3) (2002) 265–274.

[38] H.M. Berman, et al. The protein data bank, Nucleic Acids Res. 28 (1) (2000) 235–242.

[39] R.A. Laskowski, Protein structure database, Mol. Biotechnol. 48 (2) (2008) 183–198.

[40] SIB—Swiss Institute of Bioinformatics. UniProtKB/Swiss-Prot (Lausanne), 2013, http://web.expasy.org/docs/swiss-prot_guideline.html.

[41] P.W. Rose, et al. The RCSB Protein Data Bank: redesigned web site and web services, Nucleic Acids Res. 39 (Database issue) (2011) D392–D401.

[42] G. Küppers, J. Lenhard, T. Shinn, Computer simulation: practice, epistemology, and social dynamics, in: J. Lenhard, G. Küppers, T. Shinn (Eds.), Simulation: Pragmatic Construction of Reality, first ed., Springer, Dordrecht, 2006, pp. 3–22.

[43] E. Winsberg, Computer simulation and the philosophy of science, Philos. Compass. 4 (5) (2009) 835–845.

[44] W. Humphrey, A. Dalke, K. Schulten, VMD: visual molecular dynamics, J. Mol. Graph. 14 (1) (1996) 33–38.

[45] M. Johnson, et al. NCBI BLAST: a better web interface, Nucleic Acids Res. 36 (Suppl. 2) (2008) W5–W9.

[46] J. Stone, et al. Using VMD (Urbana-Champaign), 2012, http://www.ks.uiuc.edu/Training/Tutorials/vmd/vmd-tutorial.pdf.

[47] Theoretical and Computational Biophysics GroupWhat is VMD? Urbana-Champaign, Illinois, USA, 2008, http://www.ks.uiuc.edu/Research/vmd/allversions/what_is_ vmd.html.

[48] W.L. De Lano, The PyMol molecular graphics system (San Carlos), 2002, http://www.pymol.org/.

[49] J.D. Gale, Gulp: a computer program for the symmetry-adapted simulation of solids, J. Chem. Soc. Faraday Trans. 93 (1997) 629–637.

[50] G. Stockwell, PyMOL tutorial (Cambridge), 2003, http://www.ebi.ac.uk/~gareth/pymol/.

[51] R.M. Hanson, et al. JSmol and the next-generation web-based representation of 3D molecular structure as applied to proteopedia, Isr. J. Chem. 53 (3–4) (2013) 207–216.

[52] E.L. Willighagen, Processing CML conventions in Java, Internet J. Chem. 4 (2001) 4.

[53] R.M. Hanson, Jmol—a paradigm shift in crystallographic visualization, J. Appl. Cryst. 43 (5) (2010) 1250–1260.

[54] A. Herraez, Biomolecules in the computer: Jmol to the rescue, Biochem. Mol. Biol. Educ. 34 (4) (2006) 255–261.

[55] E.F. Pettersen, et al. UCSF Chimera—a visualization system for exploratory research and analysis, J. Comput. Chem. 25 (13) (2004) 1605–1612.

[56] G.S. Couch, D.K. Hendrix, T.E. Ferrin, Nucleic acid visualization with UCSF Chimera, Nucleic Acids Res. 34 (4) (2006) e29.

[57] N. Guex, M.C. Peitsch, Swiss-Model and the Swiss-PdbViewer: an environment for comparative protein modeling, Electrophoresis 18 (15) (1997) 2714–2723.

[58] W. Kaplan, T.G. Littlejohn, Swiss-PDB viewer (deep view), Brief Bioinform. 2 (2) (2001) 195–197.

[59] W. Cornell, et al. A second generation force field for the simulation of proteins, nucleic acids, and organic molecules, J. Am. Chem. Soc. 117 (3) (1995) 5179–5197.

[60] L.V. Kalé, NAMD: a case study in multilingual parallel programming, in: Z. Li, et al. (Ed.), Languages and Compilers for Parallel Computing, Springer, Heidelberg, 1998, pp. 367–381 (Chapter 26) (Lecture notes in computer science).

[61] R.J. Woods, et al. Molecular mechanical and molecular dynamic simulations of glycoproteins and oligosaccharides. 1. GLYCAM_93 parameter development, J. Phys. Chem. 99 (11) (1995) 3832–3846.

[62] C.J. Cramer, Essentials of Computational Chemistry: Theories and Models, second ed., John Wiley & Sons, Chichester, 2004, 596 p.

[63] F. Jensen, Introduction to Computational Chemistry, second ed., John Wiley & Sons, Chichester, 2007, 599 p.

[64] J.E. Koehler, W. Saenger, W.F. Van Gunsteren, A molecular dynamics simulation of crystalline alpha-cyclodextrin hexahydrate, Eur. Biophys. J. 15 (4) (1987) 197–210.

[65] L. Verlet, Computer "experiments" on Lennard-Jones Molecules. I. Thermodynamical Properties, Defense Technical Information Center, New York, 1967.

[66] C.L. Brooks, M. Karplus, B.M. Pettitt, Advances in Chemical Physics, Proteins: A Theoretical Perspective of Dynamics, Structure, and Thermodynamics, John Wiley & Sons, New York, 1990.

[67] D. Frenkel, Statistical mechanics for computer simulators. Monte Carlo and Molecular Dynamics of Condensed Matter Systems, vol. 49, 1996, pp. 3–42.

[68] J. Plato, Boltzmann's ergodic hypothesis, Arch. Hist. Exact Sci. 42 (1) (1991) 71–89.

[69] W.C. Swope, et al. A computer simulation method for the calculation of equilibrium constants for the formation of physical clusters of molecules: application to small water clusters, J. Chem. Phys. 76 (1) (1982) 637–649.

[70] J.A. Snyman, A new and dynamic method for unconstrained minimization, Appl. Math. Model. 6 (6) (1982) 449–462.

[71] J.A. Snyman, An improved version of the original leap-frog dynamic method for unconstrained minimization: LFOP1(b), Appl. Math. Model. 7 (3) (1983) 216–218.

[72] B. Alder, Methods in Computational Physics, Academic Press, New York, 1977.

[73] W.F. van Gunsteren, H.J.C. Berendsen, Algorithms for macromolecular dynamics and constraint dynamics, Mol. Phys. 34 (5) (1977) 1311–1327.

[74] D. Beeman, Some multistep methods for use in molecular dynamics calculations, J. Comput. Phys. 20 (2) (1976) 130–139.

[75] W.F. van Gunsteren, et al. Biomolecular modeling: goals, problems, perspectives, Angew. Chem. Int. Ed. 45 (25) (2006) 4064–4092.

[76] H.J.C. Berendsen, et al. Interaction models for water in relation to protein hydration, Intermol. Forces 14 (1981) 331–342.

[77] J.-P. Ryckaert, G. Ciccotti, H.J. Berendsen, Numerical integration of the Cartesian equations of motion of a system with constraints: molecular dynamics of *n*-alkanes, J. Comput. Phys. 23 (3) (1977) 327–341.

[78] G. Vannucchi, et al. Rituximab treatment in patients with active Graves' orbitopathy: effects on proinflammatory and humoral immune reactions, Clin. Exp. Immunol. 161 (3) (2010) 436–443.

[79] B. Hess, P-Lincs, A parallel linear constraint solver for molecular simulation, J. Chem. Theory Comput. 4 (1) (2008) 116–122.

[80] J.C. Shelley, M.Y. Shelley, Computer simulation of surfactant solutions, Curr. Opin. Colloid Interface Sci. 5 (1–2) (2000) 101–110.

[81] L.O. Modesto Orozco, et al. Coarse-grained representation of protein flexibility. Foundations, successes, and shortcomings, Adv. Protein Chem. Struct. Biol. 85 (2011) 183–215.

[82] A. Barducci, M. Bonomi, M. Parrinello, Metadynamics, Wiley Interdiscip. Rev. Comput. Mol. Sci. 1 (5) (2011) 826–843.

[83] A. Medarde Agustín, et al. Valve of the detection of Australia antigen in blood donors, Rev. Sanid. Hig. Publica (Madr) 47 (6) (1973) 529–536.

[84] D. Trzesniak, R.D. Lins, W.F. Van Gunsteren, Protein under pressure: molecular dynamics simulation of the arc repressor, Proteins 65 (1) (2006) 136–144.

[85] H.J.C. Berendsen, J.R. Grigera, T.P. Straatsma, The missing term in effective pair potentials, J. Phys. Chem. 91 (24) (1987) 6269–6271.

[86] W.L. Jorgensen, et al. Comparison of simple potential functions for simulating liquid water, J. Mol. Biol. 79 (2) (1983) 926–935.

[87] M.W. Mahoney, W.L. Jorgensen, A five-site model for liquid water and the reproduction of the density anomaly by rigid, nonpolarizable potential functions, J. Chem. Phys. 112 (20) (2000) 8910–8922.

[88] L. Degreve, C.A. Fuzo, Structure and dynamics of the monomer of protein E of dengue virus type 2 with unprotonated histidine residues, Genet. Mol. Res. 12 (1) (2013) 348–359.

[89] E.F. Franca, et al. Designing an enzyme-based nanobiosensor using molecular modeling techniques, Phys. Chem. Chem. Phys. 13 (19) (2011) 8894–8899.

[90] D. Frenkel, B. Smit. Understanding Molecular Simulation: From Algorithms to Applications, second ed., vol. 1, Academic Press, Florida, 2001.

[91] G.S. Oliveira, et al. Molecular modeling of enzyme attachment on AFM probes, J. Mol. Graph. Model. 45 (2013) 128–136.

[92] O.V.D. Oliveira, J.D.D. Santos, L.C.G. Freitas, Molecular dynamics simulation of the GAPDH–NAD+ complex from *Trypanosoma cruzi*, Mol. Simulat. 38 (13) (2012) 1124–1131.

[93] G.M. Morris, M. Lim-Wilby, Molecular docking, in: A. Kukol (Ed.), Molecular Modeling of Proteins, first ed., Human Press, New York, 2008, pp. 363–382.

[94] M.J. Garcia-Godoy, et al. Solving molecular docking problems with multi-objective metaheuristics, Molecules 20 (6) (2015) 10154–10183.

[95] T. Lengauer, M. Rarey, Computational methods for biomolecular docking, Curr. Opin. Struct. Biol. 6 (3) (1996) 402–406.

[96] G.M. Morris, et al. Automated docking using a Lamarckian genetic algorithm and an empirical binding free energy function, J. Comput. Chem. 19 (4) (1998) 1639–1662.

[97] P.C. Ng, S. Henikoff, SIFT: predicting amino acid changes that affect protein function, Nucleic Acids Res. 31 (13) (2003) 3812–3814.

[98] N. Brooijmans, I.D. Kuntz, Molecular recognition and docking algorithms, Annu. Rev. Biophys. Biomol. Struct. 32 (2003) 335–373.

[99] J. Ren, et al. Structural basis for the resilience of efavirenz (DMP-266) to drug resistance mutations in HIV-1 reverse transcriptase, Structure 8 (10) (2000) 1089–1094.

[100] G.M. Morris, et al. Automated docking with protein flexibility in the design of femtomolar "click chemistry" inhibitors of acetylcholinesterase, J. Chem. Inf. Model. 53 (4) (2013) 898–906.

[101] D.B. Kitchen, et al. Docking and scoring in virtual screening for drug discovery: methods and applications, Nat. Rev. Drug. Discov. 3 (11) (2004) 935–949.

[102] W.F. de Azevedo Jr., MolDock applied to structure-based virtual screening, Curr. Drug. Targets 11 (3) (2010) 327–334.

[103] G. Heberlé, W.F. de Azevedo Jr., Bio-inspired algorithms applied to molecular docking simulations, Curr. Med. Chem. 18 (9) (2011) 1339–1352.

[104] G.M. Morris, et al. AutoDock4 and AutoDockTools4: automated docking with selective receptor flexibility, J. Comput. Chem. 30 (16) (2009) 2785–2791.

[105] G. Jones, et al. Development and validation of a genetic algorithm for flexible docking, J. Mol. Biol. 267 (3) (1997) 727–748.

[106] M.C. Melo, et al. GSAFold: a new application of GSA to protein structure prediction, Proteins 80 (9) (2012) 2305.

[107] I. Halperin, et al. Principles of docking: an overview of search algorithms and a guide to scoring functions, Proteins 47 (4) (2002) 409–443.

[108] A. Warshel, M. Levitt, Theoretical studies of enzymic reactions: dielectric, electrostatic and steric stabilization of the carbonium ion in the reaction of lysozyme, J. Mol. Biol. 103 (2) (1976) 227–249.

[109] L.C.G. Freitas, R.L. Longo, A.M. Simas, Reaction-field–supermolecule approach to calculation of solvent effects, J. Chem. Soc. Faraday Trans. 88 (2) (1992) 189–193.

[110] H.M. Senn, W. Thiel, QM/MM methods for biomolecular systems, Angew. Chem. Int. Ed. Engl. 48 (7) (2009) 1198–1229.

[111] A.L. Tchougréeff, Hybrid Methods of Molecular Modeling, first ed., Springer, New York, 2008, 344 p.

[112] W. Feng, L. Pan, M. Zhang, Combination of NMR spectroscopy and X-ray crystallography offers unique advantages for elucidation of the structural basis of protein complex assembly, Sci. China Life Sci. 54 (2) (2011).

[113] J. Grotendorst, et al. Hierarchical Methods for Dynamics in Complex Molecular Systems vol. 10 IAS Series, Forschungszentrum Jülich, Germany, 2012.

[114] A. Heyden, H. Lin, D.G. Truhlar, Adaptive partitioning in combined quantum mechanical and molecular mechanical calculations of potential energy functions for multi-scale simulations, J. Phys. Chem. B 111 (9) (2007) 2231–2241.

[115] A.R. Leach, Molecular Modelling: Principles and Applications, second ed., Pearson Education, London, 2002, 744 p.

[116] R.A. Friesner, Ab initio quantum chemistry: methodology and applications, Proc. Natl. Acad. Sci. USA 102 (19) (2005) 6648–6653.

[117] R.A. Friesner, V. Guallar, Ab initio quantum chemical and mixed quantum mechanics/molecular mechanics (QM/MM) methods for studying enzymatic catalysis, Annu. Rev. Phys. Chem. 56 (2005) 389–427.

[118] D. Sholl, J.A. Steckel, Density Functional Theory: A Practical Introduction, first ed., Wiley, Hoboken, 2009, 252 p.

[119] B.R. Brooks, et al. CHARMM: a program for macromolecular energy, minimization, and dynamics calculations, J. Comput. Chem. 4 (2) (1983) 187–217.

[120] B.R. Brooks, et al. CHARMM: the biomolecular simulation program, J. Comput. Chem. 30 (10) (2009) 1545–1614.

[121] D.A. Case, et al. The Amber biomolecular simulation programs, J. Comput. Chem. 26 (16) (2005) 1668–1688.

[122] D.A. Case, et al. AMBER 2015, University of California, San Francisco, 2015.

[123] W.F. van Gunsteren, H.J.C. Berendsen, Groningen Molecular Simulation (GROMOS) Library Manual, Biomos, Groningen, The Netherlands, 1987, pp. 1–221.

[124] W.R.P. Scott, et al. The GROMOS biomolecular simulation program package, J. Phys. Chem. A 103 (1999) 3596–3607.

[125] W.L. Jorgensen, J. Tirado-Rives, The OPLS force field for proteins. Energy minimizations for crystals of cyclic eptides and crambin, J. Am. Chem. Soc. 110 (6) (1988) 1657–1666.

[126] W.L. Jorgensen, D.S. Maxwell, J. Tirado-Rives, Development and testing of the OPLS all-atom force field on conformational energetics and properties of organic liquids, J. Am. Chem. Soc. 118 (45) (1996) 11225–11236.

[127] N. Reuter, et al. Frontier bonds in QM/MM methods: a comparison of different approaches, J. Phys. Chem. A 104 (2000) 1720–1735.

[128] J. Stewart, MOPAC: a semiempirical molecular orbital program, J. Comput. Aided Mol. Des. 4 (1) (1990) 1–103.

[129] D. van Der Spoel, et al. GROMACS: fast, flexible, and free, J. Comput. Chem. 26 (16) (2006) 1701–1718.

[130] M.J. Frisch, et al. Gaussian 03, Revision C. 02, Gaussian, Inc, Wallingford CT, 2004.

[131] F. Furche, et al. Turbomole, WIREs Comput. Mol. Sci. 4 (2) (2014) 91–100.

[132] F. Neese, The ORCA program system, Wiley Interdiscip. Rev. Comput. Mol. Sci. 2 (1) (2012) 73–78.

[133] M.F. Guest, et al. The GAMESS-UK structure package: algorithms, developments and applications, Mol. Phys. 103 (6–8) (2005) 719–747.

[134] K. Coutinho, S. Canuto, DICE: A Monte Carlo Program for Molecular Liquid Simulation, University of São Paulo, Brazil, 2003 version 29.

[135] K. Coutinho, S. Canuto, DICE Manual: Version 2.9 (São Paulo, Brazil), http://fig.if.usp.br/~kaline/dice/dicemanual.pdf

[136] F. Fogolari, A. Brigo, H. Molinari, The Poisson–Boltzmann equation for biomolecular electrostatics: a tool for structural biology, J. Mol. Recognit. 15 (6) (2002) 377–392.

[137] M. Elstnera, T. Frauenheima, S. Suha, An approximate DFT method for QM/MM simulations of biological structures and processes, J. Mol. Struct. THEOCHEM 632 (1–3) (2003) 29–41.

[138] M. Karplus, G.A. Petsko, Molecular dynamics simulations in biology, Nature 347 (1990) 631–639.

[139] J.E. Koehler, W. Saenger, W.F. van Gunsteren, A molecular dynamics simulation of crystalline alpha-cyclodextrin hexahydrate, Eur. Biophys. J. 15 (4) (1987) 197–210.

[140] J.J.P. Stewart, Optimization of parameters for semiempirical methods V: modification of NDDO approximations and application to 70 elements, J. Mol. Model. D 13 (12) (2007) 1173–1213.

[141] D.R. Askeland, P.P. Phulé, The Science and Engineering of Materials, fifth ed., Thomson, London, 2006, 790 p.

[142] V. Pattabhi, N. Gautham, Biophysics, Kluwer, New York, 2002, 253 p.

[143] I.N. Serdyuk, N.R. Zaccai, J. Zaccai, Methods in Molecular Biophysics: Structure, Dynamics, Function, Cambridge University Press, New York, 2007.

[144] RCSB—Research Collaboratory for Structural Bioinformatics, Yearly growth of structures solved by X-ray (Camden), 2016, http://www.rcsb.org/pdb/statistics/contentGrowthChart.do?content=explMethod-xray&seqid=100.

[145] E. Kaplan, et al. Aminoglycoside binding and catalysis specificity of aminoglycoside 2″-phosphotransferase IVa: a thermodynamic, structural and kinetic study, Biochim. Biophys. Acta. 1860 (4) (2016) 802–8013.

[146] M. Brecher, et al. Identification and characterization of novel broad-spectrum inhibitors of the flavivirus methyltransferase, ACS Infect. Dis. 1 (8) (2015) 340–349.

[147] C. Suryanarayana, M.G. Norton, X-ray Diffraction: A Practical Approach, Plenum Press, New York, 1998.

[148] R.L. Snyder, X-ray diffraction, in: E. Lifshin (Ed.), X-ray Characterization of Materials, Wiley-VCH, New York, 1999, pp. 1–103.

[149] M. Birkholz, Principles of X-ray diffraction, in: M. Birkholz (Ed.), Thin Film Analysis by X-Ray Scattering, first ed., Wiley, Weinheim, 2006, pp. 1–40 (Chapter 1).

[150] A. Messerschmidt, X-Ray Crystallography of Biomacromolecules, first ed., Wiley-VCH, Weinheim, 2007.

[151] RCSB—Research Collaboratory For Structural Bioinformatics, Yearly growth of structures solved by NMR (Camden), 2016, http://www.rcsb.org/pdb/statistics/contentGrowthChart.do?content=explMethod-nmr&seqid=100.

[152] J.T. Rubino, et al. Structural characterization of zinc-bound Zmp1, a zinc-dependent metalloprotease secreted by *Clostridium difficile*, J. Biol. Inorg. Chem. (2015) 1–12.

[153] J. Ding, et al. HdeB chaperone activity is coupled to its intrinsic dynamic properties, Sci. Rep. 5 (16856) (2015) 1–12.

[154] M.J. Duer, Solid State NMR Spectroscopy, first ed., Blackwell Science, London, 2002.

[155] R.A. Graaf, In vivo NMR Spectroscopy: Principles and Techniques, second ed., John Wiley & Sons, Chichester, 2007.

[156] G.S. Rule, T.K. Hitchens, Fundamentals of Protein NMR Spectroscopy, Springer, Dordrecht, 2006.

[157] D.L. Pavia, et al. Introduction to Spectroscopy, fourth ed., Cengage Learning, Stamford, 2009, 752 p.

[158] M. Renault, A. Cukkemane, M. Baldus, Solid-state NMR spectroscopy on complex biomolecules, Angew Chem. Int. Ed. Engl. 49 (45) (2010) 8346–8357.

[159] M. Balci, Basic 1H- and 13C-NMR Spectroscopy, first ed., Elsevier, Amsterdam, 2005.

[160] I.P. Gerothanassis, et al. Nuclear magnetic resonance (NMR) spectroscopy: basic principles and phenomena, and their applications to chemistry, biology and medicine, Chem. Educ. Res. Pract. 3 (2) (2002) 229–252.

[161] J. Keeler, Understanding NMR Spectroscopy, second ed., Wiley, Hoboken, 2010, 526 p.

[162] J. Cavanagh, Protein NMR Spectroscopy: Principles and Practice, second ed., Elsevier, San Diego, 2007, 912 p.

[163] G. Wider, Technical aspects of NMR spectroscopy with biological macromolecules and studies of hydration in solution, Prog. Nucl. Magn. Reson. Spectrosc. 32 (4) (1998) 193–275.

[164] L. Bordoli, T. Schwede, Automated protein structure modeling with SWISS-MODEL workspace and the protein model portal, Homology Modelling: Methods and protocols, first ed., Humana Press, New York, 2012, pp. 107–136.

[165] J. Moult, Predicting protein three-dimensional structure, Curr. Opin. Biotechnol. 10 (6) (1999) 583–588.

[166] Q. Fang, D. Shortle, Prediction of protein structure by emphasizing local side-chain/backbone interactions in ensembles of turn fragments, Proteins (53) (2003) 486–490.

[167] H. Venselaar, E. Krieger, G. Vriend, Homology modelling, in: J. Gu, P.E. Bourne (Eds.), Structural Bioinformatics, second ed., Wiley-Blackwell, Hoboken, 2009, pp. 715–736.

[168] S.J. Suhrer, et al. Effective techniques for protein structure mining, in: A.J.W. Orry, R. Abagyan (Eds.), Homology Modeling: Methods and Protocols, Springer, New York, 2012, pp. 33–54 (Chapter 2).

[169] A. Bhattacharya, et al. Assessing model accuracy using the homology modeling automatically (HOMA) software, Proteins 70 (2008) 105–118.

[170] B. Alberts, et al. Molecular Biology of the Cell, sixth ed., Garland Science, New York, 2015, 1464 p.

[171] K. Arnold, The Swiss-Model workspace: a web-based environment for protein structure homology modelling, Bioinformatics 22 (2006) 195–201.

[172] B. Wallner, A. Elofsson, All are not equal: a benchmark of different homology modeling programs, Protein Sci. 14 (5) (2005) 1315–1327.

[173] S.F. Altschul, et al. Basic local alignment search tool, J. Mol. Biol. 215 (3) (1990) 403–410.

[174] C. Camacho, et al. BLAST+: architecture and applications, BMC Bioinformatics 10 (2009) 421.

[175] N. Guex, M.C. Peitsch, T. Schwede, Automated comparative protein structure modeling with SWISS-MODEL and Swiss-Pdb Viewer: a historical perspective, Electrophoresis 30 (2009) S162–S173.

[176] C. Venclovas, Methods for sequence–structure alignment, in: A.J.W. Orry, R. Abagyan (Eds.), Homology Modeling: Methods and Protocols, Springer, New York, 2012, pp. 55–82 (Chapter 3).

[177] S. Ramachandran, N.V. Dokholyan, Homology modeling: generating structural models to understand protein function and mechanism, in: N. Dokholyan (Ed.), Computational Modeling of Biological Systems: From Molecules to Pathways, Springer, Ney York, 2012, pp. 97–116.

[178] Z. Xiang, Advances in homology protein structure modeling, Curr. Protein Pept. Sci. 7 (3) (2006) 217–227.

[179] R. Rodriguez, Homology modeling, model and software evaluation: three related resources, Bioinformatics 14 (6) (1998) 523–528.

[180] R. Sanchez, A. Sali, Comparative protein structure modeling, in: D.M. Webster (Ed.), Protein Structure Prediction: Methods and Protocols, Kluwer, New York, 2000, pp. 97–130 (Chapter 6).

[181] J.C.M. Ierich, et al. A Computational protein structure refinement of the yeast acetohydroxy acid synthase, J. Braz. Chem. Soc. 26 (8) (2015) 1702–1709.

[182] T. Gonzalez, J. Díaz-Herrera, Computing Handbook: Computer Science and Software Engineering, third ed., CRC Press, Boca Raton, 2014, 2326 p.

[183] G. Palermo, M. De Vivo, Computational chemistry for drug discovery, in: B. Bhushan (Ed.), Encyclopedia of Nanotechnology, first ed., Springer, New York, 2015, pp. 1–15.

[184] A.S. Fauci, H.D. Marston, Ending AIDS—is an HIV vaccine necessary? N. Engl. J. Med. 370 (6) (2014) 495–498.

[185] S.J. Kent, et al. The search for an HIV cure: tackling latent infection, Lancet Infect. Dis. 13 (7) (2013) 614–621.

[186] C. Katlama, Barriers to a cure for HIV: new ways to target and eradicate HIV-1 reservoirs, Lancet 381 (9883) (2013) 2109–2117.

[187] W.L. Jorgensen, The many roles of computation in drug discovery, Science 303 (5665) (2004) 1813–1818.

[188] J.G. Lombardino, J.A. Lowe, The role of the medicinal chemist in drug discovery—then and now, Nat. Rev. Drug Discov. 3 (10) (2004) 853–862.

[189] R.E. Babine, S.L. Bender, Molecular recognition of protein-ligand complexes: applications to drug design, Chem. Rev. 97 (5) (1997) 1359–1472.

[190] T.P. Lybrand, Ligand-protein docking and rational drug design, Curr. Opin. Struct. Biol. 5 (2) (1995) 224–228.

[191] I. Carvalho, A.D.L. Borges, L.S.C. Bernardes, Medicinal chemistry and molecular modeling: an integration to teach drug structure—activity relationship and the molecular basis of drug action, J. Chem. Educ. 84 (4) (2005) 588–596.

[192] A.C. Anderson, The process of structure-based drug design, Chem. Biol. 10 (9) (2003) 787–797.

[193] S. Singh, B.K. Malik, D.K. Sharma, Molecular drug targets and structure based drug design: a holistic approach, Bioinformation 1 (8) (2006) 314–320.

[194] F.L. Leite, et al., Nanoneurobiophysics: new challenges for diagnosis and therapy of neurologic disorders. Nanomedicine (London) 10 (23) (2015) 3417–3419.

[195] P.S. Garcia, et al. A Nanobiosensor based on 4-hydroxyphenylpyruvate dioxygenase enzyme for mesotrione detection, IEEE Sens. J. 15 (4) (2015) 2106–2113.

[196] C. Steffens, et al. Atomic force microscope microcantilevers used as sensors for monitoring humidity, Microelectron. Eng. 113 (2014) 80–85.

[197] C. Steffens, et al. Microcantilever sensors coated with a sensitive polyaniline layer for detecting volatile organic compounds, J. Nanosci. Nanotechnol. 14 (2014) 6718–6722.

[198] A. Da Silva, et al. Nanobiosensors based on chemically modified AFM probes: a useful tool for metsulfuron-methyl detection, Sensors 13 (2) (2013) 1477–1489.

[199] C.C. Bueno, et al. Nanobiosensor for Diclofop detection based on chemically modified AFM probes, IEEE Sens. J. 14 (5) (2014) 1467–1475.

[200] D.K. Deda, et al. The use of functionalized AFM tips as molecular sensors in the detection of pesticides, Mat. Res. 16 (3) (2013) 683–687.

[201] A.S. Moraes, et al. Evidences of detection of atrazine herbicide by atomic force spectroscopy: a promising tool for environmental sensing, Acta Microscopica 24 (1) (2015) 53–63.

[202] A.M. Higa, et al. Ag-nanoparticle-based nano-immunosensor for anti-glutathione S-transferase detection, Biointerface Res. Appl. Chem. 6 (1) (2016) 1053–1058.

[203] A.M. Amarante, et al. Modeling the coverage of an AFM tip by enzymes and its application in nanobiosensors, J. Mol. Graph. Model. 53 (2014) 100–104.

[204] F.L. Leite, et al. Theoretical models for surface forces and adhesion and their measurement using atomic force microscopy, IJMS 13 (12) (2012) 12773–12856.

[205] X. Cai, W. Liu, S. Chen, Environmental effects of inclusion complexation between methylated beta-cyclodextrin and diclofop-methyl, J. Agric. Food. Chem. 53 (17) (2005) 6744–6749.

[206] Y. Lu, et al. Stereo selective behaviour of diclofop-methyl and diclofop during cabbage pickling, Food Chem. 129 (4) (2011) 1690–1694.

7

Nanocomposites of Polymer Matrices and Lamellar Clays

F.R. Passador*, A. Ruvolo-Filho**, L.A. Pessan**

*INSTITUTE OF SCIENCE AND TECHNOLOGY, FEDERAL UNIVERSITY OF SÃO PAULO, SÃO PAULO, SÃO PAULO, BRAZIL; **DEPARTMENT OF MATERIALS ENGINEERING, FEDERAL UNIVERSITY OF SÃO CARLOS, SÃO CARLOS, SÃO PAULO, BRAZIL

CHAPTER OUTLINE

7.1 **Polymer Nanocomposites** 187
7.2 **Structures of Lamellar Clays** 188
7.3 **Structure of Polymer Nanocomposites** 189
7.4 **Methods for Obtaining Polymer Nanocomposites** 191
7.5 **Compatibilization in Nanocomposites with Nonpolar Matrices** 192
7.6 **Nanocomposites with Polar Matrices** 193
    7.6.1   Structure and Properties 193
7.7 **Nanocomposites with Nonpolar Matrices** 196
    7.7.1   Structure and Properties 197
7.8 **Final Considerations** 203
References 204

7.1 Polymer Nanocomposites

The constant search for new materials that provide the properties demanded for applications motivates research in industries, universities, and technological institutions around the world. The development of polymer nanocomposites stands out among the currently most promising technologies for improving the thermal, mechanical, gas, and organic vapor barrier properties of materials.

A polymer nanocomposite is a composite material comprising a polymer matrix and an inorganic dispersive phase that has least one dimension that is nanometric in scale. Due to the large aspect ratio of the inorganic layers, large polymer–clay surface interactions can occur, which enables the improvement in the properties.

In traditional polymer composites, the addition of large micrometric inorganic loads (between 10 and 40% weight concentration) can significantly increase the mechanical properties of the polymers. These properties can also be increased by using loads with high aspect ratios (such as glass and carbon fiber). The higher the aspect ratio of the

Nanostructures
Copyright © 2017 Elsevier Inc. All rights reserved.

inorganic load is, the greater the contact area with the polymer matrix is, which may increase the reinforcement by transferring a higher voltage from the matrix to the inorganic load. Therefore, the addition of a nanometric material with a high aspect ratio and a high stiffness to a polymer matrix may improve the performance of the polymer even more. Studies of this type of composite have been reported in the literature since 1950. However, since 1990, more attention has been given to nanocomposites because researchers at Toyota presented a study of the use of polyamide 6 (PA6) with organophilic clay in the construction of timing belts for motorized vehicles. The addition of clay at 5% by weight improved the characteristics of this material significantly when compared to unmodified PA6 resin [1].

Since then, different polymer matrices and inorganic loads have been combined to obtain new materials with different properties. Among the more frequently used inorganic loads, one can highlight laminar silicates [1–6], metal nanoparticles [7], carbon nanotubes [8,9], silica [10], calcium carbonate [11], and zinc oxide [12]. This chapter discusses the production, morphology, and properties of nanocomposites of polymer matrices with lamellar clays.

7.2 Structures of Lamellar Clays

Among the inorganic loads with potential for use in the production of nanocomposites, one can highlight lamellar or layered clays, which are composed of hydrated aluminum silicate and formed by layers with nanometric thicknesses; montmorillonite is the most widely used in the production of nanocomposites. Fig. 7.1 presents the basic crystalline structure of a silicate with a 2:1 structure. It is composed of three main layers: a central octahedral layer of alumina or magnesia and two outer tetrahedral layers of silica. The

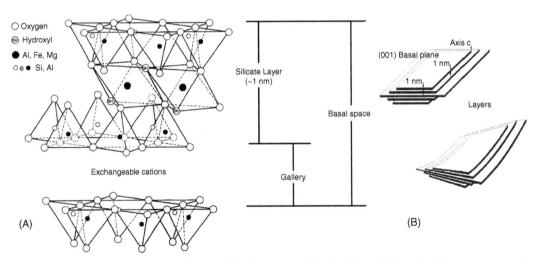

FIGURE 7.1 (A) Structure of a 2:1 phyllosilicate and **(B)** illustration of the flexibility of the lamellae [1].

layers are joined at the ends so that the oxygen ions of the octahedral layer also belong to the tetrahedral layer. The thickness of the layer is approximately 1 nm, and its lateral dimensions can vary from 30 nm to several micrometers.

Naturally, clay minerals are hydrophilic and can be dispersed in water to form thixotropic gels. To make these clay minerals compatible with hydrophobic polymers such as polyolefins, modifications of the clay surface are performed by exchanging hydrated cations with organic cations, decreasing the surface energy of the clay and increasing the distance between the layers [13]. The vast majority of organophilic clays are obtained from smectite clays; montmorillonite is the most common due to intrinsic characteristics of this clay mineral, which include the small size of its crystals and its high cation exchange capacity (CEC), which is used to characterize the degree of isomorphous substitution [1,14]. The CEC of smectites varies from 80 to 150 meq/100 g, which is much higher than those of other clay minerals (which are smaller than 40 meq/100 g).

In addition to the salt used as the organic treatment agent in clay minerals, the form in which this substitution is performed affects the formation of nanocomposites. One of the methods most frequently used to introduce alkylammonium ions between the layers is the ion exchange reaction [14]. This reaction consists of forming, in solution, the desired ion by dissolving the amine with a strong acid or with a salt that has a long alkyl chain containing atoms bonded to counterions, such as chloride and bromide, in hot water at approximately 80°C. These solutions are poured into a montmorillonite dispersion in hot water. A mixer is used to precipitate the particles, which are collected, washed, and then dried. During the drying process, the particles pile up again [14].

The surface treatment is important because it not only makes the clay organophilic and improves the characteristics of its interactions with nonpolar polymers but also increases the distance between the layers. In fact, surface treatment is used even when the polymers are polar and polarity modification of the clay is not fundamental to the production of nanocomposites.

By chemically modifying the surface of the clay, the organic salts allow a favorable penetration of the polymer precursors into the interlamellar regions. The role of the organic salt in the process of clay delamination depends on its chemical nature as well as the length and polarity of its chains [16].

7.3 Structure of Polymer Nanocomposites

To obtain nanocomposites with optimized properties, the clay lamellae should be dispersed and distributed appropriately in the polymer matrix. In addition, in several cases, a significant improvement of the diverse properties of the nanocomposites is reached when the polymer molecules between the clay lamellae are intercalated to a certain level or the lamellae are separated until they are completely dispersed and form an exfoliated structure. The nanocomposites are formed by distancing the clay lamellae because the forces that keep the lamellae together are relatively weak with subsequent penetration of the polymer chains in the interlamellar regions.

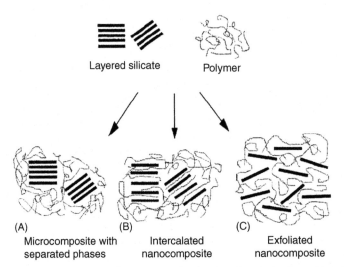

Layered silicate

Polymer

(A) Microcomposite with separated phases

(B) Intercalated nanocomposite

(C) Exfoliated nanocomposite

FIGURE 7.2 Schematic representation of the different types of composites that can be formed by mixing lamellar silicates and polymers. (A) Microcomposites (with separation of both phases), (B) intercalated nanocomposites, and (C) exfoliated nanocomposites [1].

After the separation of the clay lamellae, it is expected that polymer chains bond physically or chemically to their surfaces to form interfaces that are strong enough to maintain their bonds under high strains and flexible enough to allow the transfer of these strains from the polymer matrix to the clay lamellae.

Depending on the nature of the components used (clay, an organic modifier of the clay, a polymer matrix, and a compatibilizing agent) and on the preparation method, it is possible to obtain composites with the three main types of structure shown in Fig. 7.2.

When the polymer is not able to intercalate between the silicate layers, the structure formed is similar to that of a conventional composite, shown in Fig. 7.2A, which provides little or no improvement of the properties. The second structure, which occurs in intercalated nanocomposites, is formed when one or more extended polymer chains are intercalated between the clay lamellae, which increases the distance between them while preserving the lamellar organization, as shown in Fig. 7.2B. In a third case, that of exfoliated nanocomposites, the clay lamellae are completely separated and dispersed, and the system does not present any order, as shown in Fig. 7.2C [1].

Structural characterization of polymer nanocomposites with lamellar clay is usually performed using two main complementary techniques: high-angle X-ray diffraction (XRD) and transmission electron microscopy (TEM). The intercalation and/or exfoliation process can be verified through XRD by observing the shift of the (0 0 1) diffraction peak of clay for values smaller than 2θ (for intercalation) or the absence of this peak (for exfoliation), which implies a loss of the structural regularity of the clay layers. The dispersion state and the distribution of the inorganic load in the polymer matrix are observed using TEM.

7.4 Methods for Obtaining Polymer Nanocomposites

In general, three main strategies can be used to obtain polymer–clay nanocomposites: intercalation of the polymer in solution, in situ polymerization, and intercalation in the melted state (melted intercalation).

When a polymer is intercalated in solution, the organically modified clay and the polymer are dispersed in an organic polar solvent. The silicates in layers can be easily dispersed in an appropriate solvent. The polymer dissolves in the solvent and then adsorbs to the layers of expanded silicates. When the solvent evaporates, the layers regroup and form an intercalated structure [17,18]. The selection of an appropriate solvent is a primary criterion for obtaining the desired level of exfoliation of an organophilic clay.

In situ polymerization involves the insertion of an appropriate monomer into the clay galleries before the polymerization process. The silicate layers are "swollen" by the liquid monomer (or a solution of the monomer) to ensure that the polymer forms between the intercalated layers. Polymerization can be initiated by heat, radiation, or the diffusion of an appropriate initiator, such as an organic initiator or a fixed catalyzer, through cation exchange [1,15].

When nanocomposites are obtained by melted intercalation, the clay is directly dispersed in the melted polymer. During mixing in the melted state, the strain that the polymer exerts on the clay depends on its molecular weight. High levels of shear stress help reduce the size of the clay particles, which aids the intercalation and/or exfoliation process. The mechanism proposed for the action of the shear flow during the exfoliation of organically modified clay in melted intercalation is shown in Fig. 7.3. Initially, particles break down and form piles (tactoids) that disperse through the matrix, as shown in Fig. 7.3A. The transfer of strain from the polymer to these tactoids leads to stronger shearing, which

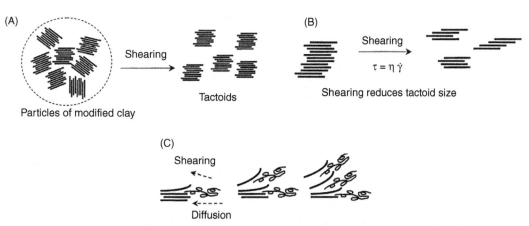

FIGURE 7.3 Effect of shearing on the intercalation and/or exfoliation of organically modified clay during melted intercalation [19].

breaks these tactoids into smaller piles (Fig. 7.3B). Finally, the individual layers are separated through a combination of shearing and diffusion of the polymer chains in the galleries; this step depends fundamentally on time and on the chemical affinity between the polymer and the clay (Fig. 7.3C) [19].

Thermodynamic models [20–22] have been used to explain the formation of nanocomposites by melted intercalation. Vaia and Giannelis [22] showed that this process occurs as a result of the concurrent action of entropic and enthalpic changes. The main factors contributing to changes in the free energy during the formation of the nanocomposite are the confinement of polymer chains between the clay layers, changes in the conformation of the surfactant molecules, and the appearance of new molecular interactions between the polymer, the surfactant, and the surface of the clay lamellae. The global decrease in the entropy of the system due to the confinement of polymer chains in silicate galleries can be compensated for by increasing the conformational freedom of the surfactant molecules and by separating the clay lamellae with a less confining medium. Therefore, these two opposing effects make the total entropic variation of the system small and negative, that is, the formation of a nanocomposite is entropically unfavorable. Therefore, the intercalation of the polymer must be managed, mostly by enthalpic variations.

The enthalpy of the mixture can be assumed to be favorable because the magnitude and number of interactions between the polymer chains and the surfactant molecules fixed on the surfaces of the clay lamellae are maximized when both contain polar groups. Therefore, intercalated nanocomposites can be obtained in systems that feature weak interactions between polymers and clays, and exfoliated nanocomposites can be produced by systems that feature strong interactions between polymers and clays, which exfoliates the layers. For nonpolar polymers, direct intercalation with clay lamellae is difficult and requires the use of compatibilizers.

7.5 Compatibilization in Nanocomposites with Nonpolar Matrices

Recent studies have shown that organically modified clays can be efficiently exfoliated in matrices of polar polymers, such as polyamides, using appropriate processing conditions and techniques. However, obtaining exfoliated nanocomposites based on the polyolefins that are more frequently used, such as polypropylene (PP) and polyethylene (PE), has proven more difficult because these materials are hydrophobic and appropriate interactions with the polar surface of the aluminosilicates in the clay do not occur [16,23,24]. The compatibility of the inorganic load with the polymer matrix is a significant challenge in the preparation of nanocomposites and can be improved by chemically modifying the surfaces of the particles in the components and by adding compatibilizers between the surfactant and the polymer matrix. Strength is promoted by the clay that sediments due to the efficiency of the compatibilization and by the restricted mobility of the polymer chains that are in contact with the clay lamellae.

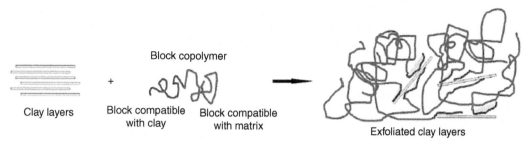

FIGURE 7.4 Schematic representation of the action of a block copolymer during exfoliation of the clay lamellae in a polymer matrix [20].

Two main and joint strategies are currently used to enable chemical and physical inter-actions between the components of nanocomposites with nonpolar matrices. The first consists of making the surfaces of the inorganic particles organophilic, and the second consists of incorporating a comonomer with hydrophilic characteristics into the polymer chains.

Hydrophilic units can be directly inserted into the polymer chain in the matrix using copolymerization techniques in the presence of monomers and using reactive extrusion in the presence of polymers grafted with maleic anhydride, acrylic acid, or another functional group. An ideal compatibilizing agent between two noncompatible components must have parts that combine thermodynamically with both components. Surfactants provide these functions only partially because their ionic parts interact favorably with the surfaces of the clay particles. The long branches of alkyl groups, however, exhibit only limited com-patibility with polymer chains. More efficient compatibilization can be obtained by using macrosurfactants, such as block or grafted copolymers, as shown in Fig. 7.4, to combine blocks that can react with the surface of a solid particle and with the polymer matrix and then improve the interaction between the surface of the clay and the polymer chains [20].

7.6 Nanocomposites with Polar Matrices

Among the various polar polymer matrices used in the preparation of nanocomposites, polyamide is the one that stands out the most; a broad range of studies of nanocompos-ites of polyamide and lamellar clay can be found in the literature [1,19,25–31]. Other polar polymer matrices that stand out are poly(ethylene terephthalate) [4,32–37], poly(vinyl chloride) [38–42], and polycarbonate [43–46].

7.6.1 Structure and Properties

In general, the properties of polymer nanocomposites are highly dependent on their microstructures. A nanocomposite with an exfoliated and well-dispersed structure fea-tures important modifications to the mechanical, physical, and chemical properties of its matrix, which are due to the stronger polymer-clay interactions that occur in these systems. The thermal stability of these polymer nanocomposites is higher than those of pure polymers because of the presence of anisotropic layers of clay in the polymer matrix, which slow or stop the diffusion of volatile products through the nanocomposite.

Nanocomposites with concentrations of clay on the order of 2–10% may show significant improvements in their properties over those of pure polymers. Among the main improvements are in mechanical properties, such as increases in the elastic modulus, the maximum tensile strength, and the flexural modulus, in barrier properties, such as a reduced permeability to gases, optical properties, and ionic conductivity.

The advantage of adding a smaller amount of clay during the formulation and preparation of a nanocomposite in the melted state causes the final density of the material to decrease, which contributes to the production of lighter components and reduces wear and tear to the equipment used to process these materials.

In addition to these properties, the preparation of polymer/lamellar clay nanocomposites includes improvements to their thermal stability and the ability to promote flame retardancy.

A good example of a nanocomposite with a polar matrix that has excellent properties is PA6 with organically modified montmorillonite clay (OMMT). Fig. 7.5 presents a TEM micrograph of a nanocomposite of PA6 with the addition of 5% OMMT by weight, which was obtained by melted intercalation in a twin-screw extruder.

The morphology of this nanocomposite comprises individual lamellae that are completely exfoliated, well dispersed and distributed, which strengthens the polymer matrix significantly.

The excellent dispersion of the clay lamellae promotes a significant increase in the elastic modulus from 2.7 GPa for pure PA6 to 4.7 GPa with the added nanoload. Oliveira et al.

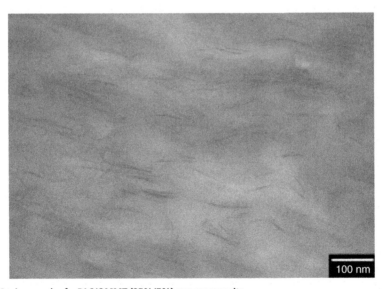

100 nm

FIGURE 7.5 TEM micrograph of a PA6/OMMT (95%/5%) nanocomposite.

[5] reported an increase of more than 50°C in the heat deflection temperature (HDT) of this nanocomposite (HDT_{PA6} = 55°C and $HDT_{nanocomposite}$ = 108°C).

The improvement in these mechanical properties has been connected to the dispersion, the degree of exfoliation, the aspect ratio of the clay, and the polymer–clay interfacial interactions. Luo and Daniel [46] studied epoxy/clay nanocomposites and observed that a high degree of dispersion increases the elastic modulus. The efficiency of the strengthening of the nanocomposite was attributed to the fraction of exfoliated material because the effective aspect ratio of the incorporated additive becomes extremely elevated in exfoliated nanocomposites. However, this effect can restrict the delamination of other layers of clay that could be exfoliated but remain aggregates. Therefore, partial intercalation may contribute positively to the increase in the elastic modulus because a higher concentration of clay may be intercalated and, therefore, may be in the structure of a material with a high aspect ratio. Therefore, the high aspect ratios of exfoliated particles and intercalated aggregates is desirable for strengthening the nanocomposites.

In addition to preparing nanocomposites using matrices that consist of only one polymer, several research groups have been developing nanocomposites with polymer matrices that consist of mixtures of two or more polymers; these are nanocomposites of polymer blends. Polymer blends are excellent alternatives for adding value to materials because the physical and chemical properties of these polymer matrices are modified to allow a vast range of applications [47]. Polymer blends are polymeric systems that originate in the physical mixture of two or more polymers and/or copolymers that do not undergo significant chemical reactions [47,48].

A balance between properties is a typical result of using a polymer blend as a nanocomposite matrix. The addition of lamellar clay significantly increases the elastic modulus of the nanocomposite, that is, the stiffness of the material increases. However, the impact resistance (IR) of the nanocomposite decreases, that is, it becomes more fragile. An alternative for improving both properties could be the addition of a second phase that has excellent resistance to impact but does not mix with the polymer matrix to balance the mechanical properties (tenacity vs. stiffness). An example of such a system is the nanocomposite of PA6 with OMMT with the addition of a second phase composed of an acrylonitrile–butadiene–styrene (ABS) blend as an impact modifier for engineered polymers.

Fig. 7.6 shows the morphology of a nanocomposite of the PA6/ABS blend with OMMT developed by Oliveira et al. [5].

The morphology features elongated ABS domains. Inside these domains, it is possible to observe polybutadiene particles with a wide range of sizes in addition to exfoliated lamellae, which are only in the matrix of PA6. In this case, the clay lamellae prevent the coalescence of the dispersed polybutadiene domains, which reduces the size of this elastomeric phase. Regarding mechanical properties, one can observe that the addition of clay does not alter the Izod IR of the system from that of the PA6/ABS blend. However, adding clay to the blend increases the elastic modulus significantly ($E_{PA6/ABS}$ = 2.8 GPa and $E_{PA6/ABS/OMMT}$ = 4.2 GPa) [5].

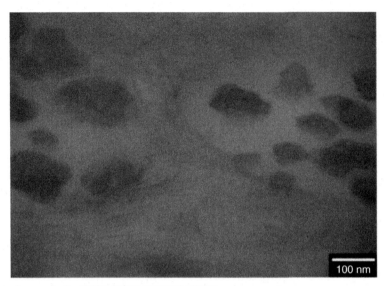

FIGURE 7.6 TEM micrograph of a PA6/ABS/OMMT (57.5/37.5/5%) nanocomposite.

Several other examples can be found in the literature on nanocomposites with polar matrices, including studies of different treatments of lamellar clays, compatibilization of the polymer matrix, and different sequences for mixing the components to identify the best relationship between structure and properties.

7.7 Nanocomposites with Nonpolar Matrices

Polyolefins are widely used for applications in the packaging, automotive, and electric industries, where thermal and mechanical resistance and gas and vapor transport properties are very important. The main representatives of this class of polymer are PE [6,17,23,24,49–56] and PP [3,21,57–60], which are the most frequently used in preparing nanocomposites with nonpolar matrices.

However, obtaining exfoliated nanocomposites from polyolefins and clay is still a significant challenge because polyolefins are hydrophobic and do not interact appropriately with the polar surface of the clay. The surfaces of clays are treated with quaternary ammonium salts to make them hydrophobic. For PE and PP, organophilization of the clay is not enough to guarantee the formation of exfoliated nanocomposites, and therefore, it is necessary to use compatibilizing agents, which are more commonly added during in situ polymerization and melted intercalation.

Different routes of compatibilization and/or types of compatibilizing agents used in the formation of nanocomposites with intercalated/exfoliated structures have been reported in the literature. Among the compatibilizing agents that are most common and most appropriate for nanocomposites are polyolefins grafted with maleic anhydride. They increase the chemical and structural affinity of the clay for the polymer matrix and

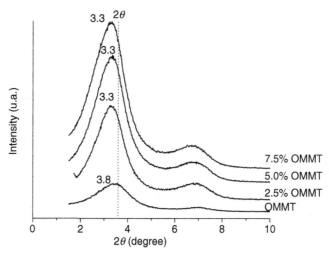

FIGURE 7.7 XRD patterns of lamellar clay (OMMT) and HDPE/LLDPE nanocomposite blends with different concentrations of lamellar clay (2.5, 5.0, and 7.5% by weight).

promote significant improvement of mechanical and barrier properties in addition to being able to mix with polymer matrices, as shown in the following sections.

7.7.1 Structure and Properties

Among polyolefins, high-density polyethylene (HDPE) has a great deal of potential for applications in the packaging and electric sectors. However, it is difficult to obtain nanocomposites of HDPE and lamellar clay because, in addition to weak interactions between the nonpolar matrix and the clay, there are difficulties in processing such materials due to their high viscosities when melted. To ease the process and obtain nanocomposites of polyolefins and lamellar clays that have good properties, Passador et al. developed [6] HDPE/linear low-density polyethylene (LLDPE) nanocomposite blends. Adding LLDPE, which has a chemical structure that is similar to that of HDPE and a lower viscosity, aids processing by modifying the morphology and the properties of the polymer blend. Fig. 7.7 shows XRD patterns of OMMT and the HDPE/LLDPE nanocomposite blends with different concentrations of clay. Fig. 7.8 shows TEM micrographs of these nanocomposites.

Analysis of the XRD patterns in Fig. 7.7 shows a small shift of the diffraction peak for smaller angles referring to the crystallographic plane (0 0 1) in relation to the OMMT, which indicates a small increase in the basal spacing ($2\theta = 3.8$degree corresponds to a basal spacing of 2.32 nm, while $2\theta = 3.3$ degree corresponds to one of 2.68 nm). This suggests that a significant fraction of the tactoids in the system remain unchanged, that is, the distance between the clay lamellae does not change. Therefore, the shear that results from mixing in the melted state is not enough to efficiently delaminate the clay mineral. There is also a secondary reflection at $2\theta = 7.4$ degree, which corresponds to crystallographic plane (0 0 2) of the nanocomposite but is shifted to a smaller angle.

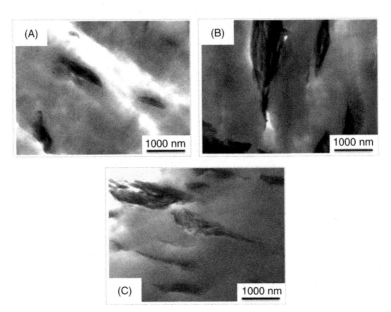

FIGURE 7.8 TEM micrographs of HDPE/LLDPE/OMMT nanocomposites (A) with 2.5% OMMT by weight, (B) with 5.0% OMMT by weight, and (C) with 7.5% OMMT by weight.

Unlike the PA6/ABS system, in which the polymer blend is immiscible, that is, phase separation occurs, the HDPE/LLDPE system presents full miscibility in the melted state, where it forms a single phase, as shown by the micrographs in Fig. 7.8. The addition of lamellar clay allows the formation of a microcomposite with a structure that is mostly composed of agglomerates or tactoids of lamellar clay. The small amount of interaction between the components (the matrix and the load) does not allow these tactoids to break down during processing, and increasing the amount of lamellar clay results in larger agglomerates in the polymer matrix.

In this system, it is necessary to add compatibilizing agents to increase the interaction between the polymer matrix and the lamellar clay. Compatibilizing agents grafted with maleic anhydride that have melt flow indices that are similar to those of materials used as polymer matrices have been studied. With these, the compatibilizing agents and polymers are expected to be miscible, and the compatibilized HDPE/LLDPE blend is expected to have a better affinity for the organic modifier of the lamellar clay.

Fig. 7.9 shows the XRD patterns and Fig. 7.10 shows the TEM micrographs of the HDPE/LLDPE/OMMT nanocomposite blend compatibilized with high-density polyethylene grafted with maleic anhydride (HDPE-g-MA).

With the addition of a compatibilizing agent (HDPE-g-MA), the diffraction peak referring to the crystallographic plane (0 0 1) undergoes a more significant shift for smaller values of 2θ, which corresponds to an increase in the interlamellar spacing in the clay mineral. This suggests that the polymer chains may have been intercalated between the lamellar silicate layers. As the lamellar clay content of the polymer matrix increases, an increase

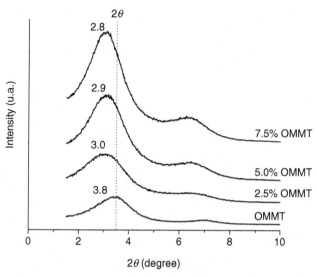

FIGURE 7.9 XRD patterns of OMMT and HDPE/LLDPE nanocomposite blends with different concentrations of lamellar clay (2.5, 5.0, and 7.5% by weight) and with the compatibilizing agent HDPE-*g*-MA.

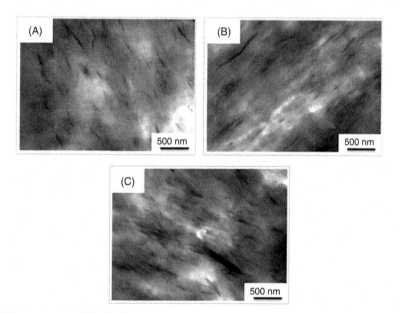

FIGURE 7.10 TEM micrographs of HDPE/LLDPE/OMMT nanocomposites with the compatibilizing agent HDPE-*g*-MA (A) with 2.5% OMMT by weight, (B) with 5.0% OMMT by weight, and (C) with 7.5% OMMT by weight.

of the basal spacing is observed: 2.96 nm for 2.5% OMMT, 3.04 nm for 5.0% OMMT, and 3.17 nm for 7.5% OMMT.

The compatibilizing agent completely modifies the morphology of the HDPE/LLDPE nanocomposites. The addition of HDPE-g-MA, which has a high viscosity, combines with the imposed shear during the extrusion process to ease the process of breaking down and reducing the size of the tactoids in the organophilic clay and, consequently, helps polymer chains intercalate between the silicate layers and the lamellar clay that is dispersed throughout the polymer matrix. The compatibilizing agent can reduce the size of the load particles due to thermodynamic factors (compatibilization due to the surfactant in organophilic lamellar clays) and kinetics (an increase in the viscosity due to the presence of clay nanoparticles). The morphology of these systems is created by intercalated lamellae; however, the presence of some exfoliated lamellae of organophilic clay and tactoids that have been significantly reduced in size compared to the morphologies of the nanocomposites shown in Fig. 7.8 are also noted.

In addition to the effects on the morphology, modifications to the mechanical and gas permeability properties are also observed. The effect that adding clay has on the elastic and flexural moduli are shown in Fig. 7.11.

The elastic modulus or Young's modulus is the ratio of stress to strain within the elastic limit, where the strain is completely reversible, proportional to the stress, and directly related to the stiffness of the material. This is a much-sought after property in studies involving polymer nanocomposites.

In general, nanocomposites have mechanical properties that are superior to those of HDPE/LLDPE blends. Comparing nanocomposites with lamellar clay contents of up to 5.0% by weight shows that nanocomposites without added compatibilizing agents present mechanical properties that are very close to those of nanocomposites compatibilized with HDPE-g-MA. This behavior, which was also observed by Spencer et al. [56] when they studied HDPE-g-MA/HDPE blends, is due to the lower crystallinity of the former with respect to the latter. However, when 7.5% lamellar clay by weight is added, the improved dispersion that results from the addition of the compatibilizing agent surpasses the properties of nanocomposites with large numbers of tactoids.

The three-point bending test measures the maximum flexural strength of a material. At the loading point, the top surface of the specimen is compressed, while the bottom surface is under tension. Because the specimen is subjected to compressive and tensile stresses when it is bent, the magnitude of its flexural strength is greater than its tensile fracture strength. Good performance of a nanocomposite in the three-point bending test is related to good dispersion and distribution of the lamellar clay in the polymer matrix. The values of the flexural modulus obtained are smaller than those of the elastic modulus; however, a significant increase in the flexural modulus is observed when organophilic lamellar clay is added; this is the result of the good interface created. In addition, the good distribution of stress observed in the nanocomposites developed, leads to an increase of 100% in the flexural modulus of the nanocomposite compatibilized with HDPE-g-MA containing 7.5% lamellar clay by weight.

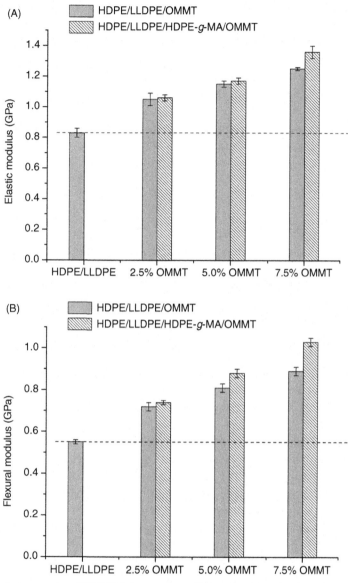

FIGURE 7.11 Mechanical properties of HDPE/LLDPE nanocomposite blends with and without compatibilizing agents. (A) The elastic modulus obtained from the uniaxial tensile test and (B) and the flexural modulus obtained from the three-point bending test.

In addition to decreasing the mechanical properties, adding lamellar clay can decrease the gas permeability of the polymer matrix.

The ability of a polymer film or membrane to transport gas and vapor molecules depends strongly on the molecular structure of the polymer matrix and is very sensitive to changes in it [61]. The addition of clay to a polymer system leads to changes in its transport properties in the vast majority of cases. The permeability to small molecules of a nano-composite that contains clay can be significantly less than that of a pure polymer because of the large aspect ratio and the impermeability of clay lamellae, which are responsible for the effects of tortuosity when they are dispersed in a polymer matrix [62–64]. Several theoretical models have been proposed in the literature to express the tortuosity factor as a function of the form, orientation, disperse state, and volume fraction of impermeable particles.

Nielsen [65] developed a permeability model for composites that includes the effect of tortuosity; for measurements in nanocomposites the following expression is used:

$$\frac{P}{P_0} = \frac{1-\phi_{NC}}{\tau} = \frac{(1-\phi_{NC})}{\left(1+\frac{L}{2W}\phi_{NC}\right)}, \tag{7.1}$$

where P and P_0 are the permeabilities of the nanocomposite and the pure polymer, respectively. The tortuosity factor $\left[\tau = (d'/d) = 1 + (L/2W)\phi_{NC}\right]$ is defined as the ratio of the distance or real path (d') that the penetrator must cross to the shorter distance (d) that it would cross if there were no silicate in the layer. It is expressed in terms of the length (L) and thickness (W) of the clay layer and the volumetric fraction of the load (ϕ_{NC}). Loads with high-form factors (L/W) lead to high tortuosities and, consequently, the permeability of such a nanocomposite is less than that of a pure polymer. Fig. 7.12 illustrates a model of the path of a penetrator diffusing through a nanocomposite.

The model proposed by Nielsen is limited because clay layers are assumed to be oriented perfectly transverse to the direction of permeation and of the same size, the diffusivity of the matrix is not changed by the presence of the clay, and preferential transport does not occur at polymer/clay interfaces.

Examples of the effects of the addition of lamellar clay on the permeability of HDPE/LLDPE nanocomposite blends to oxygen are presented in Table 7.1.

It is anticipated that the addition of an inorganic load will increase the mean free path for diffusion and decrease the permeability coefficient. In HDPE/LLDPE nanocomposite blends without compatibilizers, the agglomeration of lamellar clay and the poor dispersion of the inorganic load in the matrix help increase the number of microvoids in the polymer matrix and, therefore, aid diffusion. The oxygen permeability coefficient increases, in

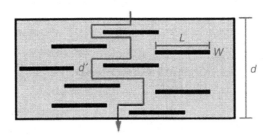

FIGURE 7.12 Model of the path of a penetrator diffusing through a nanocomposite.

Table 7.1 Oxygen Permeability of the Films Obtained Using An OX-TRAN 2/21 T Oxygen Transmission Rate System and the Degree of Crystallinity (%) Obtained from the Cooling Curve of Differential Scanning Calorimetry (DSC) Tests

Sample	P_{O_2} (Barrier)	Crystallinity (%)
HDPE/LLDPE	1.44 ± 0.22	65
HDPE/LLDPE/OMMT (2.5%)	2.28 ± 0.05	62
HDPE/LLDPE/OMMT (5.0%)	1.83 ± 0.41	63
HDPE/LLDPE/OMMT (7.5%)	0.98 ± 0.05	66
HDPE/LLDPE/OMMT/HDPE-*g*-MA (2.5%)	1.18 ± 0.17	66
HDPE/LLDPE/OMMT/HDPE-*g*-MA (5.0%)	1.16 ± 0.36	61
HDPE/LLDPE/OMMT/HDPE-*g*-MA (7.5%)	0.98 ± 0.05	57

comparison with polymer blends, for concentrations of up to 5% OMMT by weight. However, with 7.5% lamellar clay, the permeability coefficient decreases because the degree of crystallinity increases when the system's OMMT concentration increases. Therefore, in addition to the tortuosity created by the clay tactoids, there is a larger fraction of crystalline phase that contributes to the increase in the diffusion path, which decreases the effective coefficient of permeability to oxygen.

The addition of a high concentration of a compatibilizing agent to a nanocomposite can have some effects, such as a decrease in the density of the crystalline domain of the blend and an increase in the amorphous character and fluctuations of the free volume. However, it aids the process of breaking down tactoids and dispersing the clay lamellae in the polymer matrix, which increases the gas diffusion path, as observed for HDPE/LLDPE nanocomposite blends compatibilized with HDPE-*g*-MA. In addition, the oxygen permeability coefficient decreases when the OMMT concentration increases. The morphological changes due to the addition of the compatibilizing agent allow the tortuosity to increase in response to the intercalated and well-dispersed state of the clay lamellae in the polymer matrix despite lower barrier properties intrinsically presented by HDPE-*g*-MA. This decrease in the coefficient of permeability is connected to good inorganic load dispersion, high wettability by the matrix, and strong interactions at the interface that decrease the number of microvoids, which could help the diffusion process.

7.8 Final Considerations

The development of new materials that provide the properties demanded by applications, especially those in the automotive, electrical power, and packaging industries, has attracted the interest of large industries and research groups. Among the different techniques used to obtain these properties, one can highlight the preparation of nanocomposites with polymer matrices by adding lamellar clays with at least one nanoscale dimension.

Two large groups of polymer matrices can be used to prepare these nanocomposites, polar matrices, and nonpolar matrices, and it is possible to combine materials from the two groups. The nanocomposite of PA6 and organophilic lamellar clay is an excellent example of how the polarity of the polymer matrix affects the interactions between the polymer and the clay, in addition to modifying the morphology, and the thermal and mechanical properties of the nanocomposite, which can be compared to those of pure PA6. For nonpolar matrices, such as polyolefins, obtaining exfoliated nanocomposites is more difficult because polyolefins are hydrophobic and there is a lack of appropriate interactions with the polar surface of aluminosilicate clay. The compatibility of inorganic loads with polymer matrices is a big challenge in the preparation of nanocomposites that can be improved by chemically modifying the surfaces of the particles of the components and by adding compatibilizing agents between the surfactant and the polymer matrix. The strengthening promoted by the clay is determined by the efficiency of the compatibilization and the restricted mobility of the polymer chains that are in contact with the clay lamellae.

References

[1] M. Alexandre, P. Dubois, Polymer-layered silicate nanocomposites: preparation, properties and uses of a new class of materials, Mater. Sci. Eng. 28 (1–2) (2000) 1–63.

[2] F. Hussain, M. Hojjat, M. Okamoto, R.E. Gorga, Review article: polymer-matrix nanocomposites, processing, manufacturing, and application: an overview, J. Compos. Mater. 40 (17) (2006) 1511–1575.

[3] C.G. Martins, N.M. Larocca, D.R. Paul, L.A. Pessan, Nanocomposites formed from polypropylene/EVA blends, Polymer 50 (7) (2009) 1743–1754.

[4] S.E. Vidotti, A.C. Chinellato, G.H. Huo, L.A. Pessan, Preparation of poly(ethylene terephthalate)/organoclay nanocomposites using a polyester ionomer as a compatibilizer, J. Polym. Sci. Part B 45 (22) (2007) 3084–3091.

[5] A.D. Oliveira, N.M. Larocca, D.R. Paul, L.A. Pessan, Effects of mixing protocol on the performance of nanocomposites based on polyamide 6/acrylonitrile-butadiene-styrene blends, Polym. Eng. Sci. 52 (9) (2012) 3084–3091.

[6] F.R. Passador, A.C. Ruvolo-Filho, L.A. Pessan, Effect of blending protocol on the rheological properties and morphology of HDPE/LLDPE blend-based nanocomposites, Int. Polym. Proc. 27 (3) (2012) 378–385.

[7] M.U. Jurczyk, A. Kumar, S. Srinivasan, E. Stefanakos, Polyaniline-based nanocomposite materials for hydrogen storage, Int. J. Hydrogen Energy 32 (8) (2007) 1010–1015.

[8] G. Terife, K.A. Narh, Properties of carbon nanotube reinforced linear low density polyethylene nanocomposites fabricated by cryogenic ball-milling, Polym. Comp. 32 (12) (2011) 2101–2109.

[9] C.M. Chang, K.Y. Hsu, Y.L. Liu, Matrix-polymer-functionalized multiwalled carbon nanotubes as a highly efficient toughening agent for matrix polymers, J. Polym. Sci. B 50 (16) (2012) 1151–1155.

[10] M. Rahimi, I. Iriarte-Carretero, A. Ghanbari, M.C. Bohm, F. Muller-Plathe, Mechanical behavior and interphase structure in a silica-polystyrene nanocomposite under uniaxial deformation, Nanotechnology 23 (30) (2012) 305702.

[11] D. Eiras, L.A. Pessan, Mechanical properties of polypropylene/calcium carbonate nanocomposites, Mater. Res. Ibero Am. J. Mater. 12 (4) (2009) 517–522.

[12] A. Olad, R. Nosrati, Preparation, characterization, and photocatalytic activity of polyaniline/ZnO nanocomposite, Res. Chem. Intermediat. 38 (2) (2012) 323–336.

[13] C.E. Powell, G.W. Beall, Physical properties of polymer/clay nanocomposites, Curr. Opin. Solid St. M. 10 (2006) 73–80.

[14] S.C. Tjong, Structural and mechanical properties of polymer nanocomposites, Mater. Sci. Eng. 53 (3–4) (2006) 73–197.

[15] T.J. Pinnavaia, G.W. Beall, Polymer – Clay Nanocomposites, John Wiley & Sons, Inc., New York, (2000).

[16] R.A. Vaia, R.K. Teukolsky, E.P. Giannelis, Interlayer structure and molecular environment of alkylammonium layered silicates, Chem. Mater. 6 (7) (1994) 1017–1022.

[17] M. Alexandre, P. Dubois, T. Sun, J.M. Garces, R. Jérôme, Polyethylene-layered silicate nanocomposites prepared by the polimerization-filling technique: synthesis and mechanical properties, Polymer 43 (8) (2002) 2123–2132.

[18] S.S. Ray, M. Okamoto, Polymer/layered silicate nanocomposites: a review from preparation to processing, Prog. Polym. Sci. 28 (11) (2003) 1539–1641.

[19] T.D. Fornes, P.J. Yoon, H. Keskkula, D.R. Paul, Nylon 6 nanocomposites: the effect of matrix molecular weight, Polymer 42 (25) (2001) 9929–9940.

[20] H. Fischer, Polymer nanocomposites: from fundamental research to specific applications, Mater. Sci. Eng. C 23 (6–8) (2003) 763–772.

[21] Y. Wang, F.B. Chen, K.C. Wu, Effect of the molecular weight of maleated polypropylenes on the melt compounding of polypropylene/organoclay nanocomposites, J. Appl. Polym. Sci. 97 (4) (2005) 1667–1680.

[22] R.A. Vaia, E.P. Giannelis, Lattice model of polymer melt intercalation in organically-modified layered silicates, Macromolecules 30 (25) (1997) 7990–7999.

[23] S. Hotta, D.R. Paul, Nanocomposites formed from linear low density polyethylene and organoclays, Polymer 45 (2) (2004) 7639–7654.

[24] J. Heinemann, P. Reichert, R. Thomann, R. Mulhaupt, Polyolefin nanocomposites formed by melt compounding and transition metal catalyzed ethene homo- and copolymerization in the presence of layered silicates, Macromol. Rapid Comm. 20 (8) (1999) 423–430.

[25] Y. Kojima, A. Usuki, M. Kawasumi, A. Okada, T. Kurauchi, O. Kamigaito, K. Kaji, Fine structure of nylon 6-clay hybrid, J. Polym. Sci. Part B 32 (4) (1994) 625–630.

[26] T.D. Fornes, D.R. Paul, Modeling properties of nylon 6/clay nanocomposites using composite theories, Polymer 44 (17) (2003) 4993–5013.

[27] E. Picard, A. Vermogen, J.F. Gérard, E. Espuche, Barrier properties of nylon 6-montmorillonite nanocomposite membranes prepared by melt blending: influence of the clay content and dispersion state. Consequences on modeling, J. Membrane Sci. 292 (2007) 133–144.

[28] T.D. Fornes, D.L. Hunter, D.R. Paul, Effect of sodium montmorillonite source on nylon 6/clay nanocomposites, Polymer 45 (7) (2004) 2321–2331.

[29] F. Chavarria, R.K. Shah, D.L. Hunter, D.R. Paul, Effect of melt processing conditions on the morphology and properties of nylon 6 nanocomposites, Polym. Eng. Sci. 47 (11) (2007) 1847–1864.

[30] M.A. Souza, N.M. Larocca, E.M. Araújo, L.A. Pessan, Preparation and characterization of nanocomposites of polyamide 6/Brazilian clay with different organic modifiers, Mater. Sci. Forum 570 (2008) 18–23.

[31] R.A. Paz, A.M.D. Leite, E.M. Araújo, T.J.A. Mélo, R. Barbosa, E.N. Ito, Nanocomposites de Poliamida 6/Argila Organofílica: Efeito do Peso Molecular da Matriz na Estrutura e Propriedades Mecânicas e Termomecânicas, Polímeros: Ciência e Tecnologia 18 (4) (2008) 341–347.

[32] S.E. Vidotti, A.C. Chinellato, L.F. Boesel, L.A. Pessan, Poly (ethylene terephthalate)-organoclay nanocomposites: morphological, thermal and barrier properties, J. Metastable Nanocryst. Mater. 22 (2004) 47–54.

[33] L.F. Boesel, L.A. Pessan, Poly(ethylene terephthalate)-organoclay nanocomposites: morphological characterization, J. Metastable Nanocryst. Mater. 14 (2002) 89–94.

[34] Z. Ke, Y.P. Bai, Improve the gas barrier property of PET film with montmorillonite by in situ interlayer polymerization, Mater. Lett. 59 (27) (2005) 3348–3351.

[35] Y.M. Wang, J.P. Gao, Y.Q. Ma, U.S. Agarwal, Study on mechanical properties, thermal stability and crystallization behavior of PET/MMT nanocomposites, Compos. Part B Eng. 37 (6) (2006) 399–407.

[36] M.C. Costache, M.J. Heidecker, E. Manias, C.A. Wilkie, Preparation and characterization of poly(ethylene terephthalate)/clay nanocomposites by melt blending using thermally stable surfactants, Polymer Adv. Tech. 17 (9–10) (2006) 764–771.

[37] B. Wen, X.F. Xu, X.W. Gao, Y.F. Ding, F. Wang, S.M. Zhang, M.S. Yang, Highly exfoliated poly(ethylene terephthalate)/clay nanocomposites via melt compounding: effects of silane grafting, Polym. Plast. Technol. Eng. 50 (4) (2011) 362–371.

[38] T. Peprnicek, J. Duchet, L. Kovarova, J. Malac, J.F. Gerard, J. Simonik, Poly(vinyl chloride)/clay nanocomposites: X-ray diffraction, thermal and rheological behavior, Polym. Degrad. Stabil. 91 (8) (2006) 1855–1860.

[39] M.A. Souza, A. Rodolfo Jr., L.A. Pessan, Nanocomposites de poli (cloreto de vinila) (PVC)/argila organofílica, Polímeros: Ciência e Tecnologia 16 (4) (2006) 257–262.

[40] W.H. Awad, G. Beyer, D. Benderly, W.L. Lido, P. Songtipya, M.D. Jimines-Gasco, E. Manias, Materials properties of nanoclay PVC composites, Polymer 50 (8) (2009) 1857–1867.

[41] J. Pagacz, K. Pielichowski, Preparation and characterization of PVC/montmorillonite nanocomposites: a review, J. Vinyl Addit. Techn. 15 (2) (2009) 61–76.

[42] M. Mondragon, S. Sanches-Valdes, M.E. Sanches-Espindola, J.E. Rivera-Lopez, Morphology, mechanical properties, and thermal stability of rigid PVC/clay nanocomposites, Polym. Eng. Sci. 51 (4) (2011) 641–646.

[43] A.J. Hsieh, P. Moy, F.L. Beter, P. Madison, E. Napadensky, J.X. Ren, R. Krishnamoorti, Mechanical response and rheological properties of polycarbonate layered-silicate nanocomposites, Polym. Eng. Sci. 44 (5) (2004) 825–837.

[44] F.J. Carrion, A. Arribas, M.D. Bermudez, A. Guillamon, Physical and tribological properties of a new polycarbonate-organoclay nanocomposite, Eur. Polym. J. 44 (4) (2008) 968–977.

[45] J. Feng, J.W. Hao, J.X. Du, R.J. Yang, Effects of organoclay modifiers o the flammability, thermal and mechanical properties of polycarbonate nanocomposites filled with a phosphate and organoclays, Polym. Degrad. Stab. 97 (1) (2012) 108–117.

[46] J.J. Luo, I.M. Daniel, Characterization and modeling of mechanical behavior of polymer/clay nanocomposites, Compos. Sci. Technol. 63 (2003) 1607–1616.

[47] F.R. Passador, A. Rodolfo Jr., L.A. Pessan, Estado de mistura e dispersão da fase borrachosa em blendas PVC/NBR, Polímeros: Ciência e Tecnologia 16 (3) (2006) 174–181.

[48] L.A. Utracki, Polymer Alloys and Blends: Thermodynamics and Rheology, Hanser Publishers, New York, (1989).

[49] K. Chrissopoulou, I. Altintzi, S.H. Anastasiadis, E.P. Giannelis, M. Pitsikalis, N. Hadjichristidis, N. Theophilou, Controlling the miscibility of polyethylene/layered silicate nanocomposites by altering the polymer/surface interactions, Polymer 46 (26) (2005) 12440–12451.

[50] E. Jacquelot, E. Espuche, J. –F. Gérard, J. Duchet, P. Mazabraud, Morphology and gas barrier properties of polyethylene-based nanocomposites, J. Polym. Sci. B 44 (2) (2006) 431–440.

[51] A. Durmus, M. Woo, A. Kaşgöz, C.W. Macosko, M. Tsapatsis, Intercalated linear low density polyethylene (LLDPE)/clay nanocomposites prepared with oxidized polyethylene as a new type compatibilizer: structural mechanical and barrier properties, Eur. Polym. J. 43 (9) (2007) 3737–3749.

[52] M. Zanetti, L. Costa, Preparation and combustion of polymer/layered silicate nanocomposites based upon PE and EVA, Polymer 45 (13) (2004) 4367–4373.

[53] T.G. Gopakumar, J.A. Lee, M. Kontopoulou, J.S. Parent, Influence of clay exfoliation on the physical properties of montmorillonite/polyethylene composites, Polymer 43 (20) (2002) 5483–5491.

[54] J. Morawiec, A. Pawlak, M. Slouf, A. Galeski, E. Piorkowska, N. Krasnikowa, Preparation and properties of compatibilized LDPE/organo-modified montmorillonite nanocomposites, Eur. Polym. J. 41 (5) (2005) 1115–1122.

[55] S.H. Ryu, Y.W. Chang, Factors affecting the dispersion of montmorillonite in LLDPE nanocomposite, Polym. Bull. 55 (5) (2005) 385–392.

[56] M.W. Spencer, L. Cui, Y. Yoo, D.R. Paul, Morphology and properties of nanocomposites based on HDPE/HDPE-*g*-MA blends, Polymer 51 (5) (2010) 1056–1070.

[57] H.A. Patel, G.V. Joshi, R.R. Pawar, H.C. Bajaj, R.V. Jasra, Mechanical and thermal properties of polypropylene nanocomposite using organically modified Indian bentonite, Polym. Comp. 31 (2010) 399–404.

[58] M. Okamoto, P.H. Nam, P. Manti, T. Kotaka, N. Hasegawa, A. Usukli, A house of cards structure in polypropylene/clay nanocomposites under elongational flow, Nano Lett. 1 (2001) 295–298.

[59] F. Bellucci, A. Terenzi, A. Leuteritz, D. Pospiech, A. Franche, G. Traverso, G. Camino, Intercalation degree in PP/organoclay nanocomposites: role of surfactant structure, Polymer Adv. Tech. 19 (6) (2008) 547–555.

[60] Y. Dong, D. Bhattacharyya, Effects of clay type, clay/compatibiliser content and matrix viscosity on the mechanical properties of polypropylene/organoclay nanocomposites, Compos. Part A Appl. Sci. Manuf. 39 (7) (2008) 1177–1191.

[61] J. Comyn, Polymer Permeability, Elsevier Applied Science Publishers, London, (1998).

[62] J.H. Lee, D. Jung, C.E. Hong, K.Y. Rhee, S.G. Advani, Properties of polyethylene-layered silicate nanocomposites prepared by melt intercalation with PP-*g*-MA compatibilizer, Compos. Sci. Technol. 65 (13) (2005) 1996–2002.

[63] M. Kato, N. Okamoto, A. Hasegawa, A. Tsukigase, A. Usuki, Preparation and properties of polyethylene-clay hybrids, Polym. Eng. Sci. 43 (6) (2003) 1312–1316.

[64] A.R. Morales, C.V.M. Cruz, L. Peres, E.N. Ito, Nanocompositos de PEAD/PEBDL—Avaliação da esfoliação da argila organofílica pela aplicação do modelo de Nielsen e das propriedades mecânicas, ópticas e permeabilidade, Polímeros: Ciência e Tecnologia 20 (1) (2010) 39–45.

[65] L.E. Nielsen, Models for the permeability of filled polymers systems, J. Macromol. Sci. Chem. 1 (5) (1967) 929–942.

8

Mathematical Fundamentals of Nanotechnology

R. Marchiori

INTERDISCIPLINARY DEPARTMENT OF SCIENCE AND TECHNOLOGY, FEDERAL UNIVERSITY OF RONDÔNIA, ARIQUEMES, RONDÔNIA, BRAZIL

CHAPTER OUTLINE

8.1 Introduction

Classical mechanics, with its deterministic description of the behavior of matter, does not provide the correct mathematical structure to describe the physics and molecular interactions of matter at the atomic level. This representation is an average of a sufficiently large number of molecular interactions that is valid at the macroscopic scale. At the microscopic level, the laws of physics that describe the interactions between atoms, molecules, or nanoparticles can no longer be described by the deterministic laws of classical mechanics because they require a more appropriate description, which is offered by the mathematical formalism of quantum mechanics. This formalism modifies the deterministic perspective from the classical description to a probabilistic perspective of the behavior of matter. In the mathematical construction of quantum mechanics, time and space are interconnected, which is different from classical mechanics, in which they are independent.

In the space-time universe of quantum mechanics, this interdependence between the spatial and temporal dimensions has drastic consequences in the interpretation of reality at the atomic level [1,2]. The laws of quantum physics deviate from those that we are accustomed to observing, which thereby conflicts with intuitive comprehension and provides an extremely unconventional description of reality [3].

Several theories consider that quantum mechanics is also valid at a macroscopic scale, but the direct interaction of the observer with the object, which occurs at this scale, causes the "collapse" of the possible energetic configurations of the object into a specifically determined configuration, which represents the highest probability configuration. Thus, any object, such as the Moon, is visible and exists where we see it because we are looking at it. If that was not the case, the speculations of some quantum physicists would lead to the determination that it is not there and that it occupies all possible energetic configurations with some probability; that is, any position within space-time in the universe!

The essential characteristic in the advent of nanotechnology around 1985 was the development of knowledge and technology that allowed us to manipulate and analyze physical phenomena and material properties at the atomic scale, which made it possible to access knowledge about material properties based on interactions between particles at the nanometric scale. At this scale, which is the order of magnitude of interatomic distances, the energy levels that are accessible to a physical system are no longer continuous as in classical mechanics, in which energy can assume any value in a continuous manner. At the nanometric scale, wave–particle duality is manifest, which causes the discretization of energy. In other words, energy can acquire only specific, discrete values to respect the boundary conditions that emerge from the wave nature of matter; these conditions exclude certain configurations of the system and, consequently, certain relative energy values because the system is defined by atomic or molecular interactions. Thus, the mathematical formalism of quantum mechanics is needed to calculate the energy of the system at this scale.

At the nanometric scale, the classical formalism becomes an approximation that loses validity. To handle nanomaterials such as carbon nanotubes or graphene, which are currently the nanostructured materials with the greatest potential for use, quantum mechanics provides the mathematical background that describes the energy levels that are accessible to these materials.

When a material has properties due to its nanometric-scale structure and morphology, quantum theory is the only theory that can appropriately describe these properties, which indicates that the wave behavior of elementary particles becomes important at this scale. This behavior is described directly by the de Broglie equation, $\lambda = h/p$, which shows the dual wave–particle character of matter by relating the momentum p, which is a classical property that is directly proportional to the mass and velocity of the particle, to the wave property of matter, which is given by the corresponding wavelength λ, and Planck's constant h is the proportionality factor. The wavelength λ only assumes detectable values for extremely small masses, such as in the case of elementary particles and nanostructured materials.

Boundary conditions can be used to maintain a physical system in equilibrium at the nanometric scale; these conditions are related to the characteristics of the material and determine its electrical, optical, and other properties.

Therefore, quantum mechanics must be used in nanotechnology and can be summarized in the following essential points:

- The nanometric scale, which is on the order of the size of an atom, involves the wave behavior of matter and requires the use of a wave function in the Schrödinger equation to describe the movement of a particle at this scale.
- The interaction of particles with matter at the nanoscale, such as the movement of an electron in electrical conduction by a nanostructured material, is limited by the boundary conditions. These conditions limit particle movement in the crystal lattice of the material, which forces the electrons into trajectories that are defined by the periodic interaction potential of the Bravais lattice of the material. This determines the discretization of the energy, whose possible levels are calculated by the mathematical formalism of quantum mechanics.

8.2 Classical Mechanics

8.2.1 The Classical Formalism at the Nanometric Scale

Particle dynamics can be represented from the classical formalism that describes the dynamics of a physical system. However, the particles of microscopic systems follow wave movement laws; that is, at the nanometric scale, elementary particles behave like waves, and their description requires the introduction of a wave function $\psi(r,t)$ to describe their movement.

In general, the function $\psi(q)$ is a complex function, with $q = (q_1, q_2, \ldots q_n)$ being the generalized coordinates that represent the "state" of a dynamic system with n degrees of freedom. Due to the objectives of this chapter, a more extensive formalism of a quantum system will not be addressed. This chapter will only cover the aspects that are necessary to understand the relationship between quantum mechanics and nanotechnology.

In general, the evolution of a system, such as the simplest case of a particle, must be consistent with the mechanics equation [4,5]:

$$E = p^2/2m + V \tag{8.1}$$

which represents the energy E of a particle of mass m and momentum p as a function of the kinetic energy $p^2/2m$ and the potential energy V. The quantum mechanics equation that corresponds to Eq. 8.1 and involves the wave function will also have to satisfy the postulates of Einstein (1905), de Broglie (1921), and de Jammer (1966), (1974) [4,6–8]:

$$E = hv = \hbar\omega \quad \text{and} \quad p = \hbar k \tag{8.2}$$

and be valid for wave movement, which can be written in the form:

$$\hbar\omega = \hbar^2 k^2/2m + V(x,t) \tag{8.3}$$

where $\hbar = h/2\pi$.

A general mathematical procedure is available to handle the behavior of any microscopic system of particles: the Schrödinger theory of quantum mechanics [9–14].

8.3 Quantum Mechanics

The Schrödinger theory is based, among others phenomenological processes, on the laws of wave movement to express the space-time position of elementary particles in terms of probability, which the particles of any microscopic system obey; when this description is required, the system is called a quantum system.

A wave function $\Psi(r,t)$ for the general three-dimensional case of propagation with time t in the direction r can be represented using a sine wave [10]:

$$\Psi(r,t) = \sin 2\pi\left(\frac{r}{\lambda} - v \cdot t\right) \tag{8.4}$$

where $\Psi(r,t)$ is the wave function, λ is the wavelength, and v is the oscillation frequency. To write the wave function more easily, the following variables are introduced: $k = 2\pi/\lambda$ and $\omega = 2\pi v$, where k is the angular wave number, and ω is the angular oscillation frequency. The wave function then becomes:

$$\Psi(r,t) = sen(kr - \omega t) \tag{8.5}$$

The Schrödinger equation must also satisfy the linearity in $\Psi(r,t)$, which guarantees that wave functions can produce the constructive and destructive interferences that are characteristic of wave behavior. Remember that because the Schrödinger equation [9] is consistent with Eq. 8.1, it is not valid when it is applied to particles with relativistic velocities. In the case of a free particle, a potential function must satisfy the condition:

$$V(r,t) = V_0 = \text{const.} \tag{8.6}$$

This guarantees that no force is applied to the particle because the force F is equal to:

$$F = -\frac{\partial V(r,t)}{\partial x} = 0. \tag{8.7}$$

These conditions and concepts form the basis of the description of particles in matter and, in particular, the behavior of elementary particles, such as electrons in the crystal lattice of a material.

8.3.1 The Energy of Quantum Systems—Stationary Conditions

To describe the properties of nanostructured materials, it is necessary to obtain the eigenvalues of the energy of the system [9–14]. The Hamiltonian formulation of the time-independent Schrödinger equation is used to find the eigenvalues for the energy of a quantum system, as is shown in the example with carbon nanotubes at the end of this chapter. The majority of atomic and molecular systems show time-independent conditions; this means that the potential energy does not depend explicitly on time, or $V(r,t) \equiv V(r)$. Under these conditions, the wave function in Eq. 8.4 can be written by separating the variables:

$$\Psi(r,t) = \psi(r) \cdot \varphi(t) \qquad (8.8)$$

Thus, the Hamiltonian formulation of the time-independent Schrödinger equation is [10]:

$$\hat{H}\psi(r) \equiv H(\hat{q}^0, \hat{p}^0)\psi(x) = E\psi(r) \qquad (8.9)$$

where $\hat{H}$ is the Hamiltonian operator that is used in the quantum formalism, and $\hat{q}^0$ and $\hat{p}^0$ are the sets of operators that express the position and velocity of the particle, respectively. Acceptable solutions for this equation exist only for specific energy values E, which are called the eigenvalues of the potential energy $V(r)$ that is generated by the crystal lattice of the system. An eigenfunction $\psi(r)$ for the energy corresponds to each eigenvalue; using the Bohr interpretation, these eigenfunctions are wave functions whose energy is the energy of the stationary states of the system. The quantization of energy [10,11,13] emerges naturally in the Schrödinger theory due to the wave nature of matter. Only certain discrete energy values, which are obtained from the time-independent Schrödinger equation, permit solutions for the wave function. Considering the discrete set of eigenvalues E_i of the system, the Schrödinger equation can be written as [10]:

$$\hat{H}\psi_i(r) = E_i\psi_i(r) \qquad (8.10)$$

This equation allows one to obtain the eigenvalues for the energy of the system and is used for stable molecular aggregates at the nanometric scale. The analysis of the solutions to this equation provides information about the mechanical, thermal, and electrical properties of these nanomolecules, as is shown in the following paragraphs.

8.3.2 Periodic Structure of a Crystal Lattice—The Bloch Theorem

The main nanomolecules that are studied and used in nanotechnology applications have periodic structures due to their crystal lattice; therefore, the interaction potential of the electrons in the molecule is of the type [15,16]:

$$V(\vec{r}) = V(\vec{r} + \vec{R}) \qquad (8.11)$$

where $\vec{R}$ is the vector that represents the interatomic distance in the molecule. The vector notation that is used here is necessary to describe the crystal lattice of these nanomolecules. The electrons of these molecules occupy discrete energy states that are defined through quantum mechanics by the presence of a periodic potential that is related to the crystal structure of the nanomolecule. To find the eigenvalues of the energy of the wave functions of the electrons, appropriate approximations are used to simplify the problem without significantly altering the description of the system. This allows these molecular systems to be studied using various mathematical modeling techniques, such as the "tight binding" (TB) approximation. When this approximation is applied to the method known as the "linear combination of atomic orbitals" (LCAO), the monoelectronic states in a crystal approximately maintain their atomic behavior because the interactions between the electronic states of the neighboring atoms are relatively small and thus only represent system perturbations. In this approximation, the Hamiltonian operator H is equal to [17]:

$$H = H_{\text{atom}} + H_{\text{crystal}} \cong H_{\text{atom}} \tag{8.12}$$

This operator results from the interaction of the atom with the electron (H_{atom}) and the effect due to the interaction with the crystal (H_{crystal}), which is considered to be a perturbation; thus, the eigenfunction $\psi(\vec{r})$ of the electron becomes the eigenfunction of H_{atom} with the eigenvalue E_0. The reason for this is that these valence electrons are found in deep potential wells that are generated by the crystal lattice of the molecule and do not have the properties of free electrons. In other words, the potential energy that acts on the electrons due to the interaction with other electrons is much smaller than the atomic energy potential in the crystal lattice (Bravais lattice). Therefore, the Hamiltonian operator assumes the following form [18]:

$$\hat{H} \propto \sum_i \hat{T}_i + \sum_R \hat{V}(\vec{r} - \vec{R}) \tag{8.13}$$

where $\hat{T}_i$ is the kinetic energy operator for all of the electrons e_i, and $\hat{V}(\vec{r} - \vec{R})$ is the complete potential of the crystal lattice at the atomic site $\vec{R}$. The wave function $\psi(r)$ of the electron in the crystal lattice for a defined value of the wave vector $\vec{k}$ is:

$$\psi_k(r) = \sum_{\vec{R}} C_{kR} \varphi(\vec{r} - \vec{R}) \tag{8.14}$$

in which the positions $\vec{R}$ of the sites that are occupied by the atoms determine the periodicity of the Bravais lattice. The wave vector $\vec{k}$ has the direction and orientation of the wave propagation and the magnitude $/k/ = 1/\lambda$. The wave function $\varphi(\vec{r} - \vec{R})$ describes the electron that is close to the atomic site in $\vec{R}$. The sum is extended to all of the sites of the crystal lattice, so this is a LCAO. The coefficient C_{kR} is obtained from the Bloch theorem, which allows the expression of the eigenfunction of the electron in the periodic potential of the crystal lattice. The probability density of the wave function $\varphi(\vec{r})$ of the electron in the periodic potential is also periodic and is equal to:

$$\left| \varphi(\vec{r} + \vec{R}) \right|^2 = \left| \varphi(\vec{r}) \right|^2 \tag{8.15}$$

Using the normalization of the electronic distribution in the unit cell, it is equal to:

$$\int \left| \varphi(\vec{r} + \vec{R}) \right|^2 dV = \int \left| \varphi(\vec{r}) \right|^2 dV = 1 \tag{8.16}$$

Rewriting Eq. 8.16 yields the Bloch theorem [17]:

$$\varphi_k(\vec{r} + \vec{R}) = e^{ikR} \varphi_k(\vec{r}) \tag{8.17}$$

This equation represents the boundary conditions for solutions to the Schrödinger equation for a periodic potential. In this case, the eigenfunction of the electron in the Hamiltonian can be written in the form of a plane wave e^{ikR} with the periodicity of the Bravais lattice multiplied by the wave function $\varphi_k(\vec{r})$. In Eq. 8.14 for the wave function of the electron in the crystal lattice, the coefficients C_{kR} can be written as:

$$C_{kR} = N^{-1/2} e^{ikR} \tag{8.18}$$

where N is the number of atoms in the crystal. This term is used because the crystal is finite, which determines the finite number of translation vectors of the Bravais lattice, which consists of $N = N_x N_y N_z$ unit cells.

The normalization condition of the wave function in the expression of the density in the occupied space determines the application of the square root of the number of atoms N in the crystal (Eq. 8.18). Thus, Eq. 8.18 becomes:

$$\Psi_k(\vec{r}) = N^{-1/2} \sum R\, e^{ikR} \varphi(\vec{r} - \vec{R}) \tag{8.19}$$

8.3.3 Crystal Lattice and Reciprocal Lattice

The crystalline structure of a material can be accurately represented by considering only the electrons around it, which allows the formation of an indirect microscopic image of the material. Because electrons behave as waves, they can suffer reflections due to the periodicity of the crystal lattice of the material; these reflections are called Bragg reflections. The discussion of Bragg reflections leads to the notions of a "reciprocal lattice" or "reciprocal space" and "Brillouin zones." The study of X-ray diffractions that are generated by the crystal lattices of materials permits the acquisition of crystal lattice parameters from the diffraction angles θ, which form a spectrum of lines that correspond to the constructive interference of the incident rays. The reflection of the incident X-rays that is defined by Bragg's law is shown in Fig. 8.1 [19,20]:

$$2d\sin\theta = n\lambda \tag{8.20}$$

where d is the distance between two parallel atomic planes, n is a positive whole number, and λ is the wavelength of the incident electromagnetic wave. Bragg's law provides the conditions for the occurrence of constructive interference of the waves that are scattered by point charges at points in the lattice, which permits the calculation of the lattice

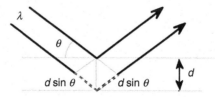

FIGURE 8.1 Reflection of X-ray radiation that is incident on the atomic planes of a material that are separated by a distance *d*.

parameters, or the spacing d between the crystal planes as a function of the diffraction angles θ. If the path difference $2d \sin\theta$ between two parallel rays is a multiple of the wavelength of the incident rays, the interference will be constructive.

To determine the intensity of each scattering that is produced by a spatial distribution of electrons within each cell of the crystal lattice, it is necessary to consider the numerical density of the electrons $n(r)$ because the intensity of the incident light that is scattered by a volume element is proportional to this density. The electronic density is naturally periodic for the translation operation that is defined by the vector $\vec{T}$ in the crystal lattice of the material. A diffraction figure of a crystal can be considered to be a representation of the reciprocal lattice of the crystal in contrast to the direct image of the real crystal lattice. Each crystal structure has two lattices: the crystal lattice and the reciprocal lattice. The concept of a reciprocal lattice is essential to describe nanomolecules, such as, carbon nanotubes or graphene. The definition of the vectors of the reciprocal lattice contributes to the calculation of the spectrum of eigenvalues of the energy of these nanostructures.

Using the concepts that will be discussed in the following paragraphs to define the "Brillouin zone," it is possible to define the structure of the valence and conduction energy bands in nanostructures, the number of electrons that belong to the Brillouin zone and, thus, the occupation of these bands by electrons. These factors characterize the material as a conductor, semiconductor or insulator and describe the behavior of the material in the presence of external fields.

8.3.3.1 Reciprocal Lattice

If $\vec{a}_1$, $\vec{a}_2$ and $\vec{a}_3$ are primitive vectors of the crystal lattice, the translation vectors of the reciprocal lattice b_1, b_2, and b_3 must satisfy the condition [17]:

$$a_i \cdot b_j = 2\pi\delta_{ij} \qquad (8.21)$$

where δ_{ij} is the Kronecker delta ($\delta = 1$ for $i = j$, $\delta = 0$ for $i \neq j$). The unit cell of the reciprocal lattice, with all of its symmetry properties, is called the first Brillouin zone, and its points are the wave vectors $\vec{k}$.

8.3.3.2 First Brillouin Zone

The Brillouin formulation is the most important diffraction condition for solid-state physics and is the only one that is used to describe the energy bands of electrons in addition to expressing the elementary excitations in crystals. The first Brillouin zone is the unit cell of the reciprocal lattice; this provides a geometric interpretation of the diffraction condition that is generated by the crystal lattice of the material [17]. The boundaries of the Brillouin zone of the linear lattice in one dimension are at $k = \pm \pi/a$, where a is the parameter of the crystal lattice of the material.

8.3.4 Electronic Structure

As was already emphasized, a material's electronic structure can be analyzed to study its properties and structure. This analysis of materials, primarily nanostructured materials, is an area of theoretical investigation in which quantum mechanics is applied to describe the spatial distribution and energy levels of electrons that make up the studied system. Through the analysis of the density of states (DOS) of the electrons, the electronic structure determines the electrical conduction properties of the material, which define whether it is an electrical insulator, semiconductor, or conductor.

8.3.4.1 Density of States—Discretization of Energy in Quantum Systems

The DOS is a property that is used extensively in quantum systems in condensed matter physics; it refers to the energy level of the electrons, photons, or phonons in a solid crystal. The electronic DOS quantifies how "packed" the electrons in a quantum mechanical system are in energy levels. The DOS can vary from zero for an energy level that is inaccessible to the electrons, with no space occupied by them, to a defined occupation value at a specific energy level that is accessible to the electrons of the material. There is a direct correlation between the concept of quantized energy levels that is described by quantum mechanics and the DOS. The DOS is an energetic configuration due to the wave property of matter. In some systems, the interatomic spacing, the crystal structure, and the atomic charge of the material only allow electrons of certain wavelengths to propagate in the system, which also limits the possible directions of wave propagations. Each wave occupies a different mode or state that can have the same wavelength or the same quantized energy levels. This determines the degeneracy of states with the same energy and the absence of states in other energies that are incompatible with the system in which no space is occupied by the system. In the case of electronic states, the DOS permits the calculation of the number of electrons for each energy level, and the diagram of these states defines the electrical conduction properties of a material. DOS is usually denoted with one of the symbols g, ρ, n, or N. If the DOS is the function $g(E)$, one can write the expression $\Delta N = g(E) \cdot \Delta E$, which represents the number of states ΔN with energies between E and $E + \Delta E$. If the fundamental state is completely occupied by electrons (full valence band) and the first excited state overlaps the valence band (an electron-hole pair in the conduction band), the material will be a conductor. However, if the bands are separated by an excitation energy E_g, which is called a bandgap, the material will be an insulator or a semiconductor as shown in Fig. 8.2 for graphene. Thus, whether a material is an insulator or semiconductor depends

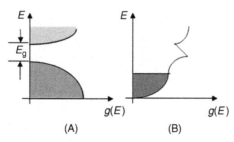

FIGURE 8.2 Occupation density $g(E)$ of the valence and conduction bands. (A) Semiconductor characteristics. (B) Metallic characteristics.

on the value of the energy gap E_g between the valence and conduction bands. In general, the bandgap E_g is on the order of eV ($E_g \sim$ eV) for insulating materials and $E_g \sim 10^{-1}$ eV for semiconductors. Several semiconducting materials, such as silicon, have a bandgap in another range of values ($E_g \sim 1.1$ eV), and materials such as ZnO, GaN, and AlN have bandgaps of a few eV in the energy band that corresponds to ultraviolet radiation.

The Fermi energy is defined as the maximum energy value in the valence band that corresponds to the energy level of the last state that was occupied by the electrons in the material. The Fermi surface is the surface in the k-space that comprises the occupied electronic states. However, in the case of materials that are "doped" with other elements, other energy levels are accessible to the system, which modifies the conductive properties of the material. In the presence of defects and impurities in the material due to doping, there is a readjustment of the effective Fermi energy to a new energetic configuration that favors electronic conduction.

8.3.4.2 Quantum Confinement of Electrons in the Volume of the Material

In a quantum system under stationary conditions, the confinement of electrons limits its energy to certain specific "discrete" levels due to the confinement in the volume of the material, which excludes specific energy values. For example, this condition determines the electrical behavior of nanomaterials, such as graphene or carbon nanotubes, as will be explained in the following paragraphs.

Consider a material with the dimensions L_x, L_y, and L_z and volume $V = L_x L_y L_z$. The stationary conditions of the system (Bloch theorem) require that the wave functions of the electrons belong to the crystal lattice of the material to respect the boundary conditions, which means that the electronic wave functions can only have certain wavelengths λ_n that correspond to the wave numbers $k_n = 2\pi/\lambda_n$. The potential of the electrons within the material does not depend on the location of the particle; that is, it is constant and can be considered null for ease of calculation.

Under these conditions, the energy levels of the electrons become "quantized" at discrete values depending on the values that are allowed by the wave vector $\vec{k}$:

$$E_n = \frac{p^2}{2m} = \frac{\hbar^2 k_n^2}{2m} \tag{8.22}$$

which is the solution for the eigenstates of the energy of the Schrödinger equation of the system. In the formalism of the quantum theory that uses the wave vector $\vec{k}$ in the energy expression, the energy of the electrons is $E = (\hbar k)^2/2m$, and the energy of the photons is $E = \hbar c k$. The periodic wave function $\psi(x)$ of the electrons, which is expressed by a sin function of the type $\psi(x) = A \sin(2\pi x/\lambda)$, will have to respect the boundary conditions of the system, which is the condition $\psi(x) = \psi(0) = \psi(L) = 0$ due to the confinement of electrons in the material. The stationary oscillations of the system are represented in Fig. 8.3, where n is the order of oscillation of the wave function of the electron and L is the distance between the walls. For $n = 1$, the wavelength is double the size of the material, and it changes phase at the edges to respect the stationary condition; $n = 2$ represents $\lambda = L$.

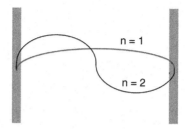

FIGURE 8.3 Quantum confinement condition of electrons with the resulting definition of the discrete energy values E_n.

Only certain oscillations of the specific wavelengths λ_n can respect the boundary conditions with the resulting discretization of the energy at defined values.

For $\psi(L) = A \cdot \sin(2\pi L/\lambda) = 0$ and $2\pi L/\lambda = n\pi$, the resulting values of λ_n are therefore $\lambda_n = 2L/n$ and $k_n = 2\pi/\lambda_n = n\pi/L$. Eq. (8.22) therefore becomes:

$$E_n = \frac{p^2}{2m} = \frac{\hbar^2 (k_n)^2}{2m} = \frac{\hbar^2 (n\pi)^2}{2m(L)^2} \tag{8.23}$$

The quantum confinement of electrons in atoms and molecules is due to the stable and stationary condition of electrons in the atomic and molecular orbitals. The theory that describes the electronic density around nuclei in a material finds an important simplification in the description of molecular orbitals in the LCAO.

The DOS leads to the concept of a "fundamental state," which is the lowest energy state of a system that contains N electrons without any type of interaction. This state is filled by the electrons of the system from the lowest energy until it is occupied by the last electron for the molecular orbital with the lowest possible energy. These orbitals represent the states that are accessible to the quantum system due to the crystal lattice, charges, and other factors. The structure of the molecular orbitals and the corresponding energies will be discussed in the following chapters. The set of orbitals that correspond to each energy level of the system determines the formation of bands of energy, the valence band and the conduction band.

8.3.5 Graphene

As an application of the concepts that were introduced in the previous paragraphs, the molecular structure of graphene is shown in real space and in reciprocal space (Fig. 8.4B) to determine the band structure that defines the electronic behavior of graphene. The Brillouin zone for graphene can be defined as the rhombus that is formed by the lattice vectors of the reciprocal space $\vec{b}_1$ and $\vec{b}_2$.

8.3.5.1 Electronic Structure of Graphene

The electronic structure of a graphene molecule that is defined by the LCAO theory is composed of a valence band and a conduction band. Fig. 8.5 shows the structure of the valence and conduction bands that are occupied by the electrons as calculated using the TB

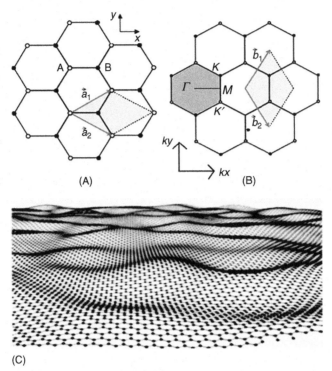

(A)

(B)

(C)

FIGURE 8.4 (A) Graphene's structure in real space. The rhombus represents the unit cell of graphene and is delimited by the network vectors $\vec{a}_1$ and $\vec{a}_2$. This area involves two atoms, which are indicated as A and B. (B) Graphene structure in reciprocal space showing the unit vectors $\vec{b}_1$ and $\vec{b}_2$ and the Brillouin zone that is delimited by them. The shaded hexagon shows the Brillouin zone that is considered because it represents all of the symmetry operations of the unit cell of graphene [21]. (C) Representation of a graphene sheet.

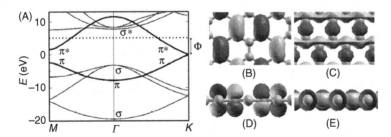

FIGURE 8.5 Electronic properties of graphene. (A) Structure of the electronic bands: band π (last occupied band) and band π^* (conduction band, first unoccupied) touch each other at the K points of the first Brillouin zone. The Fermi energy is set to zero. (B) and (D) Electronic configuration of π states at the K points in bonding orbitals. (C) and (E) σ states at Γ points far from the Fermi energy that are occupied by the electrons of the graphitic sp$_2$ bonds plan [23].

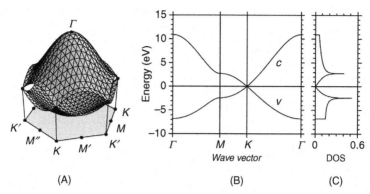

FIGURE 8.6 (A) Diagram showing the conduction and valence bands in the first Brillouin zone of hexagonal graphene [21]. The points of high symmetry Γ, K, K', M, M', and M'' are shown in the hexagonal projection. (B) Valence and conduction bands in the high-symmetry directions ΓK, KM, and $M\Gamma$ of the Brillouin zone of graphene. (C) Density of states *(DOS)* of graphene [23].

approximation and the corresponding electron cloud in the π valence orbitals (bonding) and π^* conduction orbitals (antibonding). These bands determine the electrical conduction properties of graphene as shown in the DOS diagram for graphene (Fig.8.5C). The π and π^* bands that are shown in Fig. 8.5A are also shown in three dimensions in Fig. 8.6A. These bands are close to the Fermi energy level at points of high symmetry in the Brillouin zone of graphene (points K, K', Γ and M in Figs. 8.4A–B and 8.5). These points are due to the symmetry of the hexagonal structure of graphene and represent special conditions in the description of the electrical properties of graphene [22].

The valence and conduction bands degenerate at the points K and K' of the reciprocal space, which are found at the Fermi level (Fig. 8.6C), which determines the electrical conduction properties of graphene. The energy gap between the valence and conduction bands is zero, which characterizes graphene as a zero-gap semimetallic material.

8.3.6 Carbon Nanotubes

8.3.6.1 Structure of Carbon Nanotubes

Carbon nanotubes [24] can be considered as one or several sheets of graphene that are rolled into a perfect cylinder [25,26]. The size of these materials is on the order of magnitude of an atom [diameters from 0.7–2 nm for single-walled carbon nanotubes (SWNTs) to 10 nm for multiwalled carbon nanotubes (MWNTs)]. Quantum mechanics is used to describe these nanomolecules because the movement of electrons is "confined" in the direction of the circumference of the nanotubes in a very limited number of configurations to respect the stationary conditions of the wave functions as was explained in the preceding paragraphs. A continuous number of electronic states is permitted in the direction of the tube's length because the length of the tube is much greater than the wavelength of the electronic wave functions. Placing the permitted values for the wave vector $\bar{k}$ in the first Brillouin zone of graphene generates a series of parallel lines called "cutting lines." The length, number, and orientation of these lines depend on the chiral indices (n, m) of the nanotubes, which are called "chiral vectors." For many properties, the carbon nanotube can be considered as a unidimensional crystal with a translation vector $\vec{T}$ along its main axis and a small number of carbon atoms on its circumference. The structure of a carbon nanotube can be determined unambiguously by the chiral vector $\vec{C}_h$ (Fig. 8.7A), which, when rolling the graphene sheet, coincides with the circumference of the nanotube. The chiral vector can be written as a function of the lattice vectors $\vec{a}_1$ and $\vec{a}_2$ of graphene as follow:

$$\vec{C}_h = n\vec{a}_1 + m\vec{a}_2 \tag{8.24}$$

8.3.6.2 Lattice Vectors in Real Space

To specify the symmetry properties of carbon nanotubes as a 1D system, it is necessary to express the lattice vector, which is the translation vector $\vec{T}$ (Fig. 8.7) that has the direction of the primary axis of the nanotube. The translation vector $\vec{T}$ of a nanotube can be written as a function of n and m as $\vec{T} = t_1\vec{a}_1 + t_2\vec{a}_2$, where $t_1 = (2m + n)/d_R$, and $t_2 = (2n + m)/d_R$. The length of the translation vector is $\vec{T} = \sqrt{3}\vec{C}_h/d_R$, where d_R is the maximum common divisor of $(2n + m, 2m + n)$. By defining d as the greatest common divisor of (n, m), the values of d and d_R are related by [28]:

$$d_R = \begin{cases} d \text{ if } (n-m) \text{ is not a multiple of 3d} \\ \\ 3d \text{ if } (n-m) \text{ is a multiple of 3d} \end{cases} \tag{8.25}$$

For the nanotube (6, 2) that is shown in Fig. 8.7, $d_R = d = 2$, and $(t_1, t_2) = (5, 7)$. For achiral zigzag and armchair nanotubes, $T = \sqrt{3}a$ and $T = a$, respectively. The area of the unit cell of the nanotube can be found by calculating the vector product of the two vectors $\vec{T}$ and $\vec{C}_h$, $|C_h \times \vec{T}| = \sqrt{3}\,a^2(n^2 + nm + m^2)/d_R$. Dividing this value by the area of the unit cell of graphene $|\vec{a}_1 \times \vec{a}_2| = \sqrt{3}a^2/2$ results in the number of hexagons in the unit cell of the nanotube:

$$N = \frac{2(n^2 + nm + m^2)}{d_R} \tag{8.26}$$

In the case of nanotube (6, 2), N is equal to 26, so the unit cell of this nanotube has 52 carbon atoms because there are two carbon atoms in the unit cell of graphene. For achiral nanotubes, $N = 2n$.

8.3.6.3 Application of Quantum Theory to Carbon Nanotubes

The description of carbon nanotubes is based on the analysis of graphene, whose electrons make strong covalent σ-type bonds along the graphitic plane that do not significantly affect the electronic properties of the nanotubes because they are far from the Fermi energy. The electrons that occupy the p_z atomic orbitals, which are perpendicular to the planes, interact with the p_z orbitals of neighboring atoms and generate weak interactions with the creation of π bonding molecular orbitals and antibonding π* molecular orbitals. The bond energy of these orbitals is close to the Fermi energy in the first Brillouin zone and

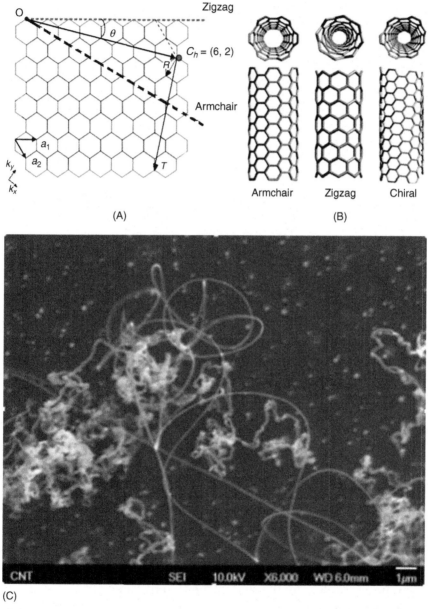

(A)

(B)

(C)

FIGURE 8.7 (A) Projection of an open nanotube on a graphene layer. When the graphene sheet is closed to form a nanotube, the chiral vector $\vec{C}_h$ becomes the circumference of the cylinder (nanotube), and the translation vector $\vec{T}$ is aligned parallel to the nanotube's axis. The vector $\vec{R}$ is the vector of symmetry, θ indicates the chiral angle, and $\vec{a}_1$ and $\vec{a}_2$ are the unit vectors of graphene. The unit cell of the nanotube is defined by the rectangle that is defined by the two vectors $\vec{C}_h$ and $\vec{T}$. (B) Helical arrangement in achiral and chiral nanotubes. (C) FEG microscopy image of carbon nanotubes [27].

is responsible for the weak interaction between the SWNTs in bundles in the same manner that it acts in graphite to generate weak bonds between the planes. The relatively simple Hamiltonian formalism is normally used to describe the π and π^* electronic bands, which is known as the TB approximation.

8.3.6.4 Application of the TB Model

By applying the TB formalism to carbon nanotubes [18], it is possible to write the Bloch states for the p_z orbital of the electron in each carbon atom in the Brillouin zone of graphene. Rewriting Eq. 8.10 yields:

$$\psi_k^A(r) = \frac{1}{(N)^{1/2}} \sum_{R_A} \exp(ik \cdot R_A)\chi(r - R_A) \tag{8.27}$$

and

$$\psi_k^B(r) = \frac{1}{(N)^{1/2}} \sum_{R_B} \exp(ik \cdot R_B)\chi(r - R_B) \tag{8.28}$$

The A and B indices refer to the two carbon atoms that occupy the primitive cell. The sums are extended to all of the sites of the crystal lattice in relation to the position of the electron in A, where the wave function is a linear combination of the wave functions relative to the atomic sites, which also applies to the electron of atom B. $\chi(r)$ represents the wave function of the p_z atomic orbital normalized to an isolated atom, while the wave function of the electron in the main cell is the linear combination of the wave functions for each atom:

$$\Psi = \psi_k^A(r) + \lambda\psi_k^B(r) \tag{8.29}$$

Considering the general expression for the Hamiltonian operator [10]:

$$H_{\mu\nu}(k) = \int \Psi^{\mu*}(k,r)\hat{H}\Psi^\nu(k,r)d^3r \tag{8.30}$$

substituting Eq. 8.29 into $H\Psi = E\Psi$ (Eq. 8.9) [10], the following components are obtained:

$$H_{AA} + \lambda H_{AB} = E \tag{8.31}$$

$$H_{BA} + \lambda H_{BB} = \lambda E \tag{8.32}$$

The coefficient λ is eliminated using the variational principle [29], which obtains the most appropriate eigenvalues for the energy. The H_{ii} terms are:

$$H_{AA} = H_{BB} = \alpha \tag{8.33}$$

With the parameter α that is defined from the expression for the Hamiltonian that is valid in the TB approximation:

$$\alpha(r) = \int |\chi_\mu(r)|^2 \left[\sum_{R\neq0} V(r - R) \right] d^3r \tag{8.34}$$

where α represents the effect of the distant potentials, or the crystal field, on the electron of the atom under consideration. The terms $H_{AB} = H_{BA}^* = \int \psi_A^*(r)\hat{H}\psi_B(r)d^3r$ are obtained from Eqs. 8.27 and 8.28, which represent the wave functions for the p_z atomic orbitals of the electrons of the primitive cell:

$$H_{AB} = \beta\left[\exp\left(ik_x\frac{1}{\sqrt{3}}\right) + 2\exp\left(-ik_x\frac{a}{2\sqrt{3}}\right)\cos\left(\frac{k_ya}{2}\right)\right]$$ (8.35)

The parameter β is defined as the Hopping term and represents the quantity that is related to the overlap between the p_z atomic orbitals via the potential:

$$\beta(R) = \int \chi_\mu^*(r)\left[\sum_R V(r-R)\right]\chi_\mu(r-R^{ll})d^3r$$ (8.36)

This term defines the width of the energy bands. The components k_x and k_y of the wave vector $\vec{k}$ are shown in Fig. 8.7A. A 2×2 matrix is thus obtained, and setting the determinant equal to zero leads to the secular Schrödinger equation:

$$\mathrm{Det}\begin{pmatrix} H_{AA}\text{-}\varepsilon & H_{AB} \\ \\ H_{BA} & H_{BB}\text{-}\varepsilon \end{pmatrix} = 0$$ (8.37)

The spectrum of eigenvalues of the graphene sheet is obtained by calculating the determinant of the matrix as a function of the components k_x and k_y of the wave vector $\vec{k}$:

$$\varepsilon(k_x,k_y) = \alpha \pm \beta\left[1 + 4\cos\left(\frac{\sqrt{3}k_xa}{2}\right)\cos\left(\frac{k_ya}{2}\right) + 4\cos^2\left(\frac{k_ya}{2}\right)\right]^{1/2}$$ (8.38)

where a is the length of the vectors $\vec{a}_1$ and $\vec{a}_2$ of the primary lattice. Eq. 8.38, which was obtained for graphene, determines the electronic structure of the carbon nanotube and defines the energy dispersion to the energy bands that is related to the π orbitals of graphene as shown in Fig. 8.5A.

By considering the structure of the carbon nanotube, two new vectors of the reciprocal lattice, the vector $K_\perp$ in the direction of the circumference and the vector $K_{//}$ along the axis of the nanotube (Fig. 8.8), can be defined from the relationship:

$$R_i \cdot K_j = 2\pi\delta_{ij}$$ (8.39)

which is translated as:

$$C_h \cdot K_\perp = 2\pi, \quad T \cdot K_\perp = 0;$$ (8.40)

$$C_h \cdot K_{//} = 0, \quad T \cdot K_{//} = 2\pi,$$ (8.41)

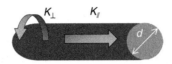

FIGURE 8.8 Representation of the wave vectors in the directions of the translational vector $\bar{T}$ *(K$_{\parallel}$)* and of the chiral vector $\bar{C}_h$ along the circumference *(K$_{\perp}$)*.

Using the expressions for the vectors t_1 and t_2 (Section 8.3.6.2), obtained due to the condition $C_h \cdot T = 0$; the following is valid:

$$K_{//} = 1/N(-t_2 b_1 + t_1 b_2) \tag{8.42}$$

$$K_{\perp} = 1/N(mb_1 - nb_2) \tag{8.43}$$

where b_1 and b_2 are the vectors of the reciprocal lattice of graphene, and N is the number of hexagons per unit cell.

The wave vector $K_{\perp}$, which is associated with the chiral vector $\vec{C}_h$, is now quantized with a finite number of states k, while the wave vector $K_{//}$, which is associated with the translation vector $\vec{T}$ along the axis of the tube, is continuous if the difference Δk between the two contiguous values of $K_{//}$ is infinitesimal ($\Delta k = 2\pi/T \to 0$), which is respected for a nanotube of length $T \to \infty$. The energy eigenvalues can therefore be written in the form:

$$\varepsilon_p^{\text{nanotube}}(k) = \varepsilon^{\text{graphene}}\left[kK_{//} + pK_{\perp} \right] \tag{8.44}$$

The indices k and p can assume the values:

$$-\pi/T < k < \pi/T; \quad p = 0, 1, \ldots, N-1 \tag{8.45}$$

For example, consider a nanotube (n, m), where $C_h = na_1 + ma_2$. The wave functions must satisfy the periodicity condition of the tube:

$$C_h \cdot k = 2\pi q \tag{8.46}$$

where q is a whole number, and $\vec{k}$ is a wave vector of graphene in the first Brillouin zone. Considering the case of an armchair-type nanotube, the indices that define the chiral vector are equal: $n = m$. The length C of the circumference of the nanotube is:

$$C = |C_h| = \sqrt{C_h \cdot C_h} = a\sqrt{n^2 + nm + m^2} \tag{8.47}$$

$$C = a\sqrt{3}n \tag{8.48}$$

Considering the quantization that is defined in Eq. 8.46, the following is valid:

$$C \cdot k = a\sqrt{3}nk_x = 2\pi q \tag{8.49}$$

where k_x is the component of the wave vector $\vec{k}$ in the direction of the chiral vector C_h, and $q = 1, 2, \ldots, 2n$.

Substituting this result into Eq. 8.38 results in:

$$\varepsilon_q^{(n,n)}(k) = \alpha \pm \beta \left[1 + 4\cos\left(\frac{q\pi}{n}\right)\cos\left(\frac{ka}{2}\right) + 4\cos^2\left(\frac{ka}{2}\right) \right]^{1/2} \tag{8.50}$$

where $-\pi/a < k < \pi/a$.

Consider a nanotube with the indices $n = 5$, $m = 5$ as an example. $C_h = 5a_1 + 5a_2$, and the spectrum to the eigenvalues of this type of carbon nanotube is:

$$\varepsilon_q^{(5,5)}(k) = \alpha \pm \beta \left[1 + 4\cos\left(\frac{q\pi}{5}\right)\cos\left(\frac{ka}{2}\right) + 4\cos^2\left(\frac{ka}{2}\right) \right]^{1/2} \tag{8.51}$$

The spectrum of the energy eigenvalues can be visualized as a function of the wave vector $\vec{k}$ as shown in Fig. 8.9. The number of hexagons N per unit cell is:

$$N = 2\cdot(n^2 + m^2 + nm)/d_R = 10 \tag{8.52}$$

Since there are 2 atoms/cell, 20 electrons occupy 10 orbitals.

The increasing energetic behavior of the orbitals in the case of the conduction band π^* from the edge ($k = -\pi/a$) to the center of the Brillouin zone indicates a greater anti-bonding state of the electrons in the molecule. The width of each band is related to the energy dispersion in the band, which is proportional to the overlap between the orbitals of neighboring atoms. Fig. 8.9 shows that some of the orbitals are degenerate; that is, they are characterized by the same binding energy. The last orbital that is occupied in the π band and the first unoccupied orbital in the π^* conduction band cross the Fermi energy, which

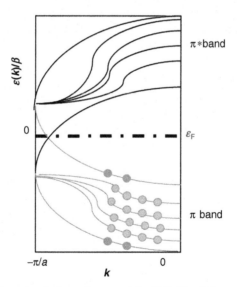

FIGURE 8.9 Occupation of the valence band π for carbon nanotube (5, 5). The circles represent the occupation of orbitals by the electrons for each allowed K-point [18].

means that there is an overlap between the orbitals in the two bands, which confers metallic conduction characteristics to the nanotube.

8.3.6.5 Electronic Structure of a Carbon Nanotube

As a first approximation, the electronic structure of a carbon nanotube can be obtained from the 2D structure of graphene. In the case of a SWNT, however, the quantum confinement of the 1D electronic states along the circumference of the nanotube must be considered. Placing the permitted values for the wave vector in the first Brillouin zone of graphene generates a series of parallel lines that indicate the wave vector values $\vec{k}$ for the considered nanotube; the length, number and orientation of these "cutting lines" depend on the chiral indices (n, m) of the nanotube. In the case of armchair-type nanotubes, such as the (5, 5) nanotube that was considered in the preceding section, the boundary conditions that are defined by the Bloch functions that are related to the chiral vector C_h are valid:

$$\psi_k\ (r + C_h) = e^{ikr} C_h \psi_k(r) \tag{8.53}$$

The value of k_x is calculated as in the previous section. Rewriting Eq. 8.49 for the wave vector k_x results in [25]:

$$k_x^v = \frac{v}{N_x} \frac{2\pi}{\sqrt{3}a} \tag{8.54}$$

where $v = 1,...,N_x$ with the number of hexagons $N_x = 5$. This shows the presence of 5 modes that are permitted for the wave vector k_x. These possible energetic configurations are normally represented using a hexagon that reproduces the first Brillouin zone (BZ) for graphene as shown in Fig. 8.10A; inserting these lines into the diagram in Fig. 8.5A, which shows the valence and conduction bands in the first hexagonal Brillouin zone of graphene, results in the configuration that is shown in Fig. 8.10B. Fig. 8.10A–B show the wave vectors k for SWNTs in the first hexagonal BZ of graphene using the cutting lines; the discrete values of the wave vector $\vec{k}$ are indicated in the direction of the vector K_1 due to the

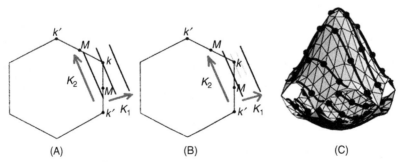

(A)   (B)   (C)

FIGURE 8.10 Illustrations of the allowed k values in the first Brillouin zone for (A) a metallic nanotube and (B) a semiconductor nanotube; (C) cutting lines in the case of a (4, 2) nanotube for illustrative purposes [21].

periodic condition of the electronic wave function in the direction along the diameter of the nanotube, while in the longitudinal direction of the nanotube for the K_2 vector, the wave vector assumes continuous values because it lacks boundary conditions that limit the discrete quantum states for the wave vectors k in this direction. In Fig. 8.10C, the lines of the values of the wave vector $\vec{k}$ cut the valence and conduction bands of a carbon nanotube with indices (4, 2). Fig. 8.10A represents a metallic nanotube because the cutting line intersects a point K on the edge of the hexagon at the Fermi energy of graphite. Fig. 8.10B refers to the case of a semiconductor nanotube because the cutting line does not pass through the point K.

The zone-folding approximation is used to represent the energy bands of carbon nanotubes in a diagram; the basic idea is that the electronic structure of a specific nanotube is determined by the overlap of the electronic bands of graphene. The cutting lines that correspond to the permitted wave vectors k that are related to the nanotube under consideration result in the diagram that is illustrated in Fig. 8.11.

The energy dispersion is obtained by substituting the values of k_x into Eq. 8.38. Fig. 8.11 shows the energy bands with degenerate and nondegenerate orbitals for 10 orbitals in the valence band and 10 in the conduction band. The DOS is shown on the right side of the figure. For zig-zag type tubes with chirality $(n, 0)$, the permitted wave vectors are:

$$k_y^v = \frac{v}{N_y}\frac{2\pi}{a} \qquad (8.55)$$

where $v = 1,...,N_y$. For example, the tube (9, 0) has nine lines of permitted wave vectors (Fig. 8.12).

The relationship of the dispersion of the energy bands and the calculation of the electronic DOS indicate that this is a metallic nanotube because the valence and conduction

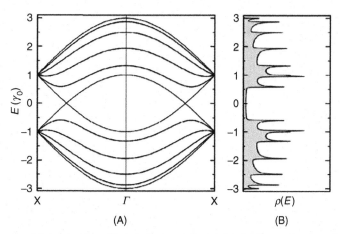

FIGURE 8.11 (A) Diagram of the molecular orbitals in the energy bands for the electronic states of the nanotube (5, 5), which was obtained using the zone-folding approximation; (B) the corresponding DOS of electrons; the Fermi energy is placed on the zero energy level [23].

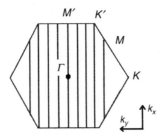

FIGURE 8.12 Illustration of the allowed values of k (cutting lines) for an armchair nanotube (9, 0) in the first Brillouin zone.

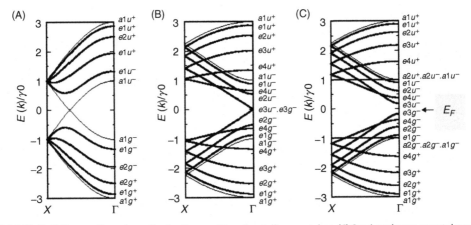

FIGURE 8.13 (A–B) Energy dispersion ratios for the considered metallic nanotubes. (C) Semiconductor nanotube with a bandgap ≠ 0 [30].

bands touch at the Fermi energy ($k = 0$) as shown in Fig. 8.13B. When a bandgap is present between the bands (Fig. 8.13C), the nanotube behaves like a semiconductor. The DOS diagrams show how the DOS between the π and π^* bands is greater than 0 for the metallic case and is equal to 0 in the semiconductor case.

Once the basic physical concepts that govern the behavior of matter at the atomic to nanometric scales have been introduced, it is possible to address the incredible properties of nanostructured materials.

Several possible applications for nanomaterials that are already being studied and used based on the exotic material properties at the nanometric scale will be described in the following chapters.

List of Symbols

E	Energy
Λ	Wavelength
H	Planck's constant
H h	Reduced Planck's constant

p	Momentum
σ	Bonding atomic orbital
π	Band of valence molecular orbitals
π^*	Band of conduction molecular orbitals
$\psi(r, t)$	Wave function
r	Position
t	Time
m	Mass
V	Potential energy
ν	Fequency
ω	Angular frequency
k	Wave number
F	Force
$\hat{H}$	Hamiltonian operator
δ_{ij}	Kronecker delta
a	Lattice parameter
$g(E)$	Density of electronic states
E_g	Gap energy
a_i	Vectors of the crystal lattice
b_i	Reciprocal space lattice vectors
C_h	Chiral vector
p_i	Atomic orbitals p $(n = 2)$
β	Hopping term
ε	Energy eigenvalue
SWNT	Single Wall Nanotube
MWNT	Multi Wall Nanotube

References

[1] E.G. Segré, Dos raios X aos quarks–Físicos modernos e suas descobertas [Modern Physicists and Their Discoveries], Editora Universidade de Brasília, Brasília, (1987).

[2] A. Einstein, Über einen die Erzeugung und Verwandlung des Lichtes betreffenden heuristischen Gesichtspunkt, Annalen der Physik[série 4] 17 (1905), 132–148.

[3] H. Everett III, The theory of the universal wave function, in: S. Bryce, DeWitt, Neill Graham (Eds.), The Many-Worlds Interpretation of Quantum Mechanics, Princeton University Press, Princeton, 1973.

[4] P.A.M. Dirac, The Principles of Quantum Mechanics, fourth ed., Oxford University Press, Ely House, London, (1958).

[5] D. Griffiths, Introduction to Quantum Mechanics, second ed., Pearson, Prentice-Hall, London, 2005.

[6] L. De Broglie, Sur La Degradation du quantum dasn les transformation sucessives des radiations de haute frequence, Compte rendus de l'Academie des Sciences de Paris 173 (1921), 1162–1165.

[7] M. Jammer, The Conceptual Development of Quantum Mechanics, McGraw-Hill, New York, 1966.

[8] M. Jammer, The Philosophy of Quantum Mechanics, Wiley, New York, 1974.

[9] E.R.J.A. Schrodinger, Quantisierung als Eigenwertproblem, Annalen der Physik 384 (4) (1926) 361–376.

[10] R. Eisberg, R. Resnick, Quantum Physics of Atoms, Molecules, Solids, Nuclei and Particles, second ed., J. Whiley and Sons, New Jersey, USA, 1985.

[11] T.S. Kuhn, Black–body Theory and the Quantum Discontinuity (1894–1912), Oxford University Press, Oxford, 1978.

[12] R.P. Feynman, The Strange Theory of Light and Matter, expanded edition, Princeton University Press, Princeton, 2006.

[13] M. Planck, Über eine Verbesserung der Wienschen Spektralgleichung, Verhandlugen der Deutschen physikalische Gesellschaft 2 (1900) 202–204.

[14] L. Van Dommelen, Fundamental Quantum Mechanics for Engineers, 2004, Copyright © 2004, 2007, 2008, 2010, 2011, and on, Leon van Dommelen.

[15] E. Wigner, F. Seitz, On the constitution of metallic sodium, Phys. Rev. 43 (1933) 804.

[16] J.C. Slater, Electronic energy bands in metals, Phys. Rev. 45 (1934) 794.

[17] C. Kittel, Introduction to Solid State Physics, eight ed., John Wiley and Sons, United States of America, 2005.

[18] J.D.M. Vianna, A. Fazzio, S. Canuto, Teoria Quântica de Moléculas e Sólidos: Simulação computacional [Solids and Molecules Quantum Theory: Computational Simulation], Livraria da Física-Brasil, São Paulo, 2004.

[19] W.H. Bragg, The nature of Röntgen rays, Transactions and proceedings and report of the Royal Society of South Australia 31 (1907) 94–98.

[20] J. Mehra, H. Rechenberg, Springer, New York, 1982–1987.

[21] M.S. Dresselhaus, et al., Raman Spectroscopy of carbon nanotubes. Phys. Rep. 409, 2005, 47–99.

[22] R. Daniel et al., Experimental review of graphene, PACS numbers: 81.05.ue, 72.80.Vp, 63.22.Rc, 01.30.Rr, 2011.

[23] J.C. Charlier, Electronic and transport properties of nanotubes, Rev. Mod. Phys. (2007) 79.

[24] S. Iijima, Helical microtubes of graphitic carbon, Nature 354 (1991) 56.

[25] P.J.F. Harris, Carbon Nanotubes and Releated Structures: New Materials for the 21st Century, Cambridge University Press, Cambridge, 2003.

[26] B. Bhushan, Springer Handbook of Nanotechnology, Springer-Verlag Berlin Heidelberg, Berlin, Heidelberg, New York, Hong Kong, London, Milan, Paris, Tokyo, 2004.

[27] R. Marchiori, Produção Por Ablação A Laser e Caracterização de Nanotubos de Carbono [Carbon Nanotubes Production by Laser Ablation and their Characterization], Doctorate Thesis, Florianópolis, 227f. Departamento de Engenharia de Materiais, Universidade Federal de Santa Catarina, Florianópolis-SC, Brazil, 2007.

[28] R. Saito, G. Dresselhaus, M.S. Dresselhaus, Physical Properties of Carbon Nanotubes, Imperial College Press, London, 1998.

[29] R.P. Feynmann, Space-Time Approach to Non-Relativistic Quantum Mechanics, Rev. Mod. Phys. 20 (1948) 367.

[30] R. Saito, et al. Electronic structure of graphene tubules based on C60, Phys. Rev. B46 (1992) 1804.

Index

9

Carbon-Based Nanomaterials

J.R. Siqueira, Jr.*, Osvaldo N. Oliveira, Jr.**

*INSTITUTE OF EXACT SCIENCES, NATURAL AND EDUCATION, FEDERAL UNIVERSITY OF TRIÂNGULO MINEIRO (UFTM), UBERABA, MINAS GERAIS, BRAZIL; **SÃO CARLOS INSTITUTE OF PHYSICS, UNIVERSITY OF SÃO PAULO (USP), SÃO CARLOS, SÃO PAULO, BRAZIL

CHAPTER OUTLINE

9.1 Introduction

Carbon is one of the most plentiful and versatile elements in the universe. Due to its allotropic characteristics, carbon forms compounds that have completely different properties depending on the arrangement of the adjacent carbon atoms. Diamond and graphite are two typical examples. Diamond is a material known to have the highest hardness, whereas graphite is fragile and brittle [1,2]. Carbon nanostructures are among the most promising in nanotechnology because they can be used in different fields. In electronics, the combination of a molecular-sized diameter (on the order of 1 nm) with a microscopic-scale length and optical and electrical properties enable carbon nanostructures to be used for the production of innovative devices [2–5]. Composites derived from these nanostructures may exhibit mechanical properties, the electrical and thermal conductivities are useful for applications, and there are desirable features for protection from corrosion in new engineering products

[2–4]. Carbon nanomaterials may also provide new functionalities in biomedical applications, such as sensing and the controlled delivery of pharmaceuticals and drugs [5,6].

For these applications to be commercially viable, there are still major challenges to overcome. Methods for large-scale synthesis, methods for incorporating and dispersing these nanomaterials, such as matrix composites, and methods for the precise control to manipulate and make electrical connections in nanocircuits [1,2] are warranted.

Among the many carbon structures, we highlight fullerenes, carbon nanotubes (CNTs), and graphene. In the following sections, we discuss each of these materials and how they have been used in nanotechnology. We hope that this chapter provides the readers a brief compendium of the properties, applications, and trends in the use of these carbon nanomaterials.

9.2 Fullerenes

Fullerene is considered to be the third allotropic form of carbon, after graphite and diamond [7,8]. Discovered in 1985 by Harold W. Kroto (University of Sussex, Brighton, England), Robert F. Curl and Richard E. Smalley (Rice University, Houston, Texas, USA) [9], fullerene has an icosahedral symmetrically closed-cage structure, which is formed by 20 hexagons and 12 pentagons in which each carbon atom is bonded to 3 other carbon atoms with sp^2 hybridization [7–9], as shown in Fig. 9.1. Because of its similarity to a building by the American architect Richard Buckminster Fuller, fullerene is called *buckminsterfullerene* [9]. It is a molecule with 60 carbon atoms (C60) that are arranged in the shape of a soccer ball with a 7 Å diameter. Due to its shape, fullerene is also known as *buckyball*. Unlike graphite or diamond crystals, which have atom arrangements, fullerene forms molecular crystals and can be considered as a zero-dimensional carbon structure [7,8]. Its discovery

FIGURE 9.1 Schematic representation of the C60 fullerene structure, which is also called *buckminsterfullerene* or *buckyball*.

revolutionized research on new allotropes of carbon and nanostructured materials, and its discoverers received recognition with the Nobel Prize in Chemistry in 1996 [7,8].

9.2.1 Fullerene Properties

In the icosahedral symmetrically closed-cage structure, there are two bond lengths in the fullerenes with "double bonds" in the 6:6 ring, which are shorter than the 6:5 bonds. The C60 is not "superaromatic" because it does not contain double bonds in the pentagonal rings, resulting in low electron delocalization. Thus, C60 behaves as an electron-deficient alkene and is therefore reactive with electron-donor species. Its molecular stability is due to its geodesic-shaped structure and its electronic bonds. In principle, there may be an infinite number of fullerenes as long as its structure forms an icosahedron with hexagonal and pentagonal rings.

In addition to the C60 molecule, fullerene may have larger structures, such as C70, C76, and C78, and smaller structures, such as C28 and C36. In the 1990s, C60 and other fullerenes began to be produced in large quantities via the condensation of soot that is generated in graphite vaporization [10]. Fullerenes may also occur naturally in materials affected by high-energy events, such as lightning and meteors, and in geological samples [7,8].

One of the most widely used processes of synthesis is based on the *Krätschmer–Huffman* method in which an electric arc is generated on graphite electrodes in a helium atmosphere at a pressure of approximately 200 torr. To separate out the fullerene, carbon soot from evaporated graphite is dissolved in a nonpolar solvent. Subsequently, the solvent is dried, and the C60 and C70 fullerenes are separated from the residue. Under an optimum current and pressure of helium flow, yields of up to 70% of C60 and 15% of C70 are achieved. The laser vaporization method is also used to produce fullerene. In a typical system, a Nd:YAG pulsed laser operating at 532 nm and 250 mJ is used as the source, and the graphite target is maintained in an oven at 1200°C. Although the yields are low, fullerenes have also been produced via deposition of soot from flames involving, for example, the combustion of benzene and acetylene [8,11].

Fullerenes have excellent mechanical properties, resisting high pressures and returning to their original shape even after subjected to more than 3000 atm. Theoretical calculations indicate that a single C60 molecule has a bulk modulus of 668 GPa when compressed to 75% of its size. This property makes fullerenes harder than steel and diamond, whose *bulk* modules are 160 and 442 GPa, respectively. Fullerenes can withstand collisions of up to 15,000 mph against stainless steel and maintain its shape, demonstrating their high stability. The optical properties of the fullerenes are also renowned for the delocalized π electrons, which generate nonlinear optical responses and intensity-dependent refractive indices [11,12].

9.2.2 Applications of Fullerenes

Fullerenes can be applied in several fields and, currently, are most notable in organic solar cells [13]. The high electron affinity and the ability to carry a charge make fullerene the best electric charge acceptor for this type of device. Fullerenes have a *lowest unoccupied molecular orbital* (LUMO) with a low energy compared with the organic charge donors that have high electron affinities [13,14]. With this LUMO energy, C60 can be reversibly reduced with up to six electrons, denoting its ability to stabilize a negative charge. Polymer blends that are combined with fullerenes exhibit ultrafast photo-induced charged transfer (at approximately 45 fs). Thus far, the greatest efficiency for a polymer solar cell is due to the polymer/fullerene heterojunction. The fullerene (most commonly C60) acts as an n-type semiconductor (electron acceptor) and is combined with a p-type polymer (electron donor), typically a polythiophene. The polymer/fullerene blend is deposited as a film to act as an active layer to create what is known as a *bulk* heterojunction [13,14], as shown in Fig. 9.2.

Solar cell efficiency steadily increases. Although the energy conversion efficiency of a polymer/fullerene heterojunction solar cell is still low (approximately 10%), compared to conventional silicon cells (20% efficiency), the production cost of this type of solar cell may be significantly lower. Furthermore, organic solar cells can be flexible, rolled up, and scattered over any surface [13–15]. Plastic photovoltaic devices can be used to cover internal and external walls of buildings. Additionally, cells of any color and texture can be manufactured. For example, a cell phone can be painted with this material, and its battery can be charged while a person walks with it on a sunny day. Organic solar cells are also expected to be used in advertising, such as on light banners, liquid crystal displays, and food packaging [13–15].

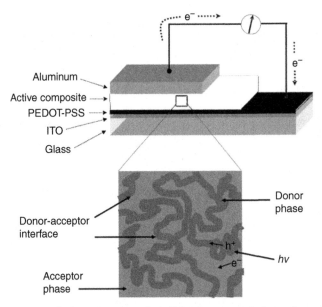

FIGURE 9.2 Schematic illustration of a heterojunction solar cell that is composed of an active polymer/fullerene layer. *Modified from B.C. Thompson, J.M. Fréchet, Polymer-fullerene composite solar cells, Angewante Chemie 47 (2008) 58–77 [13]. Copyright 2008 John Wiley & Sons, Inc.*

Fullerenes can be used in other organic electronic devices. In organic field effect transistors, they serve as an n-type semiconductor to increase charge mobility and stability [8,13]. Interdigitated capacitors with fullerenes were used in sensors that explore the electron acceptor properties of fullerene films [16]. In optical limiters, fullerenes induce a decrease in transmittance with an increase of incident light. This allows all of the light below an activation threshold to pass through, for example, maintaining the light transmission at a constant level below the damage threshold for the eye or an optical sensor [16]. Fullerene hydride shows promise for storing hydrogen gas in electrical vehicles that use a fuel cell. This is encouraging because the available hydrogen storage technologies, such as compressed gas or storage as metallic hydrides, are dangerous and/or have low hydrogen storage densities [16].

In the medical and pharmaceutical fields, fullerene derivatives can be used in photodynamic therapy for cancer, such as in antibacterial agents and in neuroprotective drugs [1,8]. Due to its ability to include atoms, fullerenes are suitable as drug carriers. Additionally, noble gases can be encapsulated in fullerenes for use in nuclear magnetic resonance. Fullerenes are also antioxidants and react with free radicals, thus preventing cell damage or death. For example, fullerene has an antioxidant performance that is 100 times more effective than vitamin E. In engineering, fullerenes are employed to strengthen metal alloys [1,8,12].

9.3 Carbon Nanotubes

CNTs are among the most studied nanostructures in recent decades [4]. They are a few nanometers in diameter and form one-dimensional structures because the length is orders of magnitude larger than the diameter. They have different properties from other allotropes of carbon, such as graphite and diamond, with unique mechanical and electrical characteristics. The discovery of CNTs has not been fully elucidated, but they have been known in the scientific community since 1991 because of the publication by Sumio Iijima [17]. That article presented a method to obtain cylindrical and concentric carbon structures, which were later called multiwalled carbon nanotubes (MWNTs) [17]. In 1993, Iijima and Donald Bethune (an IBM researcher) published separate studies that included methods to obtain single-walled carbon nanotubes (SWNTs). After these independent findings, nanotubes became an object of study in various fields of science [18].

The importance of CNTs for nanotechnology is reflected in their numerous application possibilities. In engineering, CNTs have been studied to create new composites for the aeronautics industry [19]. They have also been used in nanodevices and electronic nanocircuits to manufacture new chips for computers [20].

9.3.1 Properties of Carbon Nanotubes

The great interest in CNTs is due to their outstanding properties, such as high mechanical strength and elasticity, chemical, thermal and structural stability, and high conductivity

(A)   (B)

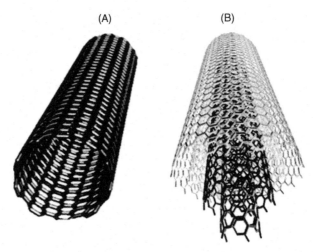

FIGURE 9.3 (A) A representation of the single-walled carbon nanotube (SWNT) structure and (B) a multiwalled carbon nanotube (MWNT). *Modified from J.J. Gooding, Nanostructuring electrodes with carbon nanotubes: a review on electrochemistry and applications for sensing, Electrochim. Acta 5 (2005) 3049–3060 [23]. Copyright 2005 Elsevier.*

[1,5,19,20]. CNTs are formed only by carbon and have a cylindrical shape. The structure can be compared to a graphene sheet rolled into a tube shape in which the ratio between the length and diameter makes it almost a one-dimensional structure [17,18]. The two most important structures are SWNTs and MWNTs, as shown in Fig. 9.3. SWNTs have a simple cylinder structure, and MWNTs consist of several concentric SWNTs. The length and diameter of the MWNT structures are very different from SWNTs, which implies differences in their properties [21,22]. Their properties depend on the atomic arrangement of how the "graphene sheet is rolled up" and on the tube diameter, length, and morphology [21–23].

The properties of SWNTs are determined by the structure formed by the bonds between the carbon atoms of the graphene sheet [21,22]. SWNT structures depend on the chirality of the tube, which is defined by the chiral vector $(\vec{C}_h)$ and the chiral angle (θ), as shown in Fig. 9.4. The vector $\vec{C}_h$ is the linear combination of the base vectors a_1 and a_2 of a simple hexagon via the relationship $\vec{C}_h = n\vec{a}_1 + m\vec{a}_2$, where n and m are integers and $\vec{a}_1$ and $\vec{a}_2$ are the single cell vectors of a two-dimensional matrix that is formed by the graphene sheet in which the direction of the nanotube axis is perpendicular to the chiral vector, as shown in Fig. 9.4A.

The chiral vector pair (n,m) indicates the direction in which the graphene sheet is rolled. The values of this pair allow three types of arrangements for SWNTs: zigzag, armchair, and chiral, as shown in Fig. 9.4B. The pair (n,m) determines the nanotube chirality and therefore its optical, mechanical and, in particular, electrical properties [21,22]. If (n,m) is a multiple of 3, the nanotubes are of the armchair type with metallic behavior, that is, they have a Fermi level in a partially filled band (Fig. 9.5A). When (n,m) is not a multiple of 3, the nanotubes are the zigzag type and have semiconductor behavior (Fig. 9.5B).

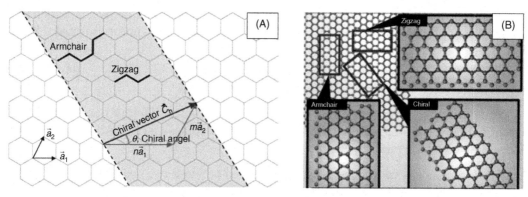

FIGURE 9.4 (A) Schematic diagram of how a graphene sheet is rolled up to form a carbon nanotube and (B) representation of the three types of SWNT structures obtained with the (*n*,*m*) pair from the chiral vector. *Modified from E.T. Thostenson, Z.F. Ren, T.W. Chou, Advances in the science and technology of carbon nanotubes and their composites: a review, Comp. Sci. Technol. 61 (2001) 1899–1912 [21]. Copyright 2001 Elsevier.*

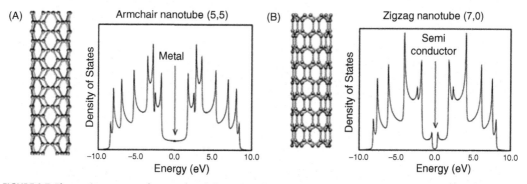

FIGURE 9.5 Electronic structure of nanotubes. (A) armchair (metallic) and (B) zigzag (semiconductor). *Reprinted with permission from J.C. Charlier, Defects in carbon nanotubes, Acc. Chem. Res. 35 (2002) 1063–1069 [22]. Copyright 2002 American Chemical Society.*

The (*n*,*m*) relationships that are formed with the nanotube structure, the chiral angle, and the type of nanotube are listed in detail in Table 9.1.

MWNTs are described by models developed from electron microscopy images. They generally have a coaxial cylindrical shape, are formed as a coaxial polygon, or are rolled-up graphene sheets. MWNTs are formed by several graphene sheets that are rolled around each other concentrically with a distance between the sheets of 3.4–3.6 Å [21,22]. The length

Table 9.1 Relationship of the (*n*,*m*) Pair with the Type of SWNT and Its Chiral Angle

(*n*,*m*)	Structure	Chiral Angle (Degree)
n = *m*	Armchair	$\Theta = 0$
m = 0	Zigzag	$\Theta = 30$
m ≠ *n*	Chiral	$0 \le \Theta \le 30$

and diameter of MWNTs are different from the length and diameter of SWNTs, which implies different properties. Another factor that defines the type of nanotube formed (SWNT or MWNT) is related to synthesis.

9.3.2 Synthesis of Carbon Nanotubes

There are many methods for synthesizing CNTs that produce different types with different properties. The three main methods are electric arc, pulsed laser, and chemical vapor deposition (CVD) [2,11,19].

- *Electric arc*: An electric arc discharge is generated between two cylindrical graphite electrodes in a steel chamber with an inert gas. The cathode and anode are kept at a distance that is less than 1 nm, allowing the current to flow so that plasma is generated between the electrodes. The temperature in the plasma region is approximately 4000 K. In this environment, the anode graphite undergoes sublimation and is deposited on the cathode or the inner walls of the chamber (soot) where the CNTs are found. To obtain MWNTs via this method, it is necessary to use catalysts to form beams of these types of nanotubes. The amount and quality of the CNTs depend on the product concentration in the chamber. This was the first method used to synthesize nanotubes and is still one of the most widely used [2].
- *Pulsed laser*: This is similar to the electric arc discharge method. The carbon is vaporized by irradiation via a laser in the presence of an inert gas. The graphite is placed inside a quartz tube and placed in a controlled-temperature tubular oven. The tube is emptied, and the temperature is increased to 1200°C, and then, the tube is filled with an inert gas (He or Ar). The laser sweeps across the graphite surface, vaporizing it and forming CNTs. During this process, it is possible to form both SWNTs and MWNTs without a catalyst [2].
- *CVD*: This involves the decomposition of a vapor or gas that contains carbon atoms, which are usually hydrocarbons in the presence of a metal catalyst, via heat treatment. The nanotubes are nucleated and formed via the decomposition of the volatile precursors and deposited on a substrate. The important parameters in this process are the type of catalyst, the substrate temperature (between 500 and 700°C) and the type of substrate, which defines the growth region of the CNTs. This technique allows CNTs to grow in an orderly manner with nanostructures designed specifically for some applications [2].

9.3.3 Applications of Carbon Nanotubes

The characteristics that result in CNTs that have different physicochemical properties allow their application in several fields. Sensors and biosensors are worth mentioning because there are hundreds of articles published from a multidisciplinary approach that involve physics, chemistry, biology, biochemistry, electronics, and materials science. The manufacturing of electronic nanodevices is the aim of this type of study, in addition to the integration of CNTs with biomolecules in nanosensors and nanoactuators. The latter applications can aid in diagnosing diseases, provide drug delivery control, and process biomaterials, for example. The sensing can be performed via conventional techniques: electrical, optical, or electrochemical [5,6,23–30].

The integration of CNTs with biological compounds (proteins, enzymes, antibodies, DNA) in electrical devices, such as biosensors, is due to the size compatibility because electronic circuit components have dimensions that are comparable to biomolecules. Electrostatic interactions and charge transfer, which are typical in biological processes, can be detected via electronic nanocircuits, which is helpful in detecting biological species [23–30]. For example, the diameter of a SWNT is comparable to the size of molecules, such as DNA (1 nm), whereas its length is greater, which provides a convenient interface for micrometric scale circuits. Additionally, the high-reactivity and effective area of CNTs provide synergy with organic molecules [23–27].

In biosensors that contain CNTs, there are typically electrodes that are modified via the immobilization of a biomolecule, such as an enzyme, as the recognition element [28,29]. CNTs are suitable for facilitating charge transfer in reactions with electroactive species in solution, in addition to increasing the electroactive surface area of the electrodes. In many cases, there is also synergy in the use of CNTs with other nanomaterials or biomolecules, for example, a more efficient catalysis of redox reactions of analytes, surface functionalization, and the compatibility of interactions with other types of nanomaterials (e.g., metal nanoparticles and quantum dots). Concerning biosensors, CNTs form suitable matrices for immobilizing biomolecules without the loss of biological activity. These biosensors that contain CNTs have been used to diagnose diseases, such as some cancers and tropical diseases (e.g., Chagas and leishmaniosis), and in traditional clinical tests, such as the measurement of glucose, urea, cholesterol, and uric acid [5,23–30]. Fig. 9.6 shows a representation of different types of sensors and biosensors that include CNTs.

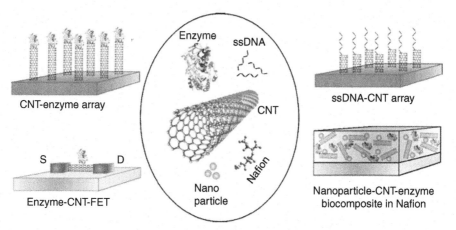

FIGURE 9.6 Representation of different types of sensors that contain CNTs. *Modified from A. Merkoçi, Nanobiomaterials in electroanalysis, Electroanalysis 19 (2007), 739–741[30]. Copyright 2007 John Wiley & Sons, Inc.*

SWNT semiconductors have been used in field-effect transistors (FET) to detect biological species under ambient temperatures and conditions. Nanotube field-effect transistors (NTFET) are obtained with the replacement of the solid-state gate with a layer of adsorbed molecules that modulates nanotube conductance. There are two classical NTFET device designs. The first design uses a single CNT as an electron channel between the source and drain electrodes. In the second type, a set of CNTs serves as a collective channel between the source and drain. The nanotube–analyte interaction may have one of two effects. The first effect is the charge transfer from the analyte molecules to the CNTs. In the second mechanism, the adsorbed analyte in the CNT walls decreases conductance [25–27]. The difference between these two types of mechanisms is obtained via measurements in a transistor. If a charge transfer occurs, the voltage threshold becomes much more positive (removing electrons) or more negative (donating electrons).

In energy storage, CNTs have been used as supercapacitors because of their high electrical conductivity, large surface area, mechanical strength, and low weight. Additionally, the possibility of modifying CNTs in chemical synthesis and their combination with other nanomaterials to improve the performance of the supercapacitors are important [31–33]. Other advantages are flexibility and optical transparency, allowing the development of flexible electronic devices, such as more pliable cell phones and computers. For this type of application, supercapacitors that contain only CNTs exhibit excellent electrical conductivity and a large surface area but do not reach the expected performance because of the contact resistance between the electrode and the current collector. Therefore, metal oxides (including those in nanoparticles) are incorporated into nanotubes to reduce the contact resistance and increase the storage density of supercapacitors [31–33].

In addition to the interest in devices, CNTs are useful as engineering materials because of their mechanical properties. CNTs have an elasticity of 1000 GPa, which is 5 times that of steel, a rupture tension of 63 GPa, and tensile strengths 50 (SWNTs) and 200 (MWNTs) times higher than that of steel. Adding to these characteristics is the lower density of CNTs compared with metals. With these properties, CNTs can be useful for aerospace engineering, such as in aircraft fuselages [19]. CNTs can form ultraresistant foams that work as high-compression springs. Threads made with CNTs are being studied to make fabrics in clothing that may be stronger than Kevlar, a material used currently in bulletproof vests [34].

Regarding products that are closer to industrialization, CNTs can be used as reinforcement in composites in which one of the phases has a nanometric size. One possibility is to apply CNTs in a polymer matrix to improve electronic properties and increase mechanical strength, for example, in plastic containers that require electrostatic shielding, such as in cell phones [35]. CNTs can also modify the mechanical properties of biodegradable polymeric nanocomposites that are used for tissue engineering and to manufacture artificial bones, cartilage, muscles, and nervous tissue. Another advantage of CNTs is the assistance they provide to bone cell proliferation and bone formation from functionalized surfaces [36–38].

9.4 Graphene

Graphene is a nanomaterial arranged in a two-dimensional layer of carbon atoms with sp^2 hybridization that are connected in a hexagonal lattice structure. Graphene was discovered in 2004 by Andre Geim and Konstantin Novoselov from the University of Manchester, England [39]. This material has driven research in nanoscience and nanotechnology because of graphene's exceptional electrical, mechanical, and chemical properties. Graphene can be used in sensors, batteries, supercapacitors, solar and fuel cells, and in biotechnology. The recognition of the importance of graphene resulted in its discoverers being awarded the Nobel Prize in Physics in 2010 [40–44].

Graphene has a remarkable band structure due to its crystalline structure. The carbon atoms form a hexagonal lattice in a two-dimensional plane in which each carbon atom is approximately 1.42 Å from its three neighboring atoms via a σ bond with each other. The fourth bond is a π bond that is oriented out of the plane. The π orbital is a pair of symmetrical lobes oriented along the z axis and centered on the core. From these π bonds, the so-called π and π^* bands form, which are responsible for most of the electronic properties of graphene [40–44].

9.4.1 Properties and Synthesis of Graphene

Graphene has many properties that are superior to any other material. For example, it has a high electron mobility of $2.5 \times 10^5 \ cm^2 \, Vs^{-1}$ at ambient temperature, a Young's modulus of 1.0 TPa, and a thermal conductivity above 3000 W/mK. Additionally, it absorbs a significant fraction of white light ($\sim$2.3%), has complete impermeability to any gas, and can sustain extremely high electrical current densities (one million times higher than copper). Furthermore, graphene is easily chemically functionalized [40–45].

The great advance in research on graphene has occurred because, among other reasons, it is cheap to obtain in a high quality. The synthesis methods can be divided into two groups, the *bottom-up* approach and the *top-down* approach [44]. The *bottom-up* approach consists of the growth of graphene sheets by combining the basic structural units. In this approach, organic synthesis methods are used, enabling graphene growth directly via molecular organic precursors. These precursors usually consist of benzene ring molecules with highly reactive functional groups, enabling controlled two-dimensional growth of the graphene sheet. Another method to grow two-dimensional graphene sheets is based on the growth in situ on catalyst substrates (e.g., Cu, Ni, Fe), such as via CVD, arc discharge, or SiC epitaxial growth. The major limitation of these techniques is that they still do not produce graphene in large amounts and uniformly, and they are expensive [44].

The *top-down* approach involves the breaking of the interplanar Van der Waals bonds of graphite. In this approach, chemical and mechanical exfoliation methods are used on graphite. These approaches have limitations in obtaining the two-dimensional graphene crystal. The mechanical or chemical graphite exfoliations do not allow complete control of the number of carbon layers in the graphene and the two-dimensional crystal lattice

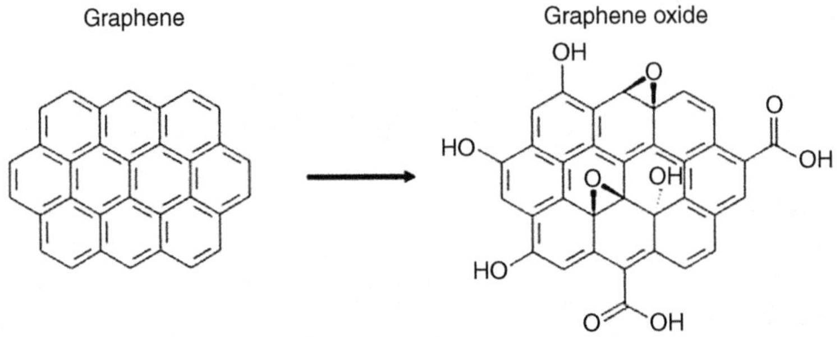

FIGURE 9.7 Simplified schematic representation of the graphene and graphene oxide structures. *Modified from Ref. [46]. Copyright 2013 Elsevier.*

quality, limiting performance. In particular, exfoliating graphite in solution requires surface modifications to break the Van der Waals bonds and the formation of structural defects on the graphene carbon lattice. An intermediate species in this type of processing is known as graphene oxide (GO) due to its high density of oxygen functional groups. With GO, it is possible to obtain colloidal solutions that are stable for long periods of time. This characteristic is important for the formation of nanocomposites by combining graphene with other types of materials, such as polymers, biomolecules, and nanoparticles. Surface charge measurements (zeta potential) of GO have indicated that GO is negatively charged when dispersed in water [44]. Fig. 9.7 shows a simplified representation of the graphene and GO structures.

GO can be converted into graphene via reduction processes, producing products with properties similar to those of graphene. However, reduced GO has significant structural differences compared to graphene [41–44]. The main purpose of the chemical reduction of GO is to obtain graphene with mechanical and conductive properties that are similar to graphene produced via the scotch tape method. GO reduction allows regeneration of the graphitic structure with deoxygenation and dehydration, restoring the conductivity of the material and restructuring the sp2 bonds in the carbon matrix [41–44].

The reduction of GO can be performed using several methods, including chemical methods with reducing agents, such as hydrazine, hydrogen plasma, dimethylhydrazine, hydroquinone, sodium borohydride, ascorbic acid, alcohols, and strong alkaline solutions, electrochemical methods, thermal methods, and ultraviolet radiation methods. Importantly, the conductivity of the reduced GO depends on both the methodology used to oxidize the graphite and the reduction methodology. The structural defects that appear during the oxidation of graphene also lead to the removal of carbon atoms from the aromatic structure of the carbon planes, creating nanometric zones of discontinuity that are impossible to recover via reduction [41–44].

9.4.2 Graphene Applications

Similar to fullerenes and CNTs, graphene also excels in applications in different fields. Graphene is employed in renewable energy sources and as transparent electrodes in dye-sensitized solar cells. Because doping can change the position of the Fermi level, graphene can act both as an electron acceptor and donor [41–44]. With the lower cost to produce graphene via liquid phase or thermal exfoliation, the increasing use of graphene can be expected in dye solar cells, especially in applications where the mechanical flexibility is crucial [41–44].

Graphene has been studied for next-generation lithium-ion batteries and in cathodes with high electrical conductivity [4,42]. With its morphology similar to a sheet, graphene can act as a conductive membrane and form a *core–shell* or nanocomposite with a sandwich-type structure. The increase in electrical conductivity of these morphologies that contain graphene can help overcome a major limitation of lithium-ion batteries, which is low power density. Additionally, the high thermal conductivity of graphene may be helpful for high current charges that generate a significant amount of heat in a battery. As anodes, the nanosheets of graphene can be used to intercalate lithium crystals in the layers [4,42].

Supercapacitors, also called electrochemical capacitors or ultracapacitors, are hosts of energy whose mechanism is associated with the electrical double layer at the interface between the electrode and the electrolyte [31,32]. Graphene is suitable for supercapacitors because of its flexibility, transparency, high electrical conductivity, large surface area, mechanical resistance, and low weight, which can allow the development of flexible electronic devices that can be incorporated in clothing and cell phones and allow the development more pliable computers. Additionally, graphene can be combined with other nanomaterials in search of synergy, as shown in Fig. 9.8 [47–52].

One challenge to produce graphene electrodes on a large scale is to avoid agglomeration of the plates because the agglomerations decrease the effective surface area, causing negative effects on the properties. To control this agglomeration, the plates are separated with nanoparticles, which also favors the electrochemical properties [47–52]. For example, metal oxides (in *bulk* or nanoparticles) are incorporated into graphene to decrease the contact resistance and increase the specific capacitance and energy density of supercapacitors. The combination of graphene and oxide in nanocomposites can generate supercapacitors with high performances, but the oxide must have a low cost and toxicity [47–52]. In this aspect, magnesium oxides may be preferable to ruthenium oxide because of their lower cost [50,51].

The unique electronic properties of graphene provide a glimpse of high-performance logic circuits of future decades, for example, in logic transistors, high-frequency transistors, flexible electronic devices, such as touch screens and electronic paper (*e-paper*), and organic light emitting diodes [42]. Graphene can also be used in photonics, photodetectors, optical modulators, optical polarization controllers, *mode-locked* solid state lasers, and insulators [42].

Like CNTs, graphene has been widely used in several types of sensors and biosensors (e.g., optical, chemical, and piezoelectric). For biosensors, in particular, graphene can be even more useful than CNTs, mainly because its two-dimensional structure facilitates functionalization for the incorporation of biomolecules and nanoparticles [53–55].

FIGURE 9.8 Schematic representation of a supercapacitor with two electrodes modified with graphene separated by a membrane. *Modified from K.S. Novoselov, V.I. Fal'ko, L. Colombo, P.R. Gellert, M.G. Schwab, K. Kim, A roadmap for graphene, Nature 490 (2012) 192–200 [42]. Copyright 2012 Nature.*

Graphene also has a higher biocompatibility and larger surface area, which can produce higher performance than CNTs or other carbon materials in biosensors [53–55].

For biomedical applications, graphene is promising due to its large surface area, chemical purity, and possibility of functionalization for drug delivery systems [42]. Its mechanical properties suggest applications in tissue engineering and regenerative medicine. Due to its single atomic layer thickness, conductivity, and mechanical resistance, graphene is suitable as a support for biomolecules to perform transmission electron microscopy [42]. Additionally, functionalized graphene can generate ultra-sensitive devices that are quick and capable of detecting biological molecules, such as glucose, cholesterol, hemoglobin, and DNA [53–55]. Graphene compounds are biocompatible with different cell types (such as mammalian and bacterial cells) both in vitro and in vivo and produce antibacterial effects [46].

9.5 Conclusions and Perspectives

Fullerenes, nanotubes, and graphene are the main nanomaterials derived from carbon, and they have unique properties that make them attractive for forming materials for use in a wide range of technologies. Although they have many common fields of application, in this chapter, we highlighted the importance of the use of fullerenes in photovoltaic solar cells, CNTs in sensors and biosensors, and graphene in supercapacitors for energy storage. Despite the significant number of studies in these areas, there is still much to investigate to fully understand the possibilities. Therefore, we described the main characteristics, properties, and applications of these carbon nanostructures, which are responsible for boosting nanotechnology.

Acknowledgments

The authors thank the National Council for Scientific and Technological Development (Conselho Nacional de Desenvolvimento Científico e Tecnológico–CNPq) (477668/2013-5), the Minas Gerais State Research Foundation (Fundação de Amparo à Pesquisa do estado de Minas Gerais–FAPEMIG) (APQ-01763-13 and APQ-01358-13), and the São Paulo Research Foundation (Fundação de Amparo à Pesquisa de São Paulo–FAPESP) (2013/14262-7) for their financial support.

References

[1] O.A. Shenderova, V.V. Zhirnov, D.W. Brenner, Carbon nanostructures critical, Rev. Solid State Mater. Sci. 27 (2002) 227–356.

[2] K. Koziol, B.O. Boskovic, N. Yahya, Synthesis of carbon nanostructures by CVD method, in: N. Yahya (Ed.), Carbon and Oxide Nanostructures, Advanced Structured Materials, 5, Springer-Verlag, Berlin, Heidelberg, 2010.

[3] A.J. Page, F. Ding, S. Irle, K. Morokuma, Insights into carbon nanotube and graphene formation mechanisms from molecular simulations: a review, Rep. Prog. Phys. 78 (2015) 38.

[4] Y. Li, J. Wu, N. Chopra, Nano-carbon-based hybrids and heterostructures: progress in growth and application for lithium-ion batteries, J. Mater. Sci. 50 (2015) 7843–7865.

[5] E. Katz, I. Willner, Biomolecule-functionalized carbon nanotubes: applications in nanobioelectronics, ChemPhysChem 5 (2004) 1085–1104.

[6] I. Willner, B. Willner, Biomolecule-based nanomaterials and nanostructures, Nano Lett. 10 (2010) 3805–3815.

[7] Rocha, R.C. Filho, Os Fulerenos e sua espantosa geometria molecular [Fullerenes and their amazing molecular geometry], Química Nova na Escola 4 (1996) 7–11.

[8] B.C. Yadav, KumarF R., Structure, properties and applications of fullerenes, Int. J. Nanotechnol. Appl. 2 (2008) 15–24.

[9] H.W. Kroto, J.R. Heath, S.C. O'Brien, R.F. Curl, R.E. Smalley, C60 Buckminsterfullerene, Nature 318 (14) (1985) 162–163.

[10] C.W.K. Isaacson, J.A. Field, Quantitative analysis of fullerene nanomaterials in environmental systems: a critical review, Environ. Sci. Technol. 43 (2009) 6463–6474.

[11] M.S. Dresselhaus, G. Dresselhaus, P.C. Eklund, Science of Fullerenes and Carbon Nanotubes, Academic Press, Boston, MA, (1996).

[12] R. S. Rouff, The bulk modulus of C60 molecules and crystals: a molecular mechanics approach, Appl. Phys. Lett. 59 (1991) 1553–1557.

[13] B.C. Thompson, J.M. Fréchet, Polymer-fullerene composite solar cells, Angew. Chem. 47 (2008) 58–77.

[14] S. Günes, H. Neugebauer, N.S. Sariciftci, Conjugated polymer-based organic solar cells, Chem. Rev. 107 (2007) 1324–1338.

[15] B.M. Alice, M. Milton, Y. Sun, Infrared spectroscopy of all-carbon poly[60] fullerene dimer model, Chem. Phys. Lett. 288 (1998) 854–860.

[16] J.C. Withers, R.O. Loutfy, T.P. Lowe, Fullerene commercial vision, Fullerene Sci. Techn. 5 (1997) 1–31.

[17] S. Iijima, Helical microtubules of graphitic carbon, Nature 354 (1991) 56.

[18] S. Iijima, T. Ichihashi, Single-shell carbon nanotubes of 1-nm diameter, Nature 363 (1993) 603.

[19] O. Gohardani, M.C. Elola, C. Elizetxea, Potential and prospective implementation of carbon nanotubes on next generation aircraft and space vehicles: a review of current and expected applications in aerospace sciences, Prog. Aerosp. Sci. 70 (2014) 42–68.

[20] U.N. Maiti, W.J. Lee, J.M. Lee, Y. Oh, J.Y. Kim, J.E. Kim, J. Shim, T.H. Han, S.O. Kim, 25th anniversary article: chemically modified/doped carbon nanotubes & graphene for optimized nanostructures & nanodevices, Adv. Mater. 26 (2014) 40–67.

[21] E.T. Thostenson, Z.F. Ren, T.W. Chou, Advances in the science and technology of carbon nanotubes and their composites: a review, Compos. Sci. Technol. 61 (2001) 1899–1912.

[22] J.C. Charlier, Defects in carbon nanotubes, Acc. Chem. Res. 35 (2002) 1063–1069.

[23] J.J. Gooding, Nanostructuring electrodes with carbon nanotubes: a review on electrochemistry and applications for sensing, Electrochim. Acta 5 (2005) 3049–3060.

[24] J. Wang, Carbon-nanotube based electrochemical biosensors: a review, Electroanalysis 17 (2005).

[25] B.L. Allen, P.D. Kichambare, A. Star, Carbon nanotube field-effect-transistor-based biosensors, Adv. Mater. 19 (2007) 1439–1451.

[26] S.N. Kim, J.F. Rusling, F. Papadimitrakopoulos, Carbon nanotubes for electronic and electrochemical detection of biomolecules, Adv. Mater. 19 (2007) 3214–3228.

[27] K. Balasubramanian, M. Burghard, Biosensors based on carbon nanotubes, Anal. Bional. Chem. 385 (2006) 452–468.

[28] O.N. Oliveira Jr., R.M. Iost, J.R. Siqueira Jr., F.N. Crespilho, L. Caseli, Nanomaterials for diagnosis: challenges and applications in smart devices based on molecular recognition, ACS Appl. Mater. Interfaces 6 (2014) 14745–14766.

[29] J.R. Siqueira Jr., L. Caseli, F.N. Crespilho, V. Zucolotto, O.N. Oliveira Jr., Immobilization of biomolecules on nanostructured films for biosensing, Biosens. Bioelectron. 25 (2010) 1254–1263.

[30] A. Merkoçi, Nanobiomaterials in electroanalysis, Electroanalysis 19 (2007) 739–741.

[31] X. Li, B. Wei, Supercapacitors based on nanostructured carbon, Nano Energy 2 (2013) 159–173.

[32] G. Yu, X. Xie, L. Pan, Z. Bao, Y. Cui, Hybrid nanostructured materials for high-performance electrochemical capacitors, Nano Energy 2 (2013) 213–234.

[33] S.W. Lee, J. Kim, S. Chen, P.T. Hammond, Y. Shao-Horn, Carbon nanotube/manganese oxide ultrathin film electrodes for electrochemical capacitors, ACS Nano 4 (2010) 3889–3896.

[34] V. Obradović, D.B. Stojanović, I. Živković, P.S. Radojević, R. Uskoković, R. Aleksić, Dynamic mechanical and impact properties of composites reinforced with carbon nanotubes, Fiber. Polym. 16 (2015) 138–145.

[35] S. Pande, A. Chaudhary, D. Patel, B.P. Singh, R.B. Mathur, Mechanical and electrical properties of multiwall carbon nanotube/polycarbonate composites for electrostatic discharge and electromagnetic interference shielding applications, RSC Adv. 4 (2014) 13839–13849.

[36] A.B. Duran, E.M. Carpenter, T.I. Malinin, J.C. Rodriguez-Manzaneque, L.P. Zanello, Carboxyl-modified single-wall carbon nanotubes improve bone tissue formation in vitro and repair in an in vivo rat model, Int. J. Nanomed. 9 (2014) 4277–4291.

[37] A. Gupta, B.J. Main, B.L. Taylor, M. Gupta, C.A. Whitworth, C. Cady, J.W. Freeman, S.F. El-Amin III, In vitro evaluation of three-dimensional single-walled carbon nanotube composites for bone tissue engineering, J. Biomed. Mater. Res. A 102 (2014) 4118–4126.

[38] H. Wang, C. Chu, R. Cai, S. Jiang, L. Zhai, J. Lu, X. Li, S. Jiang, Synthesis and bioactivity of gelatin/multiwalled carbon nanotubes/hydroxyapatite nanofibrous scaffolds towards bone tissue engineering, RSC Adv. 5 (2015) 53550–53558.

[39] K.S. Novoselov, A.K. Geim, S.V. Morozov, Electric field in atomically thin carbon films, Science 306 (2004) 666–669.

[40] K. Geim, Graphene: status and prospects, Science 324 (2009) 1530–1534.

[41] M.J. Allen, V.C. Tung, R.B. Kaner, Honeycomb carbon: a review of graphene, Chem. Rev. 110 (2010) 132–145.

[42] K.S. Novoselov, V.I. Faĺko, L. Colombo, P.R. Gellert, M.G. Schwab, K. Kim, A roadmap for graphene, Nature 490 (2012) 192–200.

[43] K. Sambasivudu, M. Yashwant, Challenges and opportunities for the mass production of high quality graphene: an analysis of worldwide patents, Nanotech Insights, 2012.

[44] V. Dhand, K.Y. Rhee, H.J. Kim, D.H. Jung, A comprehensive review of graphene nanocomposites: research status and trends, J. Nanomater. 2013 (2013) 14.

[45] R.R. Nair, P. Blake, A.N. Grigorenko, K.S. Novoselov, T.J. Booth, T. Stauber, N.M. Peres, A.K. Geim, Fine structure constant defines visual transparency of graphene, Science 320 (2008) 1308.

[46] A.M. Pinto, I.C. Gonçalves, F.D. Magalhães, Graphene-based materials biocompatibility: a review, Colloid. Surf. B 111 (2013) 188–202.

[47] X. Dong, L. Wang, D. Wang, C. Li, J. Jin, Layer-by-layer engineered Co–Al hydroxide nanosheets/graphene multilayer films as flexible electrode for supercapacitor, Langmuir 28 (2013) 293–298.

[48] Z. Niu, J. Du, X. Cao, Y. Sun, W. Zhou, H.H. Hng, J. Ma, X. Chen, S. Xie, Electrophoretic build-up of alternately multilayered films and micropatterns based on graphene sheets and nanoparticles and their applications in flexible supercapacitors, Small 8 (2012) 3201–3208.

[49] W. Liu, X. Yan, J. Lang, J. Chen, Q. Xue, Influences of the thickness of self-assembled graphene multilayer films on the supercapacitive performance, Electrochim. Acta 60 (2012) 41–49.

[50] G.D. Moon, J.B. Joo, Y. Yin, Stacked multilayers of alternating reduced graphene oxide and carbon nanotubes for planar supercapacitors, Nanoscale 5 (2013) 11577.

[51] M. Zhi, C. Xiang, J. Li, M. Li, N. Wu, Nanostructured carbon–metal oxide composite electrodes for supercapacitors: a review, Nanoscale 5 (2013) 72.

[52] W. Liu, X. Yan, Q. Xue, Multilayer hybrid films consisting of alternating graphene and titanium dioxide for high-performance supercapacitors, J. Mater. Chem. C 1 (2013) 1413.

[53] T. Kuila, S. Bose, P. Khanra, A.K. Mishra, N.H. Kim, J.H. Lee, Recent advances in graphene-based biosensors, Biosens. Bioelectron. 26 (2011) 4637–4648.

[54] W. Shao, J. Wang, W.J. Liu, I.A. Aksay, Y. Lin, Graphene based electrochemical sensors and biosensors: a review, Electroanalysis 22 (2010) 1027–1036.

[55] Y. Huang, X. Dong, Y. Shi, C.M. Li, L.J. Li, P. Chen, Nanoelectronic biosensors based on CVD grown graphene, Nanoscale 2 (2010) 1485–1488.

Index

Printed in the United States
By Bookmasters